Localization and Solitary Waves in Solid Mechanics

ADVANCED SERIES IN NONLINEAR DYNAMICS

Editor-in-Chief: R. S. MacKay *(Cambridge)*

Published

Vol. 1 Dynamical Systems
ed. Ya G. Sinai

Vol. 2 Statistical Mechanics
ed. Ya G. Sinai

Vol. 3 Topics in Bifurcation Theory & Application
G. Iooss & M. Adelmeyer

Vol. 4 Hamiltonian Systems & Celestial Mechanics
eds. J. Llibre & E. A. Lacomba

Vol. 5 Combinatorial Dynamics & Entropy in Dimension 1
L. Alseda et al.

Vol. 6 Renormalization in Area-Preserving Maps
R. S. MacKay

Vol. 7 Structure & Dynamics of Nonlinear Waves in Fluid
ed. A. Mielke et al.

Vol. 8 New Trends for Hamiltonian Systems & Celestial Mechanics
eds. J. Llibre & E. Lacomba

Vol. 9 Transport, Chaos and Plasma Physics 2
S. Benkadda, F. Doveil & Y. Elskens

Vol. 10 Renormalization and Geometry in One-Dimensional and Complex Dynamics
Y.-P. Jiang

Vol. 11 Rayleigh–Bénard Convection
A. V. Getling

Forthcoming

Symplectic Twist Maps
C. Gole

Wave Collapse
E. A. Kuznetsov & V. E. Zakharov

Positive Transfer Operators and Decay of Correlations
V. Baladi

Methods in Equivariant Bifurcation and Dynamical Systems, with Applications
P. Chossat & R. Lauterbach

ADVANCED SERIES IN NONLINEAR DYNAMICS
VOLUME 12

LOCALIZATION AND SOLITARY WAVES IN SOLID MECHANICS

A R Champneys
University of Bristol

G W Hunt
University of Bath

J M T Thompson
University College London

World Scientific
Singapore • New Jersey • London • Hong Kong

Published by

World Scientific Publishing Co. Pte. Ltd.
P O Box 128, Farrer Road, Singapore 912805
USA office: Suite 1B, 1060 Main Street, River Edge, NJ 07661
UK office: 57 Shelton Street, Covent Garden, London WC2H 9HE

The author and publisher would like to thank the following publishers of the various journals and books for their assistance and permission to include the selected reprints found in this volume:

The Royal Society (*Phil. Trans. R. Soc. Lond. A*); John Wiley & Sons, Ltd. (*Int. J. Numerical Methods in Engineering*); Elsevier Science (*Physica D, Comput. Methods Appl. Mech. Engrg.*); Pergamon Press (*J. Mech. Phys. Solids*); Springer-Verlag (*Arch. Rational Mech. Anal.*)

British Library Cataloguing-in-Publication Data
A catalogue record for this book is available from the British Library.

LOCALIZATION AND SOLITARY WAVES IN SOLID MECHANICS

ISBN 981-02-3915-7

Printed in Singapore by Uto-Print

Preface

This book comprises a collection of recent reprints and new material on the fundamentally nonlinear problems of localized responses to continuous inputs in structural systems. It has two intended audiences. First, for mathematicians and physicists, it should provide useful new insights into a classical yet rapidly developing area of application of the rich subject of dynamical systems theory. Particularly relevant is the theory of homoclinic solutions to Hamiltonian systems. Secondly, for workers in structural and solid mechanics, the book introduces a new methodology for dealing with structural localization and the related topic of the generation of solitary waves. The key to this methodology is to work with models that contain the fundamental material or geometric nonlinearities of the structure. The continuous onset of spatially-localized responses, which have been observed many times in practice, then arises naturally as solutions to such models, without the heavy reliance on imperfection effects that has traditionally been adopted to induce such behaviour. Imperfections play a secondary role in determining precisely where the localization occurs in the translational invariance associated with the homoclinic behaviour, but they do not in themselves determine the localized shape. Applications treated range from classical problems such as the buckling of cylindrical shells, twisted rods and pipelines, to the folding of geological strata, failure of sandwich structures and the propagation of solitary waves in suspended beam systems.

The book should draw attention to significant natural phenomena that often in the past, either numerically or by assumption, have been masked by co-existing spatially periodic solutions. With the growth of computer power has come the increased interest in nonlinear behaviour, and consequently the need to select physically relevant solutions from competing alternatives. Much remains to be done: the selected material should thus be seen both as a snapshot of current understanding and a pointer for future research.

Acknowledgements

The editors would like to thank each of the contributers of articles to this issue and also the various publishers who have given their permission for us to reprint their copyrighted material. We should also like to thank M. Khurram Wadee (University of Exeter) for contributing an original review article on the elastic strut model of Chapter 2. Thanks too to John Maddocks (EPFL, Lausanne) for commenting on the material pertaining to elastic rods and to Gabriel Lord (formerly University of Bristol, now NPL Limited) for supplying the cover illustration.

Contents

Chapter 1

Introduction

1.1 Overview

A response of a physical system that is localized to some portion of a distinguished co-ordinate (either space or time) is surely the next most fundamental state after homogeneous equilibrium and periodicity. This collection of original research articles is dedicated to the description and analysis of spatially localized waves in solid mechanics. The book is a greatly expanded version of the theme issue of *Phil. Trans. Roy. Soc. Lond. A* **355**(1732) 1997, with the same title. The additional material takes the form of other reprinted journal papers, this extended introduction and another original review article.

While acknowledging that sharply discontinuous localizations are also of physical importance, we concentrate here on those that arise as a balance between two conflicting influences, one pinching and the other dispersive, such that they are spread over a finite spatial domain – akin to plastic necking, for example, rather than brittle fracture. Applications range from the folding of geological strata and the buckling of cylindrical shells, twisted rods and pipelines, to the propagation of travelling solitary waves in suspended beam systems. All may be described by a new breed of mathematical theories based on the analysis of homoclinic solutions of differential equations posed on an infinite length scale. In truth no material length is infinite, but it sometimes makes sense to assume that a system sits within an infinite domain; this recognizes that if it is "long enough", the significance of different boundary conditions are often swamped by the homoclinic influence.

The key to this form of localization is that the processes in question should be nonlinear. In elastic buckling studies (see for example Chapter 5 on cylindrical shells), classical linear theory describes buckling via a critical (spatially periodic) eigenmode associated with a zero eigenvalue. However the behaviour of shells is markedly sub-critical so that buckling usually occurs via a violent jump to a finite amplitude state. Nonlinearities are then crucial in determining the shape of the post-buckled state, and a spatially-localized form is more likely to be adopted because it requires less energy than a fully periodic one (see for example the article by THOMPSON & CHAMPNEYS in Chapter 4 for a more detailed argument applied to the torsional buckling of rods). There is indeed a great deal of experimental evidence for localized buckling; see the reference lists in many of the articles that follow. A key point is that such localization does not necessarily arise from localized imperfections, but exists as a solution of the underlying (perfect) differential equation. Imperfections do however have a role in breaking the translational invariance associated with homoclinic behaviour, and hence in deciding where the localization will be centred.

A solitary wave is another form of localized phenomenon when viewed in a co-ordinate system that travels at the wave-speed. A classical example is a solution of the Korteweg de Vries (KdV) equation (Korteweg & DeVries 1895) which was first derived to describe the observation by Scott Russell in 1834 of a solitary water wave travelling along a canal. Specifically, the KdV equation may be written in terms of a dimensionless scalar wave amplitude $u(x,t)$ as

$$u_t + 6uu_x + u_{xxx} = 0. \tag{1.1}$$

Passing to a travelling co-ordinate frame by writing $\xi = x - ct$, and taking a first integral leads to the ordinary differential equation (ODE)

$$u_{\xi\xi} - cu + 3u^2 = 0 \tag{1.2}$$

which has a solitary wave solution $u = -(c/2)\text{sech}^2(\sqrt{-c}\xi/2)$ for $c > 0$. Viewed as a solution of (1.2) this is *homoclinic* solution that is bi-asymptotic to the trivial equilibrium $u = 0$ as $t \to \pm\infty$.

Loosely speaking, solitary waves are born in a balance between dispersive effects (the u_{xxx} term in (1.1), acting to spread the response in x) and those that act to (thoroughly) localize or pinch (the nonlinear, amplitude dependent, combination of u_t and uu_x terms). Solitary wave solutions of the KdV equation and the related nonlinear Schrödinger (NLS) equation have been of much interest in recent years, attracting the name "soliton" for their remarkable particle-like properties. Solitons are found to be very stable and able to pass through one another with only a shift in phase. There is now a rich theory of solitons based on the completely integrable structure of the KdV, NLS and related equations (see for example Ablowitz & Clarkson (1991)). However, such completely integrable systems are not of much concern in this book.

One of our aims is to show that solitary waves of *non-integrable* systems lead to even more interesting mathematics and physics, with many open questions. Indeed, the article by CHEN & MCKENNA in Chapter 7 suggests that solitary waves of a certain partial differential equation (PDE) share some of the remarkable stability and interaction properties of solitons, without the system appearing to be completely integrable. Soliton-like solutions can also arise in the non-integrable dynamics of rods and (see Section 1.5 below).

Note that (1.2) appears in a coordinate system that combines space and time. This ODE is in effect that of a nonlinear oscillator, other examples of which govern both spatial and dynamical problems. Consider, for example, the equation for the frictionless simple pendulum illustrated Figure 1.1(a),

$$mL\,\theta_{tt} = -mg\sin\theta, \qquad (1.3)$$

where $\theta(t)$ is the angle of swing. Solving this with initial conditions corresponding to the pendulum pointing downwards with exactly the right velocity to just reach the top, $\{\theta(0), \dot{\theta}(0)\} = \{0, 2sqrtg/l\}$, leads to a homoclinic solution with $\theta(t) \to \pi$ as $t \to \pm\infty$ (this solution may be expressed in closed form using elliptic integrals). Similarly, the spatial equilibrium configuration of the planar elastica of Figure 1.1(a), corresponding to an axially compressed elastic strut or column, can be described in terms of the angular deformation θ of the arc-length s along its centerline. A force balance leads to the equilibrium equation

$$EI\,\theta_{ss} = -P\sin\theta. \qquad (1.4)$$

Hence spatial equilibrium solutions of an infinitely long elastic strut (the elastica) are in one-to-one correspondence with motions of a simple pendu-

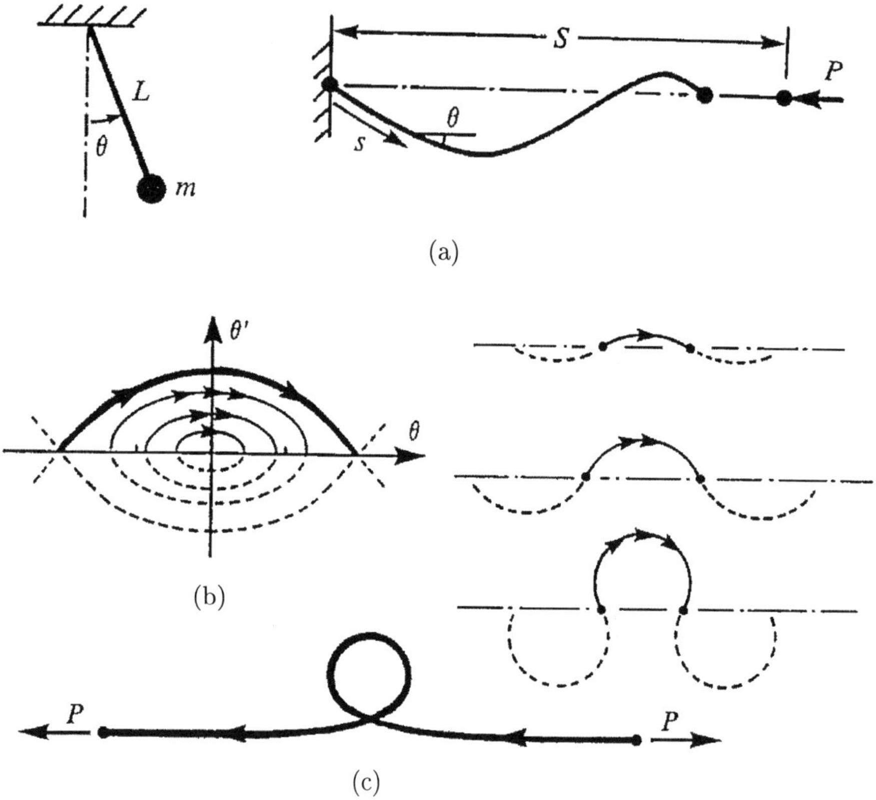

Fig. 1.1 **(a)** Illustrating the correspondence between a pendulum (where all parameters are held fixed) and an elastic strut (where we fix the compressive load P and bending stiffness EI, but vary the length S. **(b)** A single portrait of the pendulum gives the equilibrium path of the strut. For a long strut, ignoring boundary conditions, all the equilibrium solutions exist at a fixed value of the load P. **(c)** The homoclinic solution of the pendulum corresponds to a localised loop of the elastica.

lum. The simplest analogy has $mL = EI$, $mg = P$, and time t equal to the arclength s. Thinking of all other parameters as fixed and increasing the length of the strut then according to linear theory, the strut cannot buckle until

$$S_c = \pi\sqrt{EI/P}, \quad \text{which corresponds to} \quad T/2 = \pi\sqrt{L/g},$$

where T is the linear vibration period of the pendulum. The localized buckling solution (corresponding to the homoclinic orbit of the pendulum) occurs when $S = \infty$ and corresponds to pulling the two ends of the elastica through each other (see figure 1.1(c)). This analogy between spatial and temporal problems dates back to Kirchhoff (1859); it occurs also in the simplest model of torsional rod buckling, as discussed in detail by KEHRBAUM & MADDOCKS in Chapter 4 below, where the equivalent dynamical problem is that of a spinning top.

If $\sin\theta$ in (1.3) is replaced by the first two terms in its Taylor expansion then (after non-dimensionalisation) one obtains the cubic Duffing equation

$$\theta_{tt} + \theta - \frac{\theta^3}{6} = 0. \tag{1.5}$$

This has been used (with the addition of damping and forcing terms and the sometimes the inclusion of quadratic rather than cubic nonlinearity) to describe a range of nonlinear oscillations, for example describing ship capsize (Thompson ?, ?). In these studies, the bifurcations caused by the homoclinic orbit of (1.5) are shown to play an organizing role for chaotic dynamics and for fractal basins of attraction of competing stable motions. Observe however that the travelling wave KdV equation (1.2) is, up various scaling factors, identical to the quadratic Duffing equation. The variables have totally different meanings, but the homoclinic solution is important for both models, albeit for different reasons. See, for example the contribution by COLEMAN ET AL in Chapter 7 and Figure 1.6 below for similar phase plane analyses of buckling problems.

Another essential ingredient of the localization described in this book, is that it typically occurs in systems for which there is some form of energy conservation. For example, the pendulum has Lagrangian

$$\mathcal{L} = \frac{1}{2}m(L\theta_t)^2 + mgL\cos\theta,$$

while the elastic strut has the total potential energy

$$V = \int (\frac{1}{2}EI\theta_s^2 + P\cos\theta)ds.$$

Such problems can often be posed in a variational formulation, leading via Euler-Lagrange equations to differential equations which are *Hamiltonian* (the analogue of a classical mechanical system described in terms of generalized co-ordinates and momenta). Reversibility (Devaney 1976*b*) is also a common feature of many of the differential equations. A recurring theme will be homoclinic orbits of reversible Hamiltonian systems of two or more degrees of freedom. An archetype of such a system is the fourth-order equation

$$u_{xxxx} + Pu_{xx} + u = f(u), \tag{1.6}$$

where P is a parameter and $f(u)$ is a nonlinear function. Localized buckling solutions to (1.6) and their stability form the subject of Chapter 2. Also Hunt & Blackmore (1996) and the paper by CHEN & MCKENNA reprinted in Chapter 7 take $f(u)$ to be piecewise linear, respectively as a model for the equilibrium configuration of a pipeline which is free to lift off from a compressible foundation, and as a model for solitary waves of a suspended beam with zero supporting stiffness once the suspension device goes slack (see also the paper Champneys & McKenna (1997) and the contribution by HUNT & BLACKMORE in Chapter 6).

Equations such as (1.6) are known to admit infinitely many *multi-modal* homoclinic orbits, effectively comprising copies of a primary sech-like solution glued together at different separations (see Section 1.3 below and the review article by WADEE in Chapter 2 for the details). The key to the infinite multiplicity is that (for $-2 < P < 2$) the linearization at the trivial equilibrium of (1.6) has four complex eigenvalues; in this context it was proved by Devaney (1976*a*) that the existence of one non-degenerate homoclinic orbit implies chaos. For the kind of applications that we have in mind, this implies that multitudinous equilibrium solutions exist, all of which can satisfy imposed boundary conditions. Section 1.3 below outlines the mathematics behind this ‘spatial chaos’.

The main aim of this book is to show that there is now some general theory for *existence* of localized solutions and solitary waves in solid mechanics, which also suggests how one should *compute* such solutions for various mathematical models (see Section 1.4 for more details on numerics). Another important issue is the *stability* of these solutions. Apparently a general theory is much harder here, partly because stability is a more context dependent concept. For example, in Chapter 2, SANDSTEDE considers the localized bucking pulses of (1.6) as an elastic strut model; stability is then determined initially by the second variation of the energy functional. For the travelling wave problem studied by CHEN & MCKENNA in Chapter 7, it is governed by the full time-dependent PDEs. In the article by HUNT, MÜHLHAUS & WHITING in Chapter 6 on geological strata, stability must be taken in a pseudo (secular) sense appropriate for slow evolution over time. SANDSTEDE also discusses quite general conditions under which multi-modal homoclinic solutions, arising in a number of the models in this collection, may be stable (see also some references in Section 1.5 below for solitary waves in rods).

For the sake of brevity, we have concentrated in this book on localized phenomena in solid mechanics. However, localization of this kind appears important in many other disciplines as well. To mention just a few examples aside from the present applications, equation (1.6) with various nonlinear terms $f(u)$ arises in the description of solitary light pulses (e.g. Buryak & Akhmediev (1995)), of solitary water waves with surface tension (Buffoni, Groves & Toland 1996) and in the governing of pattern formation (Peletier & Troy 1996). See Champneys (1998) for a more thorough review.

The collection of articles which follow is intended to give an overall picture of why structural deformations localize and to highlight some of the surrounding mathematical issues. However, it is not a text book – rather it is a series of snapshots. Nevertheless we hope to provide a guide to current and future directions of research in the topic of localization, of use to theoretician and practitioner alike. Before proceeding with the reprinted material, the remainder of this introduction shall provide a commentary on some of the main themes to be brought out.

1.2 Structural localization

With a few notable exceptions, the phenomena described in this book were first identified in nonlinear elastic buckling problems. In the field of elastic stability, periodicity will sometimes be imposed *a priori* on the system. Useful descriptions are often found by some form of modal decomposition, with the modes themselves being picked out by linear eigenvalue analysis associated with movement away from some trivial or otherwise simple equilibrium state. Such modes frequently turn out to be periodic in one or more spatial variables. Other fields have developed other kinds of spectral breakdown, which of course can carry similar characteristics.

It is however in the localization of structural buckling that the limitations of such approaches manifest themselves. Long structures with unstable post-buckling characteristics will tend to localize, yet traditional structural analysis has often treated the deformations as periodic. This has occurred for several reasons. First, true homoclinic responses only tail away to zero in displacement and associated derivatives as $x \to \pm\infty$, and so real structural boundary conditions may only be truly satisfied at infinite length; however, although they may be denied analytically, long-but-finite systems may so nearly fulfill them as to make no practical difference. Secondly and closely linked to this is the fact that it may only be at infinite length that periodic and localized solutions emerge from the same critical state; typically, for long-but-finite systems, the latter are born at secondary bifurcation points on the periodic post-buckling path that is close to, but not coincident with, the initial bifurcation point. Periodicity and localization are thus inherently interlinked, with a strong possibility of the former masking the presence of the latter.

The contrasting influences at work on structures that experience such effects is perhaps most easily visualized by examples of simple but inherently nonlinear link models. Comprising just rotating rigid bodies and linear springs, separately and together these illustrate the underlying mechanisms in an uncluttered way. Trends towards periodicity and localization can be identified in different cases, and the ideas are readily extended to describe cellular buckling that can be seen as the transition between the two. The models can be seen as useful caricatures of real situations.

We start with the simple 1 degree of freedom (d.o.f.) situations of Fig.1.2. Rotation of the two links of Fig.1.2(a), when loaded by a conservative load P, is resisted by a linear spring of stiffness k as shown. This system has the load/coordinate, $P-Q$, and load/end-shortening, $P-\Delta$, responses shown at the bottom. The post-buckling equilibrium path ($Q \neq 0$) exhibits a fall in load with growth of deflection and is unstable (Thompson & Hunt 1973). If however, as shown in Fig.1.2(b), a spring of stiffness $2k$ is added in parallel to the links, the corresponding response shows an increase of load with deflection and is stable. The difference is significant, and takes on a greater significance when individual cells are connected together to form a load-carrying chain or structure. This is shown in Fig.1.3. Cells of Fig.1.2 are connected in series to all carry the same compressive load P. In every other way the cells are unlinked, in that displacement in one is

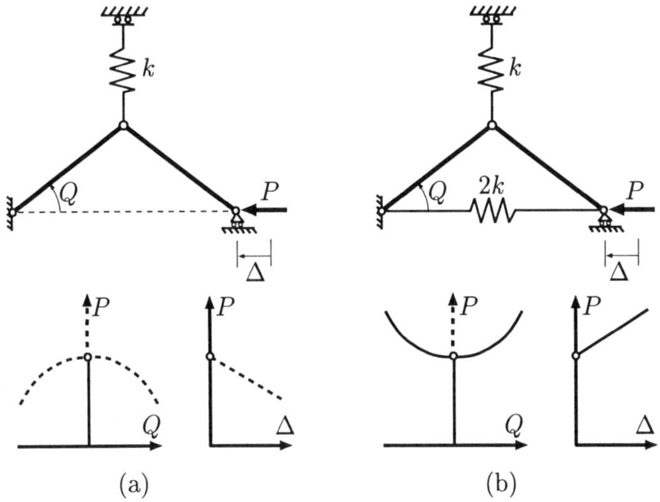

Fig. 1.2 Single degree of freedom link models and corresponding responses. Stable equilibrium is shown as unbroken lines and unstable equilibrium as broken lines. (a) Subcritical bifurcation. (b) Supercritical bifurcation.

completely unaffected by whatever happens in the others. There are thus no energy cross terms involving displacements in different cells, and in the sense of Aubry & Abramovici (1990) the system is *anti-integrable.*

As stiffnesses add according to a reciprocal law and all individual responses are locally linear in the post-buckling range, so too are combined responses, as shown. It can readily be concluded that for systems comprising the stable cells of Fig.1.2(b), the least stiff response, and thus the one that demands least work for the same end shortening, has all cells displacing the same amount (Fig.1.3(b)). On the other hand, for the unstable cells of Fig.1.2(a), minimum energy is associated with displacement in just a single cell; the area shown shaded in Fig.1.3(a) is the energy hump to be overcome in moving under constant load between pre-buckled and post-buckled states – clearly minimal when the system is localized into a single cell. With linking between cells the same trends can be observed (Hunt 1989): stable structures tend towards periodic behaviour while unstable structures want to localize.

We close this section by outlining a recent development that follows naturally from the above, but otherwise is perhaps more part future than present developments. An interesting connectivity between the two different behaviours of Fig.1.3 can be sketched by using cells that first destabilize

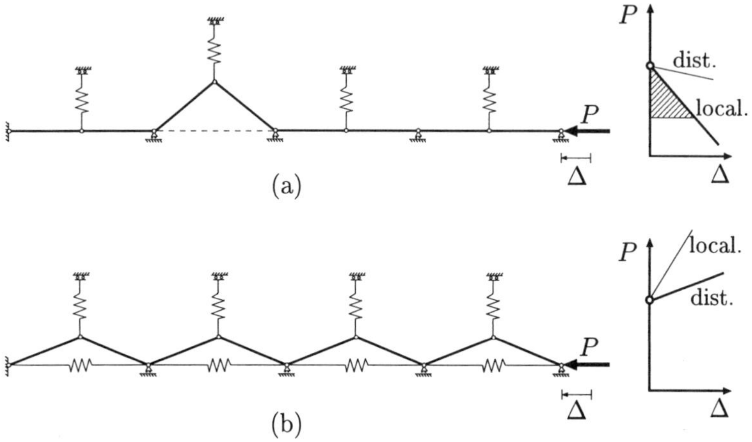

Fig. 1.3 Anti-integrable structures and corresponding responses. (a) Subcritical chain: localized response. (b) Supercritical chain: distributed response.

as in Fig.1.2(a) but subsequently restabilize as in 1.2(b). Rather than attempt to define simple spring-and-link systems which will behave in this manner, we simply consider the anti-integrable chain of linked but uncoupled elements shown schematically in Fig. 1.4(a). Individual elements are considered as black boxes each with the response of Fig. 1.4(b). As before, all elements carry the same load P, but deflection in one has no effect on, nor is it affected by, the others. Adding stiffnesses according to a reciprocal law, the combined stiffness on a rising path, R_m, when m individual elements are at stiffness k_2, is given by

$$\frac{1}{R_m} = \frac{m}{k_2} \tag{1.7}$$

and that of a falling path, F_m, when m elements are at stiffness k_2 but one is unloading at stiffness $-k_1$, is given by

$$\frac{1}{F_m} = \frac{m}{k_2} - \frac{1}{k_1} = \frac{1}{R_m} - \frac{1}{k_1}. \tag{1.8}$$

Elements are assumed to trigger one-by-one in a random order determined by small imperfections, giving the fanning sequence of equilibrium states shown as light lines in Fig. 1.4(c) on a plot of load P against overall end-shortening $\mathcal{E}$, where the load oscillates between the upper and lower loads,

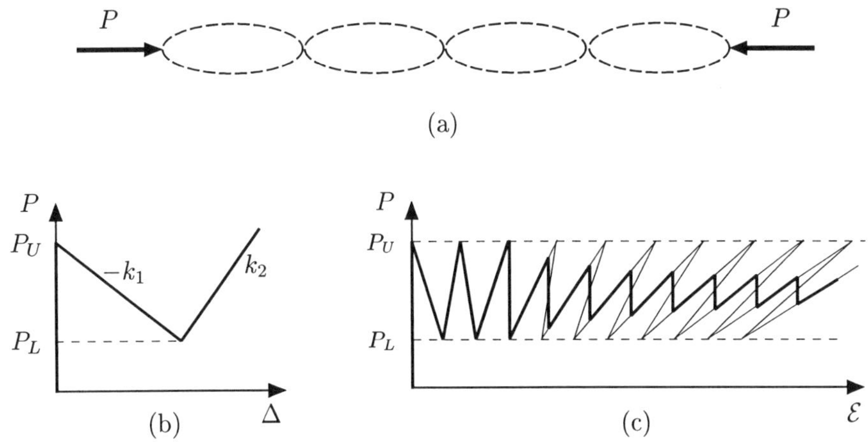

Fig. 1.4 (a) Anti-integrable chain of linked elements. (b) Tri-linear response characteristic for single element, involving regimes of infinite stiffness, negative stiffness, and restabilized positive stiffness. (c) Combined responses.

P_U and P_L, that characterize the response of a single element. In contrast, if it is assumed under the so-called Maxwell stability criterion that the system will automatically seek the global energy minimum, dynamical jumps will appear such that the thick lines of Fig. 1.4(c) are followed, with eventually all instability accumulating onto the so-called *Maxwell critical load* lying between P_U and P_L. Analytical details can be found in a recent collaborative preprint (Hunt et al. 1999).

1.3 Mathematical framework

The above discussion on mechanical reasons why the deformations of solids should localize can be complemented by a mathematical discussion. In this book, we shall not use the language and notation of classical solid continuum mechanics. Instead we shall be interested in situations (e.g. rod theories) where there is a distinguished 'long' co-ordinate direction. Regarding such a co-ordinate as time-like, one can often describe the equilibrium configuration (the 'response') of the spatial system by using an analogy with equations of motion in dynamics (the *Kirchoff* analogy illustrated in Section 1.1 for the pendulum-elastica). Hence our mathematical discussion, and also the methodological standpoint of the analysis presented in this book, shall use the language of dynamical systems theory. A note of caution, however, is that care must be taken with boundary conditions when exploiting the dynamical analogy. Structural systems naturally lend themselves to two-point boundary-value problems, whereas a dynamical system is usually solved as an initial value problem. This distinction is less important when dealing with localised solutions which decay to zero at both ends, but for spatial problems one should constantly be aware of the different possible forms of boundary condition (as for example the difference between *dead* or *rigid* loading).

Typically dynamical problem encountered as an analogy of a spatial problem will be dissipationless, time-reversible and can be expressed in classical Hamiltonian form. The simple analogy treated earlier between the elastica and the pendulum is an example where the problem has one degree of freedom (d.o.f.) in the dynamical sense. The localized buckling mode corresponded to a unique, single looped homoclinic solution. Here we shall introduce the reader to some of the extra complexity that arrives when the corresponding dynamical problem has two or more degrees of freedom. The key extra ingredient is that 1 d.o.f Hamiltonian systems are necessarily completely integrable, whereas this is not true in general for higher d.o.f systems.

An example is (1.6), which may be expressed as a reversible Hamiltonian system

$$\dot{q} = \partial H(q,p)/\partial p, \quad \dot{p} = -\partial H(q,p)/\partial q, \qquad q,p \in \mathbb{R}^2, \quad H(q,-p) = H(q,p), \tag{1.9}$$

by setting

$$t = x, \quad q_1 = u, \quad q_2 = u_{tt} + \frac{P+2}{2}u, \quad p_1 = \frac{P-2}{2}u_t + u_{ttt}, \quad p_2 = u_t,$$

and

$$H = u_t u_{ttt} + \frac{P}{2}u_t^2 - \frac{1}{2}u_{tt}^2 + G(u), \quad \text{where} \quad F(u) = \int_0^u f(v)\,dv.$$

In this section we shall consider two ingredients of localization for 2-degree of freedom reversible Hamiltonian systems: Hamiltonian Hopf bifurcation and existence of multi-modal homoclinic solutions. For more details of how these theories work in examples, see the contribution by KHURRAM WADEE in Chapter 2 for the strut model (1.6) and Section 1.5 below (also Chapter 4) for twisted rods. These notions apply in higher-dimensions also, due in part to the 'Homoclinic Centre Manifold Theorem' (Sandstede 1995). In Chapter 5 an infinite degree-of-freedom static model for cylindrical shell buckling is investigated numerically. There is some theory available for how such problems may be treated like finite d.o.f systems (see e.g. Mielke (1991), Peterhof et al. (1997)).

Consider a 4-dimensional (2 d.o.f.) reversible Hamiltonian system which we write as

$$\dot{x} = f(x, \mu), \quad x \in \mathbb{R}^4, \tag{1.10}$$

where μ is a parameter and reversibility acts according to a linear transformation R that fixes half of phase space

$$\exists R, \quad R^2 = \mathrm{Id}, \quad \mathcal{S} = \mathrm{fix}(R) \cong \mathbb{R}^2, \quad Rf(x;\mu) = -f(Rx;\mu).$$

Here $\mathcal{S}$ is referred to as the *symmetric section* of the reversibility. For classical Hamiltonian systems, reversibility usually acts to reverse the sign of the momentum variable, as in (1.9). By Liouville's Theorem (see for example Goldstein (1980)) the set of eigenvalues of the linearization around equilibria of (1.10) are symmetric with respect to the imaginary axis. A *Hamiltonian Hopf bifurcation* occurs when, upon (say) decreasing μ, two pairs of pure imaginary eigenvalues coalesce on the imaginary axis and become a complex quadruple (see, for example Figure 2 of the paper by THOMPSON & CHAMPNEYS in Chapter 4). We are interested in this bifurcation occuring at a symmetric equilibrium (i.e. one that is invariant under R) which we shall henceforth take to be at the origin of phase space. If in structural models this trivial equilibrium corresponds to an unbuckled state, a Hamiltonian Hopf bifurcation corresponds to the linear buckling load. But note (see again Figure 2 in THOMPSON & CHAMPNEYS) that the notions of stability for the structural and dynamical systems are precisely opposite.

Using double scale analysis or normal form theory close to Hamiltonian Hopf points at symmetric equilibria, one can reduce the dynamics to that of a fast sinusoidal oscillation coupled with a nonlinear oscillation in the slow time scale (see Chapter 2 for a detailed calculation on the strut model and Figure 1.6 below for twisted rods). Given an appropriate sign of a nonlinear coefficient (***subcriticality***, see Iooss & Peroueme (1993)), then the phase portrait of the nonlinear oscillator has a pair of homoclinic orbits (like the pendulum or cubic Duffing equation). In the case where the Hamiltonian system is completely integrable then the fast and slow systems completely decouple and one has the small amplitude bifurcation of a continuum of homoclinic connections. In the general, non-integrable case only two homoclinic orbits persist, which are symmetric under R (Iooss & Peroueme 1993). Thus we have that localized buckling modes (homoclinic orbits) are naturally born at zero-amplitude via sub-critical Hamiltonian Hopf bifurcation points of nonlinear systems. One does not need large-amplitude theory to obtain localization.

Now consider what happens as the parameter is reduced further so that the linearisation about 0 has four complex eigenvalues of the form $\pm\lambda \pm i\omega$ for $\lambda \neq 0$ (i.e. the origin is a *saddle focus*). First note that any given orbit lies in the 3D invariant manifold H = const. Supposing there is a homoclinic orbit connecting the saddle-focus to itself, then the 2D stable and unstable manifolds of the origin must intersect along this 1D orbit, and this intersection will generically be transversal in the 3D surface $H = H(0)$. Hence a homoclinic connection such as this will persist after its creation at a Hamiltonian Hopf bifurcation. (Note that this transversality is necessarily broken by completely integrable systems.) Devaney (1976*a*), see (Wiggins 1988, p. 275ff), used a *Shil'nikov*-type argument to show that, given a transversal homoclinic orbit, then within $H(0)$ one can construct a two-dimensional Poincaré section the return map which possesses Smale horseshoe dynamics. This shows the existence of chaotic dynamics. More precisely, there are periodic and aperiodic orbits associated with shift dynamics on an infinite set of symbols. Loosely speaking, by introducing the symbol ∞ one can also find an infinite set of homoclinic orbits which are of multi-modal type (Belyakov & Shil'nikov 1990).

A sketch of this Shil'nikov-type argument is given Figure 1.5, which is a simplification since the level set $H(0)$ near the origin is not really 3D Euclidean space, but is more like the surface of a four-dimensional cone. Let us briefly summarise the steps in this construction; the reader is referred to (Devaney 1976*a*, Wiggins 1988, Buffoni, Champneys & Toland 1996) for more technical discussions. Assume that there exists a transverse homoclinic orbit γ (Figure 1.5(a)). The flow near γ is decomposed into two bits, using Poincaré sections Σ^s, which is transverse to the local stable

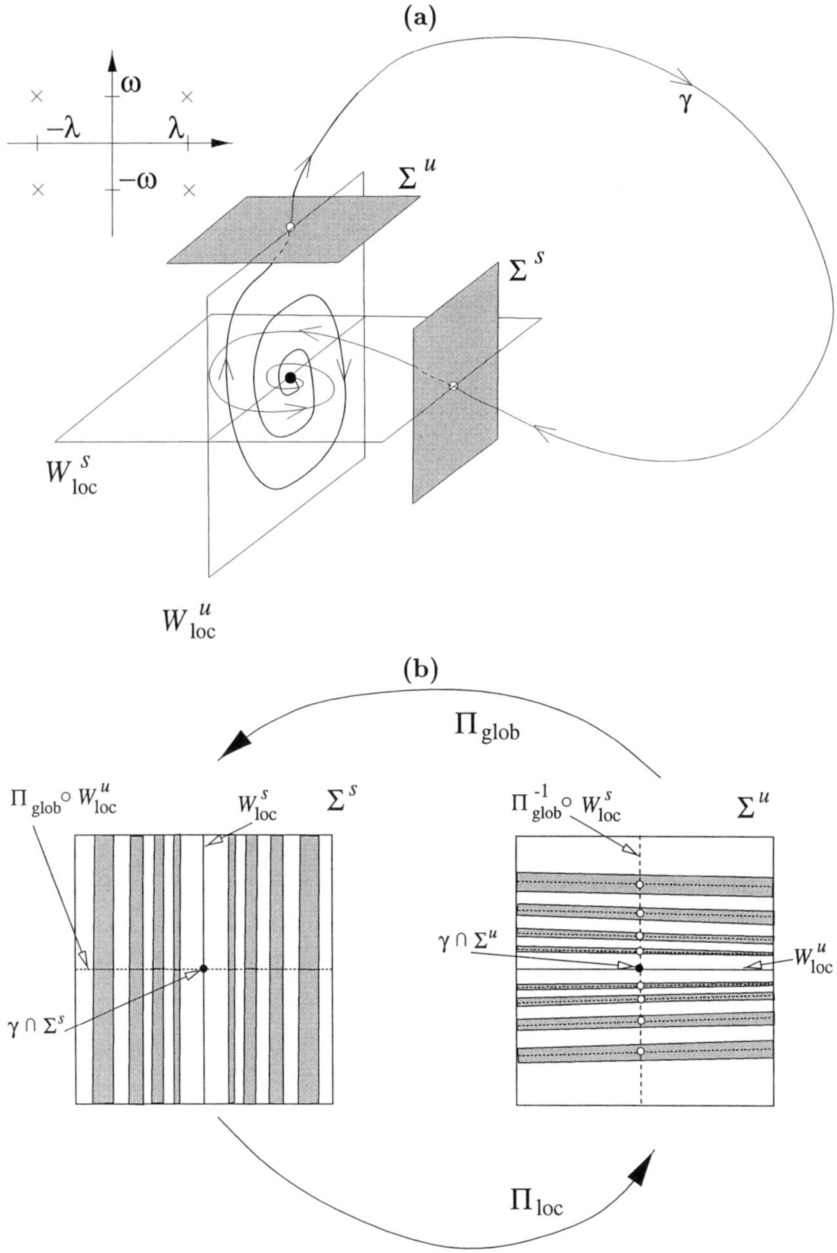

Fig. 1.5 A schematic representation of the geometrical construction behind the existence of multi-modal homoclinic solutions in the neighbourhood of a primary homoclinic orbit to a saddle-focus in a four-dimensional Hamiltonian system. See text for details.

manifold W^s, and Σ^u transverse to the unstable manifold W^u near the origin.

The local Poincaré map Π_{loc} from Σ^s to Σ^u can be approximated by solving the linearization of the local equations

$$\begin{aligned}
\dot{x}_1 &= \partial H/\partial x_3 = \lambda x_1 - \omega x_2 + P(x_1, x_2, x_3, x_4), \\
\dot{x}_2 &= \partial H/\partial x_4 = \omega x_1 + \lambda x_2 + Q(x_1, x_2, x_3, x_4), \\
\dot{x}_3 &= \partial H/\partial x_1 = -\lambda x_3 + \omega x_4 + R(x_1, x_2, x_3, x_4), \\
\dot{x}_4 &= \partial H/\partial x_2 = -\omega x_3 - \lambda x_4 + S(x_1, x_2, x_3, x_4).
\end{aligned}$$

Under this map, solutions close the stable manifold within Σ^s spend a long time before they leave a neighbourhood of the origin. In so doing solutions oscillate with angular velocity approximately ω, and hence some initial conditions are thrown out in the vicinity of Σ^u. In this manner, a thin vertical strips in Σ^s (such as one of those depicted in Figure 1.5(b)) is mapped into one of infinitely many thin vertical strips in Σ^u.

Trajectories under the global map Π_{glob} from Σ^u to Σ^s follow a small neighbourhood of γ. Since all trajectories spend a finite time, assuming Σ^s and Σ^u are defined to be small enough, then we may approximate this map as a simple linear transformation. Straight lines are mapped to straight lines (in Figure 1.5(b) the action of Π_{glob} is represented by the identity map). A complete description of solutions which remain for all time in a neighbourhood of γ can then be obtained by studying the composed Poincaré map $\Pi := \Pi_{glob} \circ \Pi_{loc}$ from $\Sigma^s \to \Sigma^s$. By construction we then have a countable family of vertical 'boxes' (the shaded regions of Σ^s in Figure 1.5(b)) that are mapped into countably many horizontal boxes (the shaded regions of Σ^u in Figure 1.5(b)). We have thus constructed a Smale horseshoe map, or more precisely shift dynamics on a countable set of symbols (see, for example Wiggins (1988)).

One consequence of the existence of the Smale horseshoe is the presence of chaotic dynamics, including *infinitely many periodic orbits.* Of more importance in the present context is that the construction also implies the existence of infinitely many 'two-modal' homoclinic orbits. Such solutions start out life in the local unstable manifold W^s_{loc} get mapped by Π_{glob} into the dashed line in Σ^s (represented as horizontal only for artistic convenience) in Figure 1.5(b). The solid portions of this line survive under the action of the local map Π_{loc}, to become the dotted lines inside each horizontal segment in Σ^u. Each of these line segments is forced by the geometry to have an intersection with the set that forms the local stable manifold under one further iteration of Π_{glob}. These points of intersection (represented by hollow circles in the figure) are by construction in both the stable and unstable manifolds, and hence are homoclinic orbits. Also, they are mapped twice under Π_{glob} and so are two-modal as they resemble two copies of the primary orbit γ separated by a passage close to the origin. Note that we have constructed an infinite family of such solutions. The ones with hollow circles nearer to W^u_{loc} in Σ^u have the property that there is more time, and hence more small-amplitude oscillation, between the two copies of γ

Iterations to this argument give the existence of homoclinic orbits with more than two humps. Following (Buffoni, Champneys & Toland 1996) these may be labeled by a string of integers $\mathbf{n}(m_1, m_2, \ldots, m_{n-1})$ where $\mathbf{n}$ measures the number of humps (copies of the primary homoclinic) and the m_i count the number of half-oscillations (with period π/ω) in between each hump. See the contribution by KHURRAM WADEE in Chapter 2 for various mode shapes of such responses for the strut equation (1.6).

Finally, note that the only essential hypothesis in the whole of the above argument was that W^u and W^s intersect transversally along γ (so that we could draw lines intersecting at right-angles in Figure 1.5(b)). However, away from Hamiltonian Hopf points it is in general a hard task for a given Hamiltonian system to prove categorically that a homoclinic orbit exists and is transverse. Sometimes one can make use of variational or analytical shooting methods; see Champneys (1998) for further discussion on this.

1.4 Numerical methods

One of the practical messages of this book concerns numerical computation. We argue that if localized solutions are required, numerical methods specifically designed for such purposes should be used. Straightforward 'black box' finite element analysis for example is likely to miss the solutions of practical significance, in the chaotic melee we have just outlined. The contribution by HUNT, LAWTHER & PROVIDÊNCIA E COSTA in Chapter 3 deals with this issue for a simplified link model of a strut, which leads to a discrete dynamical system in the dynamical analogy. Spatial chaos here is caused by homoclinic tangencies, and it is shown that a finite-element style solution to the problem has difficulty in coping with the inherent multiplicity of paths.

The key to computing localized solutions with minimum error over a truncated interval is to prescribe boundary conditions that force decay at $x = \pm\infty$ with the correct asymptotics (Beyn 1990, Friedman & Doedel 1991). Using these ideas Champneys & Spence (1993) derived a shooting method designed for computing the multiplicity of localised solutions in reversible systems. Also the latest release of the continuation code AUTO (Doedel et al. 1997) contains routines and an example (of the form (1.6)) of continuation of such solutions. This approach is adopted in many of the articles included in this collection.

Another interesting aspect of the numerical computation of localization is the possibility of the destruction or spurious creation of extra localized solutions under numerical discretisation. The article by DOMOKOS in Chapter 3 discusses this topic applied to the classical Euler strut problem (see also Fiedler & Scherle (1996) for some interesting theoretical results).

We also note that often when investigating the statics or dynamics of some of the structural systems under consideration in this issue, it is advantageous for both numerical and analytical purposes to exploit the special Hamiltonian or symmetry structure of the model. For example in rod theories see Simo et al. (1988) or Dichmann, Li & Maddocks (1996), the second of which also includes a description of visualization techniques applied to a bifurcation problems in rod mechanics.

Clearly finite element analysis and other 'black box' numerical techniques for analysing structural mechanics are extremely useful and their use has greatly enhanced practitioners' capabilities to understand real world configurations. However, it is our view that designers and users of such packages need to be aware of the underlying theory of the properties of the systems to be solved. For example, for problems with the propensity to localize via subcritical Hamiltonian Hopf bifurcations, one may need to to choose specific boundary conditions or initial mode shapes for path following, in order to avoid the spatial complexity and multiplicity of solutions.

1.5 Rod mechanics

The largest chapter of this book (Chapter 4) is dedicated to a description of complex localized equilibria of torsionally loaded rods. Rod mechanics as a subject has a long history, and a good state of the art description of the fundamentals can be found in the book by Antman (1995). Once buckled from a straight configuration, the problem of determining equilibria or dynamical states of rods is fundamentally nonlinear for geometric reasons. There are (at least) two kinds of interesting boundary value problems for rods: ones where the stress free configuration is infinitely long and asymptotically straight; and those where the ros is bent, e.g. loop boundary value problems pertinent for DNA mini-circles See for example Manning et al. (1996) and the paper by STUMP ET AL in Chapter 4 for examples of the latter, (see also Schlick (1995) and Olson (1996) for more biologically oriented reviews). For the most part, in this collection we shall be interested mainly in the former case for rods whose natural shape is straight. The kinds of application we have in mind are to the torsional buckling of structural members, and the formation of loops and kinks in underwater pipelines, cables and jacketed optical fibres.

The study of the three-dimensional spatial equilibria of thin elastic rods dates at least as far back as Kirchhoff (1859). It was he who showed the direct analogy between the equations describing equilibrium positions of an infinite rod and the dynamics of a spinning top. In fact the analogy is subtle as the statics of a rod is usually posed as a boundary-value problem whereas the dynamics of a top is an initial-value problem. Also the centerline $r(s)$ of the rod does not explicitly appear in the analogy and yet may appear explicitly in the boundary conditions. See the article by KEHRBAUM & MADDOCKS for more details.

Love (1927) noticed that an ordinarily straight rod whose principal bending stiffnesses are equal can be bent via end forces and moments into the form of a helical wave, corresponding to a periodic orbit of the spinning top. The key to this analysis was the recognition that the Kirchhoff-Love equations are *completely integrable* when regarded as a Hamiltonian system, but only in the special case of a naturally straight, circular-cross-section rod with uniform properties along its length.

More recently, Coyne (1990) showed that the same integrability implies the existence of a localized mode corresponding to a homoclinic orbit of the dynamical system. An analysis of the comparative bifurcational behaviour of these two classical buckling modes is given in the contribution by THOMPSON & CHAMPNEYS and is taken up in more generality via a dynamical instability analysis by GORIELY & TABOR.

The modern nonlinear theory of elastic rod buckling began with the work of Antman and co-workers (e.g. Antman & Kenney (1981)). They considered a more general *Cosserat* formulation of general *hyperelastic* rods that allows quite general constitutive relations including the effects of shear deformations and axial extensibility of the material. Such models can be written in a Hamiltonian form which has a special symmetry (Poisson) structure that means it is effectively a 2 d.o.f. problem; see the discussion in KEHRBAUM & MADDOCKS and the review Dichmann, Li & Maddocks (1996) for more details. Using this Hamiltonian structure it is known that for quite general Cosserat models, circular symmetry in the cross-section of a uniform rod always implies complete integrability. In the present setting, this means that the initial buckling of a long straight rod under decreased tension or increased torsion is at an integrable Hamiltonian Hopf bifurcation. That is, all solutions are equivalent to a 1 d.o.f. equivalent nonlinear oscillator (see Figure 1.6). It is argued in THOMPSON & CHAMPNEYS that the bifurcating homoclinic solution is the physically prefered postbuckling mode for long rods under *rigid* loading.

Some other discrete symmetry properties of rods with polygonal cross-sections are taken up in Buzano et al. (1985) and Pierce (1991), but do not concern us here.

A key result by Mielke & Holmes (1988) uses Melnikov's method to show that destroying the complete integrability associated with circular symmetry (e.g. by considering a tape, rather than a tube) breaks the non-transverse intersection of stable and unstable manifolds along the manifold of localized solutions. However, transverse homoclinic solutions survive this perturbation and hence the theory of Section 1.4 applies to give spatially chaotic solutions. Davies & Moon (1993) used a somewhat artificial model of a twisted rod subject to no external forces, but with a periodic non-uniformity along its length, to compute beautiful 3D pictures of some spatially chaotic equilibria. Champneys & Thompson (1996) (see Figure 1.8) computed some of the primary and multi-modal localised buckling modes of a uniform rod loaded by a wrench. These solutions were extensively path-followed in van der Heijden et al. (1998) in order to compute their load-deflection characteristics. Figure 1.7 gives a brief summary. The contribution by VAN DER HEIJDEN & THOMPSON contains the necessary normal form calculations to show the Hamiltonian Hopf bifurcation to be subcritical and also outlines a new 'Hamiltonian pitchfork' buckling behaviour that can happen for tape-like rods.

Another perturbation that breaks complete integrability of the Kirchoff-Love equations is present in a rod with an intrinsic curvature in its unloaded state. This effect is explored in CHAMPNEYS, VAN DER HEIJDEN

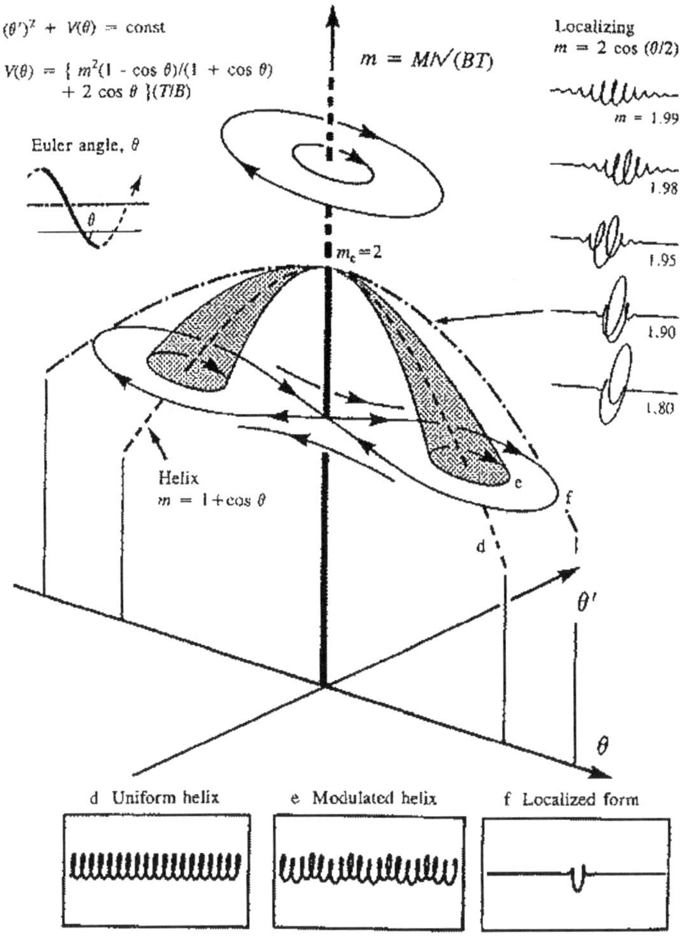

Fig. 1.6 Equivalent oscillator for a twisted circular rod, undergoing a Hamiltonian Hopf bifurcation at $m = m_c = 2$. The notation is that of THOMPSON & CHAMPNEYS in Chapter 4, and the detailed calculations are available in the book (Antman 1995, p. 331).

& THOMPSON, whose results explain the puzzling experimental observations in THOMPSON & CHAMPNEYS of a pre-buckling *one-twist-per-wave* deformation.

The paper by STUMP ET AL concerns the post-buckling behaviour of

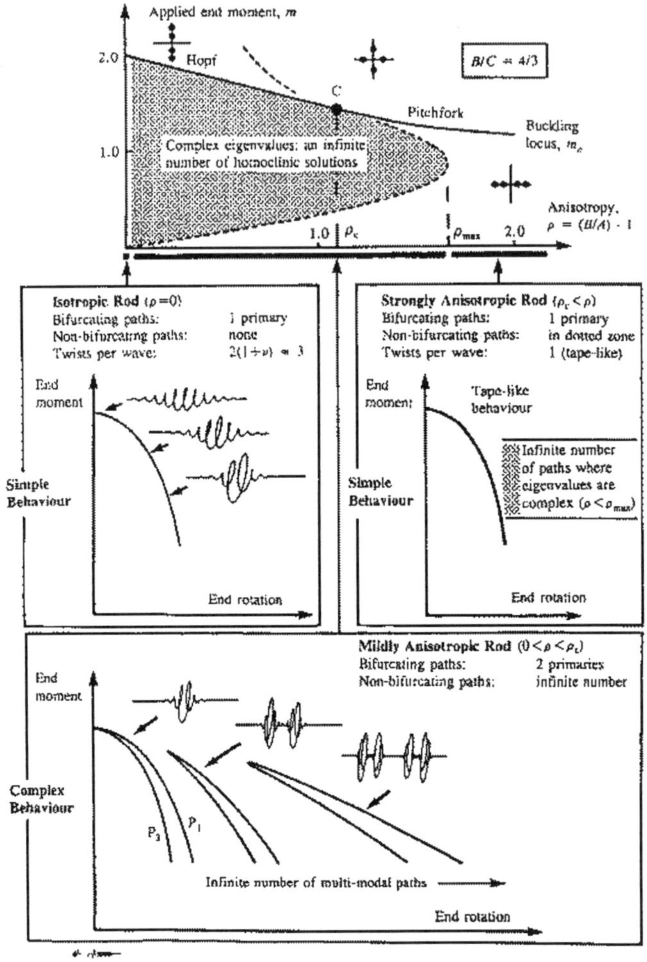

Fig. 1.7 A summary of the localized postbuckling paths for a long twisted rod with principal bending stiffnesses A and B, and torsional stiffness C. See Chapter 4 below for more details, in particular Figures 1 and 10 of the article by VAN DER HEIJDEN & THOMPSON.

circular, naturally straight, rods after the localised buckling mode self intersects and writhes into a double helix with a looped end. Their analysis also applies to loops of DNA. The paper by GORIELY & TABOR considers the buckling of finite length rods as an instability in a dynamical formula-

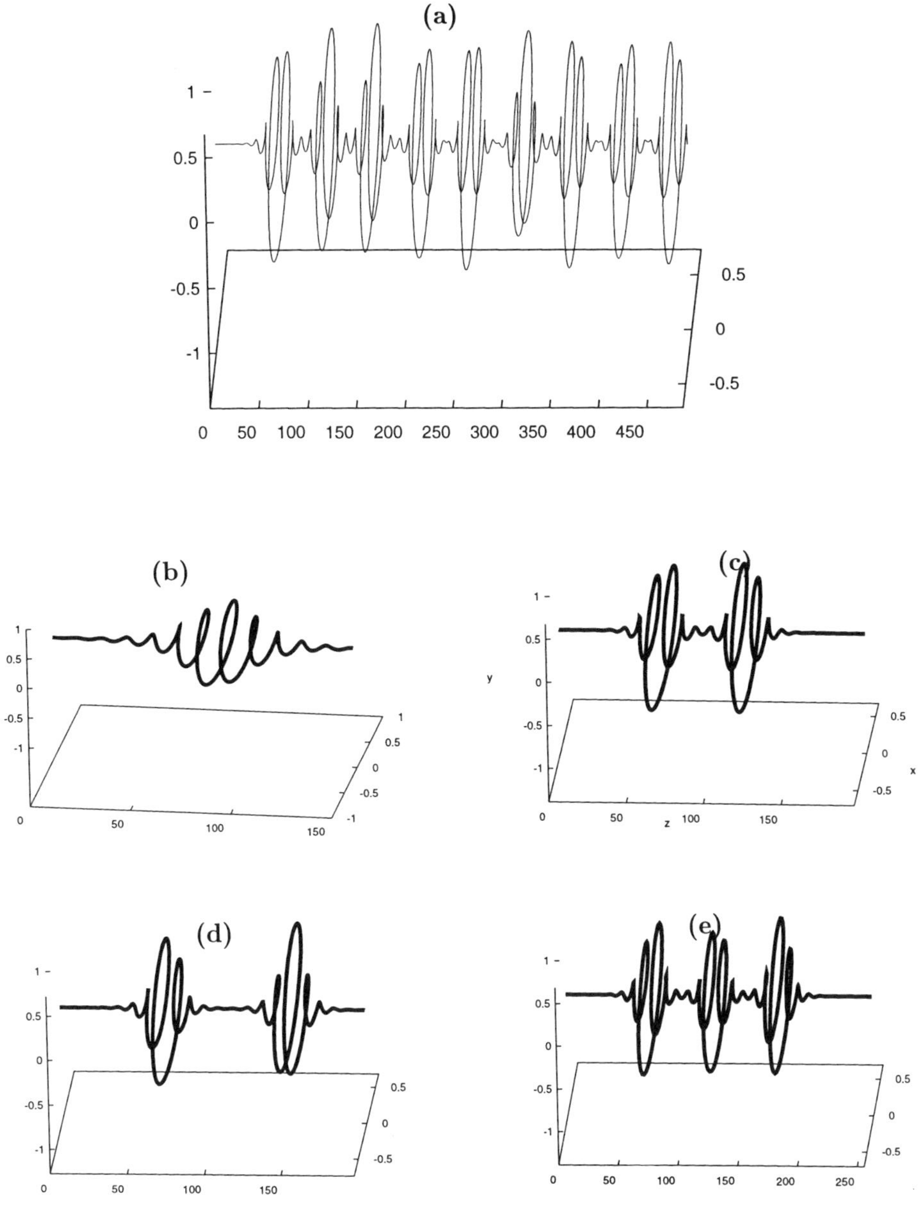

Fig. 1.8 Solutions to the Kirchoff-Love equations after Champneys & Thompson (1996) for $\rho = 0.5$, $m = 1.6$ in their notation: (a) spatial chaos; (b) a primary localized buckling mode; (c),(d) a bi-modal; and (e) a tri-modal homoclinic solution.

tion of the problem. They find that such rods initially buckle into a helical periodic waveform, which then localizes at a slightly higher value of a loading parameter. In the limit of long rods these two loads converge to one another, recovering the Hamiltonian Hopf bifurcation.

This collection deals mostly with the statics rather than the dynamics of rods. In the literature there are also results on the existence of solitary waves in rod problems. For example, the paper by COLEMAN ET AL, reproduced here in Chapter 7, which constructs explicit flexural solitary waves of extensible and shearable rods. Coleman & Xu (1993) have computed such loop-wave solutions numerically and found that their interaction is more complicated than just a shift in phase as it would be for solitons of integrable systems. We also remark on the work by Dichmann, Maddocks & Pego (1996) which shows the *stability* of solitary wave solutions in an inextensible rod model.

In general, the stability question is much harder than existence of solitary waves. Sometimes one can use the Hamiltonian structure of the (infinite d.o.f) time-dependent problem, in particular its conserved quantities, and careful variational arguments. The issue of stability is not dealt with comprehensively in this book. See Maddocks & Dichman (1984) and references therein (including to the thesis of Max Born (1906)) for the general setting for discussing the stability (in the sense of energy minimisation) of equilibrium states of rods. See also (Maddocks & Dichman 1994, Maddocks & Sachs 1993) for some recent results in this direction.

For solitary waves, if it is known that a primary wave is stable, then the techniques introduced by SANDSTEDE in Chapter 2 *may* in some circumstances be applicable to show stability of multi-modal solitary waves also. Such stable multi-modal solitary waves would seem to occur for the suspended beam model investigated by CHEN & MCKENNA in Chapter 7 (further numerical evidence on this will appear in a forthcoming article by Champneys et al. (1998)).

Bibliography

Ablowitz, M.J. & Clarkson, P.A. (1991), *Solitons, Nonlinear Evolution Equations and Inverse Scattering*, London Mathematical Society Lecture Notes, 149, Cambridge University Press.

Antman, S. S. (1995), *Nonlinear Problems of Elasticity*, Springer-Verlag, Berlin. Apllied Mathematical Sciences 107.

Antman, S. S. & Kenney, C. S. (1981), 'Large buckled states of nonlinearly elastic rods under torsion, thrust and gravity', *Arch. Rat. Mech. Anal.* **84**, 289–338.

Aubry, Serge & Abramovici, G. (1990), 'Chaotic trajectories in the standard map. the concept of anti-integrability', *Physica D* **43**, 199–219.

Belyakov, L. A. & Shil'nikov, L. P. (1990), 'Homoclinic Curves and Complex Solitary Waves', *Selecta Mathematic Sovietica* **9**, 219–228.

Beyn, W.-J. (1990), 'The numerical computation of connecting orbits in dynamical systems', IMA *J. Num. Anal.* **9**, 379–405.

Born, M (1906), Untersuchungen über die Stabilität der elastichen Linie in Ebene und Raum, unter verschiedenen Grenzbedingungen, PhD thesis, Dieterich Univ. Buchdruckerei.

Buffoni, B., Champneys, A.R. & Toland, J.F. (1996), 'Bifurcation and coalescence of a plethora of homoclinic orbits for a Hamiltonian system', *J. Dyn. Diff. Eqns.* **8**, 221–281.

Buffoni, B, Groves, M.D. & Toland, J.F. (1996), 'A plethora of solitary gravity-capillary water waves with nearly critical bond and froude numbers.', *Phil. Trans. R. Soc. Lond.* ***A*** **354**, 575–607.

Buryak, A.V & Akhmediev, N.N (1995), 'Stability criterion for stationary bound states of solitons with radiationless oscillating tails', *Physical Review E* **51**, 3572–3578.

Buzano, E., Geymonat, G. & Poston, T. (1985), 'Post-buckling behaviour of a non-linearly hyperelastic thin rod with cross-section invariant under the dihedral group D_n', *Arch. Rat. Mech. Anal.* **89**, 307–388.

Champneys, A. R. (1998), 'Homoclinic orbits in reversible systems and their applications in mechanics, fluids and optics', *Physica D* **112**, 158–186.

Champneys, A. R. & Thompson, J. M. T. (1996), 'A multiplicity of localised buckling modes for twisted rod equations', *Proc. Roy. Soc. Lond. A* **452**, 117–138.

Champneys, A. R., McKenna, P.J. & Zegeling, P.A. (1998), 'Numerical computation of solitary waves for nonlinear beam equations; stability fission and fusion'. Submitted to *Nonlinear Dynamics.*

Champneys, A.R. & McKenna, P.J (1997), 'On solitary waves of a piece-wise linear suspended beam model', *Nonlinearity* **10**, 1763–1782.

Champneys, A.R. & Spence, A. (1993), 'Hunting for homoclinic orbits in reversible systems: a shooting technique', *Advances in Computational Mathematics* **1**, 81–108.

Coleman, B.D. & Xu, J.M. (1993), 'On the interaction of solitary waves of flexure in elastic rods', *Arch. Rat. Mech. Anal.* **121**, 339–359.

Coyne, J (1990), 'Analysis of the formation and elimination of loops in twisted cable', *IEEE J. Oceanic Eng.* **15**, 71–83.

Davies, M. A. & Moon, F. C. (1993), '3-D spatial chaos in the elastica and the spinning top: Kirchhoff analogy', *Chaos* **3**, 93–99.

Devaney, R. L. (1976*a*), 'Homoclinic orbits in Hamiltonian systems', *J. Diff. Eqns.* **21**, 431–438.

Devaney, R. L. (1976*b*), 'Reversible diffeomorphisms and flows', *Trans. Amer. Math. Soc.* **218**, 89–113.

Dichmann, D. J., Li, Y. & Maddocks, J. H. (1996), 'Hamiltonian formulations and symmetries in rod mechanics', *IMA Volumes Appl. Math.* **82**, 77–113. Eds. Mesirov, Schulten & Summers.

Dichmann, D.J., Maddocks, J.H. & Pego, R.L. (1996), 'Hamiltonian dynamics of an elastica and the stability of solitary waves', *Arch. Rat. Mech. Anal* **135**, 357–396.

Doedel, E.J, Champneys, A.R, Fairgrieve, T.R., Kuznetsov, Yu.A., Sandstede, B. & Wang, X.J. (1997), 'AUTO97 continuation and bifurcation software for ordinary differential equations'. Available by anonymous ftp from FTP.CS.CONCORDIA.CA, directory PUB/DOEDEL/AUTO.

Fiedler, B. & Scherle, J (1996), 'Discretization of homoclinic orbits, rapid forcing and "invisible" chaos', *Memoirs of the AMS* **119**, R 8.

Friedman, M. J. & Doedel, E. J. (1991), 'Numerical computation and continuation of invariant manifolds connecting fixed points', SIAM *J. Numer. Anal.* **28**, 789–808.

Goldstein, H. (1980), *Classical Mechanics (2nd Ed.)*, Addison-Wesley, Reading, Mass.

Hunt, G. W. (1989), 'Bifurcation of structural components', *Proc. Instn. Civ. Engrs. Part 2* **87**, 443–467.

Hunt, G. W. & Blackmore, A. (1996), 'Principles of localised buckling for a strut on an elastoplastic foundation', *Trans. ASME* **63**, 234–239.

Hunt, G. W., Peletier, M. A., Champneys, A. R., Woods, P. D., Wadee, M. Ahmer, Budd, C. J. & Lord, G. J. (1999), Cellular buckling in long structures, (preprint).

Iooss, G. & Peroueme, M.C. (1993), 'Perturbed homoclinic solutions in reversible 1:1 resonance vector fields', *J. Diff. Eq.* **102**, 62–88.

Kirchhoff, G. (1859), 'Über das Gleichgewicht und die Bewegung eines unendlich dünnen elastischen Stabes', *J. für Mathematik (Crelle)* **56**, 285–313.

Korteweg, D.J. & DeVries, G. (1895), 'On the change of form of long waves advancing in a rectangular channel, and on a new type of long stationary waves', *Phil. Mag.* **39**, 422–443.

Love, A. E. H. (1927), *A Treatise on the Mathematical Theory of Elasticity*, 4th ed. edn, Cambridge University Press, Cambridge.

Maddocks, J. H. & Sachs, R.L. (1993), 'On the stability of KdV multi-solitons', *Comm. Pure Appl. Math.* **46**, 867–901.

Maddocks, J.H. & Dichman, D.J. (1984), 'Stability of nonlinearly elastic rods', *Arch. Rat. Mech. Anal.* **85**, 311–354.

Maddocks, J.H. & Dichman, D.J. (1994), 'Conservation laws in the dynamics of rods', *J. Elasticity* **34**, 83–96.

Manning, R. S., Maddocks, J.H. & Kahn, J.D. (1996), 'A continuum model of sequence dependent DNA structure', *J. Chem. Phys.* **105**, 5626–5646.

Mielke, A. (1991), *Hamiltonian and Lagrangian Flows on Center Manifolds with Applications in Elliptic Variational Problems*, Springer-Verlag, New York. Lecture Notes in Mathematics.

Mielke, A. & Holmes, P. (1988), 'Spatially complex equilibria of buckled rods', *Arch. Rat. Mech. Anal.* **101**, 319–348.

Olson, W. K. (1996), 'Simulating DNA at low resolution', *Current Opinions in Structural Biology* **6**, 242–256.

Peletier, L.A. & Troy, W.C. (1996), 'Spatial patterns described by the extended Fisher-Kolmogorov (EFK) equation: kinks', *Differential and Integral Equations* **8**, 1279–1304.

Peterhof, D., Sandstede, B. & Scheel, A. (1997), 'Exponential dichotomies for solitary-wave solutions of semilinear elliptic equations on infinite cylinders', *Journal of Differential Equations* **140**, 266–308.

Pierce, J. (1991), *Singularity theory and the buckling of rods*, Springer-Verlag, Berlin.

Sandstede, B. (1995), Center manifolds for homoclinic solutions, Technical report, Weierstraß Institut für Angewandte Analysis und Stochastik. Preprint No. 186, to appear in *J. Dyn. Diff. Eqns.* (1998).

Schlick, T. (1995), 'Modelling superhelical DNA: Recent analytical and dynamical approaches', *Current Opinions in Structural Biology* **5**, 245–262.

Simo, J. C., Marsden, J. E. & Krishnaprassad, P.S (1988), 'The Hamiltonian structure of nonlinear elasticity: The material and convective representation of solids, rods and plates', *Arch. Rat. Mech. Anal.* **104**, 125–183.

Thompson, J. M. T. & Hunt, G. W. (1973), *A general theory of elastic stability*, Wiley, London.

Thompson, J.M.T. (1996), Global dynamics of driven oscillators; fractal basins and intermediate bifurcations, *in* P. Aston, ed., 'Nonlinear Mathematics and its Application', Cambridge University Press, Cambridge.

van der Heijden, G.H.M., Champneys, A.R. & Thompson, J.M.T. (1998), 'The spatial complexity of localised buckling in rods with non-circular cross-section'. To appear in *SIAM J. Appl. Math.*

Wiggins, S. (1988), *Global Bifurcations and Chaos : Analytical Methods*, Springer-Verlag, New York, U.S.A.

Chapter 2

The strut on an elastic foundation

This chapter introduces many of the concepts that shall reappear later in the book by presenting material on the simple conceptual model of a strut resting on a nonlinear foundation. The first article by KHURRAM WADEE reviews what is known about existence and multiplicity of localized buckling modes for this equation. The second paper by SANDSTEDE gives new information on stability of these modes. This analysis shows that only the simplest of these modes may be physically stable (in the sense of minimising potential energy) and then only under rigid loading.

The elastic strut on an elastic foundation: A model localized buckling problem

M. Khurram Wadee*

Abstract

The paper summarizes a number of recent contributions on the existence and multiplicity of localized equilibrium solutions to a simplified model of a compressed strut supported by a nonlinear (softening) elastic foundation, which has played an important historical rôle in the general development of the subject of localization and spatial chaos in solid mechanics. Perturbation and other methods for single and multi-hump solutions are briefly outlined, and their results compared with numerics.

1 Introduction

The elastic strut on an elastic foundation has acted, and is still acting, as an exemplar in the study and general development of localization phenomena. So much has been reviewed with this as the basic description it is only possible to summarize the major contributions here. One of the main motivations for studying buckling of the strut model was that it was seen to be phenomenologically related to the classical problem of the buckling of the axially compressed cylindrical shell. The problem has been attacked from the viewpoints of applied mechanics and mathematics. The interaction and cross-fertilization of well-established disciplines has proved to be most fruitful in this particular case.

The buckling of structures was initially given a firm mathematical underpinning by Euler (1744). He studied the buckling of an axially-loaded pin-ended strut of length L with uniform bending stiffness EI and derived the value of the critical loads

$$P_i^{\mathrm{C}} = \frac{\pi^2 i^2 EI}{L^2}. \tag{1}$$

Each load is identified with a unique mode of buckling. If we concentrate on the lowest critical load, P_1^{C}, it is evident that as the length of the strut increases then this load decreases monotonically towards zero. Euler went on to perform a post-buckling analysis allowing large (elastica) deflections of the strut and found that it undergoes a stable form of buckling. This means that the buckled solution is stable even though the effective post-buckling stiffness of the strut is much lower than on the pre-buckling branch. The post-buckling analysis requires nonlinear terms to be taken into account whereas the critical loads in (1) can be derived from a linearized form of analysis. Study of elastic localization has proceeded along a similar path. Physical experience has preceded a successful mathematical description by some way.

Non-periodicity in the buckle patterns of uniform structures was largely neglected in favour of harmonic disturbances (e.g. Thompson & Hunt (1973)) until relatively recently,

*School of Engineering and Computer Science, University of Exeter, North Park Road, Exeter, Devon EX4 4QF, UK. E-mail: m.k.wadee@exeter.ac.uk

primarily owing to the lack or ignorance of techniques and tools of investigation. In certain types of structures the periodic assumption is at odds with physical experiments and among the first attempts to account for localization was work by Tvergaard & Needleman (1980) who noted the tendency of localization in structures with an unstable buckling behaviour. The treatment, however, appealed to physical arguments albeit using a version of the structural model which has proved so successful in developing our understanding of the phenomenon of localization. The equations derived were linear and so the results were of a qualitative nature. Later, Potier-Ferry (1983) used methods which allowed a periodic response to be modulated by an amplitude envelope. In this case, the author was primarily interested in hardening rather than softening behaviour in which localization is not favoured. Work combining a nonlinear softening characteristic and usage of appropriate mathematical tools was reported by Hunt *et al.* (1989).

The behaviour of a strut supported by a Winkler elastic foundation is subtly different from the simply-supported case. The critical loads and associated mode shapes (see below) are dependent on the stiffness of the foundation and the bending stiffness of the strut. However, they are *independent* of the length of the structure. Also, given different types of nonlinear behaviour the model can be made to have either stable or unstable post-buckling behaviour. Here we will concentrate our interest on the latter case as that is where localization of buckle patterns is favoured. When elastica deflections are allowed the strut can localize even when resting on a linear elastic foundation (Hunt *et al.*, 1993). Most of the work, however has been done with a softening nonlinearity in the foundation with linear bending and end-shortening terms in the strut itself (Hunt *et al.*, 1989) because the governing equation is easier to manipulate. In what follows, we shall assume that the model is of indefinite length and so the effects of boundary conditions can be neglected.

In modern terms the analysis of elastic buckling can be successfully performed by using the concept of total potential energy. The governing differential equation is derived from a variational principle. The exploration has placed great importance on *Kirchhoff's dynamical systems analogy* with the axial coordinate being viewed as time (Kirchhoff, 1859). Among other areas of mechanics which have had a direct impact on the buckling problem, most notable contributions have come from fluid mechanics; in particular from existence theory applied to the nonlinear phenomena of solitary waves in shallow water (e.g. Amick & Toland (1992a; 1992b)). Also asymptotic techniques have been taken from the fluids field and applied successfully to the description of solutions close to the critical buckling load (Newell & Whitehead, 1969; Stewartson & Stuart, 1971). More recently variational methods (e.g. Buffoni & Séré (1996)) and the analysis of appropriate normal forms (Iooss & Pérouème, 1993) have also been brought to bear. Together, these methods have extended our ability to discover localized solutions where only periodic ones had previously been sought.

As an important example of cross fertilization, various non-integrable extensions of the celebrated Korteweg–de Vries (KdV) equation (see, for example Drazin & Johnson (1989)) describe the profile of gravity–capillary solitary waves. One such extension is the so-called fifth-order KdV equation which may be written in dimensionless form as

$$u_t + u_{xxxxx} + u_{xxx} + 6uu_x = 0 \tag{2}$$

(see Grimshaw *et al.* (1994) and references therein). When the system is cast in an appropriate frame of reference (moving at the wave speed) the equations are transformed from partial differential equations (PDEs) to ordinary differential equations (ODEs). Specifically upon setting $u(x,t) = y(\xi)$ in (2), where $\xi = x - ct$ and c is the wave speed, and taking one integration yields the ODE

$$y'''' + y'' - c\,y + 3y^2 = 0 \tag{3}$$

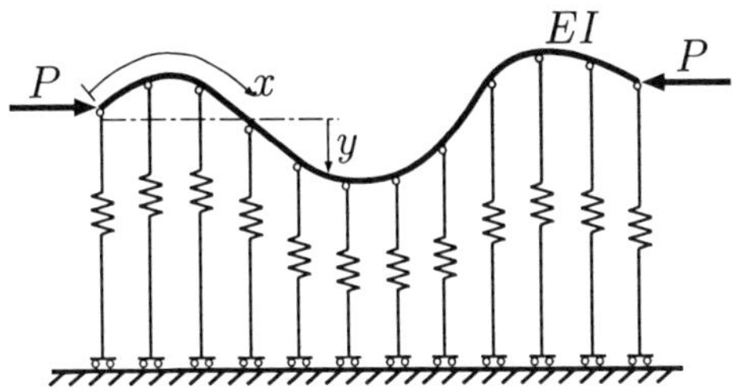

Figure 1: The elastic strut resting on an elastic foundation.

which is equivalent up to a scaling of variables and parameters to (7) below. Quite a large number of authors have considered solitary wave solutions to (3), see Champneys (1998) for a review, and also its relation to exact formulations for water waves in the presence of surface tension (Iooss & Kirchgassner, 1992; Buffoni *et al.*, 1996). In these fluids systems, interaction of *wavepackets*[1] is known to occur (Yang & Akylas, 1997) and an equivalent phenomenon which has come to be called multi-hump localization has been found in the structural system. This phenomenon has been known to exist since the study of Hunt & Wadee (1991) but a full understanding of its significance is only now being realized (Buffoni *et al.*, 1997; Wadee & Bassom, 1998).

Much of the work relating the structural model to a wider context has been reported elsewhere (see again Champneys (1998)). The present review is limited to work directly related to the structural system itself. It is set out as follows. The next section introduces the model to be studied. Preliminary analyses are given with results from linearization of the system being stated followed by a non-dimensionalization of the equation of motion. Key parameter values are identified ready for the initial post-buckling analysis in §3. This following section covers most of the early perturbation results pertaining to localized buckling, the extensions to the theory and ends with some recent findings relating to the existence of multi-hump solutions. Strategies suitable for detecting numerical solutions to the strut equation are described in §4. In §5 we review some work on more elaborate strut models which have been studied. These more sophisticated systems attempt to describe more realistic properties found in structures of practical significance e.g. elastica nonlinearities and foundations with restabilizing characteristics.

2 Basic analysis

The system under consideration is schematically shown in figure 1. The nonlinearity is assumed to lie solely in the foundation response, F per unit length, which has the form

$$F = ky + c_1 y^2 + c_2 y^3 + \cdots . \tag{4}$$

Because we are considering conservative loading and only elastic behaviour, the system is fully describable in terms of its total potential energy function, V, where

$$V(y, \dot{y}, \ddot{y}) = \int_{-\infty}^{\infty} \left(\frac{1}{2} EI\ddot{y}^2 - \frac{1}{2} P\dot{y}^2 + \frac{1}{2} ky^2 + \frac{1}{3} c_1 y^3 + \frac{1}{4} c_2 y^4 + \cdots \right) \mathrm{d}x. \tag{5}$$

[1]A wavepacket is taken to be an identifiable localized disturbance whose amplitude may modulate a faster oscillation but which is distinguishable from other disturbances.

The independent variable x is distance measured along the strut, y is the lateral deflection, P is an externally-applied compressive load, EI, k, c_1 and c_2 etc. are constants and a dot over a variable denotes differentiation with respect to x. The first and second terms are linearized forms of strain energy of bending and work done by load, respectively. The remainder are the energy stored in the foundation, the first of these being associated with the linear term in the foundation stiffness. Studies have focussed on the case where the foundation stiffness changes monotonically from the flat state as importance was placed on the post-buckling behaviour in the vicinity of criticality. In what follows we will concentrate on the specific case $F = ky + c_1y^2$ ($c_1 < 0$) with some additional results being presented for the case $F = ky + c_2y^3$ ($c_2 < 0$). In §5, we shall review recent results with more sophisticated foundation models which are more relevant for studies in the far post-buckling regime.

In dynamical systems terminology, the integrand in (5), $\mathcal{L}$, is a Lagrangian of the system and the governing ODE is found by applying the variational principle. The Euler–Lagrange equation for such a Lagrangian is

$$\frac{\partial \mathcal{L}}{\partial y} - \frac{\mathrm{d}}{\mathrm{d}x}\left(\frac{\partial \mathcal{L}}{\partial \dot{y}}\right) + \frac{\mathrm{d}^2}{\mathrm{d}x^2}\left(\frac{\partial \mathcal{L}}{\partial \ddot{y}}\right) = 0. \tag{6}$$

Explicitly, the equation of motion is

$$EI \ddddot{y} + P\ddot{y} + ky + c_1y^2 = 0. \tag{7}$$

The system is autonomous and any solutions are only defined up to translation in x. It is also *hamiltonian* and *reversible*. The well-known *symmetric section* defined by $\dot{y} = \dddot{y} = 0$ arises because of the invariance of (7) under the transformation

$$R_1 : (x, y, \dot{y}, \ddot{y}, \dddot{y}) \mapsto (-x, y, -\dot{y}, \ddot{y}, -\dddot{y}). \tag{8}$$

Furthermore, there is a second reversibility if the foundation response is symmetric with respect to y e.g. if $F = y - y^3$. A solution may then pass through the *antisymmetric section* defined by $y = \ddot{y} = 0$ and the solution is invariant under the transformation

$$R_2 : (x, y, \dot{y}, \ddot{y}, \dddot{y}) \mapsto (-x, -y, \dot{y}, -\ddot{y}, \dddot{y}). \tag{9}$$

Reversibility of the system is fundamental to finding localized solutions.

2.1 Linearization

The loss of stability of the trivial solution, $y = 0$, is of immediate interest. As P is parametrically varied there is a value at which the trivial solution becomes unstable at a bifurcation point. This point can most be easily found by linearizing equation (7). The governing equation becomes

$$EI \ddddot{y} + P\ddot{y} + ky = 0. \tag{10}$$

By feeding in the assumed form of solution $y = A\mathrm{e}^{\lambda x}$ the auxiliary equation is found to be

$$EI\lambda^4 + P\lambda^2 + k = 0. \tag{11}$$

For different values of P, five distinct forms of eigenvalues exist (figure 2).

For $P > P^{\mathrm{C}} = 2\sqrt{kEI}$, the eigenvalues are imaginary and occur as two pairs symmetrically placed about the real axis (figure 2(a)). The form of solution of (10) is sinusoidal. At $P = P^{\mathrm{C}}$ (figure 2(b))the eigenvalues either side of the real axis coalesce on the imaginary axis and the system undergoes a Hamiltonian Hopf bifurcation and for $P < P^{\mathrm{C}}$ (figure 2(c)),

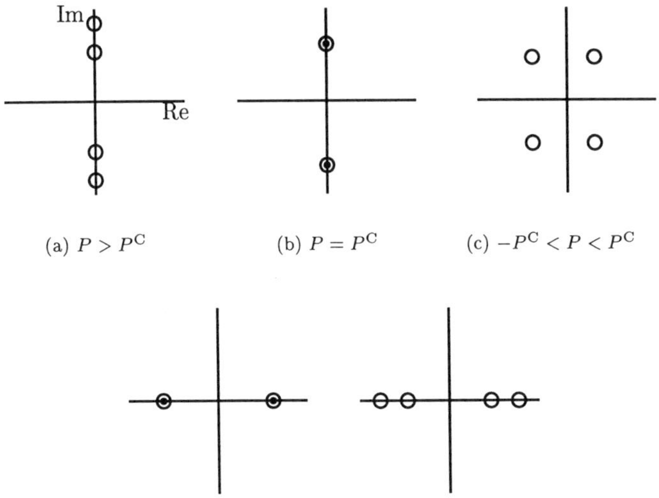

(a) $P > P^{\mathrm{C}}$ (b) $P = P^{\mathrm{C}}$ (c) $-P^{\mathrm{C}} < P < P^{\mathrm{C}}$

(d) $P = -P^{\mathrm{C}}$ (e) $P < -P^{\mathrm{C}}$

Figure 2: Eigenvalues of (10) for various values of P.

which is the range of interest for the nonlinear system, the eigenvalues split into the four quadrants of the complex plane. They have the value $\lambda = \pm\alpha \pm \mathrm{i}\omega$ where

$$\alpha = \sqrt{\frac{P^{\mathrm{C}}}{4EI} - \frac{P}{4EI}}, \qquad \omega = \sqrt{\frac{P^{\mathrm{C}}}{4EI} + \frac{P}{4EI}}. \tag{12}$$

For this range of P the phase space has a two-dimensional unstable manifold and a two-dimensional stable manifold. The behaviour is sinusoidal with wavenumber α modulated by exponential growth/decay $\mathrm{e}^{\pm\alpha x}$.

Whereas we are interested in the structural system i.e. compressive loads (positive P) we should complete the picture of the linearized system because many of the results pertaining to the problem (from fluid mechanics in particular) have been derived in the range of negative P. This range corresponds to a grossly buckled strut where the foundation is pushing away from the equilibrium position in one region of the strut and is sustained there by a *tensile* axial force. The eigenvalues of the form in (12) persist down to the second critical value of load at $P = -P^{\mathrm{C}}$ where the pairs either side of the imaginary axis coalesce on the real axis (figure 2(d)). For lower values of P the eigenvalues split into four distinct real values (figure 2(e)). The solution to (10) in this range of P is exponential only with no oscillatory component.

2.2 Non-dimensionalization

To clarify the manipulation of equations and to emphasize the relative importance of terms, equation (7) may be made dimensionless by the following scalings

$$x \to x\sqrt[4]{\frac{k}{EI}}, \qquad P \to \frac{P}{\sqrt{kEI}}, \qquad y \to \frac{y}{|c_1|} \tag{13}$$

This is equivalent to setting $EI = k = 1$ and $c_1 = -1$.

$$\ddddot{y} + P\ddot{y} + y - y^2 = 0. \tag{14}$$

This form of the equation will be used henceforth. For future reference, the non-dimensional critical load is $P^{\mathrm{C}} = 2$.

3 Post-buckling analysis

For the softening nonlinearities chosen, conventional post-buckling analysis reveals that the bifurcation point at $P = P^{\mathrm{C}}$ is unstable-symmetric and a sinusoidal buckle pattern is expected at least at criticality (Thompson & Hunt, 1973). Like most problems in post-buckling theory, there exist many solutions to the governing equation. Early analyses showed that finite-amplitude periodic solutions existed in the post-buckling range. However localization of buckle patterns has been identified as being more severe in terms of loss of stability (Hunt *et al.*, 1989). The dynamical systems analogy paved the way to a thorough understanding of localization in such uniform long structures.

3.1 Localized buckling modes

Localization is characterized by the existence of large deflections only over a small region of the strut. The deflection approaches zero at the boundaries. Hence we look for solutions of (14) which are *homoclinic* to the origin of phase space. Because the system is conservative, the Hamiltonian, $\mathcal{H}$, is conserved for all x. The system is of fourth order which corresponds to a two-degree-of-freedom hamiltonian system with

$$\mathcal{H} = \dot{y}\,\dddot{y} - \frac{1}{2}\ddot{y}^2 + \frac{1}{2}P\dot{y}^2 + \frac{1}{2}y^2 - \frac{1}{3}y^3 \tag{15}$$

and so it is clear that any solution asymptotically approaching the flat state must have zero energy or $\mathcal{H} = 0$. For regular behaviour to be expected, there must exist as many constants of motion as there are degrees of freedom (e.g. see Lichtenberg & Lieberman (1992)). Motion in phase space would then be limited to a manifold the dimension of which is the same as the number of degrees of freedom. However no constant other than $\mathcal{H}$ exists in this case which means that the system is non-integrable with the possibility of the manifestation of spatially chaotic forms. The rich structure of the solutions described below arises because of this lack of integrability.

In *path* space (the y–$\ddot{y}$ plane), the contour

$$-\frac{1}{2}\ddot{y}^2 + \frac{1}{2}y^2 - \frac{1}{3}y^3 = 0 \tag{16}$$

is of significance (Toland, 1986) (figure 3). For $P < -P^{\mathrm{C}}$, Amick and Toland (1992a) were able to use properties of this set to show that there exists a *unique* homoclinic solution of (14) which varies monotonically either side of the symmetric section (the primary orbit).

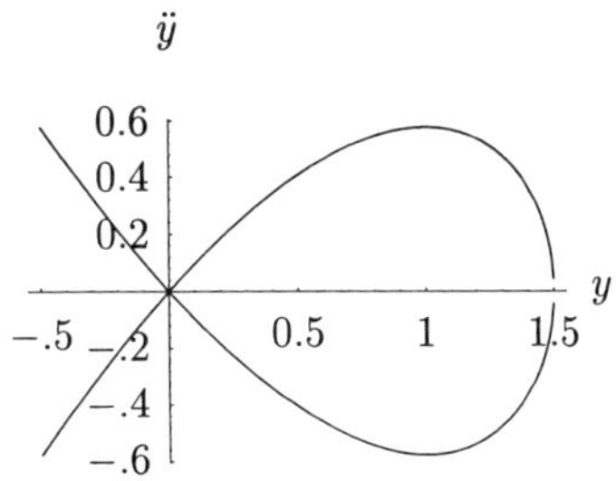

Figure 3: Zero energy contour in y–$\ddot{y}$ space.

Furthermore, at $P = -P^{\mathrm{C}}$ a bifurcation occurs at which an infinite set of *multi-humped* homoclinic trajectories come into existence. This has been proved rigorously for $-P^{\mathrm{C}} < P < -P^{\mathrm{C}} + \delta$ for small δ (Champneys & Toland, 1993; Buffoni *et al.*, 1997). There is also strong numerical and analytical evidence that the primary solution forms a continuous branch for $-P^{\mathrm{C}} < P < P^{\mathrm{C}}$ and that for each P in this interval there are always an infinite number of multi-humped solutions. See §4, specifically figure 9 below for more details. Note that in the region $P > -P^{\mathrm{C}}$, figure 2 indicates that the trajectory is no longer monotonic but oscillates with wavenumber given by the imaginary part of the linear eigenvalue.

When viewed in path space, homoclinic trajectories for $P > -P^{\mathrm{C}}$ emerge from the origin and spiral out to hit the above contour at the symmetric section whereupon they follow the same path in this projection back to the origin (see figure 4(a), (b)). The solution goes from the unstable and stable manifold and generically forms a transverse intersection. When a non-transverse intersection exists there arises the possibility of bifurcation of other types of symmetric or asymmetric homoclinic orbits as we shall explain later. Note that in addition to the primary orbit, multi-humped orbits may also be constructed analytically in path space (see figure 4(c), (d)). At the same time, the orientation of solutions in phase space (i.e. $(y, \dot{y}, \ddot{y}, \dddot{y})$-space) can underly strategies for finding even quite complex multi-humped solutions numerically (see §4).

In the meantime, let us concentrate on what may be gleaned in the vicinity of $P = +P^{\mathrm{C}}$.

3.2 In the vicinity of (positive) criticality: the onset of localization

From the structural viewpoint, our interest lies in the region $P \approx P^{\mathrm{C}}$ but the above analysis concentrated primarily on a range of values of P far from that region. In the vicinity of $P = P^{\mathrm{C}}$, the first analyses were based on a double-scale perturbation technique (Hunt *et al.*, 1989; Hunt & Wadee, 1991).

In fact, this method is equivalent to that of normal form analysis at Hamiltonian Hopf bifurcations. See Iooss & Pérouème (1993), who show rigorously the bifurcation of two small amplitude homoclinic solutions for any system satisfying a *subcriticality* inequality, and Grimshaw *et al.* (1994) who calculate the appropriate coefficients for the fifth-order KdV equation. Recently, Woods & Champneys (1998) have also utilized this method to analyse the degenerate Hamiltonian Hopf that can be used to explain the behaviour of a strut system which restabilizes after the initial unstable behaviour (see §5).

We shall concentrate here on the double-scale approach. Such a method begins with the introduction of a *slow-space* variable X

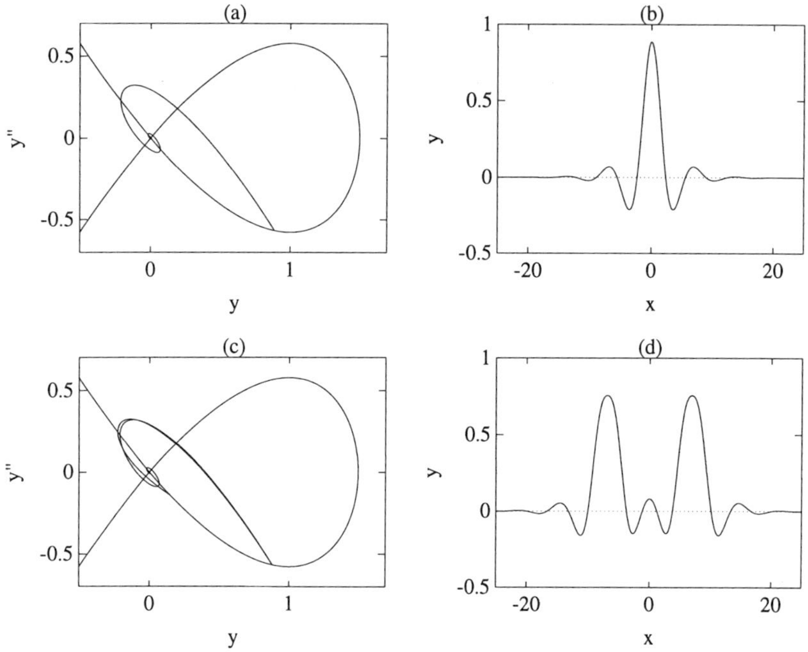

Figure 4: Illustrating the correspondence between paths and graphs for numerically calculated homoclinic solutions to (14) with $P = 1.5$. (a,b) The primary orbit; (c,d) a two-humped orbit labelled $\mathbf{2}(4)$ in the notation of Buffoni *et al.* (1997).

$$X = \varepsilon x \tag{17}$$

where ε is a perturbation parameter measuring distance from $P = P^{\mathrm{C}}$. The analysis assumes that P falls off quadratically in ε in this vicinity:

$$P = P^{\mathrm{C}} - \varepsilon^2. \tag{18}$$

The solution is assumed to be a Fourier series in x

$$y = \sum_{i=0}^{\infty} \{A_i(X) \cos i(\omega x + \phi_0) + B_i(X) \sin i(\omega x + \phi_0)\}, \tag{19}$$

with coefficients which have the ability to vary slowly with respect to x. Note the undetermined phase ϕ_0 which determines the shift of the fast oscillation with respect to the slowly-varying amplitude. Expression (19) is fed into the governing equation. By also expressing the amplitudes as formal series in ε i.e.

$$A_i = \sum_{j=1}^{\infty} \varepsilon^j A_i^{(j)} \tag{20}$$

$$B_i = \sum_{j=1}^{\infty} \varepsilon^j B_i^{(j)} \tag{21}$$

and systematically extracting terms of progressively higher order, the first order equation governing the behaviour of the amplitude of the critical mode is found to be

$$4A_1^{(1)\prime\prime} - A_1^{(1)} + \frac{19}{18} A_1^{(1)3} = 0 \tag{22}$$

where a prime denotes differentiation with respect to X (Wadee *et al.*, 1997).

There are four types of solutions to this equation (see figure 5). These are the trivial and finite-amplitude periodic solutions: respectively $A_1^{(1)} = 0$ or $A_1^{(1)} = 6/\sqrt{38}$ (marked as d in figure 5); the localized solutions, centred for clarity at $X = 0$ (denoted as f in the figure),

$$A_1^{(1)} = \frac{6}{\sqrt{19}} \operatorname{sech} \frac{1}{2} X \tag{23}$$

and the modulated solutions in between the last two (denoted e). The form of localization emerging from the critical state is thus

$$y = \varepsilon A_1^{(1)} \cos(\omega x + \phi_0) + \cdots. \tag{24}$$

The reversibility R_1 suggest that either $\phi_0 = 0$ or $\phi_0 = \pi$ giving rise to symmetric primary orbits and this is backed up by numerical evidence (figure 6). Wadee *et al.* (1997) extended the approximation in (24) by taking the analysis to higher order. However, no explicit reason for the restriction on ϕ_0 can be found using this form of analysis. In order to do so, terms *beyond all orders* of a power series expansion must be taken into account (Yang & Akylas, 1997) with the added bonus that this unlocks the restriction of only being able to find primary orbits using this perturbation method.

A detailed description of this *exponential asymptotics* method is beyond the scope of this review. We shall close the section by highlighting how the method can also reveal the complex behaviour of multi-humped solutions for P close to P^{C}.

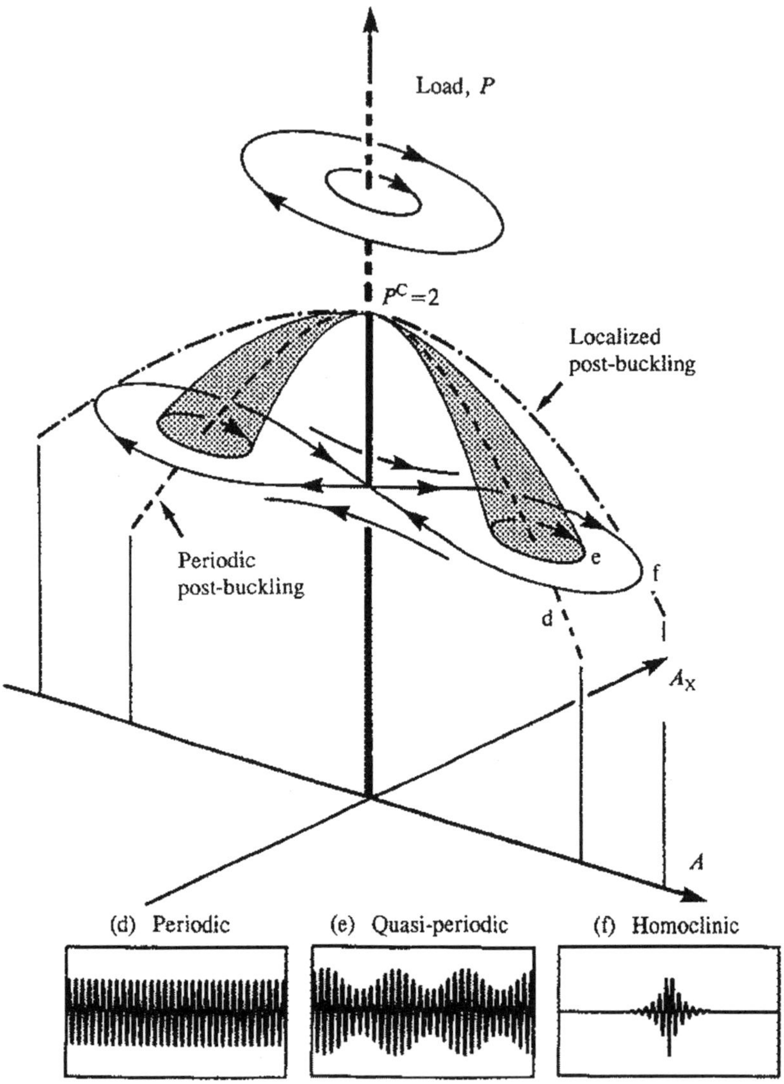

Figure 5: Emergence of slow-space solutions from the critical load.

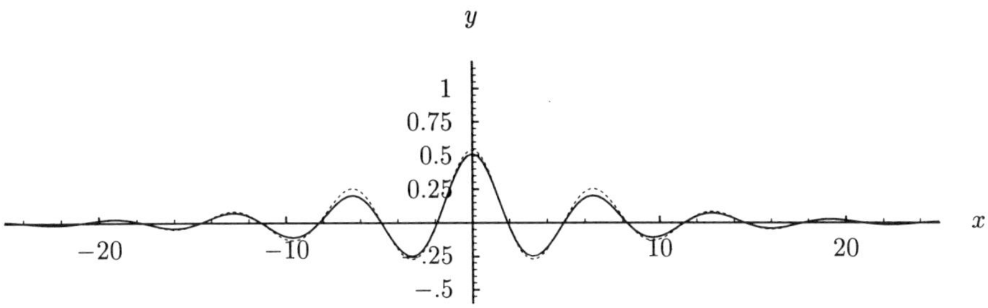

(a) $P = 1.9$

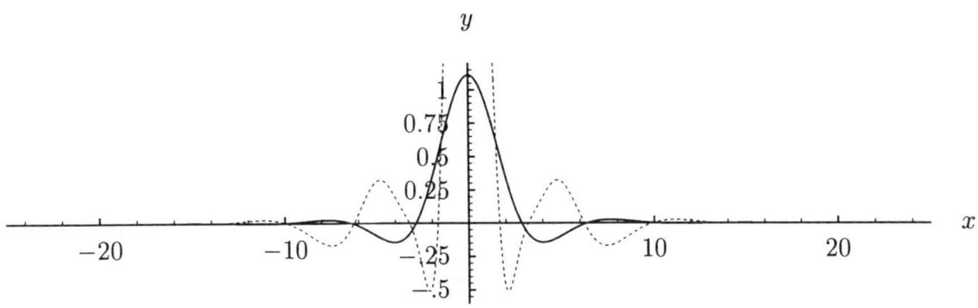

(b) $P = 1.0$

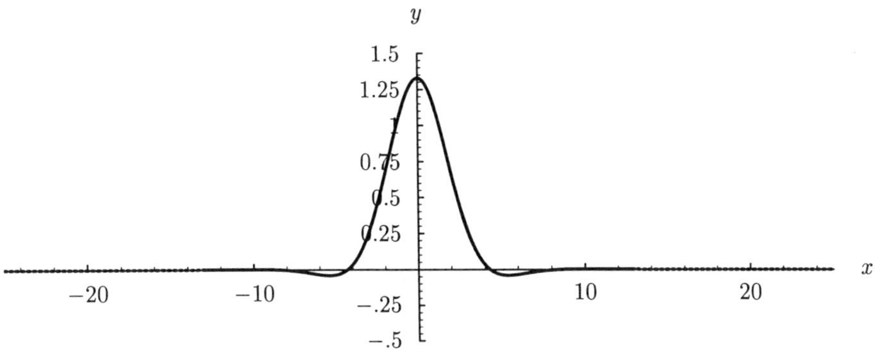

(c) $P = 0.0$

Figure 6: Primary numerical localized solutions to (14) for various axial load values. Extended perturbation approximations are shown as dashed lines in cases (a) and (b).

3.3 Multi-humped solutions as bound states of wavepackets

The method of Yang & Akylas (1997) can be further used to calculate double-hump solutions in a manner reminiscent of interactions of solitons; see also Gorshkov & Ostrovsky (1981) for a general method in the same spirit. These workers were interested in the existence of asymmetric gravity–capillary solitary waves or optical solitons with exponentially decaying oscillating tails. Starting with the results from the slow-space approach Wadee & Bassom (1998) have utilized the theory of hyperasymptotics to show which of the soliton interactions occur when the wavepackets are well separated (i.e. the interaction is weak). They worked on the "cubic" foundation, $F = y - y^3$, and found that unless ϕ_0 is restricted to the values 0, $\frac{1}{2}\pi$, π or $\frac{3}{2}\pi$ then decaying terms of the localized solution are accompanied by others which are formally exponentially small in ε but which grow exponentially large in x. The extra two particular solutions for ϕ_0 arise because of the extra reversibility, R_2, of this system (9). Despite the apparent limitation on the phase angle, it in fact allows the construction of multi-modal solutions as described by Buffoni *et al.* (1997). Situations can arise when the exponential growth in the tail of a wavepacket caused by the divergent term matches up seamlessly with that of another remote phase-shifted wavepacket. This gives rise to a double-hump solution. Furthermore, Wadee & Bassom showed that for a fixed separation of humps (in terms of wavelengths) there exist up to sixteen different combinations of solutions, which can help us to understand some of the simpler structure of figure 9 in terms of the wavepacket phase angle ϕ_0.

Consider the examples of numerically computed double humped localizations shown in figure 7. Depending on the form of each of the wavepackets, the overall solution can be symmetric, antisymmetric or asymmetric. For the (anti)symmetric case, the solution on each one branch is made up of a pair of approximate symmetric or antisymmetric wavepackets. As P is reduced along the branch, each wavepacket becomes more distinctly like a primary orbit (be it symmetric or antisymmetric). In figure 7(a)–(d) double-hump solutions are shown which are some way from the coalescence point of that branch. As we track up the branch, however, the clear distinction of the wavepackets is lost and the phase shift of the fast oscillation of each wavepacket recedes from the special values of, say, $\phi_0 = 0$ or $\frac{1}{2}\pi$ and at the coalescence point, a localized solution exists which is precisely half way between the special values (e.g. figure 7(e)). Thus we find that a transition occurs at the coalescence point where the wavepackets comprising the double-hump solution change from being both symmetric to antisymmetric while the overall (anti)symmetry is maintained. Near the limit point, there is asymptotic and numerical evidence of the existence of a symmetry-breaking bifurcation into fully asymmetric forms. Asymmetric solutions, in the case of the cubic foundation, occur when one hump is symmetric and the other is antisymmetric as seen in figure 8 (Wadee & Bassom, 1998).

While the multi-hump solutions appear to be of mathematical significance, a paper by Sandstede (1997b) has shown that when the foundation does not restabilize then all multi-hump solutions are unstable in the structural sense. However more complicated solutions become significant for systems involving restabilization of the system after the initial loss of stability at the lowest critical load as briefly discussed in §5 as appropriate for more realistic systems.

4 Global post-buckling analysis: numerical results

Figure 9 summarizes the global information that is known concerning localized buckling solutions of (14), the information for which is taken mostly from Buffoni *et al.* (1997) via a

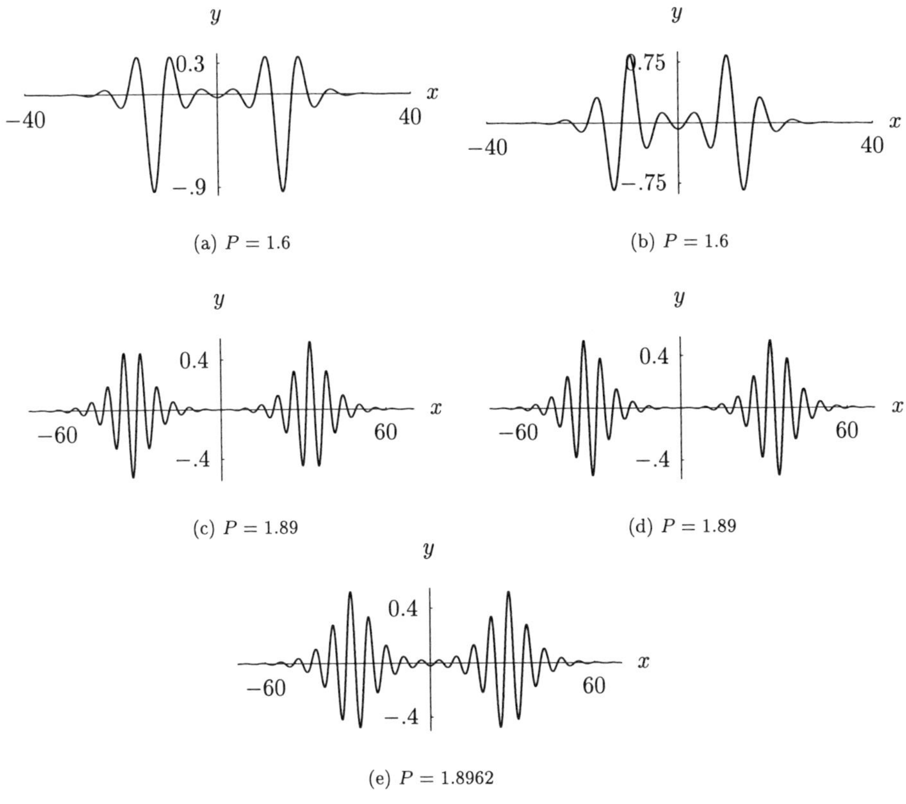

Figure 7: Symmetric and antisymmetric double-hump homoclinic solutions for the strut resting on the cubic foundation. In cases (a)–(d), each individual wavepacket is approximately a symmetric or antisymmetric primary solution whereas (e) depicts the coalescence of two branches of double-humped solutions, where the phase shift is $\phi_0 = \frac{1}{4}\pi$.

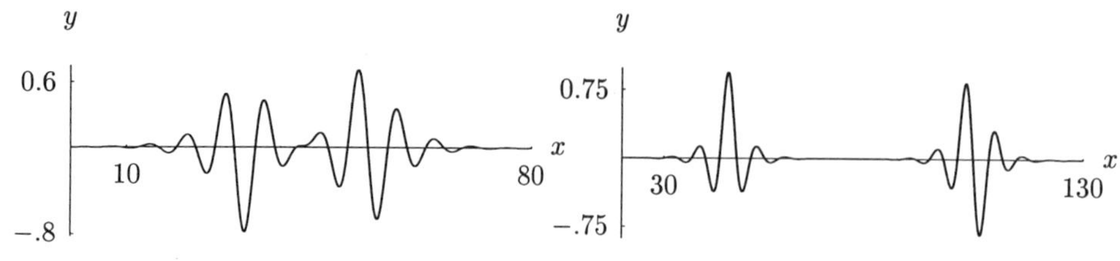

(a) $P = 1.7817$ (b) $P = 1.6078$

Figure 8: Asymmetric orbits found for the cubic foundation. Note how one wavepacket resembles a symmetric primary orbit and the other one resembles an antisymmetric orbit.

mixture of rigorous results, numerical calculations and formal arguments. Before describing some of the properties of this bifurcation diagram, we shall discuss the numerical methods used in its computation.

4.1 Numerical methods

We give a brief account of the boundary-value methods that have been used in Champneys & Toland(1993), Buffoni *et al.* (1997) and Wadee & Bassom (1998) to compute branches of homoclinic orbits as the load P varies. There are two stages; first the computation of an orbit at a fixed load P and then the continuation of the orbit as P varies. Note that a homoclinic solution satisfies a boundary-value problem on an infinite x interval. The key to performing both steps mentioned above is therefore to write a suitable boundary-value problem that approximates the full problem, but on a finite interval. The two stages of computation are then two different approaches to solving this approximate boundary-value problem. Specifically Newton-based shooting methods may be used to generate initial solutions (Champneys & Spence, 1993) and the code AUTO for continuation (see Doedel *et al.* (1997) particularly the demo `rev` in the HOMCONT section). The reader is referred to these cited works for more details on numerical implementation. We confine discussion here to the approximating boundary-value problems.

Approximate boundary-value problems for generic homoclinic orbits (which excludes the cases of homoclinic orbits in hamiltonian systems or symmetric homoclinic orbits in reversible systems) have been proposed in the broader context of computing heteroclinic connections between hyperbolic equilibria (Beyn, 1990; Friedman & Doedel, 1991). Here, to approximate the problem

$$\dot{x} = f(x, \lambda), \quad x \in \mathbb{R}^N, \quad \lambda \in \mathbb{R}, \tag{25}$$

$$x \to 0 \quad \text{as } t \to \pm\infty, \tag{26}$$

one truncates to a sufficiently large, finite time interval $t \in [-T, T]$ and solves (25) subject to

$$L_s(\lambda)x(-T) = 0, \quad L_u(\lambda)x(T) = 0, \quad \phi(x(t), \lambda) = 0, \tag{27}$$

where $L_{s,u}(\lambda)$ are projection operators onto the stable and unstable eigenspaces of $Df(0, \lambda)$, and ϕ is a scalar function used to fix the phase.

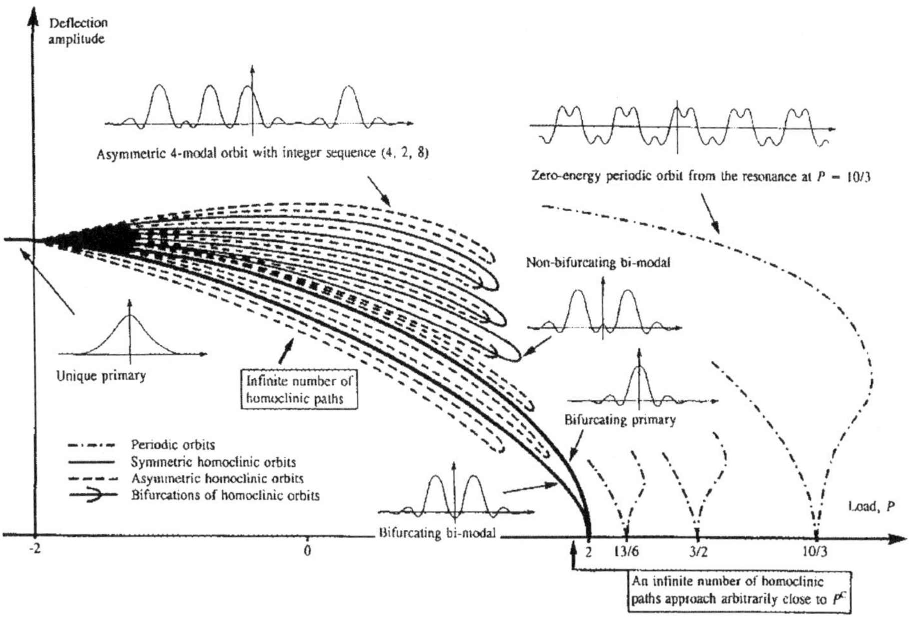

Figure 9: Schematic representation of the structure of localized and other solutions in parameter space.

Given a judicious choice of phase condition (25),(27) is a well-posed problem only if the origin is hyperbolic and homoclinic orbits exist at isolated parameter values; see Beyn (1990), Friedman & Doedel (1991) and Sandstede (1997a), where there are also results on the convergence of solutions of (25),(27) to solutions of (25),(26) as T is increased. However, homoclinic orbits to hyperbolic equilibria in hamiltonian systems typically persist as parameters are varied and so (25),(27) is not a well-posed problem in this sense for hamiltonian systems. Nevertheless, Beyn (1990) shows that if a hamiltonian system

$$\dot{q} = \frac{\partial H(q,p)}{\partial p}, \quad \dot{p} = -\frac{\partial H(q,p)}{\partial q}, \quad (q,p) \in \mathbb{R}^n \times \mathbb{R}^n \tag{28}$$

possesses a homoclinic orbit to the origin then the boundary value problem (25),(27) will be well posed if one takes $x = (q,p)$, $N = 2n$ and

$$f = \left(\frac{\partial H(q,p)}{\partial p} + \lambda \frac{\partial H(q,p)}{\partial q}, -\frac{\partial H(q,p)}{\partial q} + \lambda \frac{\partial H(q,p)}{\partial p} \right)^{\mathrm{T}}. \tag{29}$$

This extended problem is well-posed and has a regular solution with $\lambda \approx 0$, which is a good approximation to the homoclinic solution of (28).

Alternatively, a numerical method for symmetric homoclinic orbits in reversible systems was proposed in Champneys & Spence (1993). Owing to the symmetry, one only need solve an approximation to (25),(26) on the half-interval $[-T, 0]$, with a right-hand boundary condition that places the solution in the symmetric section $\mathcal{S}$ ($\{\dot{y} = \dddot{y} = 0\}$ for the present problem). Thus one solves (25) with boundary conditions

$$L_s x(-T) = 0, \quad x(0) \in \mathcal{S}. \tag{30}$$

Using this approach, one does not need a phase condition, because the phase is fixed by the right-hand boundary condition. Therefore, the parameter λ need no longer be treated as an unknown, and (25),(30) will apply to the computation of homoclinic orbits that persist as parameters are varied.

To compute asymmetric homoclinic orbits, one must use the full projection boundary conditions (27) applied to the extended system (29). For symmetric orbits, however, it is advantageous to use the symmetric section boundary condition (30), because this halves the interval of computation and leads to a more robust shooting method; see Champneys & Spence (1993). Another advantage is, using a shooting approach at fixed P, one can use an angle co-ordinate θ parametrizing solutions a fixed distance from the origin in the unstable eigenspace and the unknown time interval T as shooting parameters. Then, using the knowledge of how multi-humped solutions lie in phase space (see figure 10) and the fact that the wavelength of small amplitude oscillation is $2\pi/\omega$, one can build up accurate initial guesses for θ and T for Newton's method to compute whole sequences of multi-hump solutions systematically.

4.2 Spatial chaos and multi-humped solutions

The solutions given by the slow-space analysis are only accurate for values of P close to P^{C}. Originally when only primary solutions had been detected numerical shooting techniques suggested that there existed entire families of solutions missed by the double-scale approach (Hunt & Wadee, 1991). In fact it can be shown that there are infinitely many multi-humped solutions, argued as follows. If the primary orbit can be shown to represent the transverse intersection of stable and unstable manifolds (as numerics show to be the case for

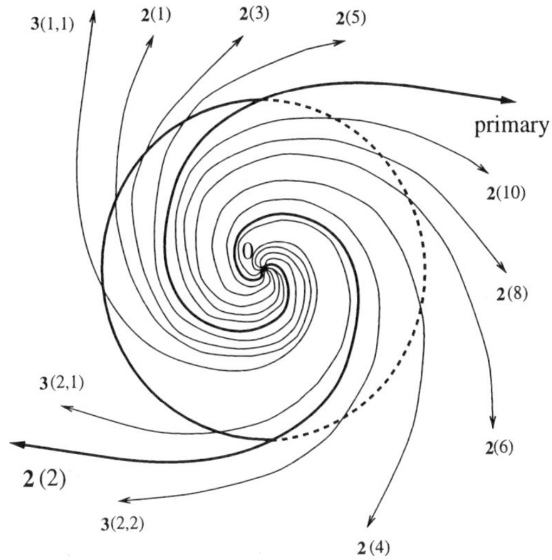

Figure 10: Schematic diagram of the ordering of the simplest multi-humped homoclinic orbits in the unstable manifold of the origin. The figure and notation for labeling of orbits is borrowed from Buffoni *et al.* (1997).

all $P \in (-P^{\mathrm{C}}, P^{\mathrm{C}})$) then the theory of Devaney (1976) shows that in the vicinity of the flat state, there exists a Smale horseshoe mapping and hence "spatial chaos". More precisely there is dynamics conjugate to a shift on infinitely many symbols. Belyakov & Shil'nikov (1990) realized that this also gives rise to a countably infinite number of multi-humped solutions. These are solutions which just miss the symmetric section but subsequently hit it close to the origin (e.g. see figure 4(c),(d) above). Buffoni *et al.* (1997) showed that it is thus possible to construct a symbolic labelling via a sequence of integers

$$\mathbf{n}(i_1, i_2, \ldots, i_{n-1})$$

of each multi-humped solution (at least for P sufficiently less than P^{C}) such that each label corresponds to a unique solution and vice versa. Here $\mathbf{n}$ counts the number of large "humps" close to the primary localized mode and the integers i_j count the number of half-oscillations between each hump. See figure 12 for some quite complex localized modes and their corresponding labels.

4.3 Coalescence and bifurcation

Further to this, Buffoni *et al.* (1997) discovered an extremely intricate bifurcation structure in the global behaviour of the different homoclinic solutions including asymmetric orbits, as depicted in figure 9. Evidence was found of the coalescence of pairs of multi-humped orbits in the range $-P^{\mathrm{C}} < P < P^{\mathrm{C}}$ with only the primary orbits associated with $\phi_0 = 0, \pi$ bifurcating at $P = P^{\mathrm{C}}$. In parameter space, numerical evidence also indicates that some asymmetric solutions bifurcate close to the limit point of double-hump solutions at a symmetry-breaking bifurcation. There are yet other such solutions which do not bifurcate from more symmetric forms. A rational explanation for these bifurcation and coalescence phenomena can be found in unfolding points of non-transverse intersections of stable and unstable manifolds along

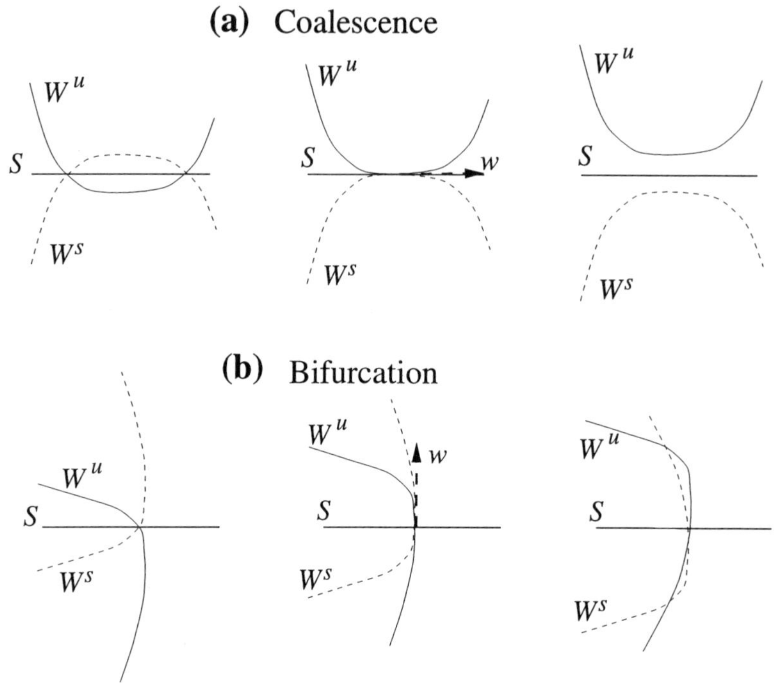

Figure 11: Unfolding under parameter variation of coalescence and bifurcation for conservative reversible systems in $\mathbb{R}^4$. Depicted are the stable and unstable manifolds W^s and W^u and the symmetric section $\mathcal{S}$ in a two-dimensional Poincaré section contained in a level set $\mathcal{H} = 0$.

homoclinic solutions (Buffoni *et al.*, 1997, Section 4). See figure 11, which has since been put on a rigorous footing by Knobloch (1997).

These coalescence and bifurcation phenomena tie in closely with the interaction calculations of the exponential asymptotics, reviewed in §3. For the quadratic foundation case Buffoni *et al.* (1997) report on numerically determined interactions involving many-humped solutions; here the interacting wavepackets can form only symmetric or asymmetric solutions. In the cubic case, as mentioned in §3.3, there is a much richer structure due to the additional reversibility.

Finally we mention that Buffoni *et al.* attempted to uncover "rules for coalescence" which govern how the labels of two symmetric orbits undergoing a coalescence may be related. As a general rule, for the quadratic nonlinearity, the process is one of an orbit with n humps gaining either 1 or 2 extra humps. For example, figure 12 shows three sets of pairs of multi-humped solutions which coalesce and annihilate each other at a slightly higher P-value.

5 Extensions of the model

The model structure which forms the basis of the work reviewed in this article gives rise to phenomena found in what may be argued to be more realistic systems. However, the initial interest was in the post-buckling behaviour of the strut in the vicinity of the critical point,

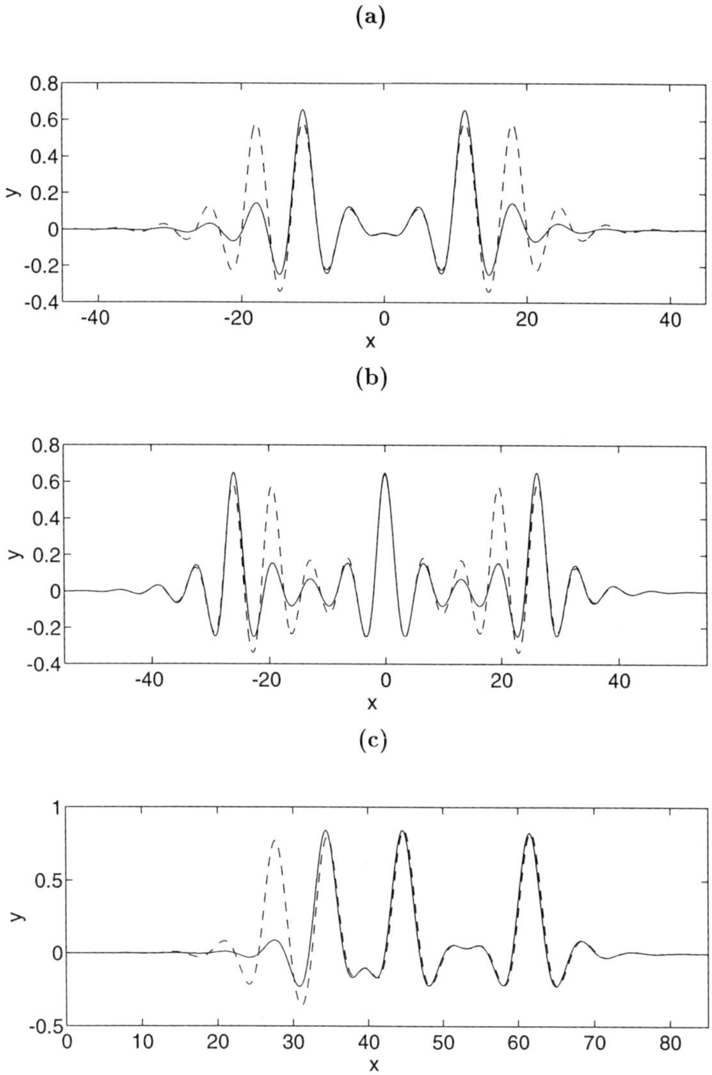

Figure 12: Graphs of pairs of multi-humped homoclinic orbits that undergo coalescence. (a) Orbits with labels $\mathbf{2}(7)$ and $\mathbf{4}(2,7,2)$ at $P = 1.8$; (b) orbits with labels $\mathbf{3}(8,8)$ and $\mathbf{5}(2,6,6,2)$ at $P = 1.8$; (c) orbits with labels $\mathbf{3}(3,5)$ and $\mathbf{4}(2,3,5)$ at $P = 1.6$.

$P = P^C$. As progress is made in the techniques of analysis, the shortcoming of the original model start to become apparent.

Firstly, the behaviour of the foundation for $y > 1$ may be argued to be of little relevance. For such large displacements, the foundation is no longer pulling back on the strut but is pushing it away. The simplified form for F thus appears to lack credibility. Even if a structure undergoes catastrophic collapse at its lowest critical load, it is likely to restabilize under the influence of higher-order terms. Indeed, if structures like the axially-compressed cylindrical shell did not have a restabilizing characteristic then their post-buckling would not be observable even under rigid loading. An obvious extension of the strut model would be to have a foundation with

$$F = y - y^2 + c_2 y^3 \tag{31}$$

with $c_2 > 0$. Such a study has been undertaken by Woods & Champneys (1998) and they use normal forms analysis to study the behaviour of the system well beyond the restabilization. For the range $c_2 < 38/27$ they show that localization is still the favoured post-buckling response. Furthermore, using a numerical solution strategy, they find a whole series of destabilization and restabilization after the first in the load–end-shortening curve. The effect is for the localization to become "smeared" out and in the limit to become periodic. In dynamical-systems terminology, a *heteroclinic* solution exists in this far post-buckling regime with the trivial solution as one asymptote and the periodic solution as the other.

We also note that such a heteroclinic analysis appears to apply to quite generally to structures which restabilize, see the forthcoming article by Hunt *et al.* (1998a). That article also indicates a connection with the Maxwell load criterion for deciding when a restabilizing structure will jump to an energetically preferred post-buckled solution; an idea which has already been successfully applied to the axially compressed cylindrical shell (Hunt & Lucena Neto, 1993; Hunt *et al.*, 1998b).

Another criticism which may legitimately be levelled at the basic model is that the nonlinearity is artificial. While this may be true, an analysis of the initial post-buckling of a strut in which elastica nonlinearities are allowed (i.e. not just truncating the strain-energy of bending and work-done-by-load terms at the first term of a Taylor series expansion as in (5)) shows that localized solutions exist even if the foundation has a linear stiffness (Hunt *et al.*, 1993). Indeed numerical experiments suggest that similar behaviour to the simpler strut model exists in the elastica model at least for the initial post-buckling regime.

Finally, we ought to mention that all of the above analyses have been on a static model. If the response of the elastic foundation is assumed to dissipate through viscous losses then the governing equation is a PDE in space and time. Localization again provides the most severe loss of stability (Wadee *et al.*, 1998). The dynamic model has been used to describe the formation of folds in geological strata (Hunt *et al.*, 1997). A recent finding is that if the end-shortening is prescribed (i.e. held constant over time) then the nonlinear integral constraint of fixed end-shortening is sufficient to cause localization—the governing equation need not be nonlinear (Budd *et al.*, 1998).

6 Conclusions

The elastic strut on an elastic foundation has proved to be a system which seems to encapsulate many of the important properties required for localized buckling. It has a simple equation which nevertheless has a great deal of complexity hiding away underneath the simple nonlinearity. It has been used as a heuristic model for systems of more direct relevance to engineers such as the axially-compressed cylindrical shell (Lord *et al.*, 1997) and the torsional buckling of the twisted rod (Thompson & Champneys, 1996). The ability to solve the system has ranged from the detection of periodic solutions a few decades ago (Thompson & Hunt, 1973) to the description of initial localized post-buckling behaviour (Hunt *et al.*, 1989), the accurate description of primary localized solutions (Wadee *et al.*, 1997) to the extremely rich structure found in parameter space including the infinitely many symmetric and asymmetric multi-hump solutions (Buffoni *et al.*, 1997). We have also said nothing in this short review about the rich topic of variational methods, which have been successfully applied to (14) to produce global existence results. The interested reader may consult Buffoni (1996) and Buffoni & Séré (1996) and references therein. There is perhaps still more that can be learnt from this little problem.

Acknowledgements

The author is grateful to Alan Champneys for his assistance and helpful discussions and in particular for drafting the section on numerical solutions. He would also like to thank Giles Hunt and Michael Thompson for kindly providing some of the figures in this review.

References

Amick, C. J., & Toland, J. F. 1992a. Global uniqueness of homoclinic orbits for a class of fourth order equations. *Z. Angew. Math. Phys.*, **43**, 591–597.

Amick, C. J., & Toland, J. F. 1992b. Homoclinic orbits in the dynamic phase space analogy of an elastic strut. *European J. Appl. Maths.*, **3**, 97–114.

Belyakov, L. A., & Shil'nikov, L. P. 1990. Homoclinic curves and complex solitary waves. *Selecta Mathematic Sovietica*, **9**, 219–228.

Beyn, W.-J. 1990. The numerical computation of connecting orbits in dynamical systems. *IMA J. Num. Anal.*, **9**, 379–405.

Budd, C. J., Hunt, G. W., & Peletier, M. A. 1998. Self-similar fold evolution under prescribed end-shortening. *J. Math. Geology*. Submitted.

Buffoni, B. 1996. Periodic and homoclinic orbits for Lorentz–Lagrangian systems via variational methods. *Nonlinear Anal. TMA*, **26**, 443–462.

Buffoni, B., & Séré, E. 1996. A global condition for quasi-random behaviour in a class of conservative systems. *Commun. Pure Appl. Math.*, **49**, 285–305.

Buffoni, B., Groves, M. D., & Toland, J. F. 1996. Plethora of solitary gravity–capillary water-waves with nearly critical Bond and Froude numbers. *Phil. Trans. R. Soc. Lond.*, A **354**, 575–607.

Buffoni, B., Champneys, A. R., & Toland, J. F. 1997. Bifurcation and coalescence of a plethora of multi-modal homoclinic orbits for a hamiltonian system. *J. Dyn. Diff. Eqns.*, **8**, 221–281.

Champneys, A. R. 1998. Homoclinic orbits in reversible systems and their applications in mechanics, fluids and optics. *Physica D*, **112**, 158–186.

Champneys, A. R., & Spence, A. 1993. Hunting for homoclinic orbits in reversible systems: A shooting technique. *Adv. Comp. Math.*, **1**, 81–108.

Champneys, A. R., & Toland, J. F. 1993. Bifurcation of a plethora of multi-modal homoclinic orbits for Hamiltonian systems. *Nonlinearity*, **6**, 665–721.

Devaney, R. L. 1976. Homoclinic orbits in hamiltonian systems. *J. Diff. Eq.*, **21**, 431–438.

Doedel, E. J., Champneys, A. R., Fairgrieve, T. F., Kuznetsov, Y. A., Sandstede, B., & Wang, X. 1997. *AUTO 97: Continuation and bifurcation software for ordinary differential equations.* Available via anonymous ftp from `ftp://ftp.cs.concordia.ca/pub/doedel/auto`.

Drazin, P. G., & Johnson, R. S. 1989. *Solitons: An introduction.* Cambridge: Cambridge University Press.

Euler, L. 1744. *Methodus inveniendi lineas curvas maximi minimive proprietate gaudentes.* Appendix: De curvis elasticis. Lausanne and Geneva.

Friedman, M. J., & Doedel, E. 1991. Numerical computation of invariant manifolds connecting fixed points. *SIAM J. Numer. Anal.*, **28**, 789–808.

Gorshkov, K. A., & Ostrovsky, L. A. 1981. Interactions of solitons in nonintegrable systems: Direct perturbation method and applications. *Physica D*, **3**, 428–438.

Grimshaw, R., Malomed, B. A., & Benilov, E. 1994. Solitary waves with damped oscillatory tails: an analysis of the fifth-order Korteweg–de Vries equation. *Physica D*, **77**, 473–485.

Hunt, G. W., & Lucena Neto, E. 1993. Maxwell critical loads for axially-loaded cylindrical shells. *Trans. ASME J. Appl. Mech.*, **60**(3), 702–706.

Hunt, G. W., & Wadee, M. K. 1991. Comparative lagrangian formulations for localized buckling. *Proc. R. Soc. Lond.*, A **434**, 485–502.

Hunt, G. W., Bolt, H. M., & Thompson, J. M. T. 1989. Structural localisation phenomena and the dynamical phase-space analogy. *Proc. R. Soc. Lond.*, **425**, 245–267.

Hunt, G. W., Wadee, M. K., & Shiacolas, N. 1993. Localized elasticæ for the strut on the linear foundation. *Trans. ASME J. Appl. Mech.*, **60**(4), 1033–1038.

Hunt, G. W., Mühlhaus, H.-B., & Whiting, A. I. M. 1997. Folding processes and solitary waves in structural geology. *Phil. Trans. R. Soc. Lond.*, A **355**, 2197–2213.

Hunt, G. W., Champneys, A. R., Woods, P. D., Peletier, M. A., Wadee, M. A., Budd, C. J., & Lord, G. J. 1998a. *Cellular buckling in long structures.* Submitted to *Nonlinear Dynamics.*

Hunt, G. W., Lord, G. J., & Champneys, A. R. 1998b. Homoclinic and heteroclinic orbits underlying the post-buckling of axially-compressed cylindrical shells. *Comp. Methods Appl. Mech. Eng.* To appear.

Iooss, G., & Kirchgassner, K. 1992. Water waves for small surface tension: An approach via normal form. *Proc. Roy. Soc. Edin. A*, **112**, 267–200.

Iooss, G., & Pérouème, M. 1993. Perturbed homoclinic solutions in reversible 1:1 resonance vector fields. *J. Diff. Eq.*, **102**, 62–88.

Kirchhoff, G. R. 1859. On the equilibrium and the movements of an infinitely thin bar. *Crelles J. Math.*, **56**.

Knobloch, J. 1997. Bifurcation of degenerate homoclinics in reversible and conservative systems. *J. Dyn. Diff. Eqns.*, **9**, 472–494.

Lichtenberg, A. J., & Lieberman, M. A. 1992. *Regular and chaotic dynamics*. New York: Springer-Verlag.

Lord, G. J., Champneys, A. R., & Hunt, G. W. 1997. Computation of localized post buckling in long axially compressed cylindrical shells. *Phil. Trans. R. Soc. Lond.*, A **355**, 2137–2150.

Newell, A. C., & Whitehead, J. A. 1969. Finite bandwidth, finite amplitude convection. *J. Fluid Mech.*, **38**, 279–303.

Potier-Ferry, M. 1983. Amplitude modulation and localization of buckling patterns. *In:* Thompson, J. M. T., & Hunt, G. W. (eds), *Collapse: The buckling of structures in theory and practice*. Cambridge: Cambridge University Press.

Sandstede, B. 1997a. Convergence estimates for the numerical approximation of homoclinic solutions. *IMA J. Numer. Anal.*, **17**, 437–462.

Sandstede, B. 1997b. Instability of localized buckling modes in a one-dimensional strut model. *Phil. Trans. R. Soc. Lond.*, A **355**, 2083–2097.

Stewartson, K., & Stuart, J. T. 1971. A non-linear instability theory for a wave system in plane convection. *J. Fluid Mech.*, **48**, 529–545.

Thompson, J. M. T., & Champneys, A. R. 1996. From helix to localized writhing in the torsional post-buckling of elastic rods. *Proc. R. Soc. Lond.*, A **452**, 117–138.

Thompson, J. M. T., & Hunt, G. W. 1973. *A general theory of elastic stability*. London: Wiley.

Toland, J. F. 1986. Existence of symmetric homoclinic orbits for systems of Euler–Lagrange equations. *Proc. Symp. Pure Math.*, **45,** Part 2, 447–459.

Tvergaard, V., & Needleman, A. 1980. On the localization of buckle patterns. *J. Appl. Mech.*, **21**, 613–619.

Wadee, M. K., & Bassom, A. P. 1998. Effects of exponentially small terms in the perturbation approach to localized buckling. *Proc. R. Soc. Lond.,* A. To appear.

Wadee, M. K., Hunt, G. W., & Whiting, A. I. M. 1997. Asymptotic and Rayleigh–Ritz routes to localized buckling solutions in an elastic instability problem. *Proc. R. Soc. Lond.*, A **453**, 2085–2107.

Wadee, M. K., Higuchi, Y., & Hunt, G. W. 1998. Galerkin approximations to static and dynamic localization problems. *Int. J. Solids Struct.* Submitted.

Woods, P. D., & Champneys, A. R. 1998. Heteroclinic tangles and homoclinic snaking in the unfolding of a degenerate reversible Hamiltonian Hopf bifurcation. *Physica D*. Submitted.

Yang, T.-S., & Akylas, T. R. 1997. On asymmetric gravity–capillary solitary waves. *J. Fluid Mech.*, **330**, 215–232.

Instability of localized buckling modes in a one-dimensional strut model

BY BJÖRN SANDSTEDE
WIAS, Mohrenstraße 39, 10117 Berlin, Germany

Stability of localized solutions arising in a fourth-order differential equation modelling struts is investigated. It was shown by Buffoni *et al.* in 1996 that the model exhibits many multimodal buckling states bifurcating from a primary buckling mode. In this article, using analytical and numerical techniques, the primary mode is shown to be unstable under dead loading for all axial loads, while it is likely to be stable under rigid loading for small axial loads. Furthermore, for general reversible or conservative systems, stability of the multimodal solutions is established assuming stability of the primary state. Since this hypothesis is not satisfied for the buckling mode arising in the strut model, any multimodal buckling state will be unstable under dead and rigid loading.

1. Introduction

In this article, localized solutions of the fourth-order ordinary differential equation

$$u_{xxxx} + Pu_{xx} + u - u^2 = 0, \quad x \in \mathbb{R}, \tag{1.1}$$

are investigated. Equation (1.1) describes equilibrium states of a strut resting on an elastic foundation with a nonlinear softening restoring force (see, for example, Hunt *et al.* 1989). Here, x and u are the arc length along the strut and vertical displacement, respectively. The parameter P denotes the axial load, while the bending stiffness has been rescaled to unity. The underlying geometry is illustrated in figure 1.

Note that (1.1) is Hamiltonian with energy given by

$$H(u) = u_x u_{xxx} - \tfrac{1}{2}u_{xx}^2 + \tfrac{1}{2}Pu_x^2 + \tfrac{1}{2}u^2 - \tfrac{1}{3}u^3.$$

By definition, *localized* solutions h of (1.1) satisfy

$$\lim_{x\to\pm\infty} h(x) = 0,$$

i.e. they correspond to homoclinic solutions of (1.1). It was shown by Amick & Toland (1992) that (1.1) has an even homoclinic solution h for each $P \in (-\infty, -2+\eta)$ for some $\eta > 0$. In addition, they proved uniqueness for $P \leqslant -2$, whence we refer to these localized solutions as the *primary* buckling modes. The solutions h are *transversely constructed*, i.e. stable and unstable manifolds of the zero equilibrium of (1.1) intersect transversely at $u = h(0)$ in the zero level set of the energy H (see Buffoni *et al.* 1996). Using the results of Devaney (1976), Buffoni *et al.* (1996) also proved that for any $P \in (-2, -2+\eta)$, with $\eta > 0$ sufficiently small, infinitely many buckling modes bifurcate from the primary state. The bifurcating equilibria are

Phil. Trans. R. Soc. Lond. A (1997) **355**, 2083–2097

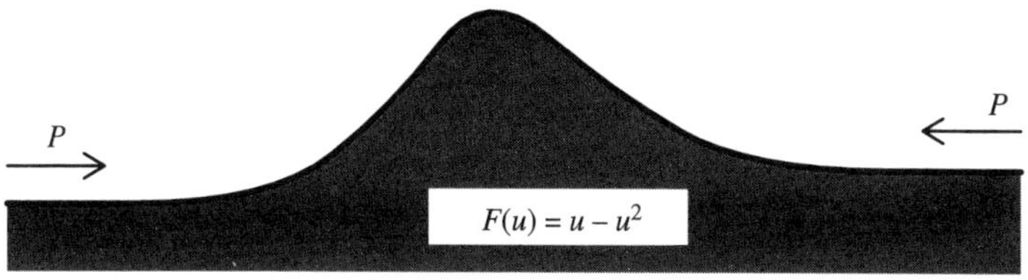

Figure 1. A strut on an asymmetric softening foundation with restoring force F under an axial load P.

multimodal solutions resembling concatenated, widely spaced copies of the primary state. There is numerical evidence that the primary buckling mode h persists up to $P = 2$ while still being transversely constructed (see Buffoni *et al.* 1996 and the references cited therein). The bifurcation to multimodal solutions mentioned above would then occur for all $P \in (-2, 2)$.

It is an interesting problem to determine whether these localized buckling modes are stable in an appropriate sense. In the present context, we adopt the notion of stability used in mechanical engineering and minimize the energy. For given axial load P, the total potential energy of localized equilibria is given by

$$W(u) = \frac{1}{2}\int_{-\infty}^{\infty} (u_{xx}^2 - Pu_x^2 + u^2 - \tfrac{2}{3}u^3)\mathrm{d}x, \tag{1.2}$$

(see, for example, Hunt *et al.* 1989). We consider two different loading devices for (1.1). Under *dead loading*, the deflection u adjusts according to (1.1). Accessing stability of an equilibrium under dead loading is then equivalent to minimizing the total energy $W(u)$ for fixed P (see Thompson & Hunt 1973). This task can be accomplished by verifying positive definiteness of the second variation $\nabla^2 W(h)$ of W given by

$$L(h)v := \nabla^2 W(h)v = v_{xxxx} + Pv_{xx} + (1 - 2h)v, \tag{1.3}$$

at a buckling state h. Note that, on account of translation invariance of $W(u)$, $L(h)$ has an eigenvalue at zero. Under *rigid loading*, the axial load P and the total displacement

$$\mathcal{E}(u) = \frac{1}{2}\int_{-\infty}^{\infty} u_x^2 \,\mathrm{d}x, \tag{1.4}$$

are fixed. We then have to minimize $W(u)$ under the additional constraint

$$\mathcal{E}(u) = \text{const.},$$

(see Thompson 1979, ch. 9). This amounts to proving positive definiteness of $L(h)$ restricted to the kernel of $\nabla\mathcal{E}(h)$.

In this paper, stability of localized solutions h of (1.1) is addressed under both dead and rigid loading by analytical and numerical techniques. In § 2, stability of the primary buckling state is investigated. It is shown by analytical means that the primary mode is unstable under dead loading for all $P < 2$. For $P \in (-\infty, -2 + \eta)$, we prove that there is precisely one negative eigenvalue. For struts under rigid loading, an integral condition for stability is derived analytically. Yet we are not able to verify this condition rigorously. However, using numerical techniques, it is likely that the primary state is unstable under rigid loading for $P < P_*$ and stable for $P > P_*$, where $P_* \approx 0.8175$. In § 3, we address the issue of stability of the multimodal states existing for $P \in (-2, 2)$. It is shown that all multimodal solutions consisting of widely spaced copies of the primary pulse are unstable under dead and rigid loading. Moreover, we determine n critical eigenvalues near zero of the second variation $\nabla^2 W$ evaluated at n-modal buckling states. They can be used to characterize obstacles for the coalescence of different multimodal states in saddle-node or pitchfork bifurcations. The results presented in § 3 hold for fairly general systems under generic assumptions on the nature of the primary mode. Thus, they may be applicable to equations for which the primary mode is stable guaranteeing existence of infinitely many stable multimodal states. In addition, existence of multimodal states is shown for $2m$-dimensional equations extending previous results obtained by Devaney (1976), Champneys (1994) and Härterich (1993). Finally, in § 4, we comment on other equations exhibiting similar phenomena.

2. Instability of the primary buckling mode

We shall determine the spectrum of the second variation

$$L(h)v = v_{xxxx} + Pv_{xx} + (1-2h)v, \tag{2.1}$$

of the total energy evaluated at a localized solution, h, of equation (1.1). The operator $L(h)$ is self-adjoint in $L^2(\mathbb{R})$ with domain $H^4(\mathbb{R})$. Therefore, its spectrum is confined to the real line. Throughout, we denote the L^2-scalar product by $\langle \cdot\,, \cdot \rangle$.

(a) *Dead loading*

In this section, we show that the primary buckling state is unstable under dead loading. As mentioned in § 1, this is tantamount to proving the existence of at least one negative eigenvalue of the second variation $L(h)$ of the total energy W. Indeed, the potential energy is then not minimized and any small imperfection will cause the mechanical system to snap into a different stable equilibrium.

Lemma 2.1. *Let $P < 2$ and h be a localized solution of (1.1).*
(i) The essential spectrum of $L(h)$ is then given by $\sigma_{\mathrm{ess}}(L(h)) = [1, \infty)$.
(ii) The operator $L(h)$ has at least one negative eigenvalue.
(iii) If the localized solution h is transversely constructed, zero is a simple eigenvalue of $L(h)$. In particular, when P is varied, the number of negative eigenvalues stays constant until h ceases to be transversely constructed.

Proof. Since $P < 2$, $h(x)$ converges to zero exponentially as $|x|$ tends to infinity. Therefore, the essential spectrum of $L(h)$ coincides with the interval $[1, \infty)$ (see Henry 1981, Appendix to § 5). We prove (ii) next. Multiplying (1.1) with h, integrating over x, and using $P < 2$, we have

$$\int_{-\infty}^{\infty} h^3 \,\mathrm{d}x = \int_{-\infty}^{\infty} (h_{xx}^2 - Ph_x^2 + h^2)\mathrm{d}x > \int_{-\infty}^{\infty} (h_{xx} + h)^2 \mathrm{d}x \geqslant 0.$$

Therefore,

$$\langle L(h)h, h \rangle = \int_{-\infty}^{\infty} (h_{xx}^2 - Ph_x^2 + h^2 - 2h^3)\mathrm{d}x = -\int_{-\infty}^{\infty} h^3 \,\mathrm{d}x < 0.$$

Thus, we conclude the existence of (at least) one negative eigenvalue of $L(h)$.

Any eigenvalue of $L(h)$ has geometric multiplicity less or equal to two. Since $L(h)$ is self-adjoint, all eigenvalues are semi-simple. If h is transversely constructed, it is clear that the eigenvalue zero with eigenfunction h_x is simple. Therefore, no eigenvalues can cross the imaginary axis upon changing P. On the other hand, since $L(h)$ is sectorial, no eigenvalues can escape to minus infinity either. ■

Next, we consider the transversely constructed primary buckling state which exists for $P < -2 + \eta$ by results of Buffoni *et al.* (1996).

Lemma 2.2. *Let $P \in (-\infty, -2+\eta)$ and consider the operator $L(h)$ evaluated at the transversely constructed primary buckling mode. There exists then precisely one negative eigenvalue λ_0 of $L(h)$.*

Proof. Due to lemma 2.1(ii), the operator has at least one negative eigenvalue. It remains to prove uniqueness. On account of transversality of the primary mode and lemma 2.1(iii), it suffices to prove the claim for $P \to -\infty$.

We exploit a coordinate transformation introduced by Amick & Toland (1992) for $P \to -\infty$. Let $-P = \sqrt{\epsilon} + (1/\sqrt{\epsilon})$ and $y = \epsilon^{1/4}x$. In the new coordinates, the localized solutions are denoted $h_\epsilon(y)$ making the dependence on the parameter ϵ explicit. By (Amick & Toland 1992), the continuous family h_ϵ converges in the sup-norm toward $h_0(y) := \frac{3}{2}\operatorname{sech}^2(\frac{1}{2}y)$ as $\epsilon \to 0$. The operator $L(h)$ transforms into

$$L_\epsilon v := \epsilon v_{yyyy} - (1+\epsilon)v_{yy} + (1 - 2h_\epsilon(y))v. \tag{2.2}$$

The eigenvalue problem for L_0 at $\epsilon = 0$ reads

$$-v_{yy} + (1 - 3\operatorname{sech}^2(\tfrac{1}{2}y))v = \lambda v, \quad v \in H^4(\mathbb{R}). \tag{2.3}$$

Changing coordinates according to $z = \tanh(\frac{1}{2}y)$ transforms (2.3) into the Legendre equation

$$(1-z^2)v_{zz} - 2zv_z + \left(12 - \frac{4(1-\lambda)}{1-z^2}\right)v = 0, \quad z \in [-1, 1].$$

Applying Abramowitz & Stegun (1972, ch. 8), it is straightforward to calculate the eigenvalues λ_n of (2.3). We obtain $\lambda_0 = -\frac{5}{4}$, $\lambda_1 = 0$, and $\lambda_2 = \frac{3}{4}$. The eigenfunctions associated with λ_0 and λ_1 are $v_0(y) = \operatorname{sech}^3(\frac{1}{2}y)$ and $v_1(y) = (\mathrm{d}/\mathrm{d}y)h_0(y)$, respectively. Since L_0 is sectorial, there exists then a constant $K > 0$ such that

$$\langle L_0 v, v\rangle \geqslant K|v|^2_{H^1(\mathbb{R})}, \quad \text{for all } v \in (\operatorname{span}\{v_0, v_1\})^\perp \cap H^1(\mathbb{R}).$$

For positive $\epsilon > 0$ sufficiently small and all $v \in (\operatorname{span}\{v_0, v_1\})^\perp \cap H^2(\mathbb{R})$, we have therefore

$$\begin{aligned}\langle L_\epsilon v, v\rangle &= \int_{-\infty}^{\infty} (\epsilon v_{yy}^2 + (1+\epsilon)v_y^2 + (1-2h_\epsilon)v^2)\,\mathrm{d}y \\ &\geqslant \epsilon(|v_{yy}|^2_{L^2(\mathbb{R})} + |v_y|^2_{L^2(\mathbb{R})}) + K|v|^2_{H^1(\mathbb{R})} - 2|h_\epsilon - h_0|_{C^0(\mathbb{R})}|v|^2_{L^2(\mathbb{R})} \\ &\geqslant \tfrac{1}{2}K|v|^2_{L^2(\mathbb{R})},\end{aligned} \tag{2.4}$$

since $|h_{\epsilon_n} - h_0|_{C^0(\mathbb{R})} \to 0$ as $n \to \infty$. We will now argue by contradiction in order to prove uniqueness of the negative eigenvalue of the operator L_ϵ for $\epsilon > 0$. Thus, assume that there exists a sequence $\epsilon_n \searrow 0$ such that L_{ϵ_n} has two or more strictly negative eigenvalues. Note that zero is always a (simple) eigenvalue. The generalized eigenspace spanned by the associated eigenfunctions has therefore dimension greater or equal to three. In particular, its intersection with the space $(\operatorname{span}\{v_0, v_1\})^\perp$ is non-trivial. For any n, choose a function w_n with $|w_n|_{L^2(\mathbb{R})} = 1$ in this intersection. Note that $w_n \in H^2(\mathbb{R})$ since eigenfunctions are actually smooth. The inequality

$$\langle L_{\epsilon_n} w_n, w_n\rangle \leqslant 0$$

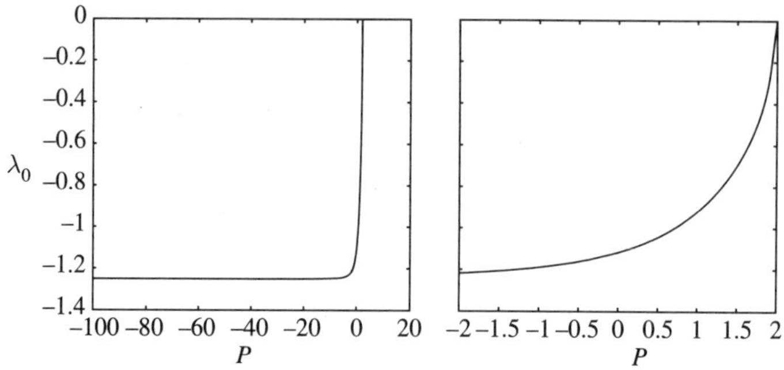

Figure 2. The negative eigenvalue λ_0 of the operator $L(h)$ at the primary buckling mode. For $P = -100$, we have $\lambda_0 = -1.24998 \approx -\frac{5}{4}$.

then holds. Substituting (2.4), we obtain

$$0 \geqslant \langle L_{\epsilon_n} w_n, w_n \rangle \geqslant \tfrac{1}{2} K |w_n|^2_{L^2(\mathbb{R})},$$

and reach a contradiction. This proves the lemma. ∎

Combining lemmata 2.1(iii) and 2.2, we see that there exists a unique negative eigenvalue of $L(h)$ as long as h is transversely constructed.

We computed the negative eigenvalue λ_0 of $L(h)$ using the driver HOMCONT (see Champneys *et al.* 1995, 1996) for the software package AUTO written by Doedel & Kernévez (1986). Projection boundary conditions with respect to the constant-coefficient operator $v_{xxxx} + P v_{xx} + (1 - \lambda) v$ are employed for the eigenfunction v_0. The initial guess $v_0(y) = \operatorname{sech}^3(\frac{1}{2} y)$, $\lambda_0 = -\frac{5}{4}$ at $P = -100$ has been used. Continuation in P shows that the negative eigenvalue persists up to $P = 2$ (see figure 2).

(*b*) *The dynamical problem*

Vibrations of the primary mode under dead loading are governed by the nonlinear equation:

$$u_{tt} + u_{xxxx} + P u_{xx} + u - u^2 = 0 \quad x \in \mathbb{R} \tag{2.5}$$

(see Lindberg & Florence 1987). We rewrite (2.5) as the first-order system

$$\begin{pmatrix} u_t \\ v_t \end{pmatrix} = \begin{pmatrix} v \\ -u_{xxxx} - P u_{xx} - u + u^2 \end{pmatrix}. \tag{2.6}$$

The linearization of (2.6) at a buckling mode $(u, v) = (h, 0)$ is given by

$$\mathcal{L}(h, 0) = \begin{pmatrix} 0 & \mathrm{id} \\ -\partial_{xxxx} - P\partial_{xx} - \mathrm{id} + 2h & 0 \end{pmatrix} = \begin{pmatrix} 0 & \mathrm{id} \\ -L(h) & 0 \end{pmatrix}.$$

Equation (2.6) generates a C^0-semiflow on the space $H^2(\mathbb{R}) \times L^2(\mathbb{R})$ (see, for example, Pazy 1983). For $P < 2$, $L(h)$ has at least one negative eigenvalue λ_0 (see lemma 2.1). Thus, the operator L has two eigenvalues $\pm\sqrt{-\lambda_0}$ on the real axis. The following lemma is then a consequence of Grillakis (1988, Appendix).

Lemma 2.3. *Any buckling mode of (1.1) under dead loading is unstable with respect to equation (2.5). In other words, there are a neighbourhood U of $(h, 0)$ and initial conditions arbitrary close to $(h, 0)$ such that the associated solutions leave the set U.*

The conclusion of lemma 2.3 is also true for parabolic or damped hyperbolic versions of equation (2.5).

(c) Rigid loading

Rigid loading is characterized by introducing an additional constraint: only imperfections with a prescribed total deflection are admissible. We therefore consider the system

$$u_{xxxx} + Pu_{xx} + u - u^2 = 0, \quad \mathcal{E}(u) = \frac{1}{2}\int_{-\infty}^{\infty} u_x^2 \,\mathrm{d}x = \text{const.}$$

Stability of the buckling mode h is equivalent to proving positive definiteness of the second variation $L(h)$ defined in (1.3) restricted to the kernel of the gradient $\nabla\mathcal{E}(h)$ of the constraint $\mathcal{E}$. The gradient $\nabla\mathcal{E}(h) : L^2(\mathbb{R}) \to \mathbb{R}$ is given by

$$\nabla\mathcal{E}(h)v = \langle h_{xx}, v\rangle = \int_{-\infty}^{\infty} h_{xx} v \,\mathrm{d}x.$$

First, we exploit that the functional $\mathcal{E}(u)$ is invariant under translation. As a consequence, $h_x \in \ker \nabla\mathcal{E}(h)$ since $\int_{-\infty}^{\infty} h_{xx}h_x \,\mathrm{d}x = 0$ by integration by parts. We may therefore restrict $L(h)$ to the space

$$H := \ker \nabla\mathcal{E}(h) \cap (\operatorname{span}\{h_x\})^{\perp} \subset L^2(\mathbb{R}).$$

Note that H is of codimension two. We have then to verify that

$$\langle L(h)v, v\rangle > 0, \quad v \in H. \tag{2.7}$$

Remark 2.4. *A buckling state h can only be stable under rigid loading if h is transversely constructed and the operator $L(h)$ has a unique negative eigenvalue.*

Proof. By lemma 2.1, $L(h)$ has at least two eigenvalues less or equal to zero. Assume that there are at least three of them. There is then a three-dimensional subspace W of $L^2(\mathbb{R})$ such that $h_x \in W$ and $\langle L(h)w, w\rangle \leqslant 0$ for all $w \in W$. Therefore, $W \cap H$ is at least one dimensional, and (2.7) is not satisfied. ■

It remains therefore to consider transversely constructed localized solutions h. We remark that whenever a localized solution h is transversely constructed at $P = P_0$ there exists a unique family $h(P)$ of buckling modes defined for P close to P_0 such that $h(P_0) = h$. With this remark in mind, we formulate the next lemma.

Lemma 2.5. *Let $P < 2$. Assume that h is transversely constructed such that $L(h)$ has precisely one negative eigenvalue. Condition (2.7) is then met if, and only if,*

$$\frac{\mathrm{d}}{\mathrm{d}P}\mathcal{E}(h(P)) < 0. \tag{2.8}$$

Here, $h(P)$ denotes the family of localized solutions mentioned before the lemma.

In other words, $h(P)$ is stable at $P = P_0$ under rigid loading if, and only if, $\mathcal{E}(h(P))$ decreases strictly near P_0.

Proof. Since zero is a simple eigenvalue of $L(h)$ by assumption, the operator $L(h)$ is invertible on $(\operatorname{span}\{h_x\})^{\perp}$. It is a consequence of Alexander *et al.* (1996), proof of lemma 2 (see also Maddocks (1985)), that $L(h)$ is positive on H if, and only if,

$$\langle L(h)^{-1}h_{xx}, h_{xx}\rangle < 0. \tag{2.9}$$

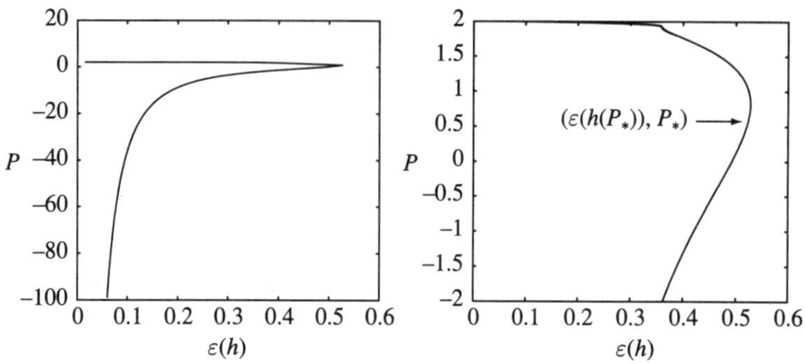

Figure 3. The integral constraint $(\mathcal{E}(h(P)), P)$. The function $\mathcal{E}(h(P))$ is increasing for $P < P_*$ and decreasing for $P > P_*$. At $P = P_* \approx 0.8175$ an exchange of stability occurs. There is an anomaly on the top of the right-hand curve which can also be found in Wadee *et al.* (1996, figure 6).

On the other hand, $h(P)$ satisfies (1.1) for all P close to P_0. Differentiating (1.1) with respect to P, we see that $(\mathrm{d}/\mathrm{d}P)h(P_0)$ satisfies $L(h)(\mathrm{d}/\mathrm{d}P)h(P_0) = -h_{xx}$. Hence, $L(h)^{-1}h_{xx} = -(\mathrm{d}/\mathrm{d}P)h(P_0)$, and (2.9) reads

$$\left\langle \frac{\mathrm{d}}{\mathrm{d}P}h(P_0), h_{xx} \right\rangle > 0.$$

Differentiating $\mathcal{E}(h(P)) = \int_{-\infty}^{\infty} h_x^2(P)\,\mathrm{d}x$ with respect to P and integrating by parts, we conclude that

$$\frac{\mathrm{d}}{\mathrm{d}P}\mathcal{E}(h(P)) = -\left\langle \frac{\mathrm{d}}{\mathrm{d}P}h(P_0), h_{xx} \right\rangle < 0.$$

Thus, (2.7) and (2.8) are equivalent. ■

We were not able to verify condition (2.8) rigorously. Numerical simulations using HOMCONT suggest that the primary buckling state is stable under rigid loading for $P \in (P_*, 2)$, while it is unstable for $P \in (-\infty, P_*)$ (see figure 3). Here, $P_* \approx 0.8175$.

3. Stability of multimodal solutions

Consider equation (1.1) and the eigenvalue problem for (1.3)

$$u_{xxxx} + Pu_{xx} + u - u^2 = 0, \tag{3.1}$$

$$L(u)v = v_{xxxx} + Pv_{xx} + (1 - 2u)v = \lambda v. \tag{3.2}$$

We assume that a transversely constructed localized solution h_1 of (3.1) has been found. Then, localized *multimodal* states h_n are sought which resemble n copies of h_1 widely spaced in x. Alexander *et al.* (1990) proved that $L(h_n)$ has n eigenvalues counted with multiplicity near any eigenvalue of $L(h_1)$. In particular, invoking lemma 2.1(iii), there are precisely n eigenvalues of $L(h_n)$ close to zero. In § 3 *a*, we will calculate these eigenvalues. To achieve this, we rewrite (3.1) and (3.2) as first-order systems

$$U_x = f(U), \tag{3.3}$$
$$V_x = (Df(U) + \lambda B)V, \tag{3.4}$$

where $U, V \in \mathbb{R}^4$, $f(U) := (U_2, U_3, U_4, -PU_3 - U_1 + U_1^2)$ and $BV := (0,0,0,V_1)$. Section 3 *b* contains stability results for the strut equation under both dead and rigid loading as well as some remarks on exclusion principles for coalescence of multimodal states.

(*a*) *The linearized eigenvalue problem*

We will consider

$$u_x = f(u), \tag{3.5}$$

$$v_x = (Df(u) + \lambda B)v, \tag{3.6}$$

for $u, v \in \mathbb{R}^{2m}$ and $\lambda \in \mathbb{C}$. Here, $B \in \mathbb{R}^{2m \times 2m}$, and $f : \mathbb{R}^{2m} \to \mathbb{R}^{2m}$ is smooth with $f(0) = 0$.

Hypothesis 3.1. *Assume that the spectrum $\sigma(Df(0))$ contains simple eigenvalues $\pm\alpha \pm \mathrm{i}\beta$ for some $\alpha, \beta > 0$. The modulus of the real part of any other eigenvalue of $Df(0)$ is strictly larger than α.*

We assume that q_1 is a transversely constructed localized solution of (3.5). As a consequence, the equation

$$w_x = -Df(q_1)^* w, \tag{3.7}$$

has, up to constant multiples, a unique bounded solution $\psi(x)$ (see Sandstede 1997). In fact,

$$\psi(x) \perp (T_{q_1(x)}W^s(0) + T_{q_1(x)}W^u(0)).$$

Suppose now that at least one of the following two hypotheses is satisfied:

Hypothesis 3.2. (Reversible systems).
(i) Suppose that $R : \mathbb{R}^{2m} \to \mathbb{R}^{2m}$ is a linear involution with $\dim \operatorname{Fix} R = m$ such that $f(Ru) = -Rf(u)$ for all u. Furthermore, assume that q_1 is symmetric, i.e. $q_1(x) = -Rq_1(-x)$.
(ii) Assume that

$$\lim_{x\to\infty} \mathrm{e}^{2\alpha x} |q_1(x)||\psi(x)| > 0.$$

or

Hypothesis 3.3. (Conservative systems).
(i) Suppose that $H : \mathbb{R}^{2m} \to \mathbb{R}$ is a smooth function such that $\nabla H(q_1(0)) \neq 0$ and $\langle f(u), \nabla H(u)\rangle = 0$ for all u.
(ii) In addition, suppose that

$$\lim_{x\to\infty} \mathrm{e}^{2\alpha x} |q_1(x)||q_1(-x)| > 0.$$

Hypotheses 3.2 and 3.3 are satisfied for generic reversible or conservative systems. Note that any Hamiltonian system is in particular conservative as energy is preserved. In this case, the solution $\psi(x)$ of (3.7) is readily computed.

Remark 3.4. *Suppose that hypothesis 3.3 is satisfied. It is then straightforward to check that $\psi(x) = \nabla H(q_1(x))$. In particular, we obtain the asymptotic expansion $\psi(x) = \nabla^2 H(0) q_1(x) + O(|q_1(x)|^2)$.*

Finally, we assume a Melnikov-type condition.

Hypothesis 3.5. $M := \int_{-\infty}^{\infty} \langle \psi(x), B(\mathrm{d}/\mathrm{d}x) q_1(x)\rangle \,\mathrm{d}x \neq 0.$

We then define the sets

$$\mathcal{R} = \left\{ \exp\left(-\frac{2\pi\alpha}{\beta} n \right); n \in \mathbb{N}_0 \right\} \cup \{0\}, \quad \mathcal{A} = \left\{ \exp\left(-\frac{\pi\alpha}{\beta} k \right); k \in \mathbb{N}_0 \right\}. \quad (3.8)$$

Note that $\mathcal{R}$ is a closed metric space. These sets will be used to characterize multi-modal solutions. We need one more definition. Suppose that hypothesis 3.1 is satisfied. A sequence $(x_j)_{j=1,\dots,k}$ is then called *admissible* if, and only if, $x_{k+1-j} = x_j$. In case hypothesis 3.2 holds, *any* sequence is called admissible. The main theorem can now be stated.

Theorem 3.6. *Let q_1 be a transversely constructed localized solution of (3.1). Assume that hypotheses 3.1 and 3.5, and, in addition, 3.2 or 3.3 (or both) are satisfied. There exists then a $\delta > 0$ such that for any $n \geqslant 2$ the following holds.*

For any admissible sequence $a_j^0 \in \mathcal{A}$ with $j = 1, \dots, n-1$ and $a_i^0 \in \{1, \exp(-\pi\alpha/\beta)\}$ for some i, there exists an $r_0 \in \mathcal{R}$, $r_0 \neq 0$, with the following property.

(i) There are C^0-functions $a_j(r) \in \mathbb{R}$ for $r \in \mathcal{R}$, $r \leqslant r_0$, with $a_j(0) = a_j^0$ for $j = 1, \dots, n-1$.

(ii) For any $r \in \mathcal{R}$ with $0 < r \leqslant r_0$, there exists an n-modal solution q_n. The distances between consecutive copies of q_1 in the n-modal solution q_n are given by

$$L_j(r) = -\frac{1}{\alpha} \ln(a_j(r) r) + \tilde{L}, \quad j = 1, \dots, n-1,$$

for some constant $\tilde{L}$.

(iii) The n-modal orbits satisfying (ii) are unique. If hypothesis 3.2 holds, they are in addition symmetric.

Denote the natural numbers associated with the $a_j^0 \in \mathcal{A}$ chosen above by k_j^0. There are then precisely n solutions λ_j of (3.6) evaluated at q_n inside a ball of radius δ around zero, and

(iv) for $M > 0$ ($M < 0$), we have

$$\#\{j; 1 \leqslant j \leqslant n-1, \operatorname{Re} \lambda_j < 0\} = \#\{j; 1 \leqslant j \leqslant n-1, k_j^0 \text{ is odd (even)}\}$$
$$\#\{j; 1 \leqslant j \leqslant n-1, \operatorname{Re} \lambda_j > 0\} = \#\{j; 1 \leqslant j \leqslant n-1, k_j^0 \text{ is even (odd)}\}$$

counted with multiplicity. Moreover, $\lambda_n = 0$ is a simple eigenvalue.

Note that the natural numbers $k_j^0 + n^0(r)$ associated with each a_j^0 and r, and hence with any n-modal state described in theorem 3.6, can be interpreted as the number of half-twists the n-modal state makes near the zero equilibrium.

Existence results of multimodal solutions in four dimensions have been first proved by Devaney (1976), Champneys (1994) and Härterich (1993). There are no rigorous stability results for multimodal solutions. Buryak & Akhmediev (1995) argued formally for a coupled nonlinear Schrödinger equation that multimodal solutions should be unstable; however, they used a criterion which is only necessary but not sufficient for instability.

Applying theorems 1 and 3 of Sandstede (1995) provides another way of extending the existence results from four- to higher-dimensional systems. In addition, Sandstede (1995, theorem 1) shows that the recurrent dynamics near the primary mode q_1 is confined to a four-dimensional locally invariant and normally hyperbolic 'centre' manifold W^c_{hom} containing q_1.

Before we prove theorem 3.6 in § 3 *c*, some consequences for the strut model are presented.

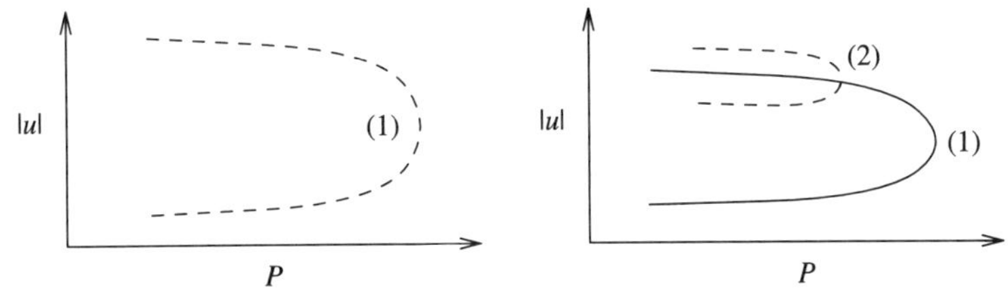

Figure 4. Coalescence of symmetric (solid) and asymmetric (dashed) multimodal solutions via saddle-node (1) and pitchfork (2) bifurcations (see Buffoni *et al.* 1996, figure 24).

(*b*) *Consequences for the strut model*

We have the following instability result for multimodal states.

Lemma 3.7. *Suppose that h is a localized solution of (1.1). Any multimodal state composed of copies of h is then unstable under dead and rigid loading.*

Proof. Note that $L(h)$ has at least one negative eigenvalue due to lemma 2.1(ii). We denote the multimodal state by h_n and assume that consecutive humps are widely separated. It follows then from Alexander *et al.* (1990) that $L(h_n)$ has at least n negative eigenvalues. Therefore, any multimodal state is unstable under dead loading. That it is also unstable under rigid loading follows from remark 2.4. ∎

For $P \in (-2, -2+\eta)$ with $\eta > 0$ sufficiently small, Buffoni *et al.* (1996) proved that the primary buckling mode is transversely constructed and that hypothesis 3.1 is met. Hypotheses 3.2 and 3.3 are also satisfied. Note that hypothesis 3.2(ii) and 3.3(ii) are automatically met since (1.1) is four dimensional. Finally, hypothesis 3.5 is true. Indeed, using remark 3.4, and the definitions $q = (h, h_x, h_{xx}, h_{xxx})$ and $Bq_x = (0, 0, 0, h_x)$, we obtain

$$M = \int_{-\infty}^{\infty} \langle \psi, Bq_x \rangle \mathrm{d}x = \int_{-\infty}^{\infty} \langle \nabla H(h(x)), (0, 0, 0, h_x(x)) \rangle \mathrm{d}x = \int_{-\infty}^{\infty} h_x^2 \, \mathrm{d}x > 0.$$

Therefore, theorem 3.6 applies to (1.1). Buffoni *et al.* (1996) observed numerically the coalescence of symmetric and asymmetric multimodal buckling states (see figure 4). Generically, coalescence is expected to occur via saddle-node or pitchfork bifurcations in the underlying partial differential equation.

Such bifurcations are related to an exchange of stability of the contributing n-modal localized solutions. In particular, generically, precisely one eigenvalue should cross the imaginary axis at zero; remember that the operators involved are self-adjoint. Therefore, the indices, i.e. the number of negative eigenvalues, of the involved n-modal solutions should differ by one. Note that we have to take the n negative eigenvalues near $\lambda_0 < 0$ into account. That immediately prevents coalescence of n_1-modal and n_2-modal solutions whenever $n_1 > 2n_2$. Furthermore, observe that the sequence $(k_j^0)_{j=1,\ldots,n-1}$ associated with each n-modal symmetric state is admissible, i.e. $k_j^0 = k_{n-j}^0$ holds for all j. In particular, eigenvalues of n-modal symmetric states come in pairs except for the eigenvalue zero and possibly another eigenvalue if n is even. Thus, it is likely that just before two symmetric states coalesce one eigenvalue has to cross the imaginary axis in a pitchfork bifurcation. No such obstacle exists for asymmetric states. A more refined criterion than just counting unstable eigenvalues is to include the symmetry of the corresponding eigenfunctions. Indeed, both, the number of even and odd unstable eigenfunctions, have to coincide for coalescence of

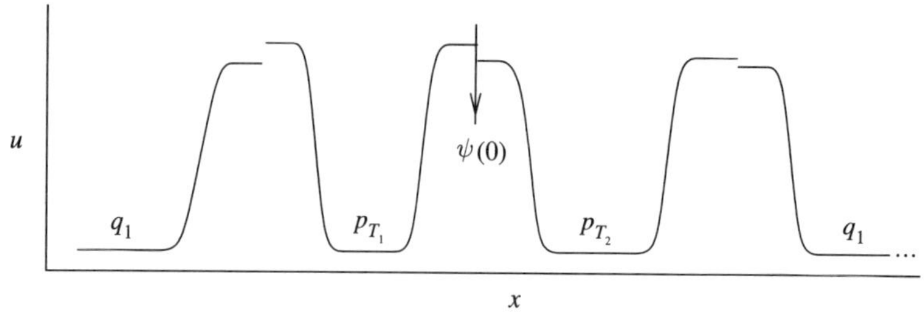

Figure 5. The discontinuous solution for $N=3$ with distances T_1 and T_2. The discontinuities are contained in the line spanned by $\psi(0)$.

two multimodal states to occur. For each particular sequence (k_j^0), these numbers can be calculated by computing the eigenvectors of the matrix A_0 appearing in the proof of theorem 3.6.

(c) *The proof of theorem 3.6*

We employ homoclinic Ljapunov–Schmidt reduction as developed in Lin (1990) and Sandstede (1993) for existence, and in Sandstede (1997) for stability. Since this method is rather technical, we shall illustrate it for two-dimensional systems, i.e. $u \in \mathbb{R}^2$, and refer to the articles mentioned above for the details in several dimensions.

Consider therefore a two-dimensional conservative system, say, and assume that q_1 is a homoclinic solution. It is then accompanied by a family p_T of periodic solutions parametrized by their period $2T$. We may assume that $p_T(0) = q_1(0) + \delta\nabla H(q(0))$ for some small δ depending on p_T. Then p_T converges to h uniformly on the interval $(-T, T)$ as $T \to \infty$. We seek multimodal solutions (which, of course, do not exist in two dimensions). The idea is then to use the distances between consecutive copies of q_1 as variables. Therefore, choose a sequence T_j for $j = 1, \ldots, n-1$ with $T_j > 0$ large such that $2T_j$ is a candidate for the distance between the jth and the $(j+1)$th copy of q_1 in the n-modal solution we want to construct. We may then concatenate the homoclinic orbit q_1 restricted to $\mathbb{R}^-$ with the solutions p_{T_j} defined on $(-T_j, T_j)$ for $j = 1, \ldots, n-1$, and finally with the other tail of q_1 defined on $\mathbb{R}^+$ (see figure 5 for $n = 3$). The resulting function looks like an n-modal solution; it is, however, discontinuous whenever it hits the line

$$\{q_1(0) + \delta\nabla H(q_1(0)); \delta \in \mathbb{R}\} = \{q_1(0) + \delta\psi(0); \delta \in \mathbb{R}\}.$$

Therefore, in order to show existence of multimodal orbits, it suffices to prove existence of numbers T_j such that all these discontinuities disappear. Since the discontinuous solution meets the line n times, there are n equations which have to be solved. If we seek symmetric solutions, we only need to solve the first $\frac{1}{2}n$ equations corresponding to the interval $(-\infty, 0]$. Indeed, we may then extend the solution to $(-\infty, \infty)$ using the symmetry. Similarly, for conservative systems, the last equation is always satisfied provided the first $n-1$ equations are: the remaining discontinuity occurs in the direction of increasing or decreasing energy, and is therefore zero since both tails have the same energy by construction.

A similar technique works in higher dimensions. As a result, we obtain a solution with n discontinuities which are contained in the line spanned by $\psi(0)$. If the discontinuities disappear, a multimodal solution is constructed. It is then possible, though a non-trivial exercise, to compute these discontinuities. Indeed, they are given by

$$\langle\psi(T_j), q_1(-T_j)\rangle - \langle\psi(-T_{j-1}), q_1(T_{j-1})\rangle - R_j((T_i)_{i=1,\ldots,n-1}) = 0, \qquad (3.9)$$

for $j = 1, \dots, n$ and large T_j (see Sandstede 1993, Satz 3). The remainder terms R_j are of higher order. Existence of n-modal solutions is equivalent to solving (3.9). We then have the following proposition.

Proposition 3.8. *Fix $n \geqslant 2$. There exists an n-modal homoclinic orbit q_n of (3.5) close to q_1 in phase space if, and only if,*

$$\left.\begin{aligned} a_1 \sin(-(\beta/\alpha)\ln(a_1 r)) &= R_1(a,r), \\ a_{j-1}\sin(-(\beta/\alpha)\ln(a_{j-1}r)) - a_j \sin(-(\beta/\alpha)\ln(a_j r)) &= R_j(a,r), \\ a_{n-1}\sin(-(\beta/\alpha)\ln(a_{n-1}r)) &= R_n(a,r), \end{aligned}\right\} \tag{3.10}$$

is satisfied where $j = 2, \dots, n-1$. In fact, if hypothesis 3.2 holds, it suffices to solve (3.10) for $j = 1, \dots, [\frac{1}{2}n]$, where $[x]$ denotes the largest integer smaller than x; if hypothesis 3.3 is satisfied, it is sufficient to solve (3.10) for $j = 1, \dots, n-1$. The remainder terms $R_j(a,r)$ are smooth in $a = (a_j)$ for $a_j \in (0,1]$ up to $r = 0$ and

$$R_j(a,r) = O(r^\gamma), \quad \frac{\mathrm{d}}{\mathrm{d}a_i} R_j(a,r) = O(r^\gamma), \tag{3.11}$$

for some $\gamma > 0$.

Proof. As mentioned above, Sandstede (1993, Satz 3) implies that existence of n-modal states is equivalent to solving (3.9). In order to derive (3.10) from (3.9), we proceed as in Sandstede (1997, § 6) and refer to that article for the details. The statement about the number of equations which need to be solved is proved in Sandstede *et al.* (1997, lemmata 3.1 and 3.2), respectively. In addition, we exploit the following symmetries. In the conservative case, hypothesis 3.3, we have

$$\begin{aligned} \langle \psi(x), q_1(-x)\rangle &= \langle \nabla^2 H(0) q_1(x), q_1(-x)\rangle + O(|q_1(x)|^2 |q_1(-x)|) \\ &= \langle q_1(x), \nabla^2 H(0) q_1(-x)\rangle + O(|q_1(x)|^2 |q_1(-x)|) \\ &= \langle q_1(x), \psi(-x)\rangle + O(|q_1(x)||q_1(-x)|(|q_1(x)| + |q_1(-x)|)) \end{aligned}$$

by remark 3.4. Under hypothesis 3.2, we have $\langle \psi(x), q_1(-x)\rangle = \langle \psi(-x), q_1(x)\rangle$ using Sandstede (1997, lemma 5.3). ■

We consider the conservative case first and comment later on the changes necessary for the reversible case. So, assume that hypothesis 3.3 is met. The proof is similar to Sandstede (1997, proof of theorem 3), where saddle-focus bifurcations for generic, non-reversible systems have been studied.

By proposition 3.8, we have to solve

$$\begin{aligned} a_1 \sin(-(\beta/\alpha)\ln(a_1 r)) &= R_1(a,r) \\ a_{j-1}\sin(-(\beta/\alpha)\ln(a_{j-1}r)) - a_j \sin(-(\beta/\alpha)\ln(a_j r)) &= R_j(a,r), \end{aligned}$$

for $j = 2, \dots, n-1$. Inserting $r = \exp(-(2\pi\alpha/\beta)n)$ with $n \in \mathbb{N}$, we obtain

$$\left.\begin{aligned} a_1 \sin(-(\beta/\alpha)\ln a_1) &= R_1(a,r), \\ a_{j-1}\sin(-(\beta/\alpha)\ln a_{j-1}) - a_j \sin(-(\beta/\alpha)\ln a_j) &= R_j(a,r), \end{aligned}\right\} \tag{3.12}$$

for $j = 2, \dots, n-1$. This allows us to take the limit $r \to 0$ yielding

$$\begin{aligned} a_1 \sin(-(\beta/\alpha)\ln a_1) &= 0, \\ a_{j-1}\sin(-(\beta/\alpha)\ln a_{j-1}) - a_j \sin(-(\beta/\alpha)\ln a_j) &= 0, \end{aligned}$$

for $j = 2, \ldots, n-1$. However, these equations are satisfied by the chosen sequence $a^0 = (a_j^0)$ with $a_j^0 \in \mathcal{A}$. Moreover, the Jacobian of (3.12) with respect to a at $(a_j, r) = (a_j^0, 0)$ is lower triangular and the entries on the diagonal are given by

$$\left(\sin\left(-\frac{\beta}{\alpha} \ln a_j \right) - \frac{\beta}{\alpha} \cos\left(-\frac{\beta}{\alpha} \ln a_j \right) \right) \Bigg|_{a_j = a_j^0 = \exp(-(\pi\alpha/\beta) k_j^0)} = (-1)^{k_j^0 + 1} \left(\frac{\beta}{\alpha} \right), \tag{3.13}$$

and thus are non-zero. Therefore, since the remainder terms are differentiable due to proposition 3.8, an application of the implicit function theorem proves (i)–(iii) of the theorem in the conservative case.

It remains to prove (iv) which addresses the stability properties of the n-modal orbits. The general approach is very similar to the existence part. In fact, any eigenfunction is an n-modal solution with respect to the primary mode $(\mathrm{d}/\mathrm{d}x)q_1$ of the linear non-autonomous equation, (3.6). In Sandstede (1997), a general procedure has been developed which reduces the calculation of the critical eigenvalues to the computation of eigenvalues of an $n \times n$ matrix. Let q_n be an n-modal solution given by $(a(r), r)$. Then, proceeding as in Sandstede (1997, § 6), all eigenvalues $\lambda(r)$ close to zero of (3.6) evaluated at q_n are given by $\lambda(r) = r\nu(r)$ for some continuous function $\nu(r)$ such that $\nu(0)$ is an eigenvalue of the matrix A_0 given by

$$(A_0)_{ij} = \begin{cases} b_j + b_{j-1}, & j = i, \\ -b_{j-1}, & j = i-1, \\ -b_j, & j = i+1, \\ 0, & \text{otherwise}, \end{cases}$$

with $b_j = (-1)^{k_j^0} c \operatorname{sgn}(M)$ for some positive constant $c > 0$ and $j = 1, \ldots, n-1$; we set $b_0 = 0$. Note that A_0 is tridiagonal, symmetric and the sum of its entries in each row vanishes. The number of positive and negative eigenvalues for such matrices has been determined in Sandstede (1997, lemma 5.4). Hence, statement (iv) of theorem 3.6 is proved for conservative systems.

In case (3.5) is reversible, we proceed as above. The only difference is that we solve (3.12) for $j = 1, \ldots, [\frac{1}{2}n]$ instead for $j = 1, \ldots, n-1$. This completes the proof of the theorem.

4. Discussion

There are several other fourth-order equations for which multimodal solutions do exist. Consider, for instance,

$$u_{tt} = -(u_{xxxx} + \mathrm{e}^u - 1), \tag{4.1}$$

$$u_t = -(\epsilon^2 u_{xxxx} - u_{xx} - u + u^3). \tag{4.2}$$

Equation (4.1) is known as the suspended beam model (see McKenna & Chen 1997), while (4.2), the extended Fisher–Kolmogorov equation, arises in the study of so-called Lifschitz points in phase transitions. Both equations exhibit similar features to (2.5) and therefore the results of §§ 2 and 3 are likely to apply to them as well. For instance, Peletier & Troy (1995*a*, *b*) proved existence of kinks for equation (4.2). Since these kinks bifurcate from the stable kinks of the Nagumo equation at $\epsilon = 0$, they are presumably also stable. Employing theorem 3.6 gives the existence of infinitely many stable multimodal kinks of (4.2) once its hypotheses are met. Note that these assumptions are generic within the class of reversible Hamiltonian equations.

Recently, Lord *et al.* (this volume) investigated the von Kármán–Donnell equations

$$\kappa^2\nabla^4 w + \lambda w_{xx} - \rho\phi = w_{xx}\phi_{yy} + w_{yy}\phi_{xx} - 2w_{xy}\phi_{xy},$$
$$\nabla^4\phi + \rho w_{xx} = (w_{xy})^2 - w_{xx}w_{yy},$$

with $(x, y) \in \mathbb{R} \times S^1$. They discretized the elliptic system in y and obtained a large system of ordinary differential equations in the unbounded variable x. Numerically, they found then a primary localized solution with oscillating tails. At least on the discretized level, theorem 3.6 can be used to confirm existence of multimodal solutions. In work in progress, Peterhof *et al.* (1997) currently extend theorem 3.6 to elliptic equations on unbounded domains and justify the numerical techniques used in Lord *et al.* (this volume).

I thank Alan Champneys for explaining the strut model as well as the mechanics behind it. Furthermore, I am grateful to one of the referees who showed me lemma 2.1(ii) and its proof. This work was partially supported by a Feodor–Lynen-Fellowship of the Alexander von Humboldt Foundation.

References

Abramowitz, M. & Stegun, I. A. 1972 *Handbook of mathematical functions*, 9th edn. New York: Dover.

Alexander, J. C., Gardner, R. A. & Jones, C. K. R. T. 1990 A topological invariant arising in the stability analysis of travelling waves. *J. Reine Angew. Math.* **410**, 167–212.

Alexander, J. C., Grillakis, M. G., Jones, C. K. R. T. & Sandstede, B. 1997 Stability of pulses on optical fibers with phase-sensitive amplifiers. *Z. Angew. Math. Phys.* **48**, 175–192.

Amick, C. J. & Toland, J. F. 1992 Homoclinic orbits in the dynamic phase space analogy of an elastic strut. *Eur. J. Appl. Math.* **3**, 97–114.

Buffoni, B., Champneys, A. R. & Toland, J. F. 1996 Bifurcation and coalescence of a plethora of homoclinic orbits for a Hamiltonian system. *J. Dynam. Differ. Equat.* **8**, 221–279.

Buryak, A. V. & Akhmediev, N. N. 1995 Stability criterion for stationary bound states of solitons with radiation-less oscillating tails. *Phys. Rev.* E **51**, 3572–3578.

Champneys, A. R. 1994 Subsidiary homoclinic orbits to a saddle-focus for reversible systems. *Int. J. Bifurcation Chaos* **4**, 1447–1482.

Champneys, A. R., Kuznetsov, Y. A. & Sandstede, B. 1995 HOMCONT: An AUTO86 driver for homoclinic bifurcation analysis. Technical report, CWI Amsterdam.

Champneys, A. R., Kuznetsov, Y. A. & Sandstede, B. 1996 A numerical toolbox for homoclinic bifurcation analysis. *Int. J. Bifurcation Chaos* **6**, 867–887.

Devaney, R. L. 1976 Homoclinic orbits in Hamiltonian systems. *J. Differ. Equat.* **21**, 431–438.

Doedel, E. J. & Kernévez, J. P. 1986 AUTO: Software for continuation problems in ordinary differential equations with applications. Technical report, California Institute of Technology.

Grillakis, M. G. 1988 Linearized instability for nonlinear Schrödinger and Klein–Gordon equations. *Commun. Pure Appl. Math.* **XLI**, 747-774.

Härterich, J. 1993 Kaskaden homokliner Orbits in reversiblen dynamischen Systemen. Diploma thesis, University of Stuttgart.

Henry, D. 1981 *Geometric theory of semilinear parabolic equations.* Springer Lecture Notes in Mathematics, vol. 804. Berlin: Springer.

Hunt, G. W., Bolt, H. M. & Thompson, J. M. T. 1989 Structural localization phenomena and the dynamical phase-space analogy. *Proc. R. Soc. Lond.* A **425**, 245–267.

Lin, X.-B. 1990 Using Melnikov's method to solve Shilnikov's problems. *Proc. R. Soc. Edinb.* A **116**, 295–325.

Lindberg, H. E. & Florence, A. L. 1987 *Dynamic pulse buckling.* Dordrecht: Martinus Nijhoff.

Maddocks, J. H. 1985 Restricted quadratic forms and their application to bifurcation and stability in constrained variational principles. *SIAM J. Math. Analys.* **16**, 47–68. (1988 Errata **19**, 1256–1257.)

McKenna, P. J. & Chen, Y. 1997 Numerical variational methods for approximating traveling waves in a nonlinearly suspended bridge. *J. Differ. Equat.* **136**, 325–355.

Pazy, A. 1983 *Semigroups of linear operators and applications to partial differential equations.* Springer.

Peletier, L. A. & Troy, W. C. 1995 Spatial patterns described by the extended Fisher–Kolmogorov (EFK) equation: kinks. *Differ. Int. Equat.* **8**, 1279–1304.

Peletier, L. A. & Troy, W. C. 1996 A topological shooting method and the existence of kinks of the extended Fisher–Kolmogorov equation. *Top. Meth. Nonlin. Analys.* **6**, 331-355.

Peterhof, D., Sandstede, B. & Scheel, A. 1997 Exponential dichotomies for solitary-wave solutions of semilinear elliptic equations on infinite cylinders. *J. Differ. Equat.* (In the press.)

Sandstede, B. 1993 Verzweigungstheorie homokliner Verdopplungen. Doctoral thesis, University of Stuttgart.

Sandstede, B. 1995 Center manifolds for homoclinic solutions. WIAS Preprint, 186.

Sandstede, B. 1997 Stability of multiple-pulse solutions. *Trans. Am. Math. Soc.* (In the press.)

Sandstede, B., Alexander, J. C. & Jones, C. K. R. T. 1997 Existence and stability of multiple pulses on optical fibers with phase-sensitive amplifiers. *Physica* D **106**, 167–206

Thompson, J. M. T. 1979 Stability predictions through a succession of folds. *Phil. Trans. R. Soc. Lond.* A **292**, 1–23.

Thompson, J. M. T. & Hunt, G. W. 1973 *A general theory of elastic stability.* Chichester: Wiley.

Wadee, M. K., Hunt, G. W. & Whiting, A. I. M. 1996 Asymptotic and Rayleigh–Ritz routes to localized buckling solutions in an elastic instability problem. Preprint.

Chapter 3

Numerics and discretization

This chapter presents two papers on the interplay between numerical discritisation, structural discretisation (in the sense of link models) and possible spatially chaos. The ideas contained have profound implications for those performing numerical calculations using 'black box' routines. In particular care should be taken in interpreting whether spatially chaotic solutions and multiplicities of localized buckling modes are really part of the underlying model or whether they are merely an artifact of numerical approximation. The first paper, by G. DOMOKOS, looks at a discritisation of the classical Euler column and finds that discrete structural models have a class of localized solutions which is not present in the continuous limit. This revisits in a structural engineering context the concepts of spurious or parasitic solutions that have been known for some time in the numerical analysis literature. The second paper, by HUNT *et al.* considers a link model which when viewed as a discrete dynamical system is chaotic due to homoclinic tangencies. This is then compared with a standard finite-element path-following approach to analyse the problem. Here the lesson the dynamical analogy gives to the finite element analyst is that care must be taken in interpreting computations of the 'fundamental path' owing to the presence of so many other candidate solutions.

Static solitary waves as limits of discretization: a plausible argument. DOMOKOS, G. *Phil. Trans. Roy. Soc. A* **355** (1997) 2099-2116.

Finite element modelling of spatially chaotic structures. HUNT, G.W., LAWTHER, R. & PROVIDÊNCIA E COSTA, P. *Int. J. Num. Meth. Eng* **40** (1997) 2237-2256.

Static solitary waves as limits of discretization: a plausible argument

By Gábor Domokos
Department of Strength of Materials, Technical University of Budapest, H-1521 Budapest, Hungary

We investigate special limits of the classical Euler buckling problem, arguing that the solution-family 'imitating' (without mass forces) the propagation of a dynamic solitary wave cannot be obtained as a limit from Euler's problem, only from its discretized version. Although we can not prove this claim rigorously, we prove other related statements that make the conjecture strongly plausible. Our results yield access to some open questions related to the discretized problem; also they show some new aspects of spatially chaotic behaviour.

1. Introduction

In this paper we will investigate special limits of the classical planar Euler buckling, a boundary value problem (BVP) describing the behaviour of an elastic beam under axial static force, discussed by Euler (1744) (cf. Love 1927). It is well-known that the homoclinic solution of the corresponding initial value problem (IVP) provides a spatially localized configuration (loop) that also solves the *dynamical* equation of the elastic line and proves to be a solitary wave. The dynamics of flexible rods, yielding solitary waves as solutions, is studied in detail by Coleman & Dill (1992), Maddocks & Dichmann (1994) provide a variational characterization. Naturally, Euler's BVP is not capable of describing the propagation of this wave. In one possible BVP interpretation (for alternative see § 2) the homoclinic solution corresponds to infinite force (instead of infinite length, as in the IVP) and the loop, shrinking to infinitesimal size, is located exactly at the middle of the beam (at least for the first mode solution). However, there exists a family $\mathcal{W}$ of BVP solutions, corresponding to the arbitrary location of the infinitesimal loop, 'imitating' the propagation of the dynamical wave. In this case the dynamics evolves amongst the set $\mathcal{W}$ of non-isolated static shapes, generated by a continuous shift transformation. The only member of $\mathcal{W}$ that can be obtained as a limit from Euler's problem is the homoclinic solution itself. The paper presents the conjecture, supported by a plausible argument, that all of $\mathcal{W}$ can be reached as a limit of the *discretized* BVP.

The fact that the continuous BVP does not converge to $\mathcal{W}$, pointing out the degenerate character of this BVP, is not very surprising. As demonstrated by Domokos & Holmes (1993*b*), there exist arbitrarily small smooth perturbations of the bending stiffness that result in drastic changes in the global bifurcation diagram, including the appearance of arbitrarily high numbers of 'parasitic' solutions. These solutions were already noticed in discrete analogues of nonlinear BVPs more than 15 years ago, (cf. Bohl 1979; Beyn 1982; Peitgen & Schmitt 1981). Peitgen *et al.* (1981) call

Phil. Trans. R. Soc. Lond. A (1997) **355**, 2099–2116

these solutions 'numerically irrelevant' (NIS). It is true that from the point of view of numerical computations they merely present an embarrassment, however, Domokos & Holmes (1993*a*) showed that parasitic solutions can be important and relevant in mechanical systems, and not only in discrete ones (Domokos & Holmes 1993*b*). This idea is carried further in this study: by using some results of Bohl (1979) and Peitgen *et al.* (1981) we investigate the relationship between limits corresponding to parasitic solutions and show the direct mechanical meaning of these limits.

The degeneracy (or structural instability) of the Euler BVP is a result of a high degree of symmetry. By perturbing the BVP one can obtain more generic cases, which, in turn, might prove to be vastly more complex than the original, classical problem. A special perturbation is discretization, either in the sense of discretizing the ordinary differential equation (ODE) by a step-by-step integrator, or, in the sense of discretizing the mechanical problem to a sequence of equal rigid links, coupled by linear torsional springs. The investigation of Domokos & Holmes (1993*a*) showed that the two discretizations can be made equivalent, moreover, the resulting map is none other than the *standard map*. Domokos & Holmes (1993*a*) investigated the extremely rich structure of the discrete BVP, proving the existence of parasitic solutions (not present in the continuous model) and characterizing them.

In this paper this study will be pursued further with the specific goal of explaining the solution family $\mathcal{W}$ as a limit. The key idea is that in contrast to the continuous BVP which possesses one parameter (the force), the discrete one has a second one: the number of links. With suitable adjustment of these two parameters one can obtain *several different* limits, two of which, $\mathcal{H}$ and $\mathcal{C}$, will be explicitly demonstrated. Moreover, we will prove that $\mathcal{H} \subset \mathcal{W} \subset \mathcal{C}$, suggesting in a plausible way that $\mathcal{W}$ itself is a special limit of the discrete BVP.

Besides supporting this argument, the tools and results of the paper help to answer a long-standing question: as already reported by Gáspár & Domokos (1989), the discrete BVP exhibits different behaviour for *odd* and *even* numbers of links. Apparently the investigation of the limits provides the key to this question.

After reviewing earlier results and stating the problem in § 2 we proceed to the new material. In § 3 we introduce special limit sets (not discussed earlier) and formulate our conjecture about wave-like solutions. The discrete problem is scrutinized in § 4, leading to observations supporting the conjecture. Finally, in § 5 we use these observations to introduce the labelling of branches and to explain the odd–even paradox. We also point out the relationship to Markov processes and draw conclusions.

This paper is aimed towards engineers, the arguments appeal to mechanical intuition and are not rigorous in the strict analytical sense. We do not claim mathematical originality for the 'observations', however. We believe that they are relevant from the engineering point of view. Whenever possible we will refer to mathematical results and discuss their connection to the investigated specific mechanical problem. The emphasis is on the mechanical interpretation of non-trivial mathematical concepts, such as chaotic and random behaviour, convergence to degenerate limits. We hope that this mechanical illustration can contribute to deeper understanding.

2. Euler buckling and its limit set $\mathcal{H}$

We recall the classical planar Euler buckling problem (cf. Euler 1744; Love 1927), corresponding to a simply supported slender linearly elastic rod of length L with uniform cross section and bending stiffness EI (E denoting Young's modulus and I

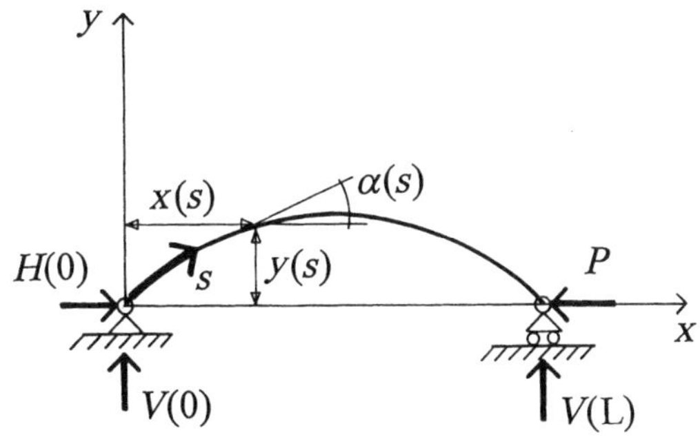

Figure 1. Euler's buckling problem

denoting the moment of inertia of the cross section), infinite axial and shear stiffness, loaded by an axial force P. Euler (1744) provided an essentially complete analysis of this classical problem, which we summarize below. As is customary, we model the (deformed) shape by a plane curve $(x(s), y(s))$ parametrized by the arc length $s \in [0, L]$ (figure 1).

If the load line passes through the supports then there always exist solutions for which the vertical reaction forces vanish identically, although the degenerate case $x(L) = 0$ (discussed by Maddocks (1984)) admits others. We will return to this special case in § 5. By restricting ourselves temporarily to the former non-degenerate solutions ($V(0) = V(L) = 0$), the equilibrium equations may be reduced to the single second order ODE in the slope $\alpha = \arctan(\mathrm{d}y/\mathrm{d}x)$:

$$\alpha'' + \lambda \sin \alpha = 0, \tag{2.1}$$

where $()'$ denotes $(\mathrm{d}/\mathrm{d}s)$ and $\lambda = (P/EI)$. The appropriate boundary conditions, for non-zero load λ, are zero moments

$$\alpha'(0) = 0 = \alpha'(L). \tag{2.2}$$

As Kirchhoff (1859) pointed out, the elastica equilibrium problem is analogous to the pendulum equation (replace $\lambda = P/EI$ by g/l in (2.1)); in fact the three-dimensional elastica is analogous to the more general heavy rigid body (top) problem in dynamics (cf. Love 1927; Mielke & Holmes 1988), of which the pendulum is a special case. This analogy suggests that results for the dynamic initial value problem can be applied to the static boundary value problem and vice versa: a strategy which has been used in studying continuous models (cf. Mielke & Holmes 1988; Thompson & Virgin 1988; ElNaschie 1990). We remark that the $\lambda \to \infty$ limit is equivalent to the $L \to \infty$ limit in the BVP. The latter one is less natural from the engineer's point of view (in a physical experiment λ is more easily varied than L), however, mathematically the two limits coincide. The $L \to \infty$ limit is adopted by Beyn (1990). This limit can be obtained by rescaling the arclength s by $S = as$ and letting $a \to \infty$. By introducing the new variable S into (2.1) and using $(\mathrm{d}\alpha/\mathrm{d}S) = (\mathrm{d}\alpha/\mathrm{d}s)(\mathrm{d}s/\mathrm{d}S) = (\alpha'/a)$; $(\mathrm{d}^2\alpha/\mathrm{d}S^2) = (\alpha''/a^2)$, one arrives at $\alpha''/a^2 + \lambda \sin \alpha = 0$, which, in turn, is equivalent to the $\lambda \to \infty$ limit. The latter will be investigated below.

Equations (2.1) and (2.2) can be solved exactly using Jacobian elliptic functions. For each fixed load $P > n^2\pi^2 EI/L^2$, in addition to the trivial solution $\alpha = 0$, there are n pairs of buckled states containing $1, 2, \ldots, n$ half waves, respectively, which may be uniquely characterized by the angles $\alpha(0) = \pm\alpha_1(P), \pm\alpha_2(P), \ldots, \pm\alpha_n(P)$ with $0 < \alpha_n < \cdots < \alpha_2 < \alpha_1 < \pi$. This statement is perhaps more easily understood

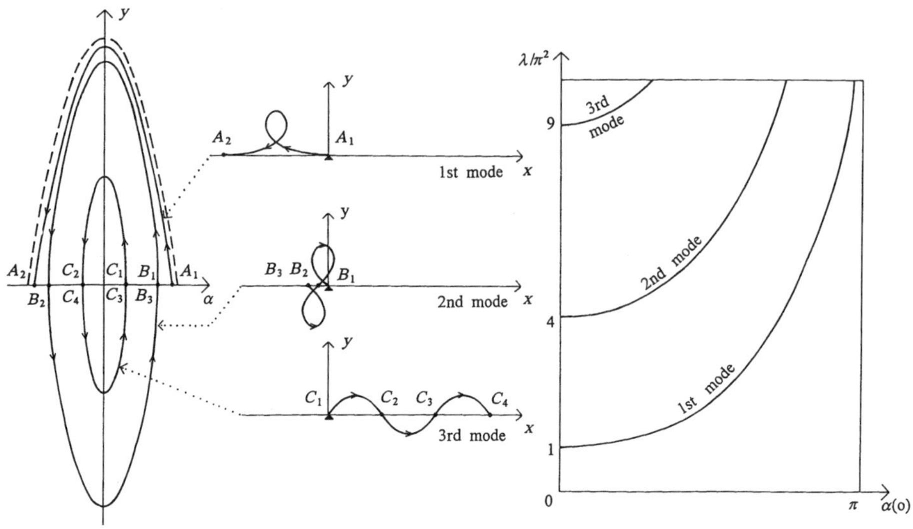

Figure 2. The solutions of the BVP in the physical and phase space and on the global bifurcation diagram.

via the phase portrait of the associated first order system, obtained by introducing the auxiliary variable $y = -\alpha'/\lambda$

$$\begin{aligned} \alpha' &= -\lambda y, \\ y' &= \sin\alpha. \end{aligned} \tag{2.3}$$

This system is Hamiltonian and possesses the first integral

$$-\tfrac{1}{2}\lambda y^2 + \cos\alpha = \cos\alpha(0) = \text{const.} \tag{2.4}$$

The level sets of (2.4) are nested periodic orbits of (2.3) whose period T increases monotonically from $2\pi/\sqrt{\lambda}$ to infinity as $\alpha(0)$ varies from 0 to π. (In fact $T = (4/\sqrt{\lambda})K(\sin(\frac{1}{2}\alpha(0)))$, where K is the complete elliptic integral of the first kind: recall that $K(0) = \frac{1}{2}\pi$ and $K(k) \approx \frac{1}{2}\ln(16/1-k^2)$ as $k \to 1$.) Solutions to the boundary value problem (2.1)–(2.2) are arcs of length L contained in those level sets having periods which are divisors of $2L$. From the monotonicity of T it is clear that these $2n+1$ orbits, including the trivial one, are the only solutions to the BVP. This fact, although with different tools, was already demonstrated by Euler. Figure 2 demonstrates three buckled shapes with the corresponding arcs in the phase plane and the global bifurcation diagram for $\alpha(0) > 0$.

The *homoclinic* solution (dashed line), connecting the two hyperbolic fixed points, is reached in the BVP as $\lambda \to \infty$. As intuitively expected, one can show by manipulating complete and incomplete elliptic integrals that $x(L) \to -L$ as $\lambda \to \infty$. (The limited length of this paper prohibits the demonstration of these calculations, however, they do not support any new claim.) Thus the n loops of the nth buckling mode each shrink to infinitesimal size, the beam becomes 'straight' in the everted limit state. Simultaneously, in the nth mode, as $\lambda \to \infty$, $\alpha'(iL/2n) \to \infty$, $(i = 1, 3, \ldots, 2n-1)$, resulting in n discontinuities of magnitude 2π in $\alpha(s)$. Consequently, the homoclinic solution can be regarded as a discrete family H_n of non-continuous solutions, the nth member H_n of the family being the $\lambda \to \infty$ limit of the nth buckled mode of the continuous problem

$$H_n(s) = k_n(s)\pi, \quad s \in [0, L], \tag{2.5}$$

where

$$k_n(s) = \begin{cases} -1, & \text{if } 1 < [(2ns/L) \bmod 4] \leqslant 3, \\ 1, & \text{otherwise.} \end{cases} \tag{2.6}$$

We remark that the functions k_n are very similar to the Walsh series w_n introduced by Walsh (1923). In fact, for $n = 2^i$ $(i = 0, 1, \ldots)$ $k_n(s) = w_n(s/L)$. The functions H_n form the homoclinic solution set $\mathcal{H} = \{H_n\}$ $(n = 1, 2, \ldots)$.

3. The solution sets $\mathcal{C}$ and $\mathcal{W}$

Let us now turn back to equations (2.1) and (2.2). Substituting $EI = 0$ (equivalent to $\lambda = \infty$) yields for $P > 0$ the trigonometric equation

$$P \sin \alpha(s) = 0, \tag{3.1}$$

to which the solution set $\mathcal{C}$ belongs

$$\mathcal{C} = \{k(s)\pi\}, \tag{3.2}$$

where

$$k(s) \in \{0, \pm 1, \pm 2, \ldots\}, \quad s \in [0, L]. \tag{3.3}$$

In plain English, the elements of $\mathcal{C}$ are $\alpha(s)$ functions that can have an arbitrary number of discontinuities in $\alpha(s)$, at arbitrary locations, the only restriction being that, away from discontinuities, $\alpha(s)$ remains an integer multiple of π. This huge set, of cardinality greater than the continuum (its cardinal number is $\aleph_2$, as Fraenkel (1923) demonstrates (p. 47, p. 86), cf. also § 5 *c*) contains many interesting subsets. One of them is the already discussed homoclinic solution set $\mathcal{H} \subset \mathcal{C}$; another interesting one-parameter subset $\mathcal{W}_1 \subset \mathcal{C}$ is given by (3.2) and

$$k(s) = \begin{cases} 1, & \text{if } 0 \leqslant s \leqslant s_0, \\ -1, & \text{if } s_0 < s \leqslant L. \end{cases} \tag{3.4}$$

The solution set $\mathcal{W}_1$ could be regarded as the quasi-static image of a single propagating solitary wave since the elements of $\mathcal{W}_1$ represent the (infinitesimal) homoclinic loop at arbitrary locations. Similar sets $\mathcal{W}_n$ (with n parameters) can be constructed in a similar manner as straightforward generalizations of the functions H_n, for the representation of n alternating, simultaneous loops at arbitrary locations. The union of the solution sets $\mathcal{W}_n$ will be denoted by $\mathcal{W}$.

The interest in the set $\mathcal{W}$ is not purely academic. Imperfect subsets of $\mathcal{W}_1$ can be observed in a physical experiment, as we stretch an elastic wire, enforcing a loop. Although this configuration is not stable in the plane (cf. Maddocks 1984), it can be stabilized by the contact force. One can observe that the location of the loop is not uniquely determined; as long as it is sufficiently small compared to the length of the wire, it can be moved back and forth in a range around the midpoint, without violating the boundary conditions. The same phenomenon can be observed in numerical experiments: here the continuous beam is replaced by a sequence of N rigid links coupled by linear springs (for details see the next section). In this model, instead of having a unique loop fixed at the middle, large numbers of adjacent

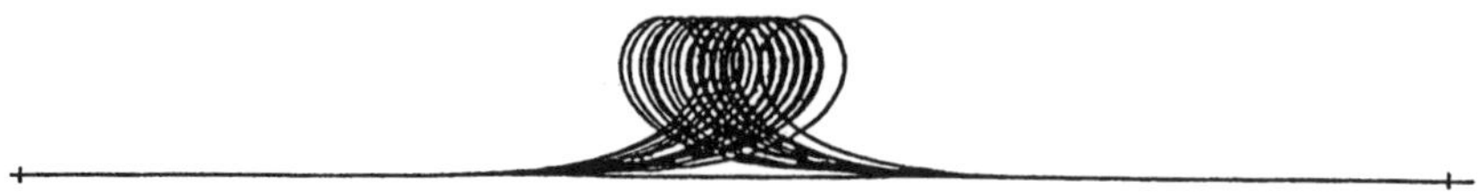

Figure 3. Imperfect subset of $\mathcal{W}_1$ emerging in a numerical experiment ($P = 493.48$, $N = 120$ elements, 59 solutions).

similar solutions appear, which also satisfy the boundary conditions. An example is illustrated in figure 3.

As we can observe, the solutions in $\mathcal{W}$ are physically and numerically relevant. The illustrated numerical example presents a version of $\mathcal{W}_1$ that is imperfect in the sense that both λ and N are finite (but large) instead of infinite, as in the perfect case. These imperfections are manifested in three ways:

(i) instead of a continuum of solution it contains only a finite number;

(ii) the distance between neighbouring solutions is finite instead of infinitesimal; and

(iii) the shape of the solutions is not identical (only very similar) and not identical with the homoclinic loop, either.

We remark that the asymmetry of the illustrated shapes is also a consequence of the imperfection: λ is finite.

It is natural to conjecture that these imperfections can be reduced as the parameter λ and the number N of links increases, thus, in the limit $\mathcal{W}_1$ can be obtained. This prompts the formulation of

Conjecture 3.1. *There exist two functions $\lambda(t)$ and $N(t)$ describing the discretized model with solution set $\mathcal{D}(\lambda, N)$, such that $t \to \infty$ implies $\lambda, N \to \infty$ and $\mathcal{D}(\lambda, N) \to \mathcal{W}$.*

In the next section we will not prove exactly this claim, however, we will show that the larger set $\mathcal{C}$ ($\mathcal{W} \subset \mathcal{C}$) and the smaller set $\mathcal{H}$ ($\mathcal{H} \subset \mathcal{W}$) can both be obtained as special limits of $\mathcal{D}$: in this way we make our conjecture plausible (Polya 1968).

4. The limit sets of the discrete model

In the first part of this section we will define and describe the discrete model, reviewing briefly parts of Domokos & Holmes (1983), to which we refer for more details. The discrete model can be defined either as a discretization of the ODE (2.3), or as a mechanical discretization of the beam, resulting in a sequence of rigid links coupled by linear torsional springs. In the former case a simplectic integrator is preferable (cf. Marsden *et al.* 1991), defining the area preserving map

$$\alpha_{i+1} = \alpha_i - \lambda l y_{i+1}, \quad y_{i+1} = y_i + l \sin \alpha_i. \tag{4.1}$$

Besides being perhaps the simplest such recursive algorithm for numerical solution of the initial value problem associated with (2.3), (4.1) also proves to be the equilibrium condition for the *mechanical* discretization (see figure 4). Each of the N rigid links are of length $l = L/N$, the torsional springs have stiffness EI/l at each joint. Solutions of (4.1) satisfying the discrete boundary conditions

$$y_0 = 0 = y_N, \tag{4.2}$$

not only provide approximations to the exact solutions of the continuum BVP (2.1)–(2.2), but are themselves exact solutions of the discrete mechanical problem with N links.

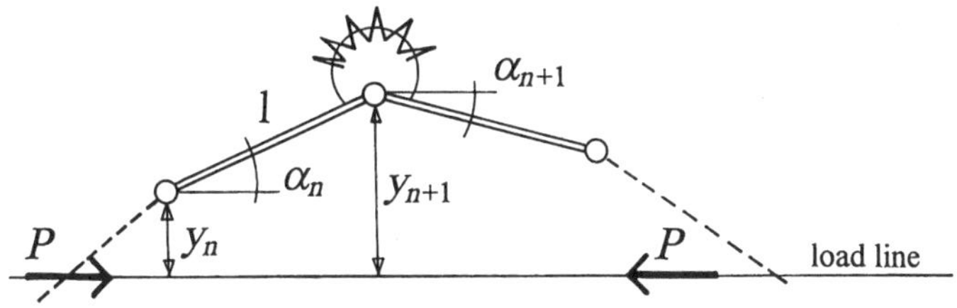

Figure 4. The mechanical interpretation of the discretization.

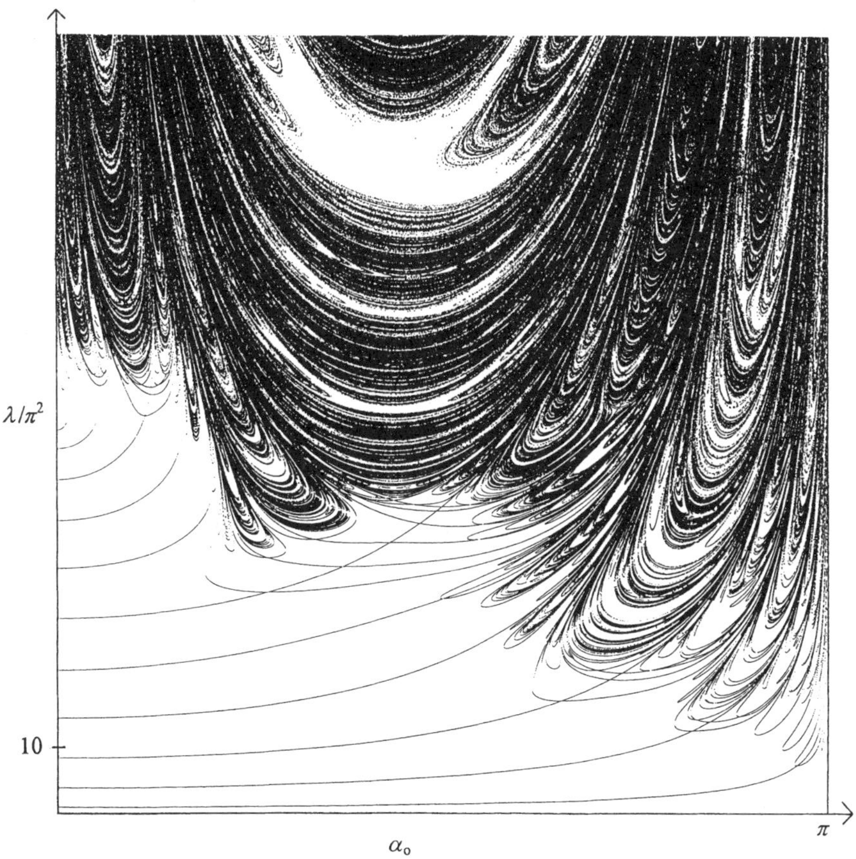

Figure 5. Bifurcation diagram of the discrete model for $N = 12$.

Equation (4.1) can be turned by a simple linear transformation into the *standard map* which has been extensively studied as being perhaps the simplest two-dimensional map providing chaotic solutions. The bifurcation diagrams for the discrete model (4.1)–(4.2) display a radically different structure from that of the continuous problem (2.2)–(2.3). As an example we present the $N = 12$ case in figure 5.

As it is apparent from the figure, besides the primary buckled modes (corresponding to the modes of the continuous model), a large number of other solutions appear in the discrete problem which we call *parasites*. It is a natural question to ask from where these parasites originate, how can they be organized, and, as perhaps most relevant to our present subject, where do they converge as $\lambda \to \infty$.

The origin of the parasites was already clarified by Bohl (1979). Domokos & Holmes (1993*a*) showed the mechanical relevance of parasites by using properties of the standard map. In addition, we characterized asymptotic behaviour for large λ, but only for the primary branches. Finally, we showed that as λ increases, the number of parasitic solutions in each mode grows beyond any finite number.

The discrete BVP (4.1)–(4.2) has two parameters, λ and N, so the solution set can be written as $\mathcal{D}(\lambda, N)$. Fixing $\lambda = \lambda_0$ and letting $N \to \infty$ is equivalent to applying an integrator with decreasing stepsize: standard ODE theory guarantees the convergence to the continuous BVP (2.2)–(2.3). After reaching that limit one can let $\lambda_0 \to \infty$. As we saw in the previous section, the continuous model converges to the homoclinic solutions. This consideration can be summarized for later reference as follows.

Observation 4.1. $\lim_{\lambda\to\infty}\{\lim_{N\to\infty}[\mathcal{D}(\lambda, N)]\} = \mathcal{H}$.

If we interchange the two limits, the result is more surprising. We can formulate the following.

Observation 4.2. $\lim_{N\to\infty}\{\lim_{\lambda\to\infty}[\mathcal{D}(\lambda, N)]\} = \mathcal{C}$.

As a first step, we deal only with the $\lambda \to \infty$ limit at constant N. This limit is also interesting from the physical point of view: it will inform us on the asymptotic behaviour of the bifurcation diagrams like the one illustrated in figure 5. The solution set $\mathcal{C}_N$ corresponding to this limit is the discrete version of $\mathcal{C}$ and is defined as

$$\mathcal{C}_N = \{k_i\pi\}, \tag{4.3}$$

where

$$k_i \in \{0, \pm 1, \pm 2, \ldots\}, \quad i \in \{0, 1, \ldots, N-1\}. \tag{4.4}$$

Our (reduced) claim is formulated as follows.

Observation 4.3. $\lim_{\lambda\to\infty}[\mathcal{D}(\lambda, N)] = \mathcal{C}_N$.

The mechanical content of this observation is confirmed by engineering intuition: as we pull with increasing force at the endpoints of the linkage, the slope of each individual member decreases, and in the limit the links are collinear with the line of action of the acting force. The individual solutions contain springs that went through an unspecified number of $k\pi$ rotations. Visualizing this limit is also helpful in understanding the origin of parasites: at a slight release from the infinite load the springs start an unwinding process, leading to the abundant solution structure illustrated in figure 5. Observation 4.3 is not new from the mathematical point of view, the $\lambda \to \infty$ limit of the discretized BVP was investigated by Bohl (1979) and Peitgen *et al.* (1981). The fact that each (unbounded) branch of the BVP converges to an element of $\mathcal{C}_N$ is proven in remark 2.4 of the latter work. The converse, i.e. that to each function in $\mathcal{C}_N$ one branch converges, is demonstrated by Bohl (1979) (cf. also Peitgen *et al.* (1981), corollary 2.5). We remark that these results could be obtained directly by an application of the implicit function theorem to (4.1), rewritten as a three-point discretization of (2.1).

In order to support observation 4.2 one further step is required: the $N \to \infty$ limit of $\mathcal{C}_N$ is a purely geometric construction: by letting $N \to \infty$ the point set defined by

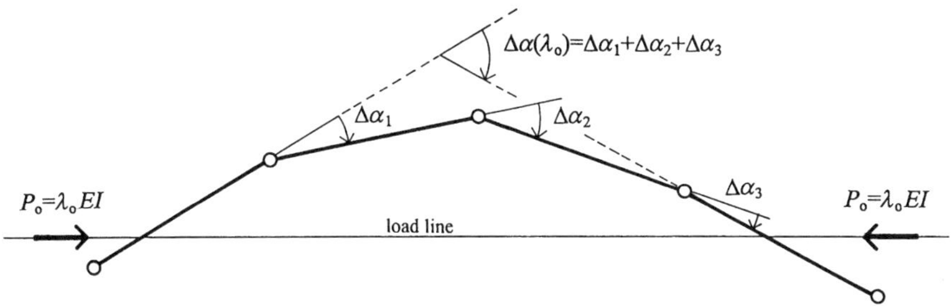

Figure 6. Segment between points on load line.

the discrete model becomes dense in the continuum, in this sense $\lim_{N\to\infty} \mathcal{C}_N = \mathcal{C}$ (cf. § 5 *c*).

The preceding considerations were rather simple, on the other hand, they rely on deeper mathematical results. We are led to the conclusion the sets $\mathcal{H}$ and $\mathcal{C}$ can both be obtained as special limits from the discretized BVP. The former is the set of homoclinic solutions to which the continuous model converges; the latter is a larger set, obtained by substituting $\lambda = \infty$ explicitly into the continuous model. The set $\mathcal{W}$, containing the quasi-static version of solitary wave propagation, is contained in $\mathcal{C}$; meanwhile it contains $\mathcal{H}$ as a subset: $\mathcal{H} \subset \mathcal{W} \subset \mathcal{C}$. These facts support, in a plausible way, conjecture 3.1, stating that $\mathcal{W}$ itself can be constructed as a limit of $\mathcal{D}(\lambda, N)$. (For more details see § 5 *c*.)

(*a*) *Additional mechanical arguments and generalization*

As mentioned before, the claim of observation 4.3 is supported by mathematical results. Nevertheless, we will provide an additional mechanical argument. This reasoning not only provides a better intuitive insight, it also yields access to mechanically more relevant generalizations which cannot be obtained from the cited mathematical results. The generalization to non-uniform discretizations will be used in § 5 *d*.

Observation 4.3 can be decomposed into two statements which we will prove subsequently.

(1) *Each branch of the discrete BVP (4.1)–(4.2) either terminates at finite* λ, *or converges to an element of* $\mathcal{C}_N$, *defined in (4.3)–(4.4).*

The idea of the mechanical proof is to decompose the discrete BVP into a series of segments, starting and ending where the structure intersects the load line (these points need not coincide with the joints). We will show for each of these segments that a finite upper bound valid for all relative angles exists, and via this upper bound the convergence to an element of $\mathcal{C}_N$ can be shown.

We regard a segment between two intersections with the load line, so y_i do not change sign along the segment (without loss of generality, we will assume $y_i \geqslant 0$) and let $\lambda = \lambda_0 < \infty$ (see figure 6).

According to (4.1), relative angles $\Delta\alpha_i = \alpha_i - \alpha_{i-1}$ at any intermediate hinge do not change sign along the segment, either. The total relative angle $\Delta\alpha(\lambda_0)$, measured between the two end-links of the segment can be computed as the sum of all intermediate relative angles. As we move along the branch, neither of the end-links can rotate by more than π, otherwise it would become collinear with the load line and the solution would become the trivial one. (Domokos & Holmes (1993*a*) prove that there are only $N - 1$ bifurcations from the trivial solution, corresponding to the primary branches. Consequently, none of the branches can *return* to the trivial

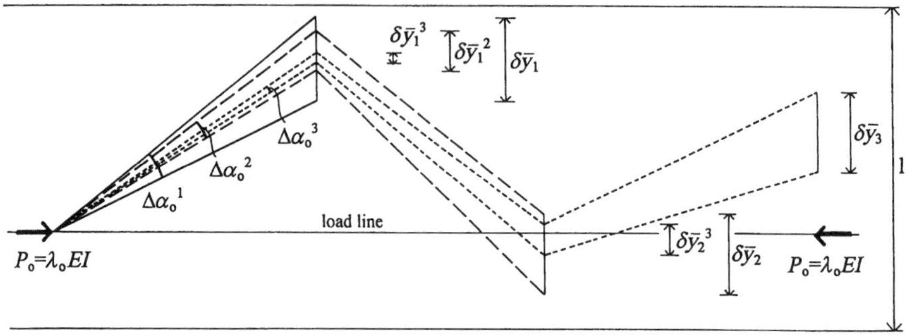

Figure 7. The strip of width l containing the intervals $\delta\bar{y}_i$.

solution.) The limited rotation at the end means that $\Delta\alpha$ is bounded from above: $\Delta\alpha \leqslant \Delta\alpha(\lambda_0) + 2\pi$. This implies the same finite upper bound for all $\Delta\alpha_i$, since we have $\Delta\alpha_i < \Delta\alpha$. As $\lambda \to \infty$, this upper bound enforces via (4.1) $y_i \to 0$, $\sin\alpha_i \to 0$, and, consequently $\alpha_i \to k_i\pi$. Since the structure can be always decomposed into a sequence of such segments, we have proved our claim for all $i \in \{0, 1, \dots, N-1\}$. (We remark that as λ increases, the length of the investigated segment might change since joints might pass through the load line. However, as they do so, the moment (and thus the relative angle) vanishes; consequently the new link belonging to the segment inherits its absolute angle from the previous one smoothly and thus our considerations are not affected.)

After proving that each branch (not terminated at finite λ) converges to an element of $\mathcal{C}_N$ we proceed to the next part of observation 4.3:

(2) *For any given function in* $\mathcal{C}_N$ *(defined in (4.3)–(4.4), there exists at least one branch of the BVP (4.1)–(4.2) that converges to the given function.*

According to (4.3)–(4.4), a function contained in the discrete limit set $\mathcal{C}_N$ is defined by a sequence of integers $k_i; i \in \{0, 1, \dots, N-1\}$, from which the new sequence $\bar{k}_i = k_{i+1} - k_i; i \in \{0, 1, \dots, (N-2)\}$ can be obtained. Let

$$k_{\max} = \begin{cases} \max\{\bar{k}_i\}, & \text{if } \max\{\bar{k}_i\} > 0, \\ 0, & \text{otherwise}, \end{cases} \tag{4.5}$$

and

$$k_{\min} = \begin{cases} \min\{\bar{k}_i\}, & \text{if } \min\{\bar{k}_i\} < 0, \\ 0, & \text{otherwise}. \end{cases} \tag{4.6}$$

We choose

$$\lambda_0 = \pi(k_{\max} - k_{\min} + 2)/l^2 \tag{4.7}$$

(recall that $l = L/N$) and define a sequence of intervals

$$\delta\bar{y}_i = [\pi(\bar{k}_i - 1)/(l\lambda_0), \quad \pi(\bar{k}_i + 1)/(l\lambda_0)]. \tag{4.8}$$

Our choice (4.7) of λ_0 ensures that all intervals lie within a strip of width l (see figure 7), containing the load line.

Now we will show that there exists a solution of (4.1)–(4.2) such that for every i, $y_i \in \delta\bar{y}_i$ and this solution converges to the prescribed limit function $\alpha_i = k_i\pi$. We construct the solution at $\lambda = \lambda_0$ by induction. Let α_0 go through a complete period: y_1 will cover the interval $[-l, l]$ twice, also it will move through $\delta\bar{y}_1$ twice. From the corresponding two intervals of α_0 we pick the one that is closer to the prescribed πk_0, let us denote this interval by $\Delta\alpha_0^1$. In the next step we let y_i travel through $\delta\bar{y}_1$: the second link will describe a complete circle, y_2 will move through $\delta\bar{y}_2$ twice. From the corresponding sub-intervals of $\delta\bar{y}_1$ we pick the one (denoted by $\delta\bar{y}_1^2$)that fixes α_1 closer to the prescribed πk_1. As we varied y_1 inside $\delta\bar{y}_1$, α_0 moved within $\Delta\alpha_0^1$. Since $\delta\bar{y}_1^2 \subset \delta\bar{y}_1$, the corresponding interval $\Delta\alpha_0^2 \subset \Delta\alpha_0^1$. As we proceed further, we obtain a nested sequence of intervals $\Delta\alpha_0^i \subset \Delta\alpha_0^{i-1}$ for α_0 and $\delta\bar{y}_j^i \subset \delta\bar{y}_j^{i-1}$ for each y_i, finally the value $y_N = 0$ is fixed by the boundary condition (4.2) and thus it specifies a single point inside the last (smallest) interval. After constructing this solution for $\lambda = \lambda_0$, we let $\lambda_0 \to \infty$. The width d of the strip can be expressed by using (4.7) $d = \pi(k_{\max} - k_{\min} + 2)/(l\lambda_0)$, and as $\lambda_0 \to \infty$, $d \to 0$ and the same holds for all $\delta\bar{y}_i$ intervals, so, finally $y_i \to 0$ and $\alpha_i \to k_i\pi$. Thus we demonstrated one branch that converges to a prescribed function in $\mathcal{C}_N$.

This mechanical argument supports observation 4.3. Now we will show that it can be readily generalized for more generic mechanical systems.

The choice of non-uniform mesh (unequal length l_i for the links) does not change the argument for statement (1). In the argument for (2) we have to consider that the stiffness c_i of the ith spring can be derived as $c_i = 2EI/(l_i + l_{i+1})$, l_i and l_{i+1} denoting the length of the adjacent links. We have to choose the spring with maximal (resp. minimal) *moment*. (The uniform mesh produced equal stiffnesses $c_i = EI/l$, so maximal moment coincided with maximal relative rotation.) Accordingly, in equations (4.5)–(4.6) and (4.8) the sequence $\bar{k}_i$ has to be replaced by $\bar{k}_i c_i$, in (4.7) l has to be replaced by $l_{\min} = \min\{l_i\}$; otherwise the argument is not affected by the choice of non-uniform discretization. This means that the same simple limit is obtained for *all* discretizations.

Physical nonlinearity of the springs can be considered in a similar manner: the sequence $\bar{k}_i$ has to be replaced by $\bar{m}_i(\bar{k}_i)$, where m_i is an *a priori* given function, denoting the moment in the ith spring as a function of the relative angle. The resulting limit behaviour is the same as in the case of linear material behaviour, as long as the functions m_i are strictly monotonic and the first derivative remains finite.

5. Related issues, discussion and conclusions

(*a*) *Branch labels*

The argument for statement (2) actually shows more than the original statement, which claimed that there exists *at least* one branch converging to a given function in $\mathcal{C}_N$. According to the argument this branch is uniquely defined; thus the map defined by our argument has a mutual one-to-one correspondence between branches going to infinity and the functions in $\mathcal{C}_N$. The latter ones are completely described by the integer string $C_N = \{k_i\}$, $i \in \{1, 2, \ldots, N\}$, which could be used to identify all infinite branches uniquely. In previous statements we excluded branches that terminate at finite λ (either at saddle-node or pitchfork critical points). However, numerical evidence indicates that *no such branches exist*. Although we do not formulate an explicit conjecture, we remark that the unique C_N labels probably apply to all branches.

From the mathematical point of view, $\mathcal{C}_N$ is an infinite set, since the constants k_i (cf. equation (4.3)) can have arbitrary integer values. However, from the physical point of view, two angles are indistinguishable if their difference is a multiple of 2π. In the case of finite λ, the distance y could serve as a distinction between two nodes whose angles differ by $2k\pi$; however, in the limit set $\mathcal{C}_N$, belonging to $\lambda = \infty$, we have $y_i = 0$ identically. Multiples of 2π could be also 'recognized' by the stored energy, however, the $\lambda \to \infty$ limit can be interpreted by letting the flexural rigidity $EI \to 0$, so the energy difference in the links also disappears in the limit. This means, that from the physical point of view we have to take all solutions in $\mathcal{C}_N$ modulo 2π. The resulting set $\mathcal{B}_N$ is defined by

$$\mathcal{B}_N = \{b_i\pi\}; \quad b_i \in \{0, 1\}; \quad i \in \{0, 1, \ldots, N-1\}. \tag{5.1}$$

where the binary integers b_i can be determined from the corresponding elements of $\mathcal{C}_N$ as

$$b_i = |k_i| \bmod 2.$$

$\mathcal{B}_N$ is (as opposed to $\mathcal{C}_N$) finite, in fact, it is equivalent to the set of all binary strings of length N, so it has exactly 2^N elements. To each element of $\mathcal{B}_N$ an infinite set of elements in $\mathcal{C}_N$ belong. We can regard the elements of $\mathcal{B}_N$ as infinite classes in $\mathcal{C}_N$. Similarly to the labels C_N we define the *binary labels* $B_N = \{b_i\}$ which will prove to be useful when organising the rather complex bifurcational structure of the BVP (4.1)–(4.2). Although the number of distinct binary labels grows rapidly with N, it is still a fixed *finite* number for *all* values of λ, as opposed to the number of branches (and the unique C_N-labels), which grow to infinity. Each physically distinguishable limit solution of the N-link discrete BVP is uniquely characterized by a binary label B_N (cf. the upper part of figure 8: here all physically relevant 16 limit solutions of the four-link BVP are characterized by 2^4 labels.)

(b) The mechanics of Markov chains

One of the most attractive points of this study is that some of the seemingly abstract limit sets have a direct mechanical analogue. This is perhaps most easily understood in the case of $\mathcal{B}_N$ that can be viewed as the solution set of a *chain* consisting of rigid elements, coupled by frictionless hinges with zero flexural rigidity. The chain is subjected to a single axial end load. In sharp contrast to the discrete elastic structure, the shape of the chain as an IVP cannot be described by a recursion formula; at each node a binary ambiguity enters. (Each element can be either in tension or in compression, *independently* of the previous one.) In other words: the shape of the chain (as an IVP) is *not* determined uniquely by the initial conditions; rather, it can be viewed as a (binary) random walk, perhaps the simplest case of a discrete Markov process, also called (independently from the present application!) a Markov-chain.

As mentioned earlier, Domokos & Holmes (1993*a*) we showed that the recursion formula for the elastic linkage is (under a linear transformation) identical with the standard map, which is known to exhibit chaotic solutions. The chaotic dynamics of the standard map lie at the core of the complex bifurcation diagrams for the BVP (cf. figure 5). We can regard the abundant appearance of parasitic solutions as a manifestation of (spatial) chaotic behaviour in the BVP. The characteristic of this behaviour is that, although deterministic (described by the recursion (4.1)), it exhibits random-like properties. The latter ones can be rigorously formulated only for infinite trajectories (IVPs); however, their presence is also apparent in the finite domain (BVP) as we look at figure 5. Our present study showed that the limit of this behaviour is a random process. Consequently, chaotic dynamics can be either viewed as an infinitely complex, deterministic system, or, equivalently, *as the imperfect (perturbed) version of a random process*. In our case the bending stiffness EI

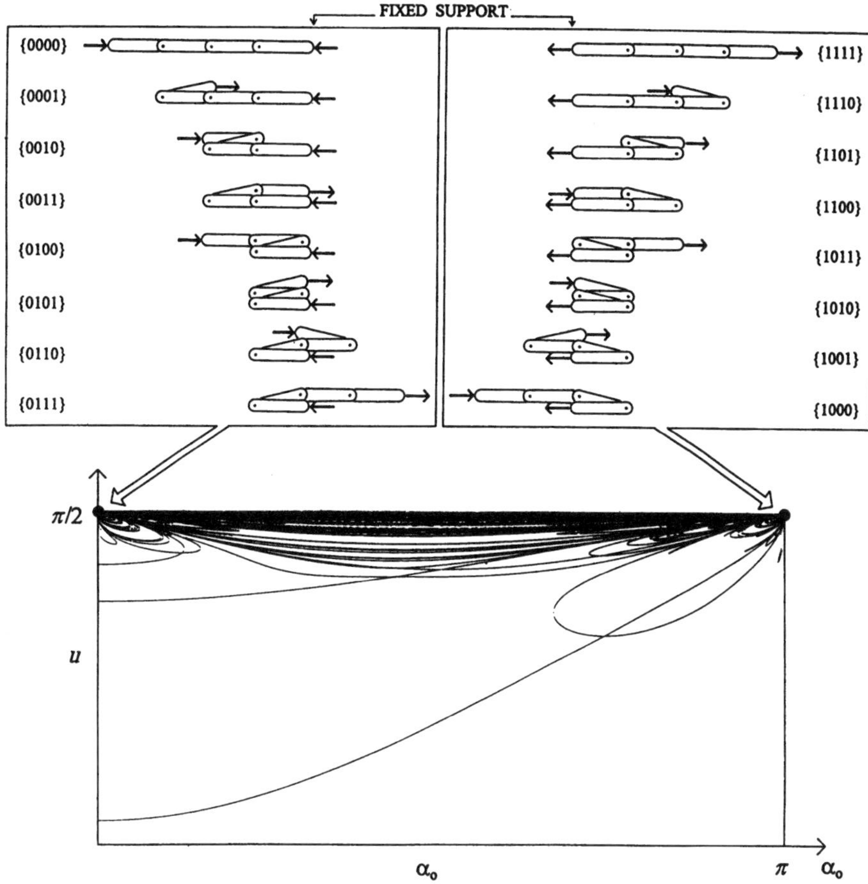

Figure 8. The limit behaviour of the $N = 4$ linkage: 2^4 binary chains.

could serve as a perturbation parameter and we were in the very fortunate situation to demonstrate this random limit as a mechanically meaningful system (chain). Similar (degenerate) random processes might be at the core of other chaotic systems; however, they might not have a direct mechanical interpretation. The appeal of this approach is that the apparently complex behaviour shown in figure 5 boils down to a startlingly simple process. This can be visualized by a rescaling of the parameter λ, bringing $\lambda = \infty$ into the finite range: we let $\lambda = \tan(u)$ ($P = \sin(u)$, $EI = \cos(u)$) and plot the $[\alpha_0, u]$ diagram: as $u \to \frac{1}{2}\pi$ we have $\lambda \to \infty$. Such a diagram for $N = 4$ is illustrated in figure 8, supplemented with the physical shape of the $2^4 = 16$ limit configurations. As mentioned before, uniqueness is violated in the $\lambda \to \infty$ limit, so 8 of the binary chains correspond to $(\alpha_0, u) = (0, \frac{1}{2}\pi)$, and other 8 to $(\alpha_0, u) = (\pi, \frac{1}{2}\pi)$. This result supplements nicely the findings of Domokos & Holmes (1993*a*). The latter referred only to the limit behaviour of the primary branches and claimed that $\alpha_0 \to 0$ or $\alpha_0 \to \pi$ as $\lambda \to \infty$, which is in perfect accordance with the concept of binary chains. Our present study shows that *not only* the primary branches (but all branches) and not only the first angle (but all angles) converge to 0 or π.

(c) Cardinal numbers and Brouwer's point of view

The construction of the limit in observation 4.2 could be mapped onto the *mechanical* construction of frictionless chains with decreasing link-size. This intuitive argument guarantees that the limit obtained by our construction falls within the frame of *intuitionistic mathematics*, pioneered by Poincaré and Brouwer. According to the latter, the continuum can be regarded only as a whole ('matrix'), and *not* as an assembly of individual elements (cf. Brouwer 1975).

As observed in § 3, the set $\mathcal{C}$ has the cardinal $\aleph_2$, which is even greater than the cardinal $\aleph_1$ of the continuum. Thus, $\mathcal{C}$ can never be reached directly in any limit starting from a finite set. The limit that we constructed is only a denumerably infinite set $\mathcal{C}^* \subset \mathcal{C}$ with cardinal $\aleph_0$, and could be regarded as the intuitive (or, intuitionistic) version of $\mathcal{C}$. One would expect $\mathcal{C}^*$ to be dense in $\mathcal{C}$, however, this is not the case. It is an interesting question from the engineer's point of view, in which sense $\mathcal{C}$ is approximated by $\mathcal{C}^*$. Similarly, since $\mathcal{W}$ has the cardinal of the continuum ($\aleph_1$), it can not be reached in any intuitive limit (starting from a finite set), only a denumerably infinite, subset $\mathcal{W}^*$. In this case the density does not present difficulty. Neither of these problems arise in the case of $\mathcal{H}$, which is by its definition, (2.5)–(2.6), denumerably infinite.

We did not introduce these considerations earlier since they do not affect the essence of our plausibility argument. The ordering $\mathcal{H} \subset \mathcal{W} \subset \mathcal{C}$ applies also for the constructed (denumerable) limits: $\mathcal{H} \subset \mathcal{W}^* \subset \mathcal{C}^*$. These set-theoretical problems related to mechanical discretization became apparent in our approach because the degree N of discretization appeared in the cardinal number of the solution sets. Otherwise, if the number of solutions does not depend on the discretization, the gap in the cardinal numbers between the approximated set (continuum) and the limit of approximation (denumerable), although still present, is less disturbing. The problems discussed here originate in the choice of the continuum as a convenient mechanical model. It serves this purpose well as long as it is not regarded as a set consisting of individual elements. This approach is not only rejected by the intuitionists, it is also alien to mechanics, a discipline based largely on intuition.

(d) Stability and the odd-even paradox

The idea of the inelastic chains casts some light onto the stability of the large number of parasitic solutions appearing in figure 5. It is a natural question to ask whether any, or how many, of these solutions are physically relevant, i.e. stable? The individual study of the solutions is out of question, however, in § 5 *a* we assigned binary labels to each solution. These labels describe the shape of the inelastic chain to which the solution converges. Without detailed argument, appealing to simple experimental intuition, we claim that such a chain cannot be in stable equilibrium if any of the links are compressed. On the other hand, if all elements are in tension, the chain is stable. This means that, at least in the $\lambda \to \infty$ limit, solutions with binary label $B_N = \{b_1, b_2, \ldots, b_N\} = \{1, 1, \ldots, 1\}$ become stable. Although this applies to only *one* of the 2^N binary classes, a single class contains infinitely many solutions.

These considerations lead further, to the (partial) answer to an old-standing paradox about the nth primary mode of the N-link discrete model: as already reported by Gáspár & Domokos (1989): secondary pitchfork bifurcations from the $N-1$ primary branches can be *always* observed if N/n is an even integer, on the other hand, *never* observed if N/n is an odd integer. In the case when N/n is not an integer, no specific conjecture was formulated (cf. figure 5: on lowest branch $N/n = \frac{12}{1} = 12$, secondary pitchfork clearly visible, on fourth mode branch $N/n = \frac{12}{4} = 3$, no secondary pitchfork can be observed). The question could be generalized in the following form: determine the number b of secondary bifurcations on the nth primary branch of the N-link model. We can not deliver a complete answer; the question still remains open, the complete solution might require deeper tools than ones applied in this paper.

On the other hand, using the concept of binary chains we can determine the *minimal* value $b_{\min}$.

Observation 5.1. *The minimal number $b_{\min}$ of secondary bifurcation on the nth primary branch of the N-link model, can be expressed as $b_{\min} = Nf(n/N)$, where*

$$f(t) = \begin{cases} (1/q), & \text{if } t = (p/q),\ p, q \text{ are relative primes and } q \text{ is even.} \\ 0, & \text{otherwise.} \end{cases}$$

Simultaneously we formulate the following.

Conjecture 5.2. $b = b_{\min}$ *for all* (N, n).

Argument for observation 5.1. In the continuous problem (2.2)–(2.3), as the nth primary branch bifurcates off the trivial solution, the configuration has $E = n - 1$ negative eigenvalues, as it can be shown by linear stability analysis. Maddocks (1984) showed that as $x(L) = 0$, the beam can perform rigid body rotations around $(x, y) = (0, 0)$, corresponding to an unstable bifurcation, the secondary branches of which *do not lie in the* α_0, λ *plane.* (This is the reason why they do not appear in figure 2: on these secondary branches the vertical force $V(0)$ (cf. figure 1) does not vanish.) This means that, as we pass this bifurcation point in each mode, the number E of negative eigenvalues is increased by one. Since no further bifurcation occurs, in the continuous model the nth primary branch has $E = n$ negative eigenvalues as $\lambda \to \infty$.

The initial count for the negative eigenvalues is identical in the discrete problem (4.1)–(4.2), including the bifurcation at $x_N = 0$ (cf. Domokos & Holmes 1993*a*), so we have $E = n$ on the nth primary branch. However, the $\lambda \to \infty$ limit is rather different. As we saw in § 5 *a*, each branch (including the primary ones) can be characterized by a binary label B_N of length N. The '0' characters correspond to *compressed* chain elements, the '1' characters to the ones in *tension.* The number, E, of negative eigenvalues on any branch in the $\lambda \to \infty$ limit is equal to the number of '0' characters in the corresponding binary string. Let us regard the first mode branch ($n = 1$) of the structures with $N = 4$ and $N = 5$ links, respectively. Contemplation of figure 9 reveals that the $B_4 = \{1111\}$, while $B_5 = \{11011\}$. This means that in the $\lambda \to \infty$ limit we have $E_4 = 0$, $E_5 = 1$.

However, as we discussed before, *initially* we had $E_4 = E_5 = 1$. From the monotonicity of the primary branches (Domokos & Holmes 1993*a*) it follows that limit points do not occur, the only kind of critical point that can change E_4 is a *a secondary bifurcation* (cf. figure 8). The same cannot be claimed for the $N = 5$ case; here E_5 is the same in the limit. (We remark that this does not imply that E_5 remains constant along the branch, however, we did not discover any *necessary* reason for E_5 to change. Conjecture 5.2 claims that the discovered necessary condition is simultaneously a sufficient one.)

The generalization of this observation to the nth primary branch of the N-link structure is purely geometrical. At the bifurcation point from the trivial equilibrium, the nth primary branch consists of n segments of length N/n, separated by intersection points with the load line (not necessarily at joints), displaying n extrema in y.

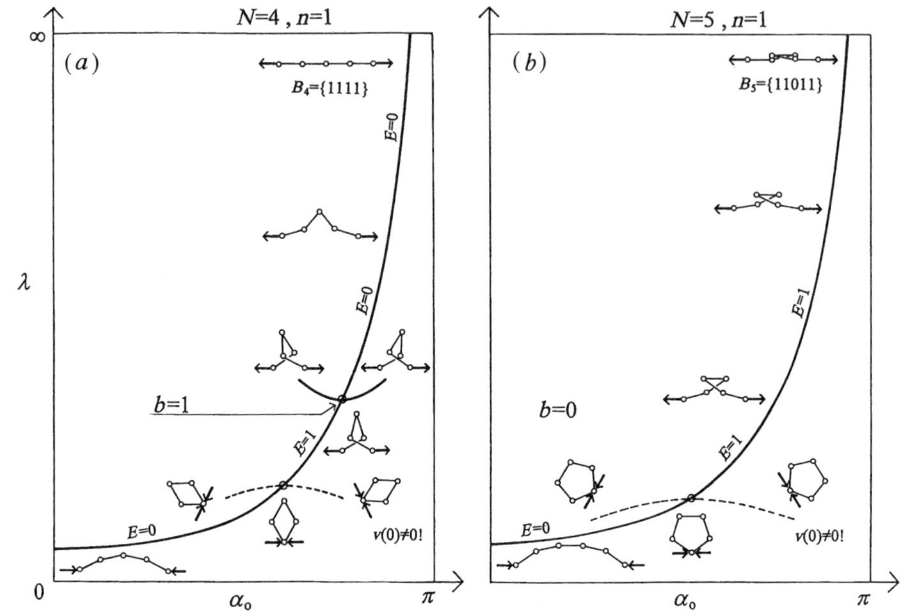

Figure 9. Global behaviour of the $n = 1$ branch for $N = 4$ and $N = 5$.

We have to count, how many *midpoints* of these segments coincide with hinges: those segments will behave similarly to the $N = 4$ branch in figure 8; thus their number is identical to the minimal change in E_N as we move along the branch. As observed before, this number is equal to $b_{\min}$. This geometric count is formulated algebraically in observation 5.1 with the aid of the function $f(t)$. (cf. figure 5: $12 * f(\frac{1}{12}) = 1$; $12 * f(\frac{2}{12}) = 2$; $12 * f(\frac{3}{12}) = 3$; $12 * f(\frac{4}{12}) = 0$; $12 * f(\frac{5}{12}) = 1$; $12 * f(\frac{6}{12}) = 6$; etc.)

We remark that (as it follows from the argument) the number b refers to secondary bifurcation in the $[\alpha_0, \lambda]$ plane and does not include the rigid body rotation at $x_N = 0$.

It seems worthwhile to describe this problem from the point of view of symmetry and symmetry breaking (cf. Golubitsky & Schaeffer 1988). The original problem (both discrete and continuous) has D_2 symmetry in the physical $[xy]$ configuration space. At the *primary* pitchforks the D_2 is factored by one of its subgroups: at *odd* modes by the C_2 (rotation), at *even* modes by the Z_2 (reflection) subgroup. As a consequence, the odd primary branches have Z_2 symmetry, the even primary branches have C_2 symmetry. (For more detail see Domokos & Holmes 1993*a*). In this subsection we dealt only with the first primary mode, so this has Z_2 symmetry, *regardless* of whether N is odd or even. Nevertheless, the $N = 2k$ (even) cases exhibit secondary pitchforks whereas the $N = 2k + 1$ (odd) cases never do so.

At this point it might appear that the reflection symmetry has little to do with the difference between odd and even discretization. This is not the case; Gáspár & Domokos (1989) investigate the behaviour of *symmetrical models*, i.e. non-uniform discretizations that share the symmetry group of the continuous model. (The results of the present paper are generalized for non-uniform discretizations in § 4 *a*.) For the first mode such discretizations are defined by $l_i = l_{N-i+1}$. Gáspár & Domokos (1989) point out that symmetric discretizations consist *typically* of odd number of links; 'even' models are non-generic. Symmetric discretizations are helpful in understanding the transition from 'even' to 'odd' models of uniform discretization. Figure 10

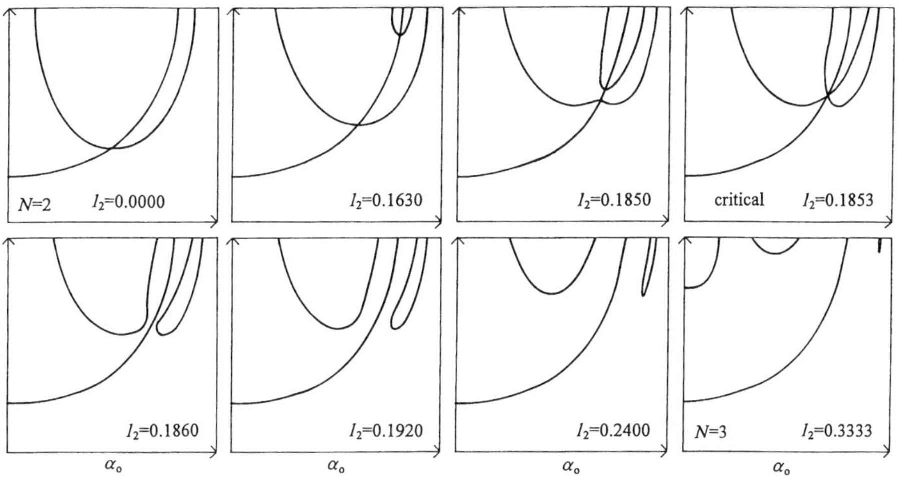

Figure 10. Bifurcation diagrams for the symmetric three-link model.

illustrates eight bifurcation diagrams from the one-parameter family of three-link symmetrical discretizations. The first and last diagram correspond to the $N = 2$ and $N = 3$ uniform discretizations, respectively. The bifurcation behaviour and stability of this structure is described by Maddocks (1985) with analytical tools. The diagrams of figure 10 are in accordance with his results, although he used different parametrization for the links.

(*e*) *Concluding remarks*

In this paper we studied special limits of Euler's buckling problem and of its discretized version. The appearance of imperfect wave-like solutions in physical and numerical experiments motivated our first conjecture, stating that the static image of solitary waves can be obtained by an appropriately constructed limit of the discretized problem. These quasi-static waves are solutions of BVPs that admit the homoclinic loop at arbitrary locations. We supported our conjecture by a plausible argument: we proved that a larger and a smaller set can be equally reached as limits.

These limits sets proved to be very helpful in organising the complex bifurcational structure of the discrete problem. In particular, we were able to assign individual unique labels to all branches in the N-link problem in the form of integer strings of length N. These labels open many interesting questions, e.g. they could be the basis of accurate estimates on the number of parasitic solutions. Branches can be ordered according to the parameter values where they emerge, or, according to their labels. The relationship between these orderings seems to be a challenging problem. We also defined binary labels, characterizing the stability at infinity. With the aid of these labels we were able to answer an old question: why does the behaviour of models with even and odd N differ? The construction of binary labels leads to the observation that the limit of this spatially complex behaviour is an appealingly simple random walk, or, conversely, spatial chaos in this BVP can be viewed as a perturbation of a Markov process. We hope that beyond making our conjecture on wave-like solutions plausible, the study of this simple mechanical example contributes to the better understanding of the relationship between discrete and continuous structures, their limits, and to the concepts relating chaotic and random behaviour.

The author highly appreciates the stimulating discussions with László Fehér, Zsolt Gáspár and Phil Holmes. The support of OTKA grants no. F021307 and no. T015851 is gratefully acknowledged. The fast and precise preparation of the figures is due to Antónia Szabados.

References

Beyn, W.-J. 1982 Spurious solutions for discrete superlinear boundary value problems. *Computing* **28**, 43–51.

Beyn, W.-J. 1990 The numerical computation of connecting orbits in dynamical systems. *J. Num. Analysis* **9**, 379–405.

Bohl, E. 1979 On the bifurcation diagram of discrete analogues for ordinary bifurcation problems. *Math. Meth. Appl. Sci.* **1**, 566–571.

Brouwer, L. E. J. 1975 *Collected works* (ed. A. Heyting), vol. I, pp. 82–83, 102–104. Amsterdam: North Holland.

Coleman, B. D. & Dill, E. H. 1992 Flexure waves in elastic rods. *J. Acoust. Soc. Am.* **91**, 2663–2673.

Domokos, G. & Holmes, P. J. 1993*a* Euler's problem, Euler's method and the standard map. *J. Nonlinear Sci.* **3**, 109–151.

Domokos, G. & Holmes, P. J. 1993*b* On non-inflexional solutions of non-uniform elasticae. *Int. J. Nonlin. Mech.* **28**, 677–685.

ElNaschie, M. S. 1990 On the suspectibility of local elastic buckling to chaos. *ZAMM* **70**, 535–542.

Euler, L. 1744 *Methodus inveniendi lineas curvas maximi minimive proprietate gaudentes Ostwald's Klassiker der Exakten Wiss.* vol. 75. Leipzig: W. Engelmann. (In German.)

Fraenkel, A. 1923 *Einleitung in die Mengenlehre.* (Grundl. der Math. Wiss., vol. IX). Berlin: Springer.

Gáspár, Zs. & Domokos, G. 1989 Global investigation of discrete models of the euler buckling problem. *Acta Technica Acad. Sci. Hung.* **102**, 227–238.

Golubitsky, M. & Schaeffer, D. 1988 *Singularities and groups in bifurcation theory*, vol. I, vol. II (with I. Stewart). Berlin: Springer.

Kirchhoff, G. 1859 Über das Gleichgewicht und die Bewegung eines unendlich dünnen elastischen Stabes. *J. für Math. (Crelle)* **56**, 285–313.

Love, A. E. H. 1927 *A treatise on the mathematical theory of elasticity* New York: Dover.

Maddocks, J. H. 1984 Stability of nonlinearly elastic rods. *Arch. Ration. Mech. Analysis* **85**, 311–354.

Maddocks, J. H. 1985 Restricted quadratic forms and their application to bifurcation and stability in constrained variational principles. *SIAM J. Math. Analysis* **16**, 47–68.

Maddocks, J. H. & Dichmann, D. J. 1994 Conservation laws in the dynamics of rods. *J. Elasticity* **34**, 83–96.

Marsden, J. E., O'Reilly, O., Wicklin, F. J. & Zombro, B. W. 1991 Symmetry, stability, geometric phases and mechanical integrators. *Nonlinear Sci. Today* **1** 4–11, 14–21.

Mielke, A. & Holmes, P. 1988 Spatially complex equilibria of buckled rods. *Arch. Ration. Mech. Analysis* **101**, 319–348.

Peitgen, H.-O. & Schmitt, K. 1981 Positive and spurious solutions of nonlinear eigenvalue problems. In *Numerical solution of nonlinear equations* (ed. E. L. Allgower, K. Glashoff & H.-O. Peitgen). Springer Lecture Notes in Mathematics, no. 878, pp. 276–324.

Peitgen, H.-O., Saupe, D. & Schmitt, K. 1981 Nonlinear elliptic boundary value problems versus their finite difference approximations: numerically irrelevant solutions. *J. Reine U. Angew. Math. (Crelle)* **322**, 74–117.

Polya, G. 1968 *Mathematics and plausible reasoning*, 2nd edn. Princeton University Press.

Thompson, J. M. T. & Virgin, L. N. 1988 Spatial chaos and localization phenomena. *Phys. Lett.* A **126**, 491–496.

Walsh, J. L. 1923 A closed set of normal orthogonal functions. *Am. J. Math.* **45**, 5–24.

INTERNATIONAL JOURNAL FOR NUMERICAL METHODS IN ENGINEERING, VOL. 40, 2237–2256 (1997)

FINITE ELEMENT MODELLING OF SPATIALLY CHAOTIC STRUCTURES

G. W. HUNT

School of Mechanical Engineering, University of Bath, Bath BA2 7AY, UK

R. LAWTHER

Department of Structural Engineering, University of New South Wales, Australia

P. PROVIDÊNCIA E COSTA

Department of Civil Engineering, Universidade de Coimbra, Portugal

SUMMARY

A simple non-linear mechanical system comprising a pin-jointed string of finite-length links, supported by elastic springs at the pins and compressed by an axial load, is viewed from two perspectives. When seen as an initial-value problem, equilibrium equations provide an iterative non-linear mapping. When seen as a boundary-value problem, it becomes a simple finite element model. At loads less than the critical buckling load, a preferred buckling configuration is found that is localized along the length. In the limit of infinite length this is described as a homoclinic connection in phase space, joining the flat equilibrium state to itself. The infinite sequence of homoclinic points thus defined embeds within the complex topological structure of a homoclinic tangle, within which also appear periodic, quasi-periodic, and chaotic spatial solutions. Implications in the finite element setting are discussed. © 1997 by John Wiley & Sons, Ltd.

KEY WORDS: localization; spatial chaos; buckling; discrete system; mapping; non-linear; finite elements; homoclinic connection

1. INTRODUCTION

By employing the *dynamical phase-space analogy*, that is by running a differential equation in a spatial dimension x as though it was a differential equation in time t, it has recently been demonstrated that equilibrium equations to non-linear structural systems can have chaotic solutions,[1,2] with all the hallmarks of extreme sensitivity to initial conditions, etc. The process, however, involves replacing a boundary-value problem with an initial-value problem, and we remain one step away from demonstrating the effects of spatial chaos on a realistically defined structural example. Here we attempt to fill the gap, in examining the multiplicity of solutions predicted by the dynamical analogy with a finite element model. Interest naturally falls on the differences between chaotic and non-chaotic regimes.

Chaotic solutions have been found for a continuous strut on a linear[3] and non-linear[1] foundation, but here we choose a simpler example, largely because it can be described by a second-order difference equation and thus can be viewed in two-dimensional phase space.[4] It has a further advantage, in that its discrete nature allows immediate translation into a finite element context; indeed, the system is so simple as to render the step trivial. The model is of a classical nature,[5] but we note that the localized response, which lies at the heart of the chaotic regime, is to the best

CCC 0029–5981/97/122237–20$17.50

Received 9 January 1996
Revised 4 August 1996

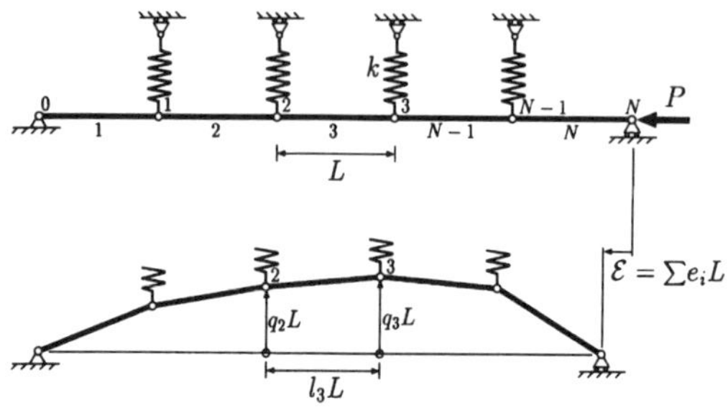

Figure 1. Spring-supported linkage: (top) undeflected system, (bottom) deflected state

of our knowledge described here for the first time; under rigid loading conditions it can be shown, either by constraint equations[6] or simple heuristic reasoning, that this is the most likely solution in practice.

We start by expressing the general equilibrium equation for the spring and rigid-bar model of Figure 1 as an iterative mapping, which if linearized at the flat fundamental state gives eigenvalues describing the stability of that state. For applied loads P less than the classical critical load $P^C = kL/4$ this is a two-dimensional saddle, with a single-dimensional stable manifold, or inset, and a single-dimensional unstable manifold, or outset. Non-linear behaviour is dominated by the presence of a homoclinic loop which has the well-known structure of a homoclinic tangle.[7–10] The fully localized solution over infinite length is expressed by an infinite set of homoclinic points, which for long-but-finite length dominates the behaviour when boundary conditions involve small-but-finite displacements at the ends. The situation is made yet more complicated by the possibility of further passages around the homoclinic loop, generating further homoclinic points and sets of intermeshing solutions. This boils down to a situation of considerable complexity, the ramifications of which are available to be explored in a finite element context over the full range of associated boundary conditions; particular emphasis is placed on solutions which are stable under the constraint of rigid loading, found with the aid of a Lagrange multiplier.

2. INITIAL VALUE PROBLEM (IVP)

2.1. Equilibrium equation

Consider the generalized structural model of Figure 1, comprising N rigid links of length L, supported by vertical linear elastic springs of stiffness k as shown, and compressed by axial load P.[5] With lateral deflection of the ith spring denoted by q_iL, and the difference between the ith and $(i+1)$th deflections denoted by v_iL, so that

$$v_i = q_{i+1} - q_i \tag{1}$$

a total potential energy function, V, can be written from the difference between strain energy stored in the springs, U, and work done by load, $P\mathscr{E}$, where $\mathscr{E}$ is the total end shortening as follows:

$$\begin{aligned} V &= U - P\mathscr{E} \\ &= \tfrac{1}{2}kL^2 \sum q_i^2 - PL\sum\left(1 \pm \sqrt{1 - v_i^2}\right) \end{aligned} \tag{2}$$

the first summation being carried out over the total number of springs, and the second over the total number of links. If link rotations, θ_i, are restricted so that $-\pi/2 < \theta_i < \pi/2$, the total potential energy becomes

$$V = \tfrac{1}{2} k L^2 \sum q_i^2 - PL \sum \left(1 - \sqrt{1 - v_i^2}\right) \tag{3}$$

and differentiation of V with respect to the generalized co-ordinate q_i, after a little manipulation[11] gives the equilibrium equation,

$$kLq_i + P\left(\frac{v_i}{\sqrt{1 - v_i^2}} - \frac{v_{i-1}}{\sqrt{1 - v_{i-1}^2}}\right) = 0 \tag{4}$$

Standard eigenvalue analysis of the flat (fundamental) state $q_i = 0$ reveals, for an infinitely long pin-ended strut, the critical bifurcation load $P^C = kL/4$ (equation (37) for $N \to \infty$). Post-buckling analysis reveals that from this point emerges a further equilibrium path that is unstable under load control and falls with respect to the load.[5] Such a path is seen emerging from point a in Figure 10.

The implicit appearance of the three adjacent generalized co-ordinates q_{i-1}, q_i and q_{i+1} means that the equilibrium equation (4) can be seen as a second-order difference equation, which can be written as a pair of first-order equations as follows:

$$\begin{aligned} q_{i+1} &= q_i + v_i \\ v_{i+1} &= \frac{\alpha_i}{\sqrt{1 + \alpha_i^2}} \end{aligned} \tag{5}$$

where

$$\alpha_i = \alpha_i(q_i, v_i) = -4\frac{q_i + v_i}{p} + \frac{v_i}{\sqrt{1 - v_i^2}} \tag{6}$$

the non-dimensional load p being defined as

$$p = \frac{P}{P^C} = \frac{4P}{kL} \tag{7}$$

This mapping can be seen as a discrete-time dynamical system.[4,10–12] Alternatively, we can write

$$a_i = (-1)^i \frac{q_{i+1} - q_i}{2}, \quad b_i = (-1)^i \frac{q_{i+1} + q_i}{2} \tag{8}$$

where $a_i L$, with alternately changing sign, measures the rotation of the $(i+1)$th link, and $b_i L$ correspondingly measures the lateral displacement of its centre. The same equilibrium equation then appears as

$$\begin{aligned} a_{i+1} &= \frac{1}{2}\frac{\beta_i}{\sqrt{1 + \beta_i^2}} \\ b_{i+1} &= a_{i+1} - a_i - b_i = \frac{1}{2}\frac{\beta_i}{\sqrt{1 + \beta_i^2}} - a_i - b_i \end{aligned} \tag{9}$$

where

$$\beta_i = \beta_i(a_i, b_i) = \frac{4}{p}(a_i + b_i) - \frac{2a_i}{\sqrt{1 - 4a_i^2}} \tag{10}$$

Here a_i can be interpreted as the amplitude of an oscillation in q_i, and b_i as a corresponding bias.[11]

2.2. Eigenvalues of the linear map

In dynamical terms, for $P < P^C$, the flat fundamental state has the form of a saddle point,[4,10] with one dimension of stable manifold (inset: convergence) and one of unstable manifold (outset: divergence). The orientation of these principal directions corresponds to the eigenvectors associated with the linear terms of the above mapping, which dominate the non-linear terms close to the fundamental state.

To show this we linearize the mapping (5) for $p > 0$ at the fundamental state, writing the result in matrix form

$$\begin{bmatrix} q_{i+1} \\ v_{i+1} \end{bmatrix} = \begin{bmatrix} 1 & 1 \\ -\dfrac{4}{p} & \left(1 - \dfrac{4}{p}\right) \end{bmatrix} \begin{bmatrix} q_i \\ v_i \end{bmatrix} \tag{11}$$

and compute the eigenvalues from $|\mathbf{A} - \lambda \mathbf{I}| = 0$, where $\mathbf{A}$ is the above square matrix. This gives the characteristic equation,

$$\lambda^2 - \left(2 - \frac{4}{p}\right)\lambda + 1 = 0 \tag{12}$$

and so

$$\lambda = 1 - \frac{2}{p} \pm \sqrt{\left(1 - \frac{2}{p}\right)^2 - 1} \tag{13}$$

For the range of interest $0 < p < 1$ this gives two real distinct eigenvalues λ_1 and λ_2 with the reciprocal property,

$$\lambda_1 \lambda_2 = 1 \tag{14}$$

indicating that the linearized map is symplectic.[10] The associated eigenvectors are

$$q_{i+1} = \lambda q_i, \quad v_{i+1} = \lambda v_i \tag{15}$$

which gives, with equation (1),

$$q_i = \frac{1}{\lambda - 1} v_i$$

The maximum absolute value of λ expresses the rate of exponential growth in amplitude, while the minimum absolute value expresses the exponential decay.

If $p = 0{\cdot}8$, for example, we have

$$\lambda = \frac{-3 \pm \sqrt{5}}{2} \tag{16}$$

which gives $\lambda_{\max} = -2{\cdot}618$ and $\lambda_{\min} = -0{\cdot}382$. The negative signs indicate that, on emergence from the flat fundamental state, q_i oscillates.

2.3. Phase portraits

When the nonlinear map is initiated from differing positions in the $a_i - b_i$ phase-plane, characteristic 2D portraits result, as shown for subcritical values of the load in Figures 2, 4 and 5 Each trajectory is associated with a set of possible equilibrium configurations of the model. These

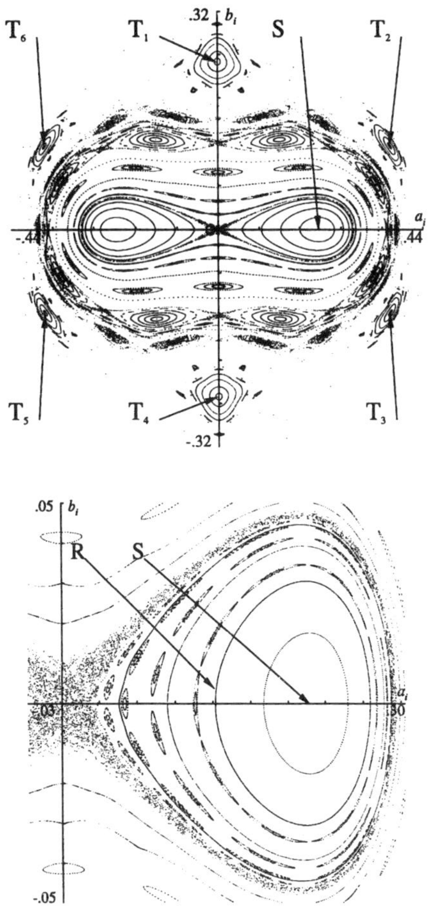

Figure 2. Trajectories in phase space at $p = 0{\cdot}9$. S is periodic of period 2, R is quasi-periodic, and T_i represent periodic solutions of period 3, as shown in Figure 3

phase portraits reveal much of the underlying topology. A periodic solution, for example, appears as a one or more invariant points, as seen in Figure 2. The point marked S, for instance, is an oscillation of period 2, the amplitude of which, being represented by a_i, appears as a single invariant point on the plot. Points marked T_i represent longer wavelength fluctuations of period 3, as illustrated in Figure 3. Four such forms are possible, although because of the change of sign in equation (8) each uses all six points $T_1 - T_6$. Similarly, the same form S is produced, with a phase difference of L, from both the position shown and its mirror image.

A regular orbit such as R of Figure 2(b) represents a non-repeating but quasi-periodic buckle pattern, involving a modulation with a wavelength that is irrational. The origin of Figure 2 is, of course, the entirely flat state, and a loop connection of the neighbourhood of this flat state with itself thus has the appearance of a localization. Bounded and unbounded (escaping) solutions lie outside the figure-of-eight formed by two such loops. As the load drops from $P^{C} = kL/4$ the

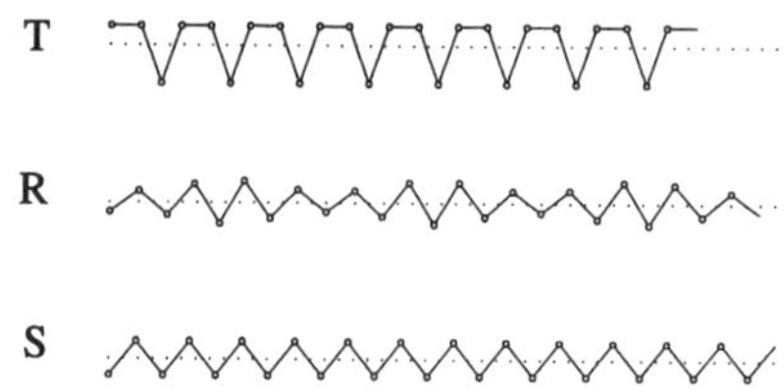

Figure 3. Some typical periodic and quasi-periodic solutions

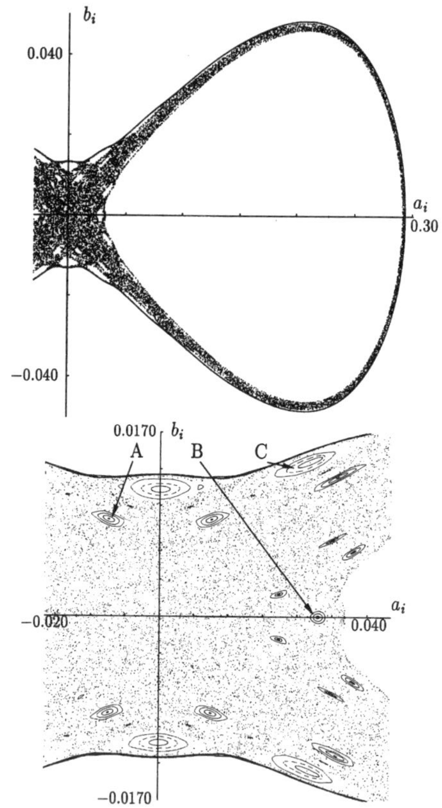

Figure 4. Chaotic trajectory at $p=0{\cdot}9$. A–C show that regular behaviour exists inside the chaotic regime

unbounded behaviour interferes with the homoclinic loop in an increasingly destructive manner, until by $P=0{\cdot}85P^{C}$ it has, to all intents and purposes, eliminated the boundaries of the loops themselves; the sharply defined externally enveloping boundaries to these loops, seen in the detailed plot of Figure 4 for $p=0{\cdot}9$, have disappeared by $p=0{\cdot}85$.

A closer examination of the homoclinic loop reveals the well-known structure of the homoclinic tangle[7,10]—a thoroughly complex affair involving periodic, quasi-periodic, homoclinic (localized)

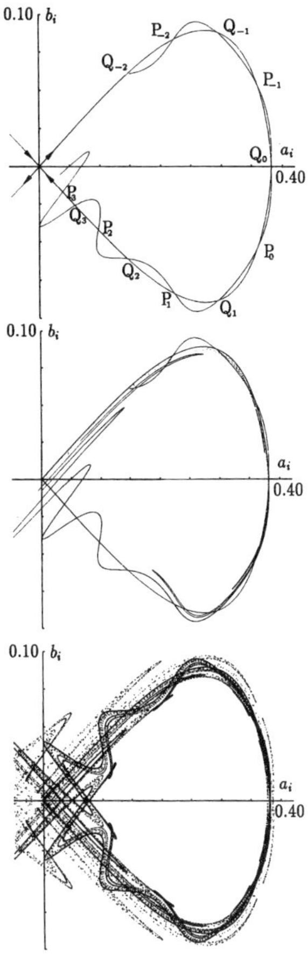

Figure 5. Stretching and folding of the evolutes at $p = 0{\cdot}8$

and chaotic trajectories. Figure 4 shows an example of the latter; typically there is a cloud, rather than clearly defined locus, of points appearing from a single trajectory. The underlying structure of this cloud is further revealed in the plots of Figure 5.

Figure 5 observes the evolution of two families of initial points supposedly lying close to one another on both the outset and the inset of the saddle. We start from very close to the flat state, and after a single iteration split the resulting interval into 585, 2340 and 1170 separate starts for parallel runs, giving, respectively, the plots of Figures 5(a), 5(b) and 5(c). Points over the interval are allowed to evolve over a finite number of iterations, increasing from (a) to (b) to (c), so that the classical stretching and folding of the evolute, first described by Poincaré,[7] is clear to see. The inset was here obtained by running 'backwards', or inverting (9):

$$\begin{aligned} a_{i-1} &= \frac{1}{2}\frac{\gamma_i}{\sqrt{1+\gamma_i^2}} \\ b_{i-1} &= -a_{i-1} + a_i - b_i = -\frac{1}{2}\frac{\gamma_i}{\sqrt{1+\gamma_i^2}} + a_i - b_i \end{aligned} \tag{17}$$

where

$$\gamma_i = \gamma_i(a_i, b_i) = \frac{4}{p}(a_i - b_i) - \frac{2a_i}{\sqrt{1 - 4a_i^2}} \tag{18}$$

In common with other mappings, the chaotic regime has a finely detailed substructure.[12] Islands of regular behaviour exist within the chaotic sea, A–C of Figure 4(b), for example. These all live in archipelagos; C for example is one of a set of 24 related islands, each of which carries a secondary family, which themselves have subfamilies and so on. As we go deeper into the substructure, longer and longer wavelength oscillations are revealed.

2.4. *Homoclinic points and localized solutions*

Part and parcel of the homoclinic tangle is the sequence of homoclinic points, where stable and unstable manifolds intersect. It is a well-known result of the non-linear theory of conservative mappings[12] that if the intersection takes place at all it takes place an infinite number of times; there are thus an infinite number of such points, stepping in towards the saddle point along both manifolds. This defines the most localized solution. Over infinite distance the amplitude steadily increases from zero, passes through the *antisymmetric section* ($b_i = 0$) at some maximum value (Q_0, for example, in Figure 5), and steadily decreases back to zero; other less localized solutions can also be found by seeking passage through the corresponding *symmetric section* where $q_{i-1} = q_{i+1}$. Such solutions are often described by simply running over half the available spatial domain, from the (anti)symmetric section to the flat state or *vice versa*.[1] (For examples of passage through the antisymmetric and symmetric sections, in a system of finite length, see, respectively, Figures 12(c) and 12(d) later.) Other localized solutions, possibly but not necessarily passing through the symmetric or antisymmetric section, also exist.

2.5. *Boundary conditions in the IVP*

The complex structure of the homoclinic tangle and its association with localized solutions naturally generate boundaries of a fractal nature, as seen in Figures 6 and 7. For a strut length of 20 links, these plots indicate the sign of q_{20} in the *space of starts* $q_0 - q_1$: if $q_{20} \geqslant 0$ the point representing the initial conditions in this space is plotted in black; if $q_{20} < 0$ it is plotted in white. The edge between black and white then relates to the single boundary condition $q_{20} = 0$, which means that a slice through the space of starts at, for example, $q_0 = 0$ provides a solution to the boundary-value problem $q_0 = q_{20} = 0$ each time it cuts a black/white interface. At infinite length there appears to be an infinite number of such edges.

Although a typical slice through the space of starts cuts across the direction of the black/white stripes, there is one exception, the black/white interface passing through the origin. At the origin this lies close to, but at a small inclination from, the inset line or stable manifold defined by the minimum eigenvalue $\lambda_{\min}$. At infinite length, and only at infinite length, do the two lines coincide. If a point on the stable manifold is forced as an initial condition, the only outcome can be a sequence of points leading inwards towards the saddle. All other starts carry a component of unstable manifold, expressed in the limit by the maximum eigenvalue, and hence are subject to divergence.

There thus apparently exists, at infinite length, an infinite number of possible equilibrium solutions to the boundary-value problem. This is further revealed in Figure 8, which shows how q_{20} varies with q_1 along the line $q_0 = 0$ in the space of starts, representing a single boundary condition. Solutions to the boundary-value problem $q_0 = q_{20} = 0$ appear each time the curve crosses

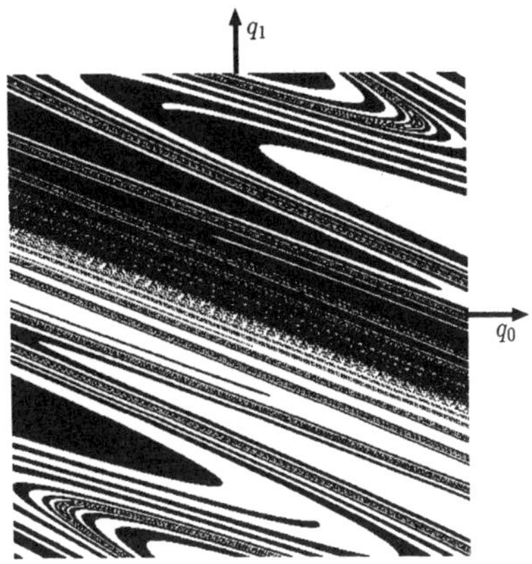

Figure 6. Fractal structure to the space of starts. Range: $q_0, q_1 = \pm 0{\cdot}05$. $p - 0{\cdot}8$; 20 links

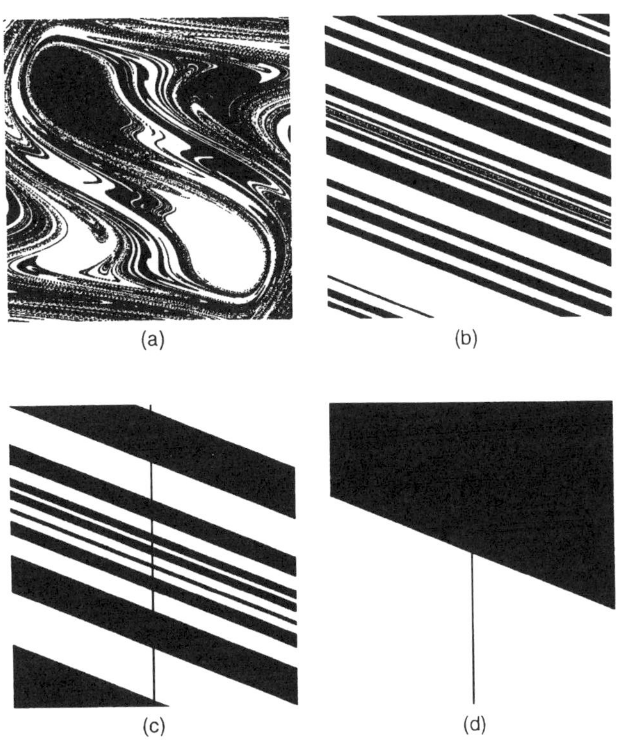

Figure 7. Space of starts $q_0 - q_1$ over four different ranges. $p = 0{\cdot}8$. 20 links: (a) q_0, q_1 in the range $\pm 0{\cdot}5$ (b) q_0, q_1 in the range $\pm 5e - 4$; (c) q_0, q_1 in the range $\pm 5e - 6$; (d) q_0, q_1 in the range $\pm 5e - 8$

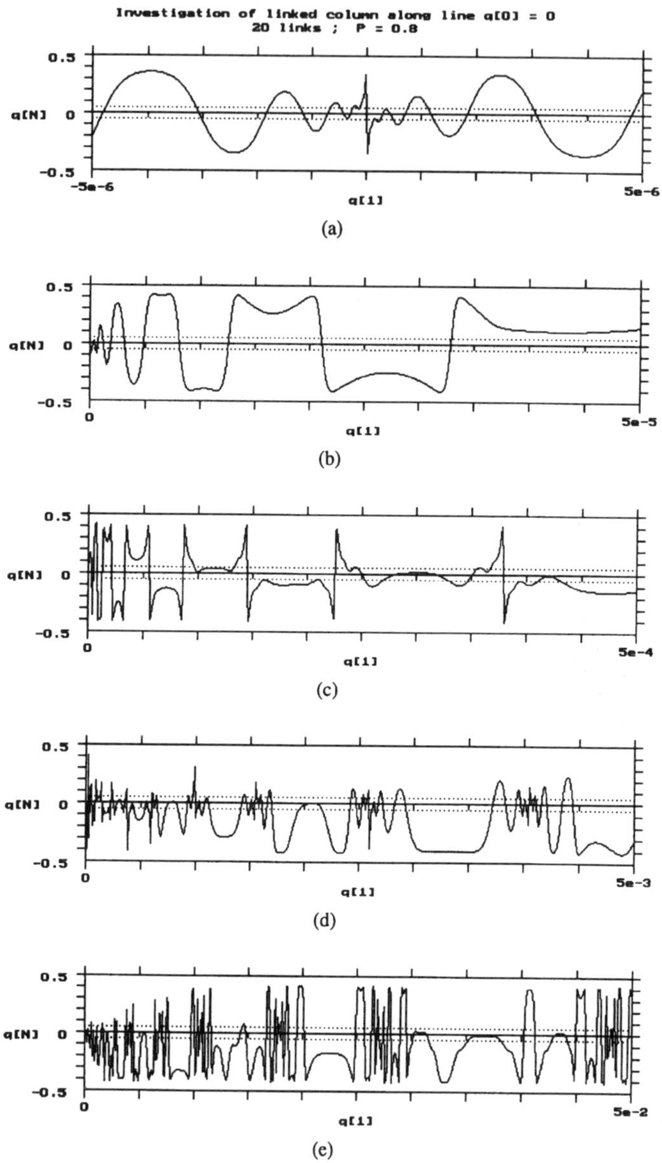

Figure 8. Chaotic fluctuation of $q_{20}(q_1)$. $q_0 = 0$; $p = 0{\cdot}8$

the horizontal axis, and similarly solutions to $q_0 = 0, q_{20} = 0{\cdot}05$ as it crosses the upper of the two broken horizontal lines. These plots again highlight the chaotic nature. The apparent discontinuity at $q_1 = 0$ in the centre of the top plot and at the left of all the remainder is a result of the starting position crossing the flat state; q_{20} suddenly jumps from positive to negative as the system swaps between branches of the figure-of-eight homoclinic loop. Interestingly, there appear similar discontinuities at different scales for non-zero values of q_{20} in the plots of Figures 8(b)–8(e), again a consequence of the fractal nature.

Wadee,[13] for this same linked system, measures the fractal dimension of the plot of Figures 6 and 7 for various load levels. For a load of $P=0{\cdot}8P^C$, for instance, he demonstrates that it is non-integer for lengths greater than about 15 links, and that it settles to a constant value of about 0·02 after some 25 to 30 links. The finite element analysis that follows is again conducted for 20 links, and not surprisingly reveals a thoroughly complex underlying bifurcational structure.

3. BOUNDARY-VALUE PROBLEM (BVP)

The following formulation of the strut system of Figure 1 as a boundary-value problem is performed in a manner that typifies a finite element system. Results are presented for a model of 20 links, which is considered sufficient to illustrate the underlying complexity. On increasing load, the flat equilibrium state initially loses its stability at an unstable-symmetric point of bifurcation,[5] a so-called subcritical phenomenon. Post-buckled states thus appear at loads below the minimum critical (bifurcation) load P^C, and are unstable under dead loading. For realistic application we thus assume that end-shortening rather than load is prescribed, thereby leaving the position of the equilibrium paths unaltered in load–deflection space, but stabilizing some of them. Equilibrium and stability are then investigated by standard variational techniques. The resulting bifurcational structure is fully analysed over the load range $0<P<P^C$, and the outcome at $P=0{\cdot}8P^C$ compared with the IVP results in phase space. A comprehensive overview begins to emerge.

3.1. Finite element formulation

Equation (3) gives the total potential energy,

$$V=U-P\mathscr{E}=\frac{1}{2}kL^2\sum_{i=1}^{N-1}q_i^2-PL\sum_{i=1}^{N}e_i \tag{19}$$

where e_iL is the end-shortening due to the ith link. Variations of V give equations for the equilibrium and stability of the strut. The structure divides naturally into discrete elements, with the springs contributing exclusively to the first (strain energy) term, and link elements contributing exclusively to the second (work done) term. The latter is the source of all non-linearity. Individual spring and link elements are shown in Figure 9.

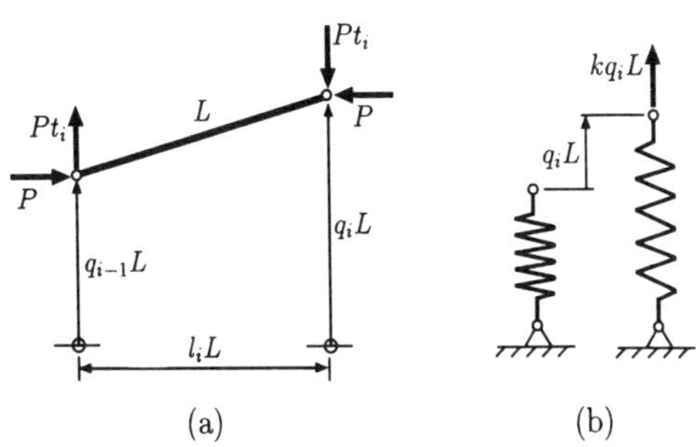

Figure 9. (a) Link element, and (b) spring element

3.1.1. Spring elements. The spring element of Figure 9(a) has a strain energy of $U_i = \frac{1}{2}kL^2q_i^2$, with the first and second variations,

$$\delta q_i \, kL^2 \, q_i \quad \text{and} \quad \delta q_i \, kL^2 \, \delta q_i \tag{20}$$

respectively.

3.1.2. Link elements. The link element of Figure 9(b) has a length of L and a horizontal projection of l_iL. This element contributes $e_iL = (1 - l_i)L$ to the end shortening, and $-PLe_i$ to the total potential. Assuming once more that $-\pi/2 < \theta_i < \pi/2$, we have $l_i = +\sqrt{1-(q_i - q_{i-1})^2}$ and the required derivatives of e_i are

$$\frac{\partial e_i}{\partial q_{i-1}} = -\frac{\partial l_i}{\partial q_{i-1}} = -\frac{q_i - q_{i-1}}{l_i} \quad \text{and} \quad \frac{\partial e_i}{\partial q_i} = -\frac{\partial l_i}{\partial q_i} = \frac{q_i - q_{i-1}}{l_i} \tag{21}$$

If the tangent of the rotation of the typical link is written $t_i = (q_i - q_{i-1})/l_i$, and $\mathbf{q} = [q_{i-1} \; q_i]^{\mathrm{T}}$, the contribution of this element to the first variation of V is

$$-\delta\mathbf{q}^{\mathrm{T}} \frac{\partial(PLe_i)}{\partial \mathbf{q}} = -\delta\mathbf{q}^{\mathrm{T}} PLt_i \begin{bmatrix} -1 \\ 1 \end{bmatrix} \tag{22}$$

and second variation similarly gives

$$-\delta\mathbf{q}^{\mathrm{T}} \frac{\partial^2(PLe_i)}{\partial \mathbf{q}^2} \delta\mathbf{q} = -\delta\mathbf{q}^{\mathrm{T}} \frac{PL}{l_i^3} \begin{bmatrix} 1 & -1 \\ -1 & 1 \end{bmatrix} \delta\mathbf{q} \tag{23}$$

3.2. Equilibrium equations

The equilibrium equations of the complete structure are given by an assembly of terms of the type of the first of equations (20) for each spring and of equation (22) for each link giving

$$kLq_i + Pw_i = 0 \tag{24}$$

for $i = 1$ to $N - 1$, where $w_i = t_{i+1} - t_i$. This is of course the same as equation (4), since $t_i = v_{i-1}/\sqrt{1 - v_{i-1}^2}$.

3.3. Stability criteria

3.3.1. Stability under load control. In a similar manner, the second term in (20) for each of the $N - 1$ springs, and the term (23) for each of the N links, assemble into a quadratic form in the δq_i, with the tridiagonal matrix,

$$\mathbf{K}_{\mathrm{lc}} = \begin{bmatrix} d_1 & s_1 & & & \\ s_1 & d_2 & s_2 & & \\ & s_2 & \ddots & \ddots & \\ & & \ddots & d_{N-2} & s_{N-2} \\ & & & s_{N-2} & d_{N-1} \end{bmatrix} \tag{25}$$

where

$$d_i = kL + P(u_i + u_{i+1}), \quad s_i = -Pu_{i+1}, \quad u_i = -\frac{1}{l_i^3} \tag{26}$$

Stability of the structure under load control is associated with positive definiteness of this matrix, which is assessed with a Gauss-type reduction without pivoting, followed by a count of the negative diagonal elements (the degree of instability,[5] or Sturm count).

3.3.2. Stability under displacement control. For displacement control, end-shortening $\mathscr{E}$ rather than load P is prescribed. We can replace the second term from the potential function (19) by $-\lambda(\mathscr{E} - c)$, such that a constraint $\mathscr{E} = c$, where c is a constant, is enforced through a Lagrange multiplier λ. This has little effect on the equilibrium equations: the contribution of the link to the first variation of the new total potential is now

$$-\delta\mathbf{q}^{\mathrm{T}} \frac{\partial(\lambda L e_i)}{\partial \mathbf{q}} - \delta\lambda \frac{\partial(\lambda L e_i)}{\partial \lambda} = -\delta\mathbf{q}^{\mathrm{T}} \lambda L t_i \begin{bmatrix} -1 \\ 1 \end{bmatrix} - \delta\lambda\, L e_i \tag{27}$$

and to the second variation is

$$-\delta\mathbf{q}^{\mathrm{T}} \frac{\partial^2(\lambda L e_i)}{\partial \mathbf{q}^2} \delta\mathbf{q} - \delta\mathbf{q}^{\mathrm{T}} \frac{\partial^2(\lambda L e_i)}{\partial \mathbf{q}\, \partial\lambda} \delta\lambda - \delta\lambda \frac{\partial^2(\lambda L e_i)}{\partial\lambda \partial\mathbf{q}} \delta\mathbf{q} - \delta\lambda \frac{\partial^2(\lambda L e_i)}{\partial\lambda^2} \delta\lambda$$

$$= -\delta\mathbf{q}^{\mathrm{T}} \frac{\lambda L}{l_i^3} \begin{bmatrix} 1 & -1 \\ -1 & 1 \end{bmatrix} \delta\mathbf{q} - \delta\mathbf{q}^{\mathrm{T}} L t_i \begin{bmatrix} -1 \\ 1 \end{bmatrix} \delta\lambda - \delta L t_i [-1 \quad 1]\, \delta\mathbf{q} + 0 \tag{28}$$

Thus, the first variation simply changes P to λ in (24), and adds the constraint equation. Second variation similarly replaces P by λ in (25), but also augments this matrix, giving

$$\mathbf{K}_{\mathrm{dc}} = \begin{bmatrix} & & & w_1 \\ & \mathbf{K}_{\mathrm{lc}}(\lambda) & & \vdots \\ & & & w_{N-1} \\ w_1 & \cdots & w_{N-1} & 0 \end{bmatrix} \tag{29}$$

A Lagrange multiplier has been added and the Sturm count adjusted accordingly.[14] This reduces the negative count by a number ranging from zero to the number of constraints, so in this problem displacement control either makes the count one less or leaves it unaltered.

3.4. Path-following algorithm

An equilibrium point is defined as any set of values (q_i, P), $i = 0, 1, \ldots, N$ satisfying the equilibrium equations (24), and an equilibrium path is a smooth continuous set of equilibrium points. In this boundary-value problem all points on a path will have the same given values of q_0 and q_N. The computational problem is stated as:

(i) Given an initial equilibrium point find a nearby one, and continue doing this, so producing a sequence of points on an equilibrium path.

Nearby is defined in terms of a distance measure. Finding a nearby point is one problem: knowing that it is connected to the original by an equilibrium path is certainly another.

Finding nearby points at a given distance can be visualized as constructing a sphere, centred at the current point and with radius equal to the specified distance, and searching for equilibrium points on the surface of this sphere. There are usually two intersections, which can be interpreted as the forward and backward directions on the current path. Further intersections will occur if the sphere contains multiple paths. If these paths are distinct then all but the current path can be eliminated by choosing the radius to be sufficiently small, but equilibrium paths can intersect, at a *bifurcation point*. In the neighbourhood of a bifurcation point there are several (commonly four) nearby points, all connected; the number of intersections with the sphere cannot be reduced by shrinking while the bifurcation point remains inside. There will however be a change in Sturm count along at least one of the individual paths.

An equilibrium path with some known points can be propagated from a current equilibrium point to a nearby point, given a definition of distance and an approximate direction, using the arc-length method.[15] The direction to an anticipated nearby point is taken as the tangent to the path at the current point, which is predicted from this and previous equilibrium points. If the current point is a bifurcation, a choice of path has to be made.

The first estimate of a new equilibrium point is the current point plus a vector of the required length in the given direction. Equilibrium of this point is checked using equation (24) defining

$$R_i = kLq_i + Pw_i \tag{30}$$

The residuals R_i will generally not be zero, and adjustments to P and q_i must be made. The differential form of equation (30) is

$$\delta R_i = \frac{\partial R_i}{\partial q_j}\delta q_j + \delta P w_i \tag{31}$$

the repeated suffix implying summation. The derivatives $\partial R_i/\partial q_j$ are the elements of the tangent stiffness $\mathbf{K}_{\mathrm{lc}}$ of equation (25).

The aim is, of course, to find adjustments $(\delta q_i, \delta P)$ such that

$$R_i + \delta R_i = 0 \quad \text{or} \quad \delta R_i = -R_i \tag{32}$$

under conditions imposed by an arc-length constraint, commonly held to be either linear or quadratic. The major difficulty of path following in the current exercise was the tendency of an algorithm to reverse out of bifurcation and limit points, and subsequently track back to the origin. The following scheme, using a linear constraint, was found to be both robust and fast.

A general linear constraint

$$c_i \delta q_i + c_0 \delta P = 0 \tag{33}$$

is considered, where suitable choices of c_i and c_0 give load, displacement, or more general arc-length, control. Premultiplication of equation (31) by $\mathbf{K}_{\mathrm{lc}}^{-1}$ gives

$$\delta q_i + q_i^w \delta P = -q_i^R \tag{34}$$

where $\mathbf{q}^{\mathrm{w}} = \mathbf{K}_{\mathrm{lc}}^{-1}\mathbf{w}$ is the displacement vector associated with the vector $\mathbf{w}$, and $\mathbf{q}^{\mathbf{R}} = \mathbf{K}_{\mathrm{lc}}^{-1}\mathbf{R}$ is similarly defined with respect to the residuals. A second premultiplication by the constraint coefficients c_i produces

$$(-c_0 + c_i q_i^w)\delta P = -c_i q_i^R \tag{35}$$

Hence,

$$\delta P = \frac{-c_i q_i^R}{-c_0 + c_i q_i^w} \tag{36}$$

with the values of δq_i being obtainable from (34). Arc-length constraint, denying the possibility of backtracking, is provided by simply choosing (c_i, c_0) to be the current path vector in (q_i, P) space, calculated at the start of the increment.

The new estimate of the nearby point will (almost certainly) not be on the required sphere, but can be adjusted by scaling the changes. Iteration then returns to equation (30), and continues until the residuals are sufficiently small—calculations have been conducted on a PC using TurboPascal with 10 byte real representation, *sufficiently small* being taken as $|R_i/kL| < 10^{-15}$. This is a reasonably stringent convergence criterion, and scaling to satisfy the distance constraint can easily destroy an otherwise good estimate. For efficiency, the distance constraint is thus relaxed to ± 5 per cent, which means that a fairly thick-walled sphere is being considered.

The path tracking is started by specifying an initial point and a direction. If the initial point is not in equilibrium, a nearby point is sought with the above algorithm, without at this stage any distance constraint.

3.5. *Bifurcations of the perfect system*

The following results are mostly given for the perfect system, defined by $q_i = 0$ for $i = 0, 1, \ldots, N$ in the unloaded state; non-zero values at the boundaries or elsewhere identify imperfect systems. This is done on the basis that if the perfect system is well-understood, imperfect behaviour is also readily dealt with; the complication of a bifurcation point is more likely to arise in a perfect than an imperfect response, so that if the former is tractable so too should be the latter.

3.5.1. Fundamental path. Thompson and Hunt[5] show that loading of the fundamental equilibrium state $q_i = 0$ results in $N - 1$ critical bifurcation loads

$$P^m = \frac{kL}{4\sin^2(m\pi/2N)}, \quad \mathbf{q} = 0, \quad m = 1 \ldots N-1 \tag{37}$$

The corresponding tangents to the new equilibrium paths are

$$\Delta P = 0, \quad \Delta q_i = \sin\frac{im\pi}{N} \tag{38}$$

The lowest of these (symmetric) bifurcations occurs at $P^{N-1} = kL/4\sin^2[(N-1)\pi/2N]$, which approaches $kL/4$ as the number of links increases. Accordingly, a non-dimensional load $p = 4P/kL$ is defined. The minimum critical load p^C for a strut of finite length is thus slightly greater than 1.

3.5.2. Post-buckled solutions. To explore fully the response of the perfect system under slowly changing load, boundary conditions $q_0 = q_N = 0$ are first imposed. The path-tracking procedure described above is then employed, special care being taken to identify bifurcation points by keeping a running check on the Sturm count. Paths emanating from the lowest two bifurcation points for $N = 20$ are shown in Figure 10 on a plot of load p against end-shortening $\mathcal{E}/L$, with the branching sequence described schematically in Figure 11 and representative deflected states shown in Figure 12.

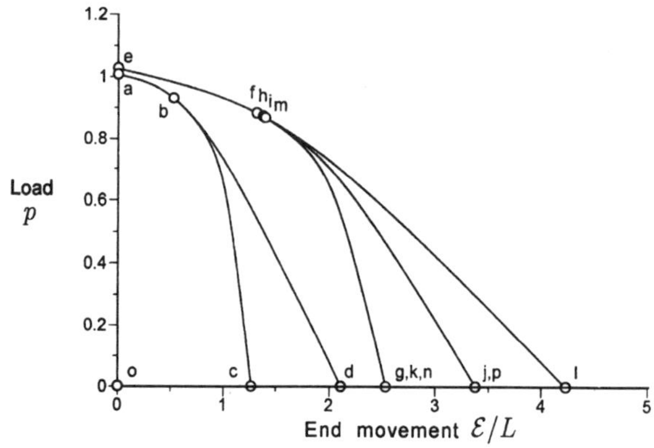

Figure 10. Some equilibrium paths of the 20 link strut

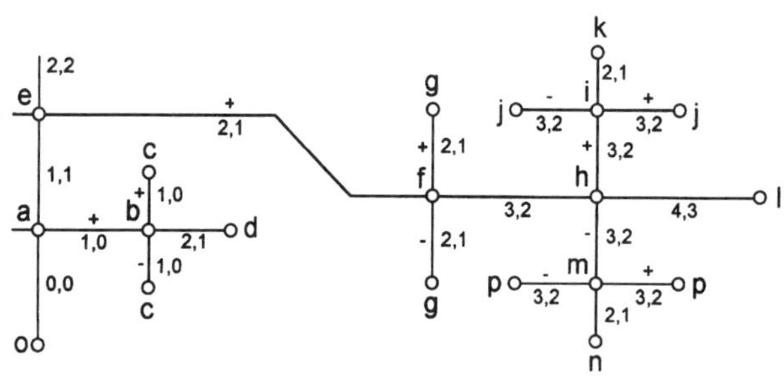

Figure 11. Bifurcation diagram. Numbers (1, 0), etc. denote Sturm counts under load and displacement control respectively. Paths out of a bifurcation are labelled + and −, according to whether q_1 tends to increase or decrease

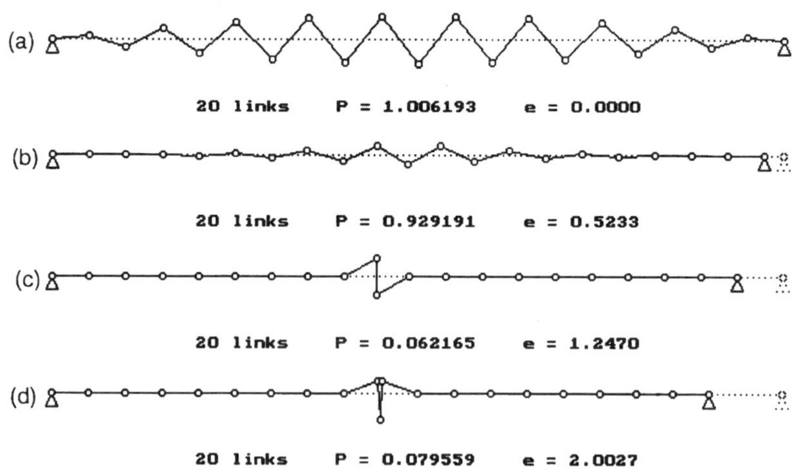

Figure 12. Some typical equilibrium states: (a) near a—vertical scale exaggerated by 1000; (b) at b; (c) near c; (d) near d

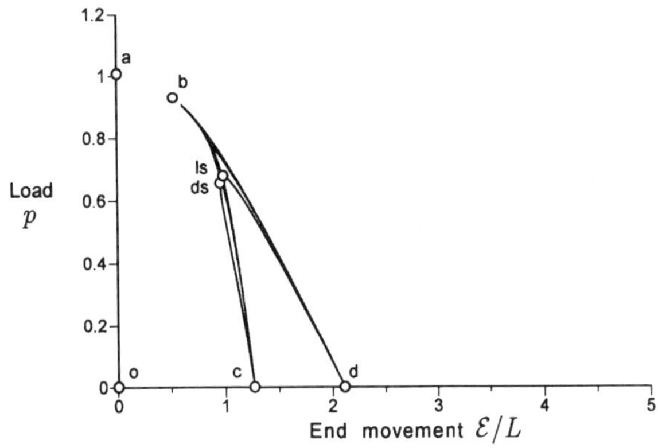

Figure 13. Equilibrium paths disconnected from the fundamental path $q_i = 0$

We see from Figure 11 that under rigid loading conditions only the path lying closest to the fundamental path in Figure 10 is stable. Localized buckling, ending up at zero load as the near-centred self-equilibriating three-link blip of Figure 12(c), thus represents the only realistic practical solution. But all the other possibilities suggested in Section 2 remain open to analytical schemes based purely on equilibrium considerations. Special path-following techniques are therefore often necessary to overcome such difficulties associated with multi-equilibria (see, for example, Reference 16).

To complicate the issue further, paths disconnected from the fundamental state, as shown in Figure 13, also occur. These unload to points c and d, which, respectively, represent blips similar to those of Figures 12(c) and 12(d), but with the localized displacement situated off-centre.

3.6. BVP solutions in IVP terms

The boundary- and initial-value formulations give the same information in somewhat different terms, and it is interesting to interpret the results of the former in the space of the latter. We concentrate in what follows only on paths that, for the BVP, are stable under displacement control—with a corresponding Sturm count of zero as indicated in the bifurcation diagram of Figure 11. Solutions to the boundary-value problem $q_0 \neq 0$, $q_{20} = 0$ are scattered in the initial space of starts $q_0 - q_1$ as shown in Figure 14. The comparison emphasizes that there are a large number of bounded but physically unstable solutions, with Sturm count greater than zero, that appear on the edge between grey and white but are not marked in black. These suggest that finite element solution routines which take no account of second-order information such as Sturm counts may be prone to tracking spurious equilibrium paths. It is clearly important in such problems to check continuously for bifurcation behaviour.

For every sufficiently small q_0 there are at least two stable BVP solutions, one close to the smooth, and one close to the oscillating, region of the loop, as seen in Figure 14. The unbroken part represents a family of solutions on stable rising paths, imperfect in the sense that they are not completely flat but nevertheless close to the fundamental path of the perfect system. The broken line is clearly influenced by the form of the homoclinic tangle. The disconnected portions all represent paths which end in the single blip of Figure 12(c), in turn (as q_0 increases from zero)

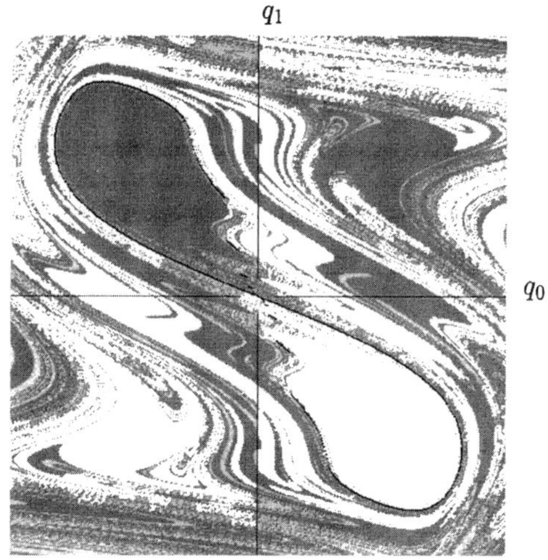

Figure 14. Stable BVP solutions under displacement control (black) compared with IVP solutions (grey/white interface), at $p = 0{\cdot}8$. Range $q_0, q_1 = \pm 0{\cdot}5$

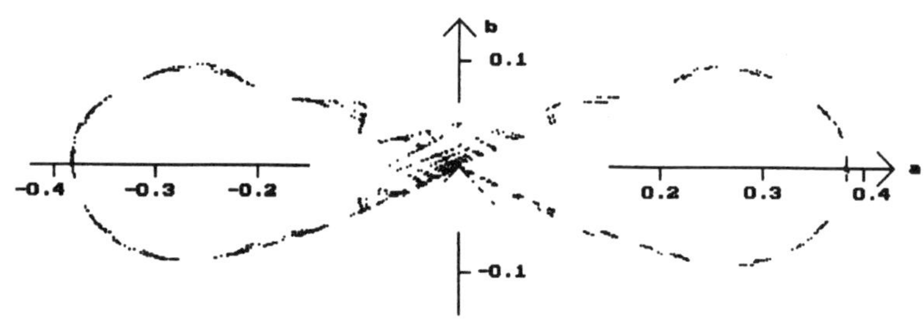

Figure 15. 152 competing 'perfect' BVP solutions in $a_i - b_i$ space. $p = 0{\cdot}8$; 20 links

in the central position, one link away from centre, two links away, and so on. For finite length the sequence is limited to a finite number of disconnected segments, which grows to infinity with the number of links. We note close similarities with the so-called 'parasitic solutions' of an elastic loop (elastica) discovered by Domokos and Holmes,[17] which are absent from the continuous system but present in its discretization.

Figure 15 shows finally a composite of 150 or so solutions to the boundary-value problem $q_0 = q_{20} = 0$, $p = 0{\cdot}8$, in the $a_i - b_i$ space of equation (8). To emphasize the underlying topology, all 21 pin displacements are shown. The development of the tangle is clearly evident, as is the tendency for solutions to cluster. In this representation, as in Figure 5, the distinction between stable and unstable manifolds has disappeared; the invertible nature means that both maps (9) and (17) are equally valid.

4. CONCLUDING REMARKS

In a world where it is often claimed that physical structural experiments are becoming less necessary, being replaced in importance by numerical modelling and experimentation, it is perhaps a salutary lesson that such a simple (non-linear) model can exhibit complex, chaotic, behaviour. And yet a laboratory experiment of such a system would not be regarded as unpredictable; although its exact position would be determined by imperfections, we would expect a single localization to develop along the length. Conventional finite element analysis based just on equilibrium considerations would be at a loss to distinguish this most significant solution from amongst less important ones. Even a check on Sturm counts may not be enough. Imperfections, for example, might give preference to one particular localized solution of Figure 14, but others can also exist that are locally if not globally stable. Some other criterion, like global minimum energy[6] may be required in addition to equilibrium.

We close with a remark, albeit an important one, on semantics. The word 'stability', indeed even the phrase 'structural stability', are subject to different interpretations. The intersecting paths of the stable or unstable symmetric (pitchfork) bifurcation, for instance, would be structurally unstable to a topologist, being readily destroyed by small imperfections or perturbations. In the sense of structural mechanics however it is the paths themselves that are described as stable or unstable; the fundamental path, for example, would be regarded as stable below and unstable above the minimum critical load. But in complete contrast, in the IVP sense of the dynamical phase-space analogy, it is the other way around. The former is an unstable saddle point, allowing the divergence necessary for localization, while the latter is stable.[18] We profess a certain sympathy for those coming new to the field.

ACKNOWLEDGEMENTS

This work was initiated while the first author was Munro Visiting Fellow to the Department of Structural Engineering, University of New South Wales.

REFERENCES

1. G. W. Hunt and M. K. Wadee, 'Comparative lagrangian formulations for localized buckling', *Proc. R. Soc. Lond.*, **434**, 485–502 (1991).
2. A. R. Champneys and J. F. Toland, 'Bifurcation of a plethora of large amplitude homoclinic orbits for Hamiltonian systems', *Nonlinearity*, **6**, 665–721 (1993).
3. G. W. Hunt, M. K. Wadee and N. Shiacolas, 'Localized elastic for the strut on the linear foundation', *ASME, J. Appl. Mech.*, **60**, 1033–1038 (1993).
4. I. Percival and D. Richards, *Introduction to Dynamics*, Cambridge University Press, Cambridge, 1982.
5. J. M. T. Thompson and G. W. Hunt, *A General Theory of Elastic Stability*, Wiley, London, 1973.
6. G. W. Hunt and G. Baker, 'Principles of localization in the fracture of quasi-brittle structures', *J. Mech. Phys. Solids*, **43**, 1127–1150 (1995).
7. H. Poincaré, *Les Methodes Nouvelles de la Mechanique Celeste*, Vol. 3, Gauthier-Villars, Paris, 1899.
8. M. V. Berry, 'Regular and irregular motion', in R. S. MacKay and J. D. Meiss (eds.), *Hamiltonian Dynamical Systems*, Adam Hilger, Bristol, 1987, pp. 27–53.
9. V. I. Arnol'd, 'Small denominators and problems of stability of motion is classical and celestial mechanics', in R. S. MacKay and J. D. Meiss (eds.), *Hamiltonian Dynamical Systems*, Adam Hilger, Bristol, 1987, pp. 250–366.
10. D. K. Arrowsmith and C. M. Place, *An Introduction to Dynamical Systems*, Cambridge University Press, Cambridge, 1990.
11. G. W. Hunt, P. Providência and K. Ikeda, Discrete lagrangian theory and phase portraits for cellular structural forms, Preprint.
12. Ian Stewart, *Does God Play Dice?: The New Mathematics of Chaos*, Blackwell, Oxford, 1989.
13. M. K. Wadee, 'Fat-fractal scaling in an area-preserving structural mapping', *Comput. Struct.*, **58**(4), 851–858 (1996).
14. F. W. Williams and M. S. Anderson, 'Incorporation of lagrange multipliers into an algorithm for finding exact natural frequencies or critical buckling loads', *Int. J. Mech. Sci.*, **25**, 579–584 (1983).

15. M. A. Crisfield, 'An arc length method including line searches and accelerations', *Int. j. numer. methods eng.*, **19,** 1269–1289 (1983).
16. M. A. Crisfield and J. Wills, 'Solution strategies and softening materials', *Comput. Methods Appl. Mech. Eng.*, **66,** 267–289 (1988).
17. G. Domokos and P. J. Holmes, 'Euler's problem, Euler's method and the standard map', *J. Nonlinear Sci.*, **3,** 109–151 (1993).
18. P. Providência, 'Post-buckling behaviour of uniform cellular structures', *Ph.D. thesis*, University of London, London, 1994.

Chapter 4

Twisted rods

This, the largest chapter of the book, deals with the three-dimensional static definition of long rods under combined tension and torsional end loading. The first article by THOMPSON & CHAMPNEYS presents some experimental evidence for localized buckling of circular-cross section, inextensible, naturally straight rods and compares using energy analysis a classical periodic helical buckling mode with a localized one. For long rods, the localized solution is shown to be energetically favourable. The next article by VAN DER HEIJDEN & THOMPSON considers the buckling of rods with non-circular cross-section. Using ideas of normal form analysis and reversing symmetries, they show that several types of behaviour are possible depending on material and cross-sectional properties of the rod (Poisson's ratio and aspect ratio). Either the wave number of the mode is uncorrelated with the applied twist rate, or for tape-like rods, the mode shape is like a tape wrapped round a cylinder. The article by CHAMPNEYS, VAN DER HEIJDEN & THOMPSON considers other perturbing influences from simple the isotropic case, namely axial extensibility, shearability and the effect of a rod that is naturally curved. The latter anisotropy is shown numerically to break integrability and so lead to spatial chaos and infinitely many multi-modal localized buckling modes.

Next, the contribution by STUMP, FRASER & GATES returns to the isotropic case but considers the shape taken by the rod after localization has caused the rod to self-contact and then writhe into a double helix with a loop at the end. Application is also given to the writhing of DNA strands. Finally, the paper by GORIELY & TABOR revisits the initial buckling and localization from a dynamical point of view. The buckling is identified with a temporal instability of the time-dependent formulation of the problem. Localization is then a secondary bifurcation superimposed on this initial instability. In the limit of long rods, the two bifurcations converge to the same value of the control parameter. These asymptotic results agree with the energetic heuristics in the first paper of the chapter.

From helix to localized writhing in the torsional post-buckling of elastic rods

By J. M. T. Thompson[1] and A. R. Champneys[2]

[1] *Centre for Nonlinear Dynamics, Civil Engineering Building, University College London, Gower Street, London WC1E 6BT, UK*

[2] *Department of Engineering Mathematics, University of Bristol, Queen's Building, University Walk, Bristol BS8 1TR, UK*

In this paper we study the competition between helical and localizing modes in the torsional buckling of stretched and twisted elastic rods. Within the Love–Kirchhoff formulation, we make comparative studies of the helical deformation of Love (1927) and the homoclinic localizing solution of Coyne (1990). Plots of the loads against their corresponding deflections allow the energetically preferred mode to be identified: it is found that preference switches from the helix to the localized mode early in the post-buckling range. These plots also allow us to predict the jumps that are observed under a variety of dead and rigid loading processes: these dynamic jumps take the rod from the spatially localized form to the familiar writhing state. Preliminary experiments confirm this preference for the localizing mode. They also reveal a second type of helical deformation, at a shorter wavelength, that is not predicted by the above long-rod analyses. A programme of further experimental and theoretical studies is suggested, and in a companion paper we lay the mathematical foundations for a numerical investigation of the complex and spatially chaotic deformations of a wider class of elastic rods.

1. Introduction

Problems associated with the large spatial deformations of thin elastic rods arise in many fields of pure and applied science. In molecular biology, the writhing or 'supercoiling' of DNA is known to effect almost every physical, chemical and biological property of the molecule. This has importance for genetic engineering, because the phenomenon of integration, in which a small piece of DNA is physically inserted into a larger DNA molecule, seems to require that the integrated element be supercoiled (Bauer *et al.* 1980; Calladine 1980; Calladine & Drew 1992). In ocean engineering, the formation of loops and kinks can seriously damage pipelines, electromechanical submarine cables and jacketed optical fibres (Coyne 1990; Tan & Witz 1992).

The *spatial elastica* of a thin rod deforming in three dimensions is, moreover, worthy of careful study as an archetypal example of spatial localization. As such, it can offer many insights into the localization of buckling patterns that is a significant feature in the nonlinear post-buckling of thin shells (Hunt & Neto 1991, 1993). Kirchhoff's identification of the nonlinear differential equations of the statical elastica with those of a spinning top leads naturally to the general dynamical phase space analogy. This analogy allows the use of modern geometrical concepts of dynamical systems

Proc. R. Soc. Lond. A (1996) **452**, 117–138

theory, and their associated computational techniques which will be the topic of a companion paper (Champneys & Thompson 1995). It identifies the buckling of a long rod under torsion as a Hamiltonian Hopf bifurcation, with the localized spatial form corresponding to a homoclinic saddle connection. The normal form of this bifurcation serves to give explicit formulae for the initial post-buckling behaviour of the elastic rod: as it does for the planar strut on a nonlinear elastic foundation. It is the homoclinic nature of the localized mode, in the dynamical analogy, that allows it to be perturbed into spatial chaos (Mielke & Holmes 1988; Thompson & Virgin 1988; Davies & Moon 1993).

In this paper we focus attention on the arbitrarily large nonlinear deformations of an elastic rod subjected to tension and torsion, using the formulation of Love (1927). For a circular cross section and linear constitutive relationships, the equations have three integrals, allowing complete analytical solutions. In particular, the solution for helical deformations was given by Love, and the (homoclinic) solution for localized deformations was recently written down in a convenient form by Coyne (1990).

These two modes of deformation are illustrated in figure 1. Here a long elastic rod is subjected to an end tension T and a twisting moment M. If M is increased under constant T, simple buckling theory predicts a subcritical bifurcation from the trivial undeflected configuration (a) into the unstable helical form of Love (b). Notice that we have here drawn a sinusoidally modulated helix, as is predicted by the linear buckling theory for a rod of great but finite length when we assume zero deflection at the boundaries. Meanwhile the homoclinic analysis shows that the helix can progressively localize, giving an alternative subcritical equilibrium path leading continuously from the initial buckling mode to the highly localized forms of (c) to (e). Under rigid loading conditions (Thompson 1979), in which the end shortening and end rotation are controlled, theory shows that the homoclinic path can be stabilized, except for a small dynamic jump from (d) to the first self-contact at (e). Our experiments described in § 3 confirm this result, showing a very well defined and continuous localization process under rigid loading, which is quasi-static from (c) to (d) and dynamic from (d) to (e): under further loading the continuous quasi-static writhing of (f) is observed.

To throw some light on the early stages of this process, we make some careful theoretical comparisons between the two solutions under different loading conditions. These include dead loading where one controls the magnitude of the loads (T and M in the present context) and rigid loading where one controls the corresponding displacements of the loads (here D and R). We show how, under dead loading conditions, their initial unstable (subcritical) post-buckling characteristics agree with the normal form of a Hamiltonian Hopf bifurcation (van der Meer 1985). By examining plots of the loads against their corresponding deflections, we identify the dynamic jumps that can be expected under a variety of dead and rigid laboratory loading arrangements. It is shown that under certain rigid loading conditions the initial post-buckling characteristics are of opposite type: the localized solution is supercritical, while the helical solution is subcritical.

Now the localized form is certainly the preferred, minimal energy, mode of deformation at large amplitude. But the above result suggests that close to the point of bifurcation, the helical form is preferred. There is thus an element of competition between the two modes of deformation, and in the presence of imperfections we might expect a delayed transition from the helical to the localized form.

The helical and localized modes that we have been discussing are those of the

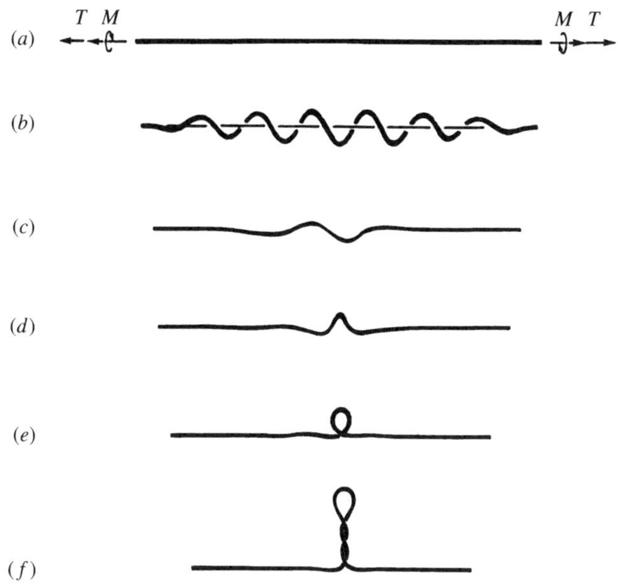

Figure 1. Spatial localization of a stretched and twisted rod: (a) the trivial undeflected straight configuration; (b) the helical form which bifurcates from the trivial solution; (c) to (d) continuous quasistatic localization, observable under rigid loading; (d) to (e) a dynamic jump, observed under rigid loading; (f) the final writhing that follows the self-contact at (e).

Love–Kirchhoff theory, and preliminary experiments on steel wires and rubber rods revealed a second type of helix with a sigificantly shorter wavelength: these have led us to the conclusion that there is a lot of further work needed in this area. Detailed numerical studies will certainly be required, and in a companion paper (Champneys & Thompson 1995) we lay down an appropriate computational framework based on the dynamical analogy and present new results using recently developed shooting techniques (Champneys & Spence 1993). We also compare and contrast our findings about spatial localization in twisted rods with recent results for the strut on a nonlinear elastic foundation (Hunt *et al.* 1989; Hunt & Wadee 1991; Champneys & Toland 1993; Buffoni *et al.* 1994).

2. Formulation and eigenvalue analysis

Consider an initially straight long elastic rod of length S, loaded at its ends by a tensile force T and a twisting moment M, as shown in figure 2. These together constitute a wrench, and it is clear that a trivial equilibrium solution will always exist in which the rod remains straight and experiences a uniform (internal) twist. This trivial solution becomes unstable at a critical point (M_c, T_c) in the (M, T) control space, which can be obtained from the linear eigenvalue analysis of Love (1927, § 272) or Timoshenko & Gere (1961, equation 2-74). Letting the length of the rod tend to infinity in these equations, we have

$$M_c^2 = 4BT_c, \quad \text{or } m_c = 2 \tag{2.1}$$

where $m \equiv M/\sqrt{BT}$. Here B is the bending stiffness which is assumed to hold for any plane of bending, as would, for example, be the case for circular and square

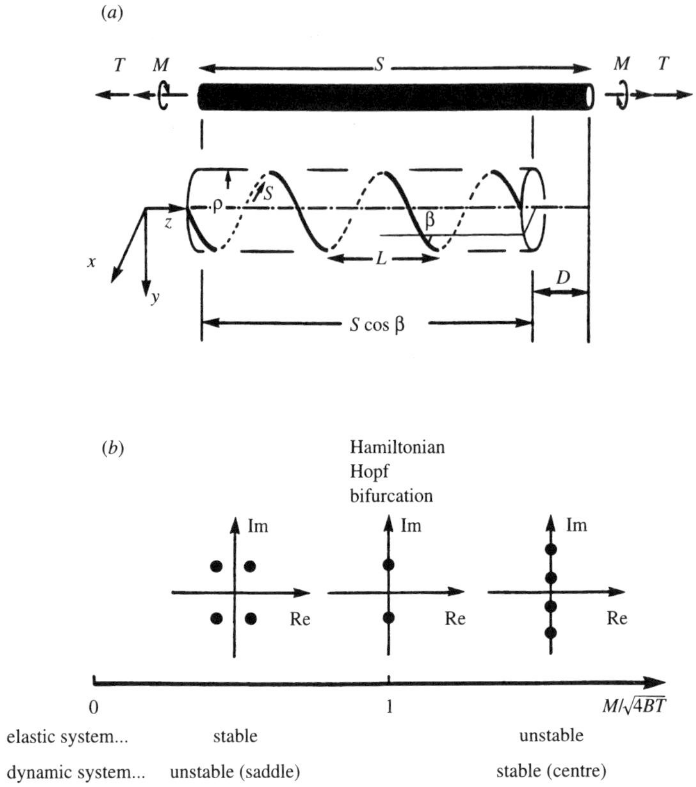

Figure 2. Torsional buckling of a rod into a helical form: (*a*) notation for the rod and the finitely deformed helical state; (*b*) movement of the eigenvalues of the linearized problem.

cross sections: it is given by the well-known formula $B = EI$ in terms of the Young's modulus of the material, E, and the second moment of area I. Notice that the critical moment (2.1) does not depend on the torsional stiffness of the rod C, which for a circular cross-section is given by GJ, where $G = E/2(1+\nu)$ is the shear modulus and $J = 2I$ is the polar second moment of area. We shall see that stability under rigid, as opposed to dead, loading conditions depends on the relative magnitudes of B and C, and for a circular rod we have $B = (1+\nu)C$ so that B is always greater than C. With Poisson's ratio ν equal to one third, appropriate for most metals, we have $B = \frac{4}{3}C$. For soft rubber, ν is approximately equal to its greatest value, 0.5, giving $B = \frac{3}{2}C$. For hard rubber, Cottrell (1964) gives Poisson's ratio as 0.4.

At the point of bifurcation at which the rod loses its elastic stability, as M is increased through its critical value (or, equivalently as T is decreased through T_c), the eigenvalues (Timoshenko & Gere 1961) move as illustrated in figure 2*b*. Two complex conjugate pairs meet on the imaginary axis to become four imaginary eigenvalues: a saddle, which is dynamically unstable, is thus followed by a centre which is dynamically stable. This corresponds to a Hamiltonian Hopf bifurcation, and is the same as for the strut on an elastic foundation (Hunt *et al.* 1989, figure 4). The companion paper (Champneys & Thompson 1995) gives the full details of this bifurcational behaviour.

In terms of the (internal) twist per unit length at buckling $\tau = \tau_c = M_c/C$, the critical condition (2.1) can be written as

$$\tau_c = 2\sqrt{BT_c}/C \tag{2.2}$$

At this critical twist, the eigenvalue theory predicts that an infinitely long rod will buckle into a three-dimensional helix with axial wavelength, $L = L_c$, given by

$$L_c = 4\pi B/M_c = 2\pi B/\sqrt{BT_c}. \tag{2.3}$$

For a long but finite rod the theory predicts a sinusoidally modulated helix, as sketched in figure 1*b*. With this well-defined finite wavelength we would not expect the boundary conditions of our long rod, $S \gg L_c$, to be important: and experimentally we can prescribe conditions that make T and M conservative. In this way we can avoid the well-known problems of non-conservative circulatory torsional loading (Ziegler 1968). However, the questions about the length and the boundary conditions require further careful investigation, and we shall make some further comments about them in the final discussion and conclusions of § 6.

A simple consequence of (2.2) and (2.3) is that the number of complete turns made by the end moment per incipient helical wave is $\tau_c L_c/2\pi = 2B/C = 2(1+\nu)$. This is a delightfully general result, depending only on Poisson's ratio and independent of the applied tension. It is nicely confirmed in our experimental work.

3. Experimental results

We have performed a number of preliminary experiments on both metallic wires and rubber rods. The qualitative behaviour seems to be similar, and we present here some observations on the response of the rubber specimens. These are easier to test and handle, and do not suffer from the irreversible plastic deformation which accompanies the writhing of a wire. The disadvantage of rubber is that gravitational effects are more severe, and the rod suffers considerable axial extension of its centre line, which is ignored in the theory. It must also be remembered that it has a hardening stress–strain curve, and at the large strains that are encountered in the tests the departure from linearity may be appreciable.

The specimens were loaded in the purpose-built testing machine shown in figure 3*a*, which allows the rigid control of the end rotation R and the end displacement D. The applied twisting moment M and the end tension T will vary passively as we adjust R or D, and the machine has, as yet, no device for measuring them. Notice that the passive variation of T will alter the predicted wavelength of the helix.

On some occasions a well developed helix was observed, as illustrated in figures 3*b* and *c*. On other occasions, there was a fairly continuous deformation into the localized mode displayed in figure 3*d*, and the process of deformation as R is increased at constant D is shown in figure 4. The black circular rod was painted half white, so that in the top untwisted state it appears as two parallel black and white strips. The pictures, traced from photographs, show the deformed shape of the rod as R is increased from 0 to 28π in steps of 4π. A short length of the rod at one end was obscured by a measuring device: which is why the turns of the end, $R/2\pi$, implied by the black and white patterns of twist, do not quite seem to be integers.

Analysis of this picture allows us to distinguish two distinct modes of deformation. In the early stages of the test, at $R = 8\pi$ for example, there is a clearly visible helical deformation (H_1) which has approximately one full turn per helical wave. Later in

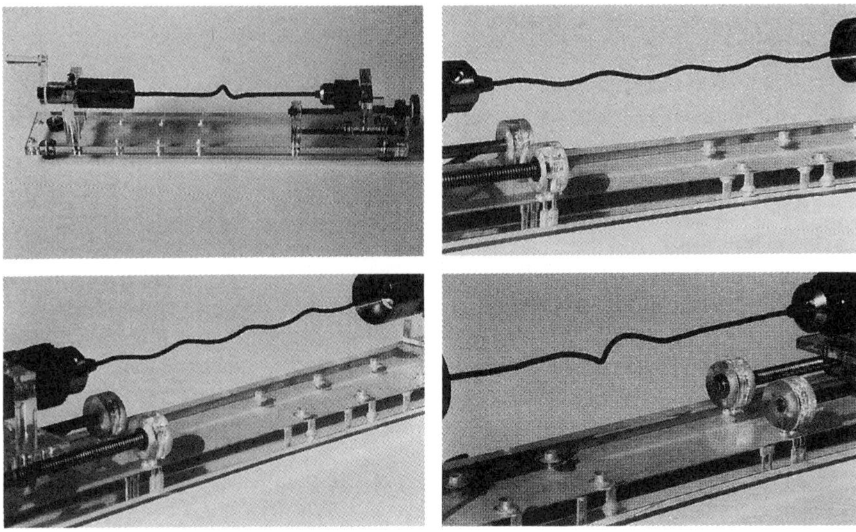

Figure 3. Photographs of the testing machine and some examples of the modes of deformation observed in a circular rubber rod.

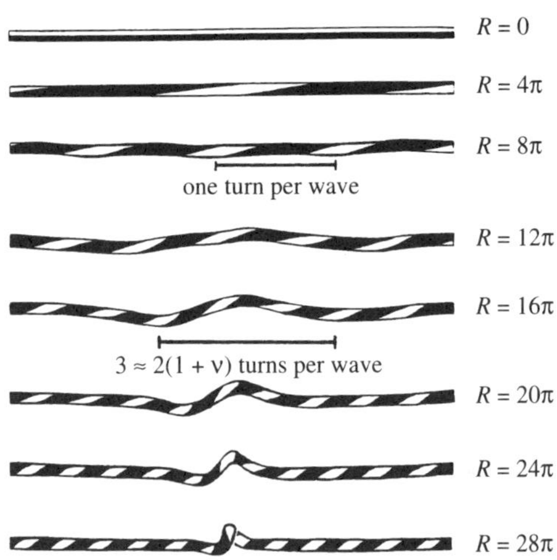

Figure 4. The deformation sequence of one specimen which has been painted to make the twist visible. There is evidence for an initial helical deformation (H_1) with approximately one twist per wave. This is followed by the characteristic localization process, the localized mode (L_3) having approximately three twists per helical wave in agreement with the prediction of Coyne's theory.

the test we see that the localized form (L_3) has roughly $3 \approx 2(1+\nu)$ turns per wave, in agreement with the eigenvalue prediction. These two modes do not seem to interact in any significant way, the mode L_3 simply superimposing itself onto the H_1 mode, as we shall see more clearly in figure 6.

Now both the helical and localized modes of the Love–Kirchhoff formulation that are analysed in this paper have $2(1+\nu)$ turns per wave at the bifurcation point, and this value changes very little along the post-buckling path. So while we can identify

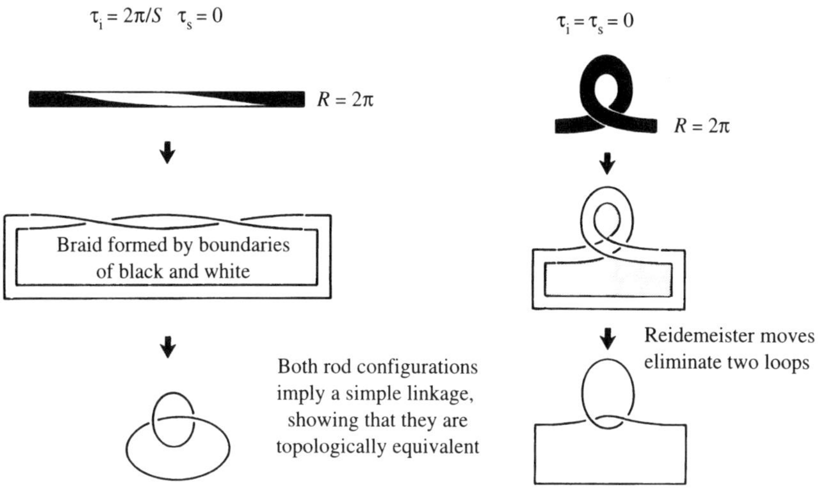

Figure 5. A short rod with one complete end turn ($R = 2\pi$) will form a single untwisted loop as the ends are allowed to approach one another. The topological equivalence of these two configurations can be established by examining the linking of the two black–white boundaries.

the theoretical localizing mode with the experimental L_3, there is no experimental observation of the theoretical helix H_3. This is in agreement with our theoretical predictions.

Meanwhile it seems that the short-wave helix H_1 either lies outside the domain of Love–Kirchhoff theory or is associated with the finite length and the boundary conditions: or perhaps with gravity and/or initial curvature of the test specimens (see § 6). This theory treats the rod as an elastic line (as opposed to an elastic solid of small but finite radius) with no axial or shear deformability, and with linear constitutive relationships. So there are plenty of reasons why it might not give a complete description of the response of our rubber specimens. However, we should emphasize that H_1 is readily observable in thin metallic wires at low elastic stress levels, as we shall see in figure 15. We leave an examination of this to further work, emphasizing here that our present theoretical work is concerned exclusively with the competition between H_3 and L_3. In this context, the fact that we never observe H_3 experimentally can be taken as confirmation of our theoretical prediction that L_3 is energetically more favourable that H_3.

The black and white colouring of figure 4 shows clearly how one complete loop removes the torsion produced by one complete end turn, as illustrated in figure 5. Here R is held constant at 2π, and the short straight specimen with one turn deforms into the right-hand loop as the ends are allowed to approach one another by varying D. Notice that the loop appears all black, showing that it has no internal torsion: and being approximately planar, its centre line has no space torsion either. The topological equivalence of the straight and looped rods is established by considering the linking of the two black–white boundaries as indicated in the figure. Notice how the two loops highlighted by the dot screen can be eliminated by the (intuitively obvious) Reidemeister move of knot theory (Kauffman 1991).

The specimens tested in the machine are relatively short, so we have also examined the torsional buckling of long rods hanging in a shallow catenary. These usually exhibit well developed H_1 deformations, as illustrated in figure 6. This mode is re-

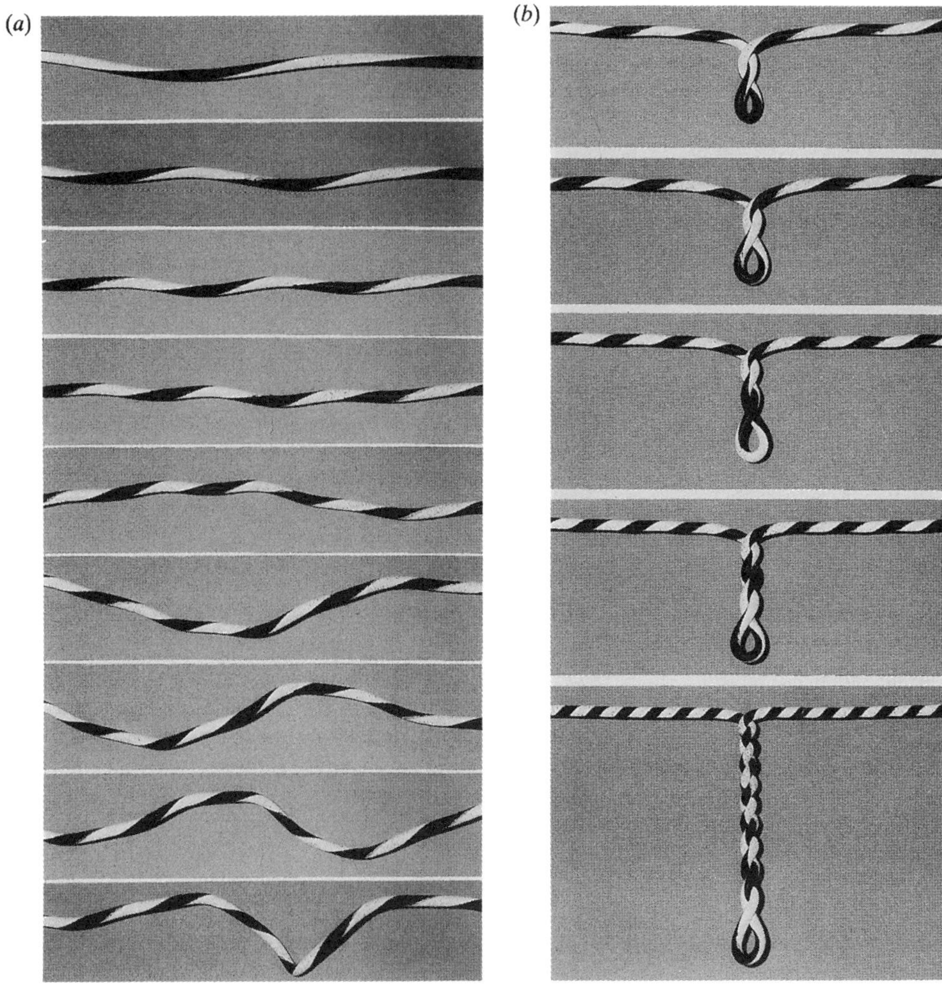

Figure 6. Two photographic sequences of a single test on a long twisted rubber rod that is allowed to hang in a shallow catenary under fixed end displacement and controlled end rotation. In the first sequence there is continuous quasi-static development of a short-wave helix H_1, onto which is superimposed the localized mode L_3. A dynamic jump takes the rod, at constant end displacement and end rotation, from the last picture of the first sequence to the first picture of the second sequence. The writhing form of the second sequence then develops quasi-statically under further slow increase in the end-rotation.

markably robust against dynamic disturbances, and the hanging catenary can be shaken vigorously without disturbing the pattern. As the end rotation is increased the amplitude of the helical wave first increases, and then seems to decrease. Eventually the L_3 mode superimposes itself on the H_1, giving finally a large dynamic jump to a highly writhed form. For a long rod, the size of this jump cannot be limited, because the flexibility of the rod itself means that even under the rigid control of R and D, a short central section of the rod will experience effectively dead loading conditions.

4. Post-buckling into a helix

To analyse the large deflection helical forms that emerge from the point of bifurcation discussed in § 2, we employ the exact nonlinear formulation of Love (1927, § 270) using the radius of the helix ρ and the angle β to describe the form of the helix shown in figure 2*a*. Here β, representing a constant value of the Euler angle θ, replaces Love's $\alpha = \frac{1}{2}\pi - \beta$.

Simple geometry gives us the axial wavelength L, the total number of helical waves N and the total axial end shortening of the deformed rod D as

$$L = 2\pi\rho/\tan\beta, \tag{4.1}$$

$$N = S\sin\beta/2\pi\rho, \tag{4.2}$$

$$d \equiv D/S = (1 - \cos\beta). \tag{4.3}$$

(*a*) *Internal and kinematic torsion*

Following Love, we must consider the total twist per unit arc-length τ decomposed into two parts, which we shall call the internal torsion τ_i and the kinematic torsion of the helical space curve τ_s, so that

$$\tau = \tau_{\mathrm{i}} + \tau_{\mathrm{s}}. \tag{4.4}$$

Here the symbol τ has the same meaning as in Love, while τ_{i} is Love's $\mathrm{d}\phi/\mathrm{d}s = \mathrm{d}f/\mathrm{d}s$. Notice that this use of τ for a kinematic twist rate (a generalized strain) follows a long geometrical tradition: since we shall never be talking about stresses, there will be no cause for confusion with its use in continuum elasticity as a shear stress. For our linearly elastic rod with torsional stiffness C, the internal twisting moment at any cross-section is given by $C\tau$.

The geometrical theory of space curves gives the kinematic torsion τ_{s} and the total curvature κ as

$$\tau_{\mathrm{s}} = \sin\beta\cos\beta/\rho, \tag{4.5}$$

$$\kappa = \sin^2\beta/\rho. \tag{4.6}$$

The total angle in radians R turned by the end moment is given by

$$r \equiv R/S = \tau_{\mathrm{i}} + 2\pi N/S = \tau_{\mathrm{i}} + \sin\beta/\rho. \tag{4.7}$$

This expression is adequate for the present investigation, being strictly correct for a long uniform helix that is supported in a manner similar to that envisaged by Love. It will be approximately correct for the central section of the slowly modulated helix of figure 1*b*. Notice that r is not given by $\tau_{\mathrm{i}} + \tau_{\mathrm{s}}$. If it were, we would find that under a prescribed end rotation $r =$ const., we would have a prescribed total torsion $\tau = r =$ const.: under these conditions a deformation into a helix would not achieve any reduction in torsional strain energy, so there would be no buckling.

To further clarify this situation, it is useful to examine two specific illustrations of helical kinematics. We consider first the tape on a wire. A springy tape slotted over a smooth rigid wire (or a straight strip of paper with a central line that is placed on the wire) is a useful device for visualising the principal axes of any space curve defined by the form of the wire. Since the tape cannot bend in its stiff direction, the normal to the tape will everywhere coincide with the principal normal of the space curve. If the wire is in the form of a helix, the tape will lie as if it were wound on the cylinder of the helix with $\tau_{\mathrm{i}} = 0$ and $\tau = \tau_{\mathrm{s}}$. Notice that it has a non-zero

τ, implying that it needs a corresponding moment on the end: but this is irrelevant because our idealized tape with an infinite bending stiffness is locked in position, and has no freedom of movement. The angle turned is given by (4.7) with $\tau_i = 0$, and for one helical wave the number of turns $R/2\pi$, is equal to 1. Sections of the tape separated by one helical wave are parallel to one another. Consider, secondly, the tube on a wire. A rubber tube (straight when free from stress) slotted freely over our smooth rigid helix illustrates the addition of τ_i and τ_s. In particular, it is instructive to consider the form that it adopts when there is no end moment applied, so that $\tau = 0$. Clearly we must then have $\tau_i = -\tau_s = -\sin\beta\cos\beta/\rho$, giving for one helical wave the number of turns as $1 - \cos\beta$.

A useful analysis of the relationship between internal torsion and kinematic torsion is given by Fuller (1971), who introduces the writhing number of a space curve: and see also the paper of Alexander & Antman (1982).

(b) *Energy analysis of the post-buckled states*

It is instructive to derive the equilibrium conditions of the large amplitude helix by an energy analysis, assuming that we start with a uniform elastic rod that is straight when free from stress. To do this we regard T and M as prescribed dead loads which do work through the corresponding deflections $-D$ and R. The total potential energy, V, can then be treated as a function of the three generalized coordinates (ρ, β, τ_i). The potential energies of T and M are V_T and V_M, given by

$$V_T = TD = TS(1 - \cos\beta), \tag{4.8}$$

$$V_M = -MR = -MS(\tau_i + \sin\beta/\rho). \tag{4.9}$$

The strain energy of bending V_B and the strain energy of torsion V_C are given by

$$V_B = \tfrac{1}{2}BS\kappa^2 = \tfrac{1}{2}BS\sin^4\beta/\rho^2, \tag{4.10}$$

$$V_C = \tfrac{1}{2}CS\tau^2 = \tfrac{1}{2}CS[\tau_i + \sin\beta\cos\beta/\rho]^2. \tag{4.11}$$

The total potential energy of the system, rod plus dead loads, is now

$$V \equiv vS = V_T + V_M + V_B + V_C, \tag{4.12}$$

and the necessary and sufficient condition for equilibrium is that $\partial v/\partial\rho = \partial v/\partial\beta = \partial v/\partial\tau_i = 0$. Setting $\partial v/\partial\tau_i = 0$ gives immediately

$$M = C\tau, \tag{4.13}$$

which is an additional condition over and above those established by Love for the equilibrium of the helix (it appears in Love's later stability analysis in § 272(d)). Whatever the (unspecified) boundary conditions are, equation (4.13) is a natural overall requirement, demanding that the end of the rod carries about its own material axis the twisting moment M. It is a simple consequence of the invariance of τ (see Champneys & Thompson 1995) for boundary conditions that involve the rod eventually lining up with the loading axis, as in the slowly modulated helix of figure 1*b* or in Coyne's localizing solution of figure 1*d*. The equation plays a vital role in the analysis, and it is nice to see it emerge so clearly from this energy formulation. Next, $\partial v/\partial\rho = 0$ gives

$$M = C\tau\cos\beta + B\sin^3\beta/\rho, \tag{4.14}$$

which is Love's equation (37): and finally $\partial v/\partial\beta = 0$ gives, using (4.14),

$$T = C\tau\sin\beta/\rho - B\cos\beta\sin^2\beta/\rho^2, \tag{4.15}$$

which is Love's equation (36). Notice that these two equations relate the applied loads (T, M) to the deformation variables which can be chosen as either (ρ, β, τ) or (ρ, β, τ_i). They are insufficient to determine the deformation under a given loading: which is why we need an additional condition such as (4.13).

(*c*) *Examination of the equilibrium equations*

Although we prescribed dead T and M for our energy analysis, the equilibrium conditions that we have established do of course hold for either rigid or dead loading. Our further discussions could indeed be made quite independent of the energy analysis by simply adopting Love's two equilibrium equations, our (4.14) and (4.15), together with the new equation (4.13), which we can presume to be an equilibrium requirement at the end of a section of helix.

Using (4.13) we can eliminate τ to get the required two equilibrium conditions relating the loads (T, M) to the generalized helix coordinates (ρ, β),

$$T = (M \sin\beta)/\rho - (B\cos\beta\sin^2\beta)/\rho^2, \tag{4.16}$$

$$M(1-\cos\beta) = (B\sin^3\beta)/\rho. \tag{4.17}$$

Note that, like the eigenvalue result, these full nonlinear equilibrium equations do not contain the torsional stiffness C. From (4.16) and (4.17) the applied loads are given by

$$M\rho/B = \sin^3\beta/(1-\cos\beta), \tag{4.18}$$

$$T\rho^2/B = \sin^2\beta. \tag{4.19}$$

Notice that under these equilibrium conditions the total number N of waves is given by $N/S = \sqrt{(T/B)}/2\pi$: this number is constant if T is constant. Eliminating ρ gives

$$m \equiv M/\sqrt{BT} = 1 + \cos\beta. \tag{4.20}$$

In addition to this final equilibrium equation, we shall also need the corresponding deflections of T and M. Equation (4.3) gives us immediately

$$d = 2 - m, \tag{4.21}$$

and using (4.4), (4.5), (4.7), (4.13), (4.19) and (4.20) we find

$$r/\sqrt{BT} = (2-m)/B + m/C. \tag{4.22}$$

These equilibrium conditions will allow us to discuss a variety of loading sequences.

(*d*) *Control parameters in physical experiments*

We shall want to imagine a physical experiment in which we observe an equilibrium response under the slow quasi-static variation of one or more control parameters, followed possibly by a rapid dynamic jump to a new remote equilibrium state. To do this we shall need to distinguish between the following.

A *fixed* quantity will not vary at all during the experiment. This category will include the obvious candidates (that we need not subsequently list) such as the length of the rod S and the stiffnesses B and C. It might also include a pre-fixed quantity, such as a prescribed amount of pre-twist that we imagine to be fixed before the experiment starts.

A *controlled* parameter (more briefly, a *control*) is assumed to be given a very

	pre-fixed	controlled	passive
dead loading (figure 7)	(none)	(T, M)	(D, R)
mixed loading (figure 8)	(T)	$(M$ or $R)$	$(D, R$ or $M)$
rigid loading (figure 14)	(R)	$(T$ or $D)$	$(M, D$ or $T)$

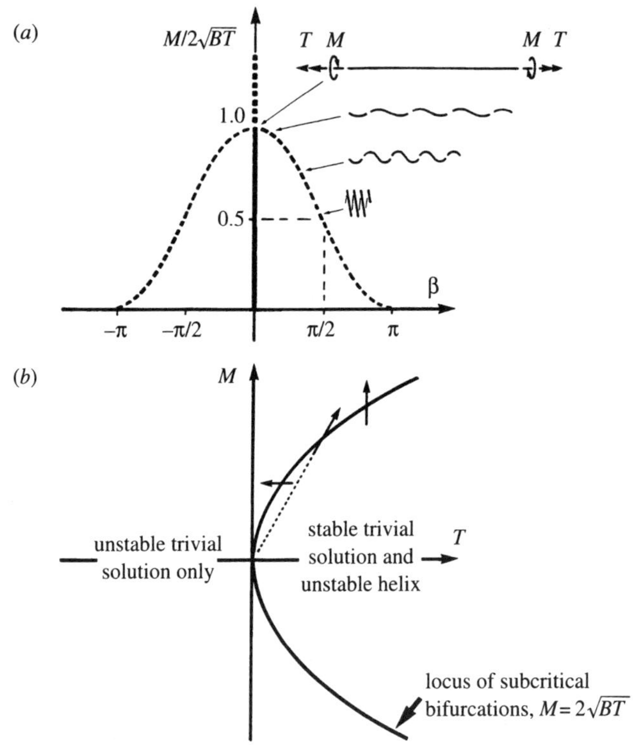

Figure 7. Helical response of the rod under dead loading: (*a*) the falling unstable subcritical path of helical equilibrium states, on a plot of a load parameter against a measure of the lateral deflection; (*b*) the locus of the subcritical bifurcation in the (T, M) control space, showing three possible loading rays across the instability threshold.

slow quasi-static variation during the experiment. The slow variation means that a control will not change its value during any stability test or dynamic jump.

A *passive* variable is one that is neither fixed nor controlled, and will normally vary both during the quasi-static progression along an equilibrium path and during a dynamic jump. For our elastic rod under end loading, we shall concentrate on the following experiments, categorizing the loads (T, M) and their associated deflections (D, R) in the table.

The appearance of the composite load parameter $m \equiv M/\sqrt{BT}$, does of course point to the equivalence of controlling M or $1/\sqrt{T}$: the precise dynamic similarities are clearly established in the non-dimensionalization employed by Champneys & Thompson (1995).

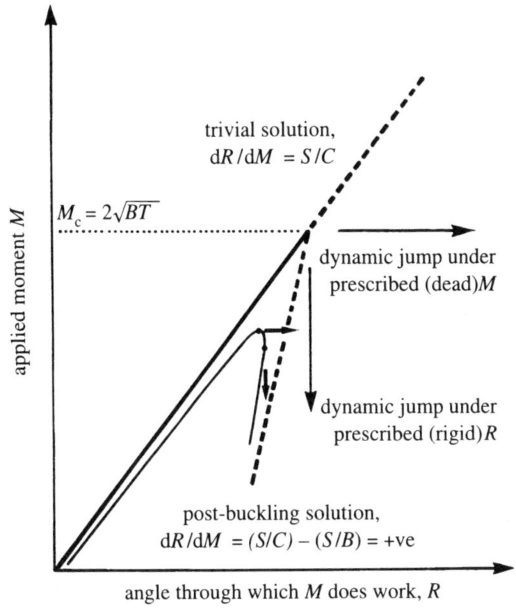

Figure 8. If T is constant, the $r(m)$ equation of the helix gives a linear post-buckling relationship between M and its corresponding angle R. The positive slope of the post-buckling line shows that under fixed and dead T, the rod will exhibit a dynamic jump under either controlled M or controlled R. The jumps of an imperfect rod would be from the horizontal and vertical tangencies shown.

(e) *Behaviour under totally dead loading*

If we first think of T as pre-fixed while M is controlled, equation (4.20) gives us the $M(\beta)$ relationship of a falling unstable post-buckling path of helical equilibrium states, shown in figure 7*a*. This path bifurcates from the trivial solution at $\beta = 0$, $m_c = 2$, in agreement with (2.1).

If we next regard T and M as independent controls, we can draw the critical parabola $T_c = M_c^2/4B$, shown in figure 7*b*. Above this parabola (high T) there exists a stable trivial solution and an unstable helix: below the parabola there is just an unstable trivial solution. As we cross the parabola from high T along any $T(M)$ control path, we encounter a subcritical bifurcation at which the rod must jump dynamically: under the helical constraint of this analysis the system apparently jumps dynamically to a flat bracelet ($\beta = \frac{1}{2}\pi$) of zero axial length.

(f) *Pre-fixed dead tension with controlled rotation*

As an intermediate form of mixed dead and rigid loading, we consider the situation in which the tension T is pre-fixed, while the end rotation R is controlled. Examining equation (4.22), we see that, since $B > C$, we have an internally unstable, backwards sloping $M(R)$ post-buckling path under pre-fixed T, as shown in figure 8. So even under our rigid control of the end rotation, we would have a dynamic jump from the bifurcation point. Within the present constraint of a helical deformation, this jump would again seem to be to a compacted bracelet ($\beta = \frac{1}{2}\pi$) of zero axial length.

(g) Behaviour under prefixed end rotation

Here we shall regard R as a pre-fixed quantity, corresponding to an internal pre-twist of $\tau_0 = R/S = r$. Substituting m from (4.21) into (4.22) we have

$$r/\sqrt{BT} = d/B + (2-d)/C. \tag{4.23}$$

This exhibits a bifurcation at $d = 0$, $T = T_c$, where $C\tau_0 = 2\sqrt{BT_c}$, and writing $P = T/T_c$ we have

$$1 = \sqrt{P}\{1 - d[1-(C/B)]/2\}, \tag{4.24}$$

giving, with $k \equiv [1-(C/B)]/2$,

$$P = 1/(1-kd)^2, \tag{4.25}$$

a result in which the dependence on τ_0 has vanished into the T_c. Since d must be positive, P must be greater than 1, showing that if we had mixed loading conditions with P controlled, we would encounter a subcritical bifurcation as we decrease P. Writing $P - 1 \equiv p$, a first-order branching solution valid for small p is

$$p \approx 2kd. \tag{4.26}$$

We shall discuss these equations in a later comparison (figure 14) with the homoclinic localizing solution under conditions of controlled R and D.

5. Comparisons with the homoclinic post-buckling solution

Coyne (1990) has made a nice study of the homoclinic localizing solution that emerges from the branching point at $m = 2$. He uses the general formulation of Love (1927, § 269), and obtains an analytical solution for the long elastic rod under T and M that concerns us here.

(a) Behaviour under totally dead loading

We can first make a direct comparison with our § 4 *e*, relevant to a discussion of dead loading. Coyne's equations (17) and (19) give in the present notation,

$$m = 2\cos(\tfrac{1}{2}\theta_{\max}), \tag{5.1}$$

where $\theta_{\max}$, the maximum central value of the Euler angle θ, can be directly compared with our helix angle β. This result is compared with our helix equation (4.20) in figure 9.

Notice that close to the branching point we have the asymptotic result $\theta_{\max} = \sqrt{2}\beta$, which is a universal feature of the Hamiltonian Hopf bifurcation. It arises, for example, in the double-scale perturbation analysis of the strut on a nonlinear elastic foundation as illustrated in figure 10, reproduced from Hunt *et al.* (1989). Here P is the axial compressive load on the strut, and in this dynamical phase space diagram we have an inverted supercritical Hopf bifurcation: as shown in figure 2, there is an inversion of the stabilities as we pass from the elastic problem to the dynamic analogy. The modulation amplitude A and its derivative A' are governed by an equation of the form

$$A'' - A + A^3 = 0, \tag{5.2}$$

which has a fixed point at $A = 1$, and a homoclinic orbit reaching to $A = \sqrt{2}$.

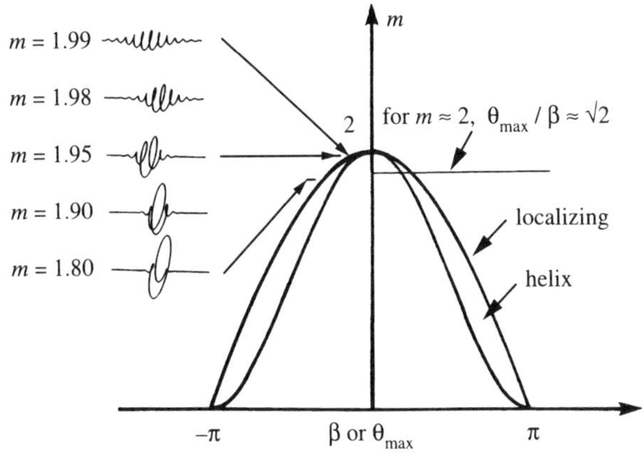

Figure 9. Comparison of the helical and localized post-buckling paths on a plot of the composite load parameter m against a measure of the lateral deflection.

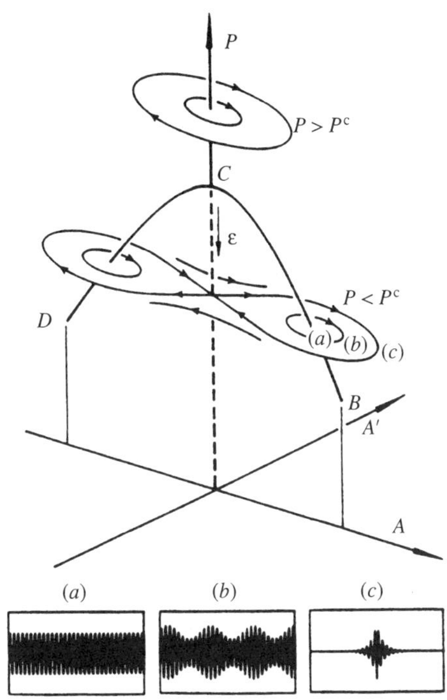

Figure 10. A schematic representation of a Hamiltonian Hopf bifurcation, as it arises, for example, in the buckling of a strut on a nonlinear elastic foundation (Hunt *et al.* 1989). Here A and A' represent the amplitude of the wave envelope and its derivative. Thus, (a) represents an unmodulated wave (such as our helix), while (c) represents a localized homoclinic solution. Periodically modulated waves such as (b) will also exist. The corresponding equation and phase portrait for our stretched and twisted rod are given by Antman (1995) (figure 3.8a, p. 331).

(b) *Pre-fixed dead tension with controlled rotation*

We can also make a direct comparison with the results of § 4 f, and we look first at Coyne's large deflection solutions for the end shortening D and the end rotation

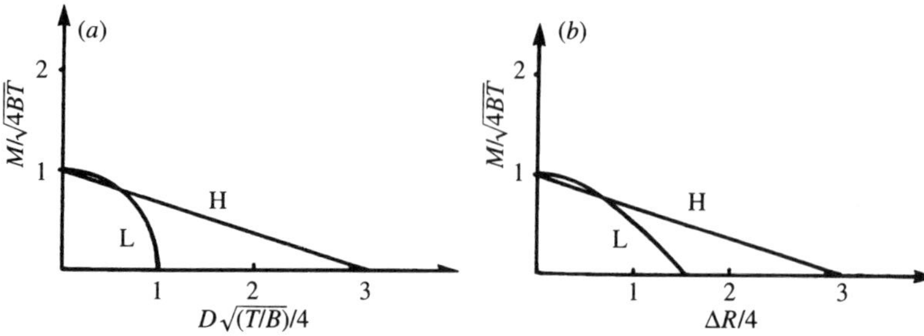

Figure 11. Comparisons of the helical and localized post-buckling solutions showing the variation of the load parameter with: (*a*) the end-shortening parameter; (*b*) the end-rotation parameter.

R. These large deflection solutions are directed towards the highly localized forms, c to e of figure 1, and may not pick up fine detail close to the bifurcation point: but they will be entirely adequate for our present investigation.

Now for a long rod, the contributions to D and R from a local deformation will not be directly comparable with those from a helical mode that extends over the whole length: the former are finite, the latter scale as S, and go to infinity with S. For this reason, we evaluate the contribution from just one helical wave, taking as our S the tension-dependent arc length of one helical wave. From the discussion below (4.19), this arc length is $S_1 = 2\pi\sqrt{B/T}$.

For the end-shortening we employ Coyne's equation (32) which relies on the assumption that the product of an amplitude parameter and the length of the rod is much greater than one. It is this type of approximation that may miss fine detail near the bifurcation, where the amplitude is vanishingly small. Using Coyne's equations (17), (21), (22) and (32), we have in our notation

$$[D\sqrt{(T/B)}/4]^2 + [\tfrac{1}{2}m]^2 = 1, \tag{5.3}$$

which gives the unit circle of figure 11*a*. This can be compared with one helical wave by setting $d = D/S_1$ in (4.21), which gives the straight line of figure 11*a*.

We write the end rotation R as the sum of the fundamental path contribution $R_\mathrm{F} \equiv MS/C$, and the buckling contribution ΔR. Then using Coyne's equations (49) and (19), and the above end-shortening results, we have for the localizing solution

$$\tfrac{1}{2}m = \cos(\tfrac{1}{4}\Delta R), \tag{5.4}$$

which is the cosine function plotted in figure 11*b*. For comparison, we set $r = R/S_1$ in (4.22), to obtain for one helical wave the straight line of figure 11*b*.

Now to discuss the response of a rod under the rigid control of R, at constant pre-fixed T, as we did in figure 8, we must include the fundamental end rotation $R_\mathrm{F} = MS/C$. We start by evaluating R_F for $S = S_1$, which gives us the diagram of figure 12 showing the fundamental path (F), the helical path (H) and the localizing path (L). Here the value 1.4 is B/C, corresponding to $\nu = 0.4$.

We can now use this diagram to discuss the behaviour of a long rod, assuming that the helical deformation extends along the full length of the rod. As we increase S, the fundamental contribution R_F and the helical ΔR both scale as S, but we can imagine this to be cancelled in the diagram by changing the horizontal scale by a factor of S so that paths (F) and (H) remain unchanged. The contribution ΔR from the localized solution has, however, no scaling with S, so as S increases and we correspondingly rescale the horizontal axis, the path (L) will be progressively flattened up against

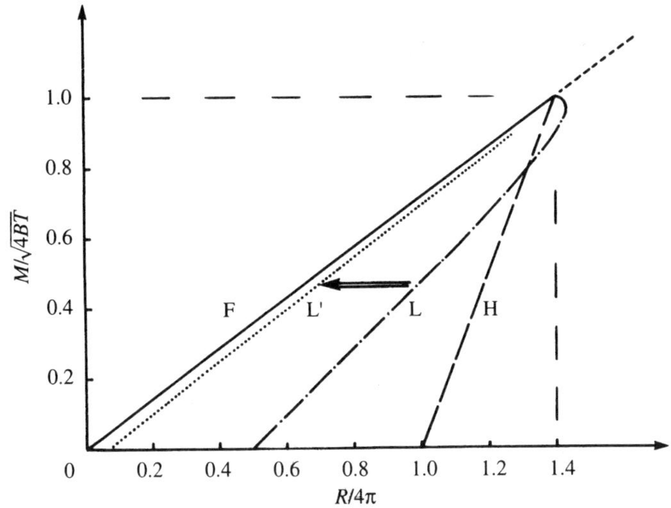

Figure 12. Comparisons of the helical and localized post-buckling solutions showing the variation of the load parameter with the total end-rotation. Here F is the fundamental straight internally twisted solution; H is the post-buckling path of helical states; L is the post-buckling path of localized states. As the length of the rod is increased, curve L will move towards F, as illustrated by L'.

(F), as shown schematically by (L′). The initial horizontal tangent of (L) at the bifurcation point will be preserved in (L′). But the length of path above the vertical tangent, which will be stable under the rigid control of R, will become vanishingly small as S increases to infinity.

With T held fixed, this diagram is a graph of the end moment against its corresponding rotation. So area under the graph is the stored energy, and we can conclude that at large amplitude the localized solution is the energetically favourable mode of deformation.

The phenomenon of localization can be further clarified by the heuristic analysis of figure 13. This follows ideas first developed for the cylindrical shell by Mansfield (1965), and pursued independently by Hunt in a number of papers (see, for example, Ikeda *et al.* 1993). Consider the chain of m, supposedly independent, helical elements, each of which has the moment-rotation relationship shown in 9*b*. Depending on how many elements buckle, we obtain the multiple paths of 9*d* for the behaviour of the chain. Of these paths, the most severe is Sv, on which only one element has buckled: this follows from the energy discussion of 9*c* which shows that the energy barrier against buckling from P to Q is given by the area A. It is also useful to consider the loading of this chain by the rigid control of R. If we assume that due to small imperfections one link has a slightly lower critical value of M, this link will jump dynamically to large amplitude, while the others will unload back along their fundamental paths.

(*c*) *A universal graph for localized post-buckling*

Since Coyne's analysis is made for a long rod, it seems natural to examine his force-displacement result (his equation (50)) in the limit as the length of the rod goes to infinity, by immediately dropping the second term of his top square bracket: this corresponds to neglecting $\arcsin^{-1}(\frac{1}{4}D/\sqrt{(B/T)})$ in comparison with $S\tau_0$. The

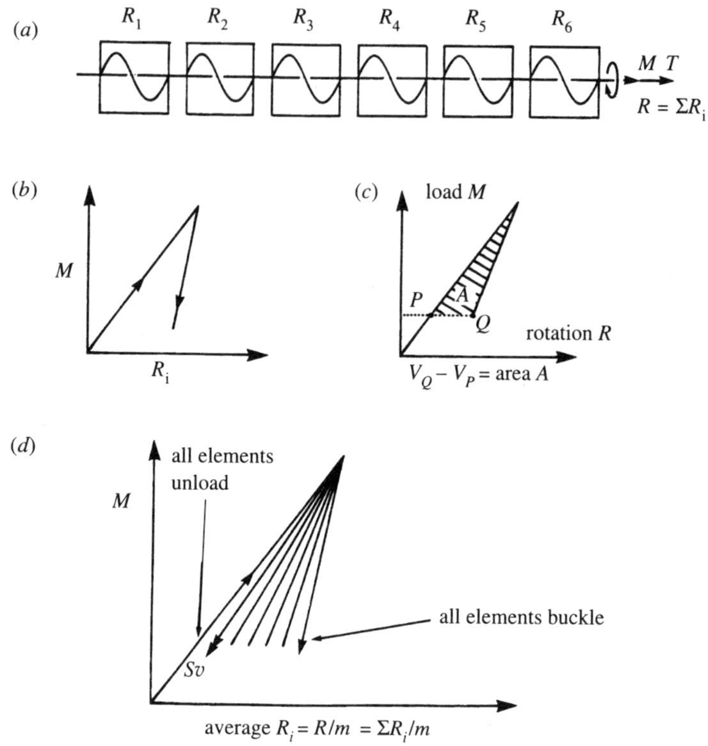

Figure 13. Heuristic explanation of spatial localization: (*a*) a chain of m independent helical elements, drawn for $m = 6$; (*b*) the response of each element; (*c*) how the energy barrier against buckling can be calculated as an area on any graph of a load against the corresponding deflection through which it does work; (*d*) the response of the chain of elements, the most severe path with the lowest energy barrier being Sv on which only one element buckles.

result is a very neat universal post-buckling curve which, like our equation (4.25), is independent of the pre-fixed twist per unit length τ_0. It does, however, miss some features of the behaviour very close to the bifurcation point, and we shall discuss this later.

In our present notation Coyne's (50) is, with the above approximation,

$$D/4\sqrt{B} = \sqrt{(T - T_c)}/T, \tag{5.5}$$

where we have used his equation (51), namely $T_c = C^2\tau_0^2/4B$. With $T/T_c = P$, and introducing the displacement parameter

$$W \equiv D\sqrt{(T_c/B)}/4, \tag{5.6}$$

we obtain the curve of localized post-buckling

$$W^2 = (P - 1)/P^2. \tag{5.7}$$

This gets all of Coyne's figure 4 onto the single universal curve shown in figure 14. To find the fold on this curve we have

$$2W\,\mathrm{d}W/\mathrm{d}P = (2 - P)/P^3 = 0, \tag{5.8}$$

giving $P = 2$, $W = \frac{1}{2}$. Notice that this value of W gives an approximation to Coyne's

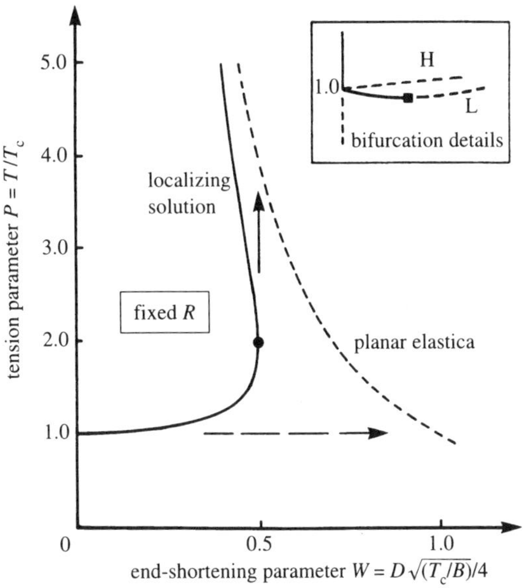

Figure 14. Universal post-buckling curve for the localizing solution under any amount of prefixed twist R. The dynamic jump under totally rigid loading (prefixed R and controlled D) is from the black circle at $(0.5, 2.0)$. Notice that the pre-fixed value of R determines the value of T_c. The inset shows bifurcational details. The very shallow helical path (H) is marginally subcritical, while the localizing path is marginally supercritical. If the localized path was followed, a jump under prefixed R and dead T would be from the black square: following roughly the path of the dashed arrow in the main diagram.

equation (70). The planar elastica, given by Coyne as $T = 16B/D^2$, becomes $W^2 = 1/P$, which we see is also the asymptotic form of the post-buckling solution (5.7) as P becomes large.

(*d*) *Behaviour under totally rigid loading*

We can finally use figure 14 to discuss the behaviour under totally rigid loading. We suppose that a prefixed amount of end rotation R has been given to the rod, the value of R influencing the diagram only through its effect on T_c. Under increasing controlled D, and ignoring the possibility that the system might prefer to follow the helical path (H) of the inset diagram, the localizing path will be followed quasi-statically up to the vertical tangent denoted by the black circle. From this point there will be a dynamic jump to the point of self-contact on the planar elastica curve. This process corresponds to the quasi-static evolution from c to d in figure 1, followed by the dynamic jump from d to e.

Some finer details of this diagram are sketched in the inset. Equations (4.25) and (4.26) show that the helical path from the branching point will have a shallow positive slope, making it subcritical under controlled T since the fundamental path is stable for $P > 1$, unstable for $P < 1$. Meanwhile the solutions of Coyne's figure 4, which have not used our approximation of § 5 c, show that the localized path falls slightly before reaching a horizontal tangent at the black square. So under decreasing controlled T, the localizing path is marginally supercritical. The suggestion, once again, is that the helical deformation might be advantageous at low amplitude, while the localizing solution is certainly advantageous at large amplitude. If the system did nevertheless follow the localizing path under controlled T, there would be a large dynamic jump from the black square, following roughly the line of the dashed arrow in the main diagram.

6. Discussion and conclusions

The theoretical comparisons between Love's helix (H_3, with approximately three twists per wave) and Coyne's localizing solution (L_3, also with approximately three twists per wave) show that for a long rod an unmodulated helix is initially the preferred energetically favourable mode of deformation. The modulated solution L_3, does, however, quickly replace it as the preferred mode, as the amplitude increases. This is confirmed by the non-appearance of H_3 in the laboratory tests. The experiments do, however, point to the H_1 helix, with approximately one twist per wave, as a commonly observed stable mode under the simple test conditions of figure 3.

Our latest tests, illustrated in figure 15, show that this H_1 mode is exhibited by a steel wire at very low stress levels within the elastic range. Such a wire should fit easily into the assumptions of the Love–Kirchhoff formulation, which treats the rod as an elastic line with no axial or shear deformability, and with linear constitutive relationships. So our search for an explanation must look beyond possible violations of the Love–Kirchhoff assumptions.

Meanwhile the H_1 mode is not observed in silicone rubber rods of square cross section when tested in water under conditions of approximate neutral buoyancy. These silicone rubber rods, kindly supplied by C. R. Calladine of the Cambridge Engineering Laboratory, were cast in a rectangular trough, and have no initial curvature in the unloaded stress-free state.

From this evidence, it seems most likely that H_1 is not a buckling mode, but rather a perturbed fundamental path generated by lateral gravitational forces and/or an initial curvature of the rod. This conclusion is reinforced by the fact that, under rigid loading, the H_1 mode seems to develop continuously from the first application of load: subsequently, there is a supercritical bifurcation at which the L_3 mode is superimposed onto the H_1 form. We leave a full investigation of mode H_1 to future work: but in the conclusions of Champneys & Thompson (1995) we draw attention to the fact that a non-trivial equilibrium state of the underlying dynamical system will be transformed into a one-twist-per-wave solution by the auxiliary Frenet–Seret equations.

Complications may also arise from the finite length and drill-chuck boundary conditions of the test specimens.

In this context it is pertinent to draw attention to the cautionary remarks of Antman & Kenney (1981). They point out that 'local behaviour, global behaviour and stability of buckled states are very sensitive to seemingly slight distinctions between boundary conditions': and they warn that mistakes can be made by a 'tacit appeal to a (non-existent) St. Venant's principle for rods'.

Many features of the experimental and theoretical studies warrant further investigation. Artificially induced initial imperfections could be used to assess the influence of material and geometrical irregularities. The effect of finite dynamic disturbances needs investigation: it is in fact remarkable just how much shaking and agitating the helical mode H_1 will tolerate, without jumping into the localized writhing form, L_3. Experiments on long rods, especially if they are made of rubber rather than metal, are of course influenced by the gravitational field, whether this be axial or

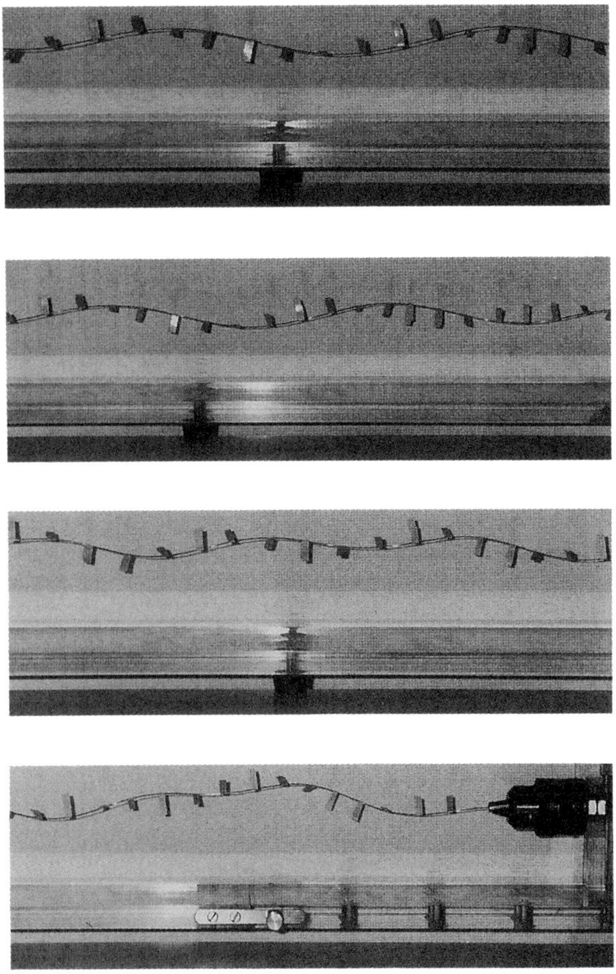

Figure 15. Experiments on a thin steel wire which was lightly stressed throughout, giving a full elastic recovery after the test. Twist can be observed by means of the flags, which were aligned in the unstressed state. The first three photographs show mode H_1. In the fourth picture the localizing mode L_3 can be seen superimposed on the H_1 mode. In this rigid loading sequence, with controlled end rotation and controlled end shortening, the secondary bifurcation into L_3 is supercritical.

lateral. Theoretical studies could throw useful light on the effect of gravity. Engineering intuition suggests that the torsional response of a long rod would not be greatly influenced by its hanging in a shallow catenary, but the sensitivities observed by Antman & Kenney (1981) might again be at work.

Looking beyond the circular rods, the complex deformations of rods with unequal principal bending stiffnesses will form a rich field of study. Here the lack of integrability in the mathematical model generates an infinity of localizing homoclinic paths, and the appearance of spatial chaos as demonstrated in our companion paper (Champneys & Thompson 1995).

References

Alexander, J. C. & Antman, S. S. 1982 The ambiguous twist of Love. *Q. appl. Math.* **40**, 83–92.

Antman, S. S. 1995 *Nonlinear problems of elasticity.* New York: Springer.

Antman, S. S. & Kenney, C. S. 1981 Large buckled states of nonlinearly elastic rods under torsion, thrust and gravity. *Arch. ration. Mech. Analysis* **76**, 289–338.

Bauer, W. R., Crick, F. H. C. & White, J. H. 1980 Supercoiled DNA. *Scient Am.* **243**, 118–133.

Buffoni, B., Champneys, A. R. & Toland, J. F. 1994 Bifurcation and coalescence of a plethora of homoclinic orbits for a Hamiltonian system. Mathematics preprint, 94/05, School of Mathematical Sciences, University of Bath.

Calladine, C. R. 1980 Toroidal elastic supercoiling of DNA. *Biopolymers* **19**, 1705–1713.

Calladine, C. R. & Drew, H. R. 1992 *Understanding DNA.* London: Academic.

Champneys, A. R. & Spence, A. 1993 Hunting for homoclinic orbits in reversible systems: a shooting technique. *Adv. comput. Math.* **1**, 81–108.

Champneys, A. R. & Thompson, J. M. T. 1995 A multiplicity of localized buckling modes for twisted rod equations. *Proc. R. Soc. Lond.* A (submitted).

Champneys, A. R. & Toland, J. F. 1993 Bifurcation of a plethora of multi-modal homoclinic orbits for autonomous Hamiltonian systems. *Nonlinearity* **6**, 665–721.

Cottrell, A. H. 1964 *The mechanical properties of matter.* New York: Wiley.

Coyne, J. 1990 Analysis of the formation and elimination of loops in twisted cable. *IEEE J. ocean. Engng* **15**, 72–83.

Davies, M. A. & Moon, F. C. 1993 3D spatial chaos in the elastica and the spinning top: Kirchhoff analogy. *Chaos* **3**, 93–99.

Fuller, F. B. 1971 The writhing number of a space curve. *Proc. natn. Acad. Sci. U.S.A.* **68**, 815–819.

Hunt, G. W., Bolt, H. M. & Thompson, J. M. T. 1989 Structural localization phenomena and the dynamical phase-space analogy. *Proc. R. Soc. Lond.* A **425**, 245–267.

Hunt, G. W. & Neto, E. L. 1991 Localized buckling in long axially-loaded cylindrical shells. *J. Mech. Phys. Solids* **39**, 881–894.

Hunt, G. W. & Neto, E. L. 1993 Maxwell critical loads for axially loaded cylindrical shells. *J. appl. Mech.* **60**, 702–706.

Hunt, G. W. & Wadee, M. K. 1991 Comparative Lagrangian formulations for localized buckling. *Proc. R. Soc. Lond.* A **434**, 485–502.

Ikeda, K., Providencia, P. & Hunt, G. W. 1993 Multiple equilibria for unlinked and weakly-linked cellular structural forms. *Int. J. Solids Struct.* **30**, 371–384.

Kauffman, L. H. 1991 *Knots and physics.* Singapore: World Scientific.

Love, A. E. H. 1927 *A Treatise on the mathematical theory of elasticity*, 4th edn. Cambridge University Press.

Mansfield, E. H. 1965 The instability of buckle pattern in cylindrical shells under axial compression. *Aeronaut. Q.* **16**, 275–281.

Mielke, A. & Holmes, P. 1988 Spatially complex equilibria of buckled rods. *Arch. ration. Mech. Analysis* **101**, 319–348.

Tan, Z. & Witz, J. A. 1992 Loop formation of marine cables and umbilicals during installation. In *BOSS 92* (ed. M. H. Patel & R. Gibbins), pp. 1270–1285. London: BPP Technical Services.

Thompson, J. M. T. 1979 Stability predictions through a succession of folds. *Phil. Trans. R. Soc. Lond.* A **292**, 1–23.

Thompson, J. M. T. & Virgin, L. N. 1988 Spatial chaos and localization phenomena in nonlinear elasticity. *Phys. Lett.* **126A**, 491–496.

Timoshenko, S. P. & Gere, J. M. 1961 *Theory of elastic stability.* New York: McGraw-Hill.

van der Meer, J. 1985 *The Hamiltonian Hopf bifurcation (Lecture Notes in Mathematics 1160).* Berlin: Springer.

Ziegler, H. 1968 *Principles of structural stability.* Waltham, MA: Blaisdell.

Received 14 December 1994; revised 13 April 1995; accepted 1 May 1995

ELSEVIER

Physica D 112 (1998) 201–224

PHYSICA D

Lock-on to tape-like behaviour in the torsional buckling of anisotropic rods

G.H.M. van der Heijden [a,*], J.M.T. Thompson [a]

[a] *Centre for Nonlinear Dynamics, University College London, Gower Street, London WC1E 6BT, UK*

Abstract

Through nonlinear normal form analysis of the equilibrium equations developed in a neighbourhood of the straight rod solution we show the existence of a 'phase transition' from mildly anisotropic to full 'tape-like' buckling in non-symmetric rods subject to terminal loads, provided the material has a sufficiently low Poisson's ratio. For a solid elliptical rod with a Poisson's ratio of 0.2 the critical point is found to occur at a cross-sectional aspect ratio of just over 3. For larger aspect ratios the rod locks on to a one-twist-per-wave buckling mode with no internal twist.

In the mildly anisotropic regime there turn out to be two physically distinct pairs of symmetric so-called primary localised buckling modes, differing in the phase of the internal twist. These four modes (homoclinic orbits) remain of the full circle of localised buckling modes existing in the symmetric case after breaking of the circular symmetry of the rod's cross-section. In the strongly anisotropic regime only the energetically favourable pair of 'flat' buckling modes survives.

Keywords: Torsional buckling; Anisotropic rods; Normal forms; Bifurcation; Homoclinic orbits

1. Introduction

In previous work [13] we numerically studied localised buckling of long, thin, initially straight, elastic rods subject to end forces and moments for the case of rods with non-symmetric cross-section. Ignoring effects of shear and axial extension and taking linear constitutive relations between (generalised) stresses and strains, we used the geometrically exact Cosserat formulation of rod mechanics [2] to derive a six-dimensional system of equilibrium equations. The subsequent assumption of an infinitely long rod gave us two advantages. Firstly, it turned our system into a dynamical system (arclength playing the role of time) for which many analytical and numerical techniques are available. Secondly, it offered us a way around the intricacies usually associated with the application of boundary conditions.

For the case of a symmetric rod it is well known that the dynamical system is completely integrable [7,23] and we have a manifold (circle) of homoclinic orbits (localised buckling solutions) to the trivial fixed point representing the straight rod. At buckling these homoclinic orbits bifurcate subcritically at a reversible Hamiltonian–Hopf bifurcation.

* Corresponding author. E-mail: g.heijden@ucl.ac.uk.

PII S0167-2789(97)00211-X

For non-symmetric rods complete integrability is lost and the manifold of homoclinic orbits splits. A Melnikov-type analysis can then be used [23] to show the persistence, amidst spatial chaos, of a finite number of isolated primary homoclinic orbits. For the D_2-symmetric problem of an initially straight and prismatic rod with linear constitutive relations but arbitrary cross-section, which we will restrict our attention to in this paper, four primary homoclinic orbits persist, representing two physically distinct localised buckling modes. Following work in [7] it was shown numerically in [13] that in addition to these primary modes, and consistent with the work of Belyakov and Šil'nikov [4], there is a countable infinity of so-called multi-modal homoclinic orbits, approximately consisting of multiple copies of primary modes.

An interesting observation in [13] was that, mathematically, buckling in anisotropic rods occurs through either of the two mechanisms. For mildly anisotropic rods the reversible Hamiltonian–Hopf bifurcation present in the isotropic case persists, while for strongly anisotropic rods, under some conditions, the reversible Hamiltonian–Hopf bifurcation disappears and buckling is ruled by a reversible Hamiltonian-pitchfork bifurcation instead. The transition is described by a codimension-two bifurcation on the buckling line. The ramifications of this codimension-two point form the subject matter of the present paper. (We should remark that although a Hamiltonian structure is known to exist in the twisted rod problem [23], we shall not use it explicitly in this paper.)

We shall perform a linear as well as a nonlinear (normal form) analysis of the equilibrium equations about the trivial straight-rod solution near the buckling line. Thus we find that for solid elliptical rods the codimension-two point occurs providing that the elastic material has sufficiently low Poisson's ratio ($< \frac{1}{3}$). Moreover, the codimension-two bifurcation has implications for the multiplicity of localised buckling solutions. Of the four homoclinic orbits present at low anisotropy only two remain in the strongly anisotropic regime. Physically, the surviving modes describe true tape-like buckling with the rod bending only in one direction with no internal twist.

The organisation of the rest of this paper is as follows. After explaining the model (Section 2), in Section 3 we use a Lagrangian formulation to linearise the equilibrium equations about the straight-rod solution. We start from Euler angle co-ordinates and, carefully dealing with the associated polar singularity, transform to a set of co-ordinates which are different from the components of force and moment used as co-ordinates in [13], but which for the present purposes give a more transparent description. An eigenvalue analysis is then carried out and local bifurcation curves are computed. After a reduction of the original six-dimensional system to a four-dimensional one a normal form analysis of the various cases follows in Section 4. In Section 5 we derive some properties of the linearised (emerging) buckling modes. Symmetry arguments are used to relate these linear modes to the full-blown localised buckling modes of the nonlinear system computed numerically in [13]. The physical behaviour of the buckled rod is further elucidated in Section 6 by looking at the cross-sectional orientation.

Although the main results of this paper are derived explicitly for solid elliptical rods, the implications are of far greater generality. This is discussed in Section 7. By tracing back our results to conditions on bending and torsional stiffnesses (thus becoming independent of any parametrisation) we show how our results on critical buckling behaviour can essentially be applied to any long, thin object with arbitrary cross-section.

We finally mention that cross-sections with less than circular symmetry have been studied before by Buzano et al. [6]. They studied compression of rods of finite length with cross-sections invariant under the dihedral group D_n with $n \geq 3$, i.e., regular polygons. Such rods will give essentially different behaviour from that of circular rods only when one deals with nonlinear constitutive relations. The post-buckling behaviour of a clamped rectangular rod under axial compression was studied qualitatively in [5] as a case of $D_4 \rightarrow D_2$ symmetry breaking. D_n-symmetric rods were also considered in [23].

2. The model

Consider a long, thin, elastic rod of non-symmetric cross-section subject to end forces and moments. As an idealisation for such objects we will assume the rod to be infinitely long. Further idealisations are made by assuming that the rod is straight and prismatic when unloaded and by ignoring effects of shear deformation and axial extension. The strain of the rod is then measured by the vector

$$\Omega(s) = \kappa_1(s)\, \boldsymbol{d}_1(s) + \kappa_2(s)\, \boldsymbol{d}_2(s) + \tau(s)\, \boldsymbol{d}_3(s),$$

where $\{\boldsymbol{d}_1(s), \boldsymbol{d}_2(s), \boldsymbol{d}_3(s)\}$ is a right-handed body-centred co-ordinate system at arclength s composed of the tangent vector $\boldsymbol{d}_3(s) = \boldsymbol{r}'(s)$ to the rod's centre line $\boldsymbol{r}(s)$, and two vectors $\boldsymbol{d}_1(s)$ and $\boldsymbol{d}_2(s)$ along the principal axes of inertia in a normal cross-section of the rod. It can be related to a fixed orthonormal system $\{\boldsymbol{e}_1, \boldsymbol{e}_2, \boldsymbol{e}_3\}$ via the Euler angles θ, ψ, ϕ as follows:

$$\begin{aligned} \boldsymbol{d}_1 &= (-\sin\psi\sin\phi + \cos\psi\cos\phi\cos\theta)\,\boldsymbol{e}_1 + (\cos\psi\sin\phi + \sin\psi\cos\phi\cos\theta)\,\boldsymbol{e}_2 - \cos\phi\sin\theta\,\boldsymbol{e}_3, \\ \boldsymbol{d}_2 &= (-\sin\psi\cos\phi - \cos\psi\sin\phi\cos\theta)\,\boldsymbol{e}_1 + (\cos\psi\cos\phi - \sin\psi\sin\phi\cos\theta)\,\boldsymbol{e}_2 + \sin\phi\sin\theta\,\boldsymbol{e}_3, \\ \boldsymbol{d}_3 &= \cos\psi\sin\theta\,\boldsymbol{e}_1 + \sin\psi\sin\theta\,\boldsymbol{e}_2 + \cos\theta\,\boldsymbol{e}_3. \end{aligned} \tag{1}$$

The κ_1 and κ_2 are the two components of curvature while τ is the twist about the body axis of the rod. Expressed in terms of Euler angles they become

$$\kappa_1 = \frac{\mathrm{d}\theta}{\mathrm{d}s}\sin\phi - \frac{\mathrm{d}\psi}{\mathrm{d}s}\sin\theta\cos\phi, \quad \kappa_2 = \frac{\mathrm{d}\theta}{\mathrm{d}s}\cos\phi + \frac{\mathrm{d}\psi}{\mathrm{d}s}\sin\theta\sin\phi, \quad \tau = \frac{\mathrm{d}\phi}{\mathrm{d}s} + \frac{\mathrm{d}\psi}{\mathrm{d}s}\cos\theta. \tag{2}$$

Making the usual assumption of a hyperelastic rod we can write the strain energy density as

$$W = \tfrac{1}{2}B_1\kappa_1^2 + \tfrac{1}{2}B_2\kappa_2^2 + \tfrac{1}{2}C\tau^2,$$

where B_1 and B_2 are the bending stiffnesses about the $\boldsymbol{d}_1$ and $\boldsymbol{d}_2$ axes, respectively, while C is the torsional stiffness of the rod. Here, we have implicitly used linear constitutive relations to express the moments m_i about the axes $\boldsymbol{d}_i$ in terms of the strains as follows:

$$m_1 = B_1\kappa_1, \qquad m_2 = B_2\kappa_2, \qquad m_3 = C\tau. \tag{3}$$

In the main body of this paper we will take, for definiteness, a uniform rod with elliptical cross-section (in Section 7 we will have something to say about the influence of the cross-section on our results). For such a rod we have (see, e.g., [27])

$$B_1 = \frac{1}{4}\pi E a^3 b, \qquad B_2 = \frac{1}{4}\pi E a b^3, \qquad C = \frac{\pi G a^3 b^3}{a^2 + b^2}, \tag{4}$$

where E is Young's modulus and G the shear modulus of the material while a and b are the semi-axes of the elliptical cross-section along $\boldsymbol{d}_2$ and $\boldsymbol{d}_1$, respectively. We shall assume, without losing generality, that $a < b$ and refer to b/a as the aspect ratio of the cross-section.

The work per unit arclength done against end tension T is $U = T(1 - \cos\theta)$. Ignoring the irrelevant constant term we can then write down the total potential energy density for the elastic rod as follows:

$$L = \tfrac{1}{2}B_1\kappa_1^2 + \tfrac{1}{2}B_2\kappa_2^2 + \tfrac{1}{2}C\tau^2 - T\cos\theta. \tag{5}$$

Note that in Kirchhoff's well-known kinetic analogy [2,20] L is the Lagrangian (density) of the heavy rigid body (or top) pivoting about a fixed point (T representing gravity). Non-dimensionalising arclength and energy according to the rescalings

$$t = \frac{M}{B_2}\, s, \qquad \mathcal{L} = \frac{B_2}{M^2}\, L, \tag{6}$$

M being the applied moment, the dimensionless Lagrangian is seen to be

$$\mathcal{L} = \frac{1}{2}\frac{\tilde{\kappa}_1^2}{1+\rho} + \frac{1}{2}\tilde{\kappa}_2^2 + \frac{\tilde{\tau}^2}{(1+\nu)(2+\rho)} - \frac{1}{m^2}\cos\theta, \tag{7}$$

where $\tilde{\kappa}_1, \tilde{\kappa}_2$ and $\tilde{\tau}$ are the dimensionless strains, i.e.,

$$\tilde{\kappa}_1 = \frac{B_2}{M}\kappa_1, \qquad \tilde{\kappa}_2 = \frac{B_2}{M}\kappa_2, \qquad \tilde{\tau} = \frac{B_2}{M}\tau, \tag{8}$$

and the parameters m, ρ and ν are, respectively, a unified load parameter, an anisotropy parameter, and Poisson's ratio (a material constant), defined by

$$m = \frac{M}{\sqrt{B_2 T}}, \qquad \rho = \frac{B_2}{B_1} - 1, \qquad 1+\nu = \frac{E}{2G} = \frac{2}{2+\rho}\frac{B_2}{C}. \tag{9}$$

3. Linearisation about the trivial solution

In order to study bifurcations of the trivial straight-rod solution of the equilibrium equations we seek to 'quadratise' the Lagrangian (7) about this solution.

The straight rod has $\theta = 0$ but, by nature of the Euler angles, the ψ and ϕ then lose their independent meaning: the rotation about the body axis of the rod is given by $\psi + \phi$. We will therefore transform to the independent co-ordinates θ, ϕ and $\chi := \psi + \phi$. The trivial solution is then given by $\theta = 0$, $\dot{\chi} = \frac{1}{2}(1+\nu)(2+\rho)$, where $\dot{\chi}$ is the fundamental twist of the unbuckled rod, an overdot denoting differentiation with respect to scaled arclength t. Linearisation about this solution is best done by introducing another co-ordinate transformation. Defining

$$x = \sin\theta\cos\phi, \qquad y = \frac{1}{\sqrt{1+\rho}}\sin\theta\sin\phi, \tag{10}$$

the Lagrangian reads

$$\begin{aligned}\mathcal{L} = {} & \frac{1}{2}(\dot{x}^2+\dot{y}^2) - \frac{\dot{\chi}}{\sqrt{1+\rho}}(x\dot{y} - (1+\rho)y\dot{x}) + \frac{1}{2}\frac{\dot{\chi}^2}{1+\rho}x^2 + \frac{1}{2}(1+\rho)\dot{\chi}^2 y^2 \\ & + \frac{\dot{\chi}^2}{(1+\nu)(2+\rho)}(1 - x^2 - (1+\rho)y^2) + \frac{\sqrt{1+\rho}}{(1+\nu)(2+\rho)}\dot{\chi}(x\dot{y} - y\dot{x}) \\ & - \frac{1}{m^2}\left(1 - \frac{1}{2}x^2 - \frac{1}{2}(1+\rho)y^2\right) + \cdots,\end{aligned} \tag{11}$$

where the ellipsis represents terms of fourth-order or higher in x, y and their derivatives. Linearisation of the equilibrium equations is now to be performed about $x = 0$, $y = 0$, $\dot{\chi} = \frac{1}{2}(1+\nu)(2+\rho)$, so after writing $\dot{\chi} = \frac{1}{2}(1+\nu)(2+\rho) + \dot{\bar{\chi}}$ we only have to retain terms in $\mathcal{L}$ that are up to combined second-order in x, y, $\bar{\chi}$ and their derivatives. The required Lagrangian thus becomes

$$\mathcal{L}_0 = \frac{1}{2}(\dot{x}^2+\dot{y}^2) - \frac{1}{2}\frac{(1+\nu)(2+\rho)}{\sqrt{1+\rho}}(x\dot{y}-(1+\rho)y\dot{x}) + \frac{1}{8}\frac{(1+\nu)^2(2+\rho)^2}{1+\rho}x^2$$
$$+\frac{1}{8}(1+\nu)^2(2+\rho)^2(1+\rho)y^2 - \frac{1}{4}(1+\nu)(2+\rho)(x^2+(1+\rho)y^2) + \dot{\chi}$$
$$+\frac{\dot{\chi}^2}{(1+\nu)(2+\rho)} + \frac{1}{2}\sqrt{1+\rho}\,(x\dot{y}-y\dot{x}) + \frac{1}{2m^2}(x^2+(1+\rho)y^2), \tag{12}$$

where an irrelevant constant term has been dropped.

From the Euler–Lagrange equations for $\mathcal{L}_0$ the linearised equilibrium equations are now found to be

$$\ddot{x}+\alpha\dot{y}+\beta_1 x=0, \qquad \ddot{y}-\alpha\dot{x}+\beta_2 y=0, \qquad \ddot{\chi}=0, \tag{13}$$

where

$$\alpha = \frac{1}{2}\sqrt{1+\rho}\left(\frac{(1+\nu)(2+\rho)}{1+\rho}+(1+\nu)(2+\rho)-2\right),$$
$$\beta_1 = -\frac{1}{4}\frac{(1+\nu)^2(2+\rho)^2}{1+\rho}+\frac{1}{2}(1+\nu)(2+\rho)-\frac{1}{m^2}, \tag{14}$$
$$\beta_2 = -\frac{1}{4}(1+\nu)^2(2+\rho)^2(1+\rho)+\frac{1}{2}(1+\nu)(2+\rho)(1+\rho)-\frac{1+\rho}{m^2}.$$

Note that for a symmetric rod (i.e., $\rho=0$) we have $\beta_1=\beta_2$, lending rotational symmetry to the oscillator system (13).

The third equation in (13) can be trivially solved to give (setting $\chi(0)=0$)

$$\chi(t)=\tfrac{1}{2}(1+\nu)(2+\rho)t, \tag{15}$$

and we are left with two coupled linear oscillators, which can be written as a first-order system

$$\dot{u}=A\,u \qquad \text{with } u=(x,\dot{x},y,\dot{y}) \quad \text{and} \quad A=\begin{pmatrix} 0 & 1 & 0 & 0 \\ -\beta_1 & 0 & 0 & -\alpha \\ 0 & 0 & 0 & 1 \\ 0 & \alpha & -\beta_2 & 0 \end{pmatrix}. \tag{16}$$

The characteristic polynomial of A is

$$P_A(\lambda)=\lambda^4+(\beta_1+\beta_2+\alpha^2)\lambda^2+\beta_1\beta_2=:\lambda^4+c_2\lambda^2+c_0. \tag{17}$$

The configuration of the four eigenvalues varies with the parameters of the system as depicted in the ρ–m plane of Fig. 1. As will be shown shortly, there are two cases: for $\nu<\frac{1}{3}$ the curves labelled m_{c_2} and m_{c_3} form a tangency, whilst for $\nu\geq\frac{1}{3}$ they do not have any point in common. The tangency has interesting physical implications which we shall discuss later. Figs. 1(a) and (b) represent the two physical extremes of a very 'stiff' ($\nu=0$) and a very 'soft' ($\nu=\frac{1}{2}$) elastic material. In diagrams to come the low-ν regime will be illustrated by taking the more realistic intermediate value of $\nu=0.2$. Experimentally, one typically finds $\nu=0.5$ for rubber, 0.3–0.4 for metals (but, e.g., 0.2 for tungsten), and 0.2–0.3 for ceramics (see, e.g., [11, Appendix H]). (Historically, from his molecular theory of matter, Poisson derived a theoretical value of 0.25, assuming a completely isotropic elastic material.)

The line denoted by m_{c_2} in Fig. 1 is the locus of Hamiltonian–Hopf bifurcations (see Section 4.3) defined by

$$c_2^2-4c_0=0, \quad c_2>0. \tag{18}$$

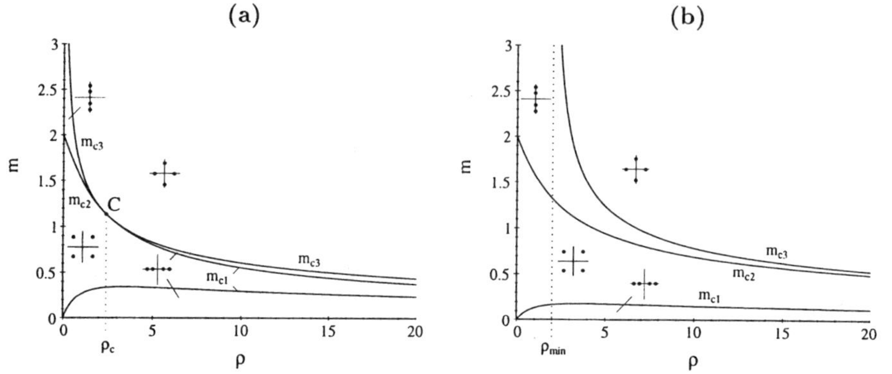

Fig. 1. Two ρ–m parameter planes with loci of bifurcation points and eigenvalue configurations to illustrate the difference between the cases $\nu < \frac{1}{3}$ and $\nu > \frac{1}{3}$: (a) $\nu = 0$ ($\rho_{\min} = 0$, $\rho_c = 2.3830$, corresponding to a critical aspect ratio of 1.8393); (b) $\nu = 0.5$ ($\rho_{\min} = 2$). In (a) the lines m_{c_2} and m_{c_3} touch each other at ρ_c and the buckling line is formed by the union of m_{c_2} and the part of m_{c_3} to the right of ρ_c; in (b) the lines do not touch and the buckling line is formed by m_{c_2}.

On m_{c_1} we have $c_2^2 - 4c_0 = 0$, $c_2 < 0$, giving double real eigenvalues. The line m_{c_3}, finally, represents Hamiltonian-pitchfork bifurcations (see Section 4.4) where two eigenvalues become zero. It is defined by

$$c_0 = 0, \tag{19}$$

which only has a real solution for $\beta_1 = 0$, giving an asymptote at $\rho_{\min} = 2\nu/(1-\nu)$. For all lines m_{c_i} we have $m \to 0$ as $\rho \to \infty$.

It was established in [13] that the buckling line in Fig. 1 is formed by the union of m_{c_2} and the large-ρ part of m_{c_3}. Thus, in particular, it is seen that for $\rho = 0$ we recover the classical critical load condition $m = 2$, valid for infinitely long symmetric rods [26, Eq. (2.74)]. At the critical point $\rho = \rho_c$ that separates the two branches of the buckling line all eigenvalues become zero. We then have $c_0 = c_2 = 0$, implying $\beta_1 = 0$, $\beta_2 = -\alpha^2$, from which ρ_c can be computed as a function of ν. A much more compact expression, however, is obtained by solving for ν in terms of ρ_c:

$$\nu = \frac{2(2+\rho_c)\sqrt{\rho_c(1+\rho_c)^3} - \rho_c^3 - 6\rho_c^2 - 12\rho_c - 8}{16 + 28\rho_c + 16\rho_c^2 + 3\rho_c^3}. \tag{20}$$

A plot is given in Fig. 2. There is an asymptote for $\nu = \frac{1}{3}$, separating the two cases illustrated in Fig. 1.

In the remaining part of this paper we will concentrate on the behaviour of the system near the buckling line. It will be seen that the critical point at ρ_c defines a transition in the type of buckling in anisotropic rods.

4. Nonlinear analysis

The nonlinear analysis consists of a centre-manifold-like reduction to a four-dimensional system followed by a normal form analysis near the relevant bifurcations. We will heavily draw on results of Iooss and co-workers [10,14,15,17,18] on normal forms in reversible dynamical systems. But before embarking on this we will have a look at the symmetry properties of our system as these play a crucial role in the analysis which follows.

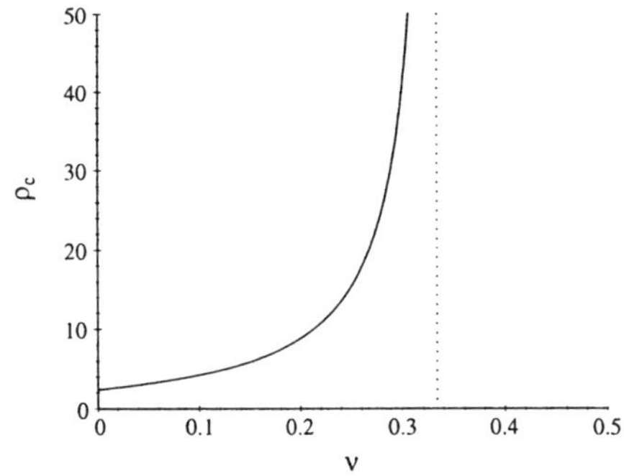

Fig. 2. The critical anisotropy parameter ρ_c as a function of Poisson's ratio ν. ρ_c tends to infinity as ν approaches $\frac{1}{3}$.

4.1. Symmetry properties

It can be verified from (7) that the full nonlinear Lagrangian is invariant under the following set of transformations:

$$\begin{aligned} R_1&: \ t \to -t, \quad \theta \to -\theta, \quad \phi \to -\phi, \ \chi \to -\chi, \\ R_2&: \ t \to -t, \quad \theta \to \theta, \quad \phi \to -\phi, \ \chi \to -\chi, \\ Z_1&: \ \theta \to -\theta, \\ Z_2&: \ \phi \to -\phi, \ \chi \to -\chi, \\ T&: \ t \to -t. \end{aligned} \tag{21}$$

R_1 and R_2 are two involutions expressing the reversibility of the system; Z_1 and Z_2 are two reflection symmetries. The symmetries Z_2 and T both correspond to applying an end moment in the opposite direction and hence can be ignored if we fix our boundary conditions (they express indifference with respect to left-handed and right-handed applied moment). The remaining symmetries R_1, R_2 and Z_1 $(= R_1 \circ R_2)$ then generate the group $D_2 \simeq \mathbb{Z}_2 \oplus \mathbb{Z}_2$.

Note that as long as we take linear constitutive relations any shape of the cross-section gives rise to a D_2-symmetric system of equations. At least, i.e., if we consider rods which are initially straight and prismatic. Indeed, allowing for initial curvature would break the D_2-symmetry and the problem would become a different one with different multiplicities of solutions. More on this case will be published elsewhere [8].

In our special case of an elliptical rod, D_2 is, of course, also the symmetry group of the cross-section. The reversibilities R_1 and R_2 then correspond to the two reflection symmetries of the cross-section. Z_1 describes the invariance of the cross-section under a 180° rotation.

Now, the Lagrangian $\mathcal{L}$ defines a vector field on a six-dimensional Euclidean phase space co-ordinated by $(x, \dot{x}, y, \dot{y}, \bar{\chi}, \dot{\bar{\chi}})$ in the obvious way:

$$\dot{w} = F(w), \qquad w = (x, \dot{x}, y, \dot{y}, \bar{\chi}, \dot{\bar{\chi}}) \in \mathbb{R}^6. \tag{22}$$

Through (21) we have a group action of D_2 on this space given by

$$R_1 = \mathrm{diag}(-1, 1, 1, -1, -1, 1), \qquad R_2 = \mathrm{diag}(1, -1, -1, 1, -1, 1), \qquad Z_1 = R_1 R_2, \tag{23}$$

and the vector field F anti-commutes with both R_1 and R_2 and commutes with Z_1:

$$F R_i = -R_i F \quad (i = 1, 2), \qquad F Z_1 = Z_1 F. \tag{24}$$

The implications of these symmetry properties are that solutions of the equilibrium equations are either invariant under the transformations in (21) or come in symmetric pairs. For Z_1 we can already remark that the only solution which is itself Z_1-invariant is the trivial straight-rod solution $\theta = 0$ (or, $x = 0$, $y = 0$). We should therefore expect to find pairs of Z_1-symmetric partners corresponding to the same rod configuration 180° rotated with respect to each other.

4.2. Reduction to a four-dimensional system

Expanding the Lagrangian in (11) up to fourth-order terms one finds that the (third-order) $\bar{\chi}$-equation becomes coupled to the x, y-equations. We can still reduce the system to a four-dimensional one, however, by exploiting a feature of the system we have not used yet, namely its conserved quantities.

Apart from the energy integral $W - U$, as a result of rotational symmetry about the loading axis e_3 we also have conservation of torque about this axis:

$$\frac{\mathrm{d}}{\mathrm{d}s}(\boldsymbol{m} \cdot \boldsymbol{e}_3) = 0, \quad \text{where } \boldsymbol{m} = m_1\,\boldsymbol{d}_1 + m_2\,\boldsymbol{d}_2 + m_3\,\boldsymbol{d}_3. \tag{25}$$

(Note that if we had a circular rod torque about the body axis $\boldsymbol{m} \cdot \boldsymbol{d}_3$ would have been the third independent integral, which would have rendered the system completely integrable.) Using the constitutive relations (3) to write the moments in terms of the strains, (25) becomes

$$B_1\kappa_1\,\boldsymbol{d}_1 \cdot \boldsymbol{e}_3 + B_2\kappa_2\,\boldsymbol{d}_2 \cdot \boldsymbol{e}_3 + C\tau\,\boldsymbol{d}_3 \cdot \boldsymbol{e}_3 = \text{const.} \tag{26}$$

Now recall that we are considering localised buckling solutions in which the rod becomes straight towards infinity. This means that $\boldsymbol{d}_3$ and $\boldsymbol{e}_3$ line up ($\theta = 0$) for large s. Since we apply an end moment M in the direction $\boldsymbol{e}_3$ it follows that the constant on the right-hand side of (26) is equal to M. Using Euler angles (2) the dimensionless form of (26) then becomes

$$-\frac{\tilde{\kappa}_1}{1+\rho}\cos\phi\sin\theta + \tilde{\kappa}_2\sin\phi\sin\theta + \frac{2\tilde{\tau}}{(1+\nu)(2+\rho)}\cos\theta = 1. \tag{27}$$

By using (2) and (10) this equation can now be used to express $\dot{\bar{\chi}}$ in terms of x, y and their derivatives. In the symmetric case, for instance, we find

$$\dot{\bar{\chi}} = -\nu(1+\nu)(x^2+y^2) + (\tfrac{1}{2}+\nu)(x\dot{y} - y\dot{x}) + \mathrm{O}(3). \tag{28}$$

Substitution into the equilibrium equations then gives a third-order expansion of the equations for x and y.

The above procedure can be carried out up to any order. Written as a dynamical system we therefore arrive at the four-dimensional reduced system

$$\dot{u} = F_r(u) = A\,u + G(u), \qquad u = (x, \dot{x}, y, \dot{y}) \in \mathbb{R}^4, \qquad G(u) = \mathrm{O}(|u|^3), \tag{29}$$

with A as defined in (16).

Obviously, this four-dimensional system inherits the symmetry properties of the original six-dimensional system (22). Explicitly, we have

$$\begin{aligned} &R_1 = \mathrm{diag}(-1, 1, 1, -1), \qquad R_2 = \mathrm{diag}(1, -1, -1, 1), \qquad Z_1 = R_1 R_2, \\ &F_r R_i = -R_i F_r \quad (i = 1, 2), \qquad F_r Z_1 = Z_1 F_r. \end{aligned} \tag{30}$$

Note that our reduced system (29) is an expansion valid only in a neighbourhood of the trivial solution $u \equiv 0$. The six-dimensional twisted rod system described by the Lagrangian (7) can actually be reduced globally to a two-degrees-of-freedom canonical Hamiltonian system [23]. Because we are only interested in behaviour near buckling a local description suffices.

In the following sections we shall simplify the vector field F_r further in a neighbourhood of the origin for parameters in various regions near the buckling line.

4.3. A normal form analysis at the reversible Hamiltonian–Hopf bifurcation

To study the possible bifurcation of homoclinic orbits (localised buckling modes) at the m_{c_2}-part of the buckling line we compute the normal form of the reduced vector field F_r at the reversible Hamiltonian–Hopf bifurcation. This treats the mildly anisotropic case ($\rho < \rho_c$) if $\nu < \frac{1}{3}$, and the full range of ρ-values otherwise. Note that the symmetric case is integrable and we already know [3,7,23] that there is a one-parameter family of homoclinic orbits.

Along the branch m_{c_2} in Fig. 1 the eigenvalues are given by $\lambda_\pm = \pm\mathrm{i}\omega$ (both with multiplicity two), where

$$\omega = \sqrt{(\beta_1^* + \beta_2^* + \alpha^2)/2}. \tag{31}$$

Here, β_1^* and β_2^* are the β_1 and β_2 in (14) taken at buckling, i.e., with m^2 solved from the Hamiltonian–Hopf condition (18). The corresponding eigenvectors v_1, $\bar{v}_1$ and generalised eigenvectors v_2, $\bar{v}_2$ are given by

$$v_1 = (A - \mathrm{i}\omega)\, v_2 = \begin{pmatrix} -\dfrac{\mathrm{i}\omega}{4\gamma}(1+\gamma^2) \\ \dfrac{\omega^2}{4\gamma}(1+\gamma^2) \\ -\dfrac{\omega}{4}(1+\gamma^2) \\ -\dfrac{\mathrm{i}\omega^2}{4}(1+\gamma^2) \end{pmatrix}, \qquad v_2 = \begin{pmatrix} \dfrac{1}{2\gamma} \\ \dfrac{\mathrm{i}\omega}{4\gamma}(1-\gamma^2) \\ -\dfrac{\mathrm{i}}{4}(1+\gamma^2) \\ 0 \end{pmatrix}, \tag{32}$$

where

$$\gamma = \frac{\omega\alpha}{\omega^2 - \beta_2^*}. \tag{33}$$

They satisfy

$$R_1 v_1 = \bar{v}_1, \quad R_1 v_2 = -\bar{v}_2, \qquad R_2 v_1 = -\bar{v}_1, \quad R_2 v_2 = \bar{v}_2. \tag{34}$$

The following facts can be checked:

$$\begin{aligned} &\beta_2^* < \beta_1^* < 0, \quad \alpha = \sqrt{-\beta_1^*} + \sqrt{-\beta_2^*} \qquad (\forall \nu, \rho) \\ &\omega = (\beta_1^* \beta_2^*)^{1/4} \geq 0, \quad 0 \leq \gamma = \left(\frac{\beta_1^*}{\beta_2^*}\right)^{1/4} \leq 1 \qquad (\forall \nu, \rho) \\ &\omega \to \tfrac{1}{2} + \nu, \quad \gamma \to 1 \qquad \text{as } \rho \to 0, \\ &\omega \to 0, \quad \gamma \to 0 \qquad \text{as } \rho \to \rho_c. \end{aligned} \tag{35}$$

Fig. 3 shows curves of ω and γ against ρ for a number of values of ν.

Using the above basis of generalised eigenvectors we can transform the linear part of the equations into (complex) Jordan normal form:

$$\dot{z}_1 = \mathrm{i}\omega z_1 + z_2 + f_1(z_1, z_2), \quad \dot{z}_2 = \mathrm{i}\omega z_2 + f_2(z_1, z_2), \qquad \text{where } f_1, f_2 = \mathrm{O}((|z_1| + |z_2|)^3). \tag{36}$$

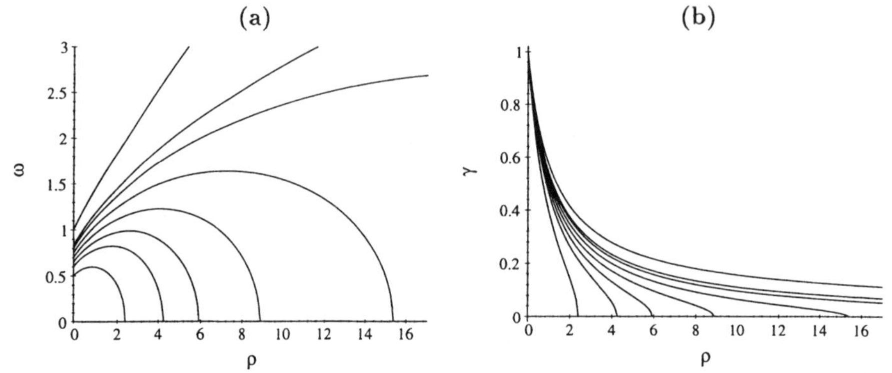

Fig. 3. ω (a) and γ (b) as a function of ρ for, from left to right, $\nu = 0, 0.1, 0.15, 0.2, 0.25, 0.3, \frac{1}{3}$ and $\frac{1}{2}$, implying, respectively, $\rho_c = 2.382976, 4.233098, 5.918595, 8.870676, 15.368498, 41.366561$, '$\infty$' and '$> \infty$'.

The reflections R_1 and R_2 act on the complex variables z_1, z_2 as follows:

$$R_1(z_1, z_2) = (\bar{z}_1, -\bar{z}_2), \tag{37}$$

$$R_2(z_1, z_2) = (-\bar{z}_1, \bar{z}_2). \tag{38}$$

It is shown in [10] (see also [9]) that a suitable normal form for the reversible $\mathbb{C}^2$-reduced system (36) is

$$\begin{aligned} \dot{z}_1 &= \mathrm{i}\omega z_1 + z_2 + \mathrm{i}z_1\, P(z_1\bar{z}_1, \tfrac{1}{2}\mathrm{i}(z_1\bar{z}_2 - \bar{z}_1 z_2)), \\ \dot{z}_2 &= \mathrm{i}\omega z_2 + \mathrm{i}z_2\, P(z_1\bar{z}_1, \tfrac{1}{2}\mathrm{i}(z_1\bar{z}_2 - \bar{z}_1 z_2)) + z_1\, Q(z_1\bar{z}_1, \tfrac{1}{2}\mathrm{i}(z_1\bar{z}_2 - \bar{z}_1 z_2)), \end{aligned} \tag{39}$$

where P and Q are real polynomials in their two arguments satisfying $P(0,0) = Q(0,0) = 0$. This normal form is valid up to arbitrary order N, where the polynomials P and Q are of correspondingly high degree.

Notice that the arguments of P and Q in (39) are invariant under both R_1 and R_2. As a result the normal form is immediately seen to preserve reversibility, i.e., the vector field defined by (39) anti-commutes with the reflections R_1 and R_2. In addition, the normal form is rotationally invariant, i.e., the vector field commutes with

$$R_\phi : (A, B) \to (A\mathrm{e}^{\mathrm{i}\phi}, B\mathrm{e}^{\mathrm{i}\phi}), \tag{40}$$

for any real ϕ. Noting that

$$R_\phi R_i = R_i R_{-\phi} \quad (i = 1, 2), \tag{41}$$

we conclude that we have the group O_2 acting on (39). As a further property, it was noted in [17] that the reversible Hamiltonian–Hopf normal form (39) is integrable (up to any fixed order). So, the original non-integrable $\mathbb{R}^4$-reduced system can be approximated as closely as we wish, near the origin, by an integrable one.

Expanding P and Q according to

$$P(u, v) = p_2\, u + p_3\, v + \mathrm{O}((|u| + |v|)^2), \qquad Q(u, v) = q_2\, u + q_3\, v + \mathrm{O}((|u| + |v|)^2), \tag{42}$$

we are now in a position to apply a result of Iooss and Pérouème [18] stating that at the Hamiltonian–Hopf bifurcation of a four-dimensional reversible dynamical system we generically have the subcritical bifurcation of a pair of symmetric homoclinic orbits provided that

$$q_2 < 0. \tag{43}$$

Moreover, these orbits persist for the full reduced system (29). Since we have a doubly reversible system we have to expect, if (43) holds, the bifurcation of four homoclinic orbits, two for each of R_1 and R_2, which are mapped into each other under Z_1.

A straightforward calculation, using Maple, yields for our system (36) in the symmetric case:

$$\begin{aligned} p_2 &= \tfrac{1}{16}(2\nu^2 + 3\nu + 1), \qquad p_3 = \tfrac{1}{12}(12\nu^3 + 16\nu^2 + 3\nu - 1), \\ q_2 &= -\tfrac{1}{8}(\nu + \tfrac{1}{2})^2, \qquad q_3 = \tfrac{1}{8}(-2\nu^2 + \nu + 1), \end{aligned} \tag{44}$$

showing that indeed $q_2 < 0$ for all acceptable values of ν.

In the general case the analogous expressions become rather forbidding and we have to content ourselves with a graphical illustration. Thus we find the four cubic normal form coefficients at the Hamiltonian–Hopf bifurcation to behave, under varying ρ, as shown in Fig. 4. The determining coefficient q_2 is negative for all $\rho < \rho_c$. All coefficients rapidly diverge as we approach the codimension-two point at ρ_c, q_2 to $-\infty$, the other three to $+\infty$.

The four expected homoclinic orbits are located numerically in [13]. Numerical continuation shows that in the ρ–m parameter plane (cf. Fig. 1) they keep existing in the entire region where we have a complex quadruple of eigenvalues, implying spatial chaos in that same region.

4.4. A normal form analysis at the reversible Hamiltonian-pitchfork bifurcation

Next we concentrate on the m_{c_3}-part of the buckling line with reversible Hamiltonian-pitchfork bifurcations. Note that a pitchfork bifurcation is not generic in a general (one-parameter) Hamiltonian system. It is, however, in our present context of a *reversible* Hamiltonian system.

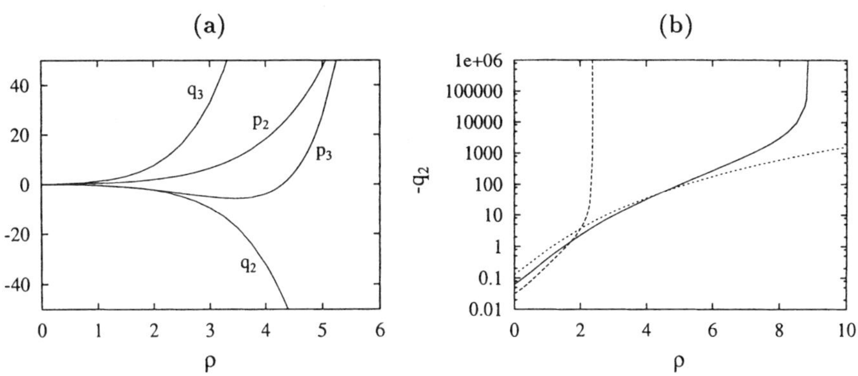

Fig. 4. (a) Cubic normal form coefficients against ρ for $\nu = 0.2$. (b) Logarithmic plot of the coefficient q_2 against ρ for $\nu = 0$ (dashed line), 0.2 (solid line) and 0.5 (dotted line).

The four-dimensional reduced system (29) now has eigenvalues

$$0,\ 0,\ \pm\lambda, \qquad \lambda = \sqrt{-\alpha^2 - \beta_2^\dagger} > 0, \tag{45}$$

where $\beta_2^\dagger$ is the β_2 defined in (14) taken at buckling, i.e., with m^2 solved from the Hamiltonian-pitchfork condition (19). A basis of generalised eigenvectors is given by

$$v_1 = A\,v_2 = \begin{pmatrix} \beta_2^\dagger \\ 0 \\ 0 \\ 0 \end{pmatrix}, \qquad v_2 = \begin{pmatrix} 0 \\ \beta_2^\dagger \\ \alpha \\ 0 \end{pmatrix}, \qquad v_{3,4} = \begin{pmatrix} \alpha \\ \pm\lambda\alpha \\ \mp\lambda \\ -\lambda^2 \end{pmatrix}, \tag{46}$$

satisfying

$$\begin{aligned} &R_1\,v_1 = -v_1, \quad R_1\,v_2 = v_2, \quad R_1\,v_3 = -v_4, \quad R_1\,v_4 = -v_3, \\ &R_2\,v_1 = v_1, \quad R_2\,v_2 = -v_2, \quad R_2\,v_3 = v_4, \quad R_2\,v_4 = v_3. \end{aligned} \tag{47}$$

Transforming to this basis normalises the first-order terms:

$$\begin{aligned} &\dot{x}_1 = x_2 + g_1(x_1, x_2, x_3, x_4), \qquad \dot{x}_2 = g_2(x_1, x_2, x_3, x_4), \\ &\dot{x}_3 = \lambda x_3 + g_3(x_1, x_2, x_3, x_4), \qquad \dot{x}_4 = -\lambda x_4 + g_4(x_1, x_2, x_3, x_4), \\ &\quad g_1, g_2, g_3, g_4 = \mathrm{O}((|x_1| + |x_2| + |x_3| + |x_4|)^3). \end{aligned} \tag{48}$$

Next we can perform a centre manifold reduction to a two-dimensional system. Because there are cubic terms in the first two equations of (48) not involving either x_3 or x_4, the tangent space approximation obtained by just putting $x_3 = x_4 = 0$ suffices. The result is

$$\begin{aligned} &\dot{x}_1 = x_2 + h_1(x_1, x_2), \qquad h_1(x_1, x_2) = g_1(x_1, x_2, 0, 0), \\ &\dot{x}_2 = h_2(x_1, x_2), \qquad h_2(x_1, x_2) = g_2(x_1, x_2, 0, 0). \end{aligned} \tag{49}$$

Reversibility can now be used to argue that we can assume $h_1 \equiv 0$. First note that reversibility with respect to R_1 implies that h_1 is even and h_2 odd in x_1, while reversibility with respect to R_2 implies that h_1 is odd and h_2 even in x_2. Now set $h_1 = x_2\,\bar{h}_1$ and transform the equations according to $x_2 \to x_2(1 + \bar{h}_1)$. This yields a system in the form

$$\dot{x}_1 = x_2, \qquad \dot{x}_2 = h(x_1, x_2), \qquad h = \mathrm{O}((|x_1| + |x_2|)^3), \tag{50}$$

where h is odd in x_1 and even in x_2. A final normalisation of the cubic terms yields the normal form

$$\dot{x}_1 = x_2, \qquad \dot{x}_2 = -p\,x_1^3 \quad (p > 0). \tag{51}$$

The dependence of the coefficient p on the parameters ν and ρ is shown graphically in Fig. 5. For fixed ν, p sharply diverges as ρ_c is approached and also diverges for large ρ.

A normal form for the centre-manifold-reduced system away from the buckling line, but still close to buckling, is obtained by inserting an unfolding parameter μ as follows:

$$\dot{x}_1 = x_2, \qquad \dot{x}_2 = \mu\,x_1 - p\,x_1^3. \tag{52}$$

Compatibility with the eigenvalue configurations of Fig. 1 requires that

$$\mu = q\,(m - m_{c_3}) + \mathrm{O}((m - m_{c_3})^2), \quad \text{with } q < 0, \tag{53}$$

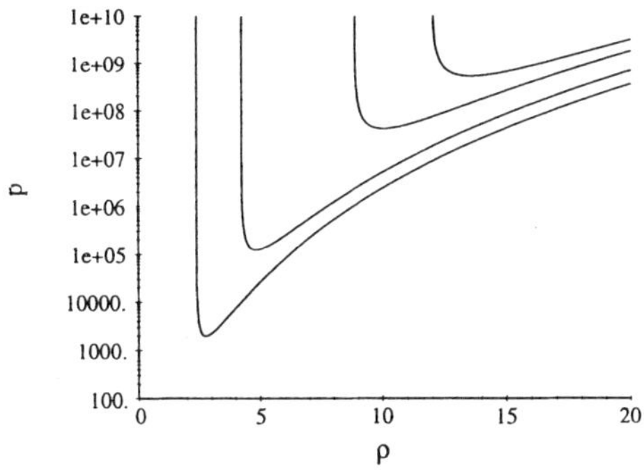

Fig. 5. Normal form coefficient p of the reduced cubic oscillator system describing the tape-like buckling across the Hamiltonian-pitchfork line m_{c_3}. The four curves are for, from bottom to top, $\nu = 0, 0.1, 0.2$ and 0.23.

where the coefficient q will still depend on ν and ρ. The system (52) is, of course, just the unperturbed Duffing oscillator which, for $\mu > 0$, is well known to have a symmetric pair of homoclinic orbits given by [12]

$$x_1^{\pm} = \pm\sqrt{\frac{2\mu}{p}}\,\frac{1}{\cosh\sqrt{\mu}\,t}, \qquad x_2^{\pm} = \mp\mu\sqrt{\frac{2}{p}}\,\frac{\tanh\sqrt{\mu}\,t}{\cosh\sqrt{\mu}\,t}. \tag{54}$$

Furthermore, the system (50) is clearly integrable with integral

$$I = \dot{x}_1^2 - 2\int_0^{x_1} h(s, x_2)\,\mathrm{d}s. \tag{55}$$

Integrability and a simple implicit function argument in [15, Section 4] suffice to prove the persistence of the homoclinic orbits (54) for the full reduced system (49). (To be sure, our doubly reversible system has more symmetry than the singly reversible system considered in [15]. Consequently, the normal forms differ in that our h in (50) does not have quadratic terms. This is why we get a symmetric pair of homoclinic orbits as opposed to the unique homoclinic orbit found in [15]. However, the persistence argument can be straightforwardly adapted to our case.)

4.5. The codimension-two point at ρ_c

From the analysis in Sections 4.3 and 4.4 it is still unclear what happens when we approach the codimension-two point at ρ_c since the coefficients of the codimension-one bifurcations all diverge. Let us, therefore, have a separate look at this case.

At $\rho = \rho_c$ along the buckling line all eigenvalues are 0. A basis of generalised eigenvectors is given by

$$v_1 = A\,v_2 = \begin{pmatrix} -\alpha^{*2} \\ 0 \\ 0 \\ 0 \end{pmatrix}, \quad v_2 = A\,v_3 = \begin{pmatrix} 0 \\ -\alpha^{*2} \\ \alpha^* \\ 0 \end{pmatrix}, \quad v_3 = A\,v_4 = \begin{pmatrix} 1 \\ 0 \\ 0 \\ \alpha^* \end{pmatrix}, \quad v_4 = \begin{pmatrix} 0 \\ 1 \\ 0 \\ 0 \end{pmatrix}, \tag{56}$$

where α^* is the α in (14) at critical ρ, i.e., $\rho = \rho_c$ with ρ_c solved from (20). On this basis the linearised vector field A and the reflections R_1 and R_2 take the form

$$A = \begin{pmatrix} 0 & 1 & 0 & 0 \\ 0 & 0 & 1 & 0 \\ 0 & 0 & 0 & 1 \\ 0 & 0 & 0 & 0 \end{pmatrix}, \qquad R_1 = \mathrm{diag}(-1, 1, -1, 1), \qquad R_2 = \mathrm{diag}(1, -1, 1, -1). \tag{57}$$

The normal form analysis of this codimension-two case was essentially carried out by Iooss [14]. (More precisely, he considered the case with only the reflection R_2 present.) He showed that the codimension-one normal forms can be rescaled such that the divergence of the coefficients is suppressed. It then follows that all the codimension-one results remain valid right up to the codimension-two point. More precisely, in the parameter plane they hold in horn-like regions about the codimension-one curves bounded by two curves tangent to the codimension-one curve at the codimension-two point.

5. Linearised buckling modes

By considering the infinitesimal buckling modes, i.e., solutions of the linearised equations (16), we identify the homoclinic orbits found in Section 4 in terms of the localised buckling modes computed numerically in [13]. Having done this we compute, in this and in Section 6, some physical properties of the rod immediately after buckling. In particular, we introduce the (average) number of twists per wave for a helically deformed rod, which might be accessible to carefully conducted experiments. For symmetric rubber rods such experiments are described in [25].

Throughout the main part of this section we will tacitly assume that $\nu < \frac{1}{3}$, i.e., that we have a critical ρ (ρ_c), although most of the analysis for $\rho < \rho_c$ also holds for $\nu \geq \frac{1}{3}$, if we pretend ρ_c to lie at infinity. There are, however, also differences, especially some limiting behaviour as $\rho \to \rho_c$. We will have more to say about this near the end of this section.

5.1. *Mildly anisotropic rods: $\rho < \rho_c$*

From the eigenvector in (32) it follows that real solutions to (16) can be written

$$\begin{aligned} x(t) &= a\cos\omega t + b\sin\omega t = r\cos(\omega t + \eta), \quad (r > 0), \\ y(t) &= a\gamma\sin\omega t - b\gamma\cos\omega t = r\gamma\sin(\omega t + \eta), \quad (r > 0) \end{aligned} \tag{58}$$

where the constants a and b, or, alternatively, r and η are to be determined by initial conditions. Note that in (58) we have ignored unbounded solutions of the form $t\cos\omega t$ and $t\sin\omega t$, which are not acceptable since we are interested in precursors to homoclinic orbits of the full nonlinear system.

As to the symmetries discussed in Section 4 we note that for the linearised system, pairs of R_1 or R_2-related solutions are simply arclength-translated copies of sines/cosines. The only solutions which are themselves reversible are given by

$$x_i = r\cos\left(\omega t + \frac{i-1}{2}\pi\right), \quad y_i = r\sin\left(\omega t + \frac{i-1}{2}\pi\right), \qquad i = 1, 2, 3, 4, \tag{59}$$

where x_1, x_3 (resp., x_2, x_4) form a Z_1-pair of R_2 (resp., R_1) reversible solutions.

In [13] we found, for mildly anisotropic rods ($\rho \leq \rho_c$), in the full nonlinear system, a Z_1-pair P_1/P_2 of R_1-reversible and a Z_1-pair P_3/P_4 of R_2-reversible single-hump homoclinic orbits constituting four so-called primary localised buckling modes for which the solutions in (59) can be regarded as the linearised versions. The precise correspondence between the (x_i, y_i) and the P_i can be ascertained by a comparison of (59) with numerically computed homoclinic orbits close to buckling. We then find the following pairing:

$$(x_1, y_1) \longleftrightarrow P_3, \quad (x_2, y_2) \longleftrightarrow P_1, \quad (x_3, y_3) \longleftrightarrow P_4, \quad (x_4, y_4) \longleftrightarrow P_2. \tag{60}$$

Given the solutions $x(t)$ and $y(t)$ of the linearised system, the transformation (10) can be used to obtain, for the incipient buckling modes, the Euler angles and their derivatives as a function of arclength t. From them other interesting quantities can be computed. For instance, the actual spatial configuration of an infinitesimal buckling mode can be obtained by integrating the (non-dimensional) centre line equation $\dot{\tilde{r}} = d_3$, which, on writing $\tilde{r} = (x_c, y_c, z_c)$ and applying (1), has the componential form

$$\begin{aligned} \dot{x}_c &= \cos\psi \sin\theta = x\cos\chi + \sqrt{1+\rho}\, y \sin\chi, \\ \dot{y}_c &= \sin\psi \sin\theta = x\sin\chi - \sqrt{1+\rho}\, y\cos\chi, \\ \dot{z}_c &= \cos\theta = 1, \end{aligned} \tag{61}$$

where χ is given by (15). For the mode P_3 the first two equations become

$$\begin{aligned} \dot{x}_c &= \tfrac{1}{2} r[(1+\gamma\sqrt{1+\rho})\cos(\dot{\chi}-\omega)t + (1-\gamma\sqrt{1+\rho})\cos(\dot{\chi}+\omega)t], \\ \dot{y}_c &= \tfrac{1}{2} r[(1+\gamma\sqrt{1+\rho})\sin(\dot{\chi}-\omega)t + (1-\gamma\sqrt{1+\rho})\sin(\dot{\chi}+\omega)t], \end{aligned} \tag{62}$$

showing that the linear buckling mode is quasi-periodic with frequencies $\dot{\chi} - \omega$ and $\dot{\chi} + \omega$. In the symmetric limit the second terms in Eqs. (62) vanish and we get a true helix of wavelength 4π described by

$$\dot{x}_c = r\cos\tfrac{1}{2}t, \quad \dot{y}_c = r\sin\tfrac{1}{2}t \qquad (\rho = 0). \tag{63}$$

In the limit $\rho \to \rho_c$ we also arrive at a true helix; this time as a result of $\omega \to 0$, rendering the two frequencies identical:

$$\dot{x}_c = r\cos\dot{\chi}t, \quad \dot{y}_c = r\sin\dot{\chi}t \qquad (\rho \to \rho_c). \tag{64}$$

For intermediate values $0 < \rho < \rho_c$ the infinitesimal buckling mode becomes modulated by the interplay of the frequencies $\dot{\chi} - \omega$ and $\dot{\chi} + \omega$. Side views of a P_1 and a P_3-type infinitesimal buckling mode are depicted in Fig. 6

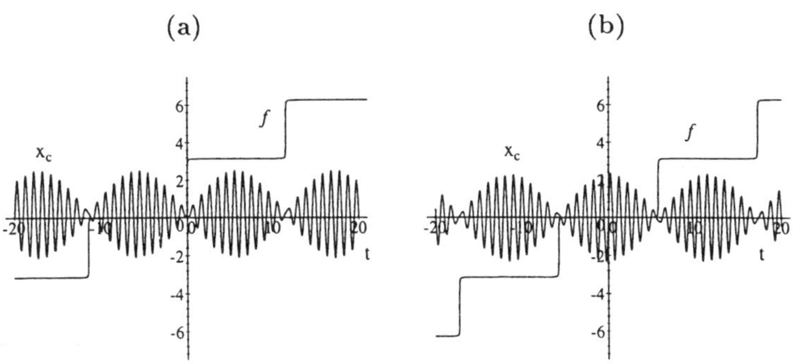

Fig. 6. A comparison of side views and internal twists f (defined in Section 6) of the buckling modes P_1 (a) and P_3 (b). ($\rho = 8.75$, $\nu = 0.2$, implying $\rho_c = 8.870676$.)

(type P_2 and P_4 shapes are obtained by reversing the signs of x_c and y_c). Close to ρ_c the buckling modes develop a pinched modulation with a wavelength that diverges as ρ_c is approached. The amplitude of the modes at the pinched points goes to zero, as follows from (62):

$$x_c^2 + y_c^2 = P^2 + Q^2 \cos^2 \omega t \quad \text{with } P \to 0, \ Q \to \frac{r}{\dot{\chi}} \qquad \text{as } \rho \to \rho_c. \tag{65}$$

We can define the (local) number of twists per wave n as the ratio of the speed of χ (the twist) and the speed of the azimuthal angle ψ (determining the 'local' wavelength):

$$n := \frac{\dot{\chi}}{\dot{\psi}} = \frac{(1+\nu)(2+\rho)}{(1+\nu)(2+\rho) - 2\dot{\phi}}, \tag{66}$$

where, from (10),

$$\dot{\phi} = \frac{\sqrt{1+\rho}\,(x\dot{y} - y\dot{x})}{x^2 + (1+\rho)y^2}. \tag{67}$$

Note that it follows from (35) that in the limit of a symmetric rod we have a helix with

$$\dot{\phi} = \tfrac{1}{2} + \nu, \ \ \dot{\psi} = \dot{\chi} - \dot{\phi} = \tfrac{1}{2} \ \text{ and hence } \ n = 2(1+\nu) \qquad (\rho = 0). \tag{68}$$

Of course, n is in general not a constant along the rod. We can, however, define its average

$$\bar{n} = \frac{\omega}{\pi} \int_{-\pi/2\omega}^{\pi/2\omega} n(t)\,\mathrm{d}t, \tag{69}$$

which is the same for all buckling modes. It can, in fact, be computed explicitly:

$$\bar{n} = \frac{1}{\pi} \int_{-\pi/2}^{\pi/2} \frac{1 - B\cos^2 t}{A + B\sin^2 t}\,\mathrm{d}t = 1 + \left(\frac{1}{\sqrt{A+B}} - \sqrt{A+B}\right)\frac{1}{\sqrt{A}}, \tag{70}$$

where

$$A = \gamma^2(1+\rho) - \frac{2\omega\gamma\sqrt{1+\rho}}{(1+\nu)(2+\rho)}, \qquad B = 1 - \gamma^2(1+\rho). \tag{71}$$

A and B are positive for all ν and ρ. Plots of $\bar{n}$ as a function of ρ for several values of ν are given in Fig. 7(a). Note that in the limit $\rho \to \rho_c$ we have $A \to 0$, $B \to 1$, and hence $\bar{n} \to 1$. Thus, the average number of twists per wave is seen to gradually change from $2(1+\nu)$ to 1 in going from the symmetric ($\rho = 0$) to the critically anisotropic ($\rho = \rho_c$) rod.

5.2. *Strongly anisotropic rods: $\rho \geq \rho_c$*

In this case the eigenvalues are $0, 0, \pm\lambda (\lambda \in \mathbb{R})$. Again demanding that solutions be bounded for all $t \in \mathbb{R}$ one deduces from the eigenvector v_1 in (46) that the only acceptable solution is the constant one

$$x(t) = a, \qquad y(t) = 0, \tag{72}$$

which immediately yields

$$\dot{\phi} = 0, \quad \text{whence } n = 1. \tag{73}$$

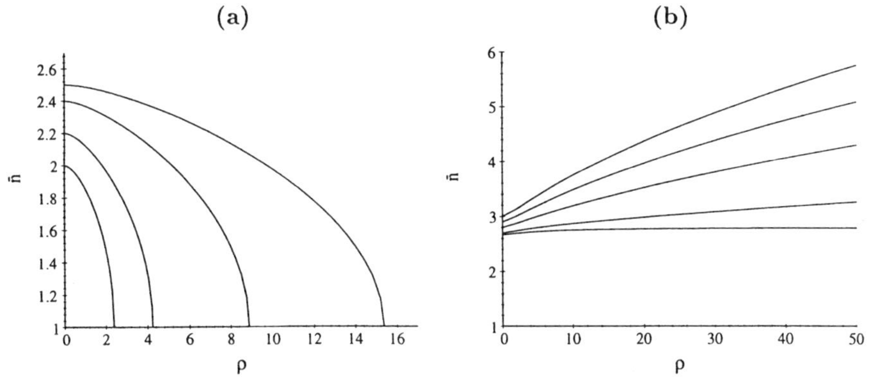

Fig. 7. Average number of twists per wave $\bar{n}$ (valid for all linear buckling modes) as a function of ρ for several values of ν. (a) $\nu < \frac{1}{3}$: $\nu = 0, 0.1, 0.2$ and 0.25, with $\rho_c = 2.382976, 4.233098, 8.870676$ and 15.368498, respectively. (b) $\nu \geq \frac{1}{3}$: $\nu = \frac{1}{3}$, 0.35, 0.4, 0.45 and 0.5 (from bottom to top).

So, n, which was found to drop to 1 at ρ_c, stays 1 at larger ρ. We conclude that if $\nu < \frac{1}{3}$ then there is a critical aspect ratio beyond which the rod locks on to a one-twist-per-wave buckling mode. This infinitesimal buckling mode is a true helix of wavelength $2\pi/\dot{\chi}$, as follows from the centre line equation (61).

An important observation to make is that for $\rho > \rho_c$ the linearised equation clearly only allows even solutions $x(t)$, thus excluding an incipient P_1 or P_2-type buckling mode. This is, of course, consistent with the change of multiplicity of homoclinic orbits found in the nonlinear analysis of Section 4. The disappearance of the modes P_1/P_2 at ρ_c can also be appreciated by imagining what happens when we let ρ in Fig. 6 tend to ρ_c, thus increasing the modulation wavelength. Whereas the mode P_3 becomes a helix, P_1 is squeezed out of existence altogether.

5.3. The case $\nu \geq \frac{1}{3}$

For $\nu \geq \frac{1}{3}$ the whole buckling line consists of Hamiltonian–Hopf bifurcations. The infinitesimal buckling modes are, therefore, given by (58) for all ρ. The behaviour of ω and γ, however, is different now. Instead of the last line in (35) we have the limiting behaviour (cf. Fig. 3):

$$\omega \to \infty, \quad \gamma \to 0, \qquad \text{as } \rho \to \infty. \tag{74}$$

More precisely, the following asymptotics can be established:

$$\begin{aligned} \nu > \frac{1}{3}: &\qquad \omega \sim \rho, \ \ \gamma \sim \frac{1}{\sqrt{\rho}} \quad (\rho \to \infty), \\ \nu = \frac{1}{3}: &\qquad \omega = \frac{1}{6}\sqrt{26}\,\sqrt{\rho} + \mathrm{O}\left(\frac{1}{\sqrt{\rho}}\right), \ \ \gamma = \frac{1}{4}\sqrt{26}\,\frac{1}{\rho} + \mathrm{O}\left(\frac{1}{\rho^2}\right) \quad (\rho \to \infty). \end{aligned} \tag{75}$$

This means that θ gets an increasingly more rapid modulation as ρ is increased. The symmetry properties of the solutions are then relatively unimportant and all infinitesimal buckling modes behave the same in the large-ρ limit, winding more rapidly while shrinking in amplitude.

The function $\dot{\phi}(t)$ also acquires unbounded frequency, while at the same time diverging for all t. As a result, for large ρ, the number of twists per wave, $n(t)$, becomes a very spiky function. Its average, $\bar{n}$, although of limited use, remains well-defined and is still given by (70). It behaves as illustrated in Fig 7(b). For $\nu > \frac{1}{3}$, A tends to 0 in the limit $\rho \to \infty$ while B remains bounded away from 1; hence $\bar{n}$ diverges. At the critical value of Poisson's ratio, $\nu = \frac{1}{3}$, $A + B$ tends to 1 as $\rho \to \infty$, rendering the second term in (70) finite. Thus, $\bar{n}$ is found to approach

$$\lim_{\rho \to \infty} \bar{n} = 1 + \tfrac{1}{2}\sqrt{13} = 2.802775\ldots \qquad (\nu = \tfrac{1}{3}), \tag{76}$$

as confirmed by Fig. 7(b).

6. Cross-sectional orientation

In Section 5 we have seen that the initial post-buckling modes P_1, P_3, P_2, P_4, in that order, have successive phase differences of 90° along the rod (cf. (59)). Only the modes P_3/P_4 were found to survive for $\rho > \rho_c$. In this section we try to explain this selection physically by looking at the cross-sectional orientation of the four modes. For this we introduce the notion of internal twist. As first noted by Love [20], the (non-dimensional) twist $\tilde{\tau}$ can be decomposed into the internal twist $\dot{f}$ (the twist of Love) and the torsion of the centre line of the rod, $\tilde{\tau}_s$:

$$\tilde{\tau} = \frac{\mathrm{d}f}{\mathrm{d}t} + \tilde{\tau}_s. \tag{77}$$

The angle f measures, in a cross-section of the rod, the relative rotation of the normal $\boldsymbol{n}$ and binormal $\boldsymbol{b}$ to the rod's centre line with respect to the local co-ordinate frame $\{\boldsymbol{d}_1, \boldsymbol{d}_2\}$. Specifically, the two co-ordinate systems are related according to

$$\boldsymbol{d}_1 = \sin f\, \boldsymbol{n} - \cos f\, \boldsymbol{b}, \qquad \boldsymbol{d}_2 = \cos f\, \boldsymbol{n} + \sin f\, \boldsymbol{b}. \tag{78}$$

The value of f at $t = 0$ (the centre of the rod) is easily obtained. Expressing the linear solutions (59) in terms of Euler angles by using (10), and substituting the result in the strain equations (2), we find

$$\begin{aligned} P_1:&\quad \tilde{\kappa}_1(0) = 0, \quad \tilde{\kappa}_2(0) = r\gamma\sqrt{1+\rho}\left(\frac{1}{2}(1+\nu)(2+\rho) - \frac{\omega}{\gamma\sqrt{1+\rho}}\right) > 0, \\ P_3:&\quad \tilde{\kappa}_1(0) = -r\left(\frac{1}{2}(1+\nu)(2+\rho) - \omega\gamma\sqrt{1+\rho}\right) < 0, \quad \tilde{\kappa}_2(0) = 0. \end{aligned} \tag{79}$$

It can be verified that for P_1, $\tilde{\kappa}_2(0)$ is positive for all values of ν and ρ. Similarly, $\tilde{\kappa}_1(0)$ is always negative for P_3. Recalling that flexure $\tilde{\kappa}_1\, \boldsymbol{d}_1 + \tilde{\kappa}_2\, \boldsymbol{d}_2$ is directed along the binormal of the rod's centre line, a comparison with (78) then establishes (also including the mirror-symmetric modes P_2 and P_4)

$$\begin{aligned} P_1:&\quad f(0) = \tfrac{1}{2}\pi, \quad \boldsymbol{d}_1 = \boldsymbol{n}, \quad \boldsymbol{d}_2 = \boldsymbol{b}, \\ P_2:&\quad f(0) = \tfrac{3}{2}\pi, \quad \boldsymbol{d}_1 = -\boldsymbol{n}, \quad \boldsymbol{d}_2 = -\boldsymbol{b}, \\ P_3:&\quad f(0) = 0, \quad \boldsymbol{d}_1 = -\boldsymbol{b}, \quad \boldsymbol{d}_2 = \boldsymbol{n}, \\ P_4:&\quad f(0) = \pi \quad \boldsymbol{d}_1 = \boldsymbol{b}, \quad \boldsymbol{d}_2 = -\boldsymbol{n}. \end{aligned} \tag{80}$$

The normal $\boldsymbol{n}$ can finally be related to the centre line $\boldsymbol{r}(t)$ by recalling that for any given space curve $\alpha(t)$ the normal is proportional to $\ddot{\alpha}(t)$, the proportionality factor being the inverse of the curvature. From the centre line equations (61) and (62) we then deduce that at $t = 0$, $\boldsymbol{n} = -\mu\, \boldsymbol{r}$ for some positive constant μ. (For a helix the normal $\boldsymbol{n}$ is pointing into the cylinder on which the helix winds *at all times*.) Consequently, projections of the modes

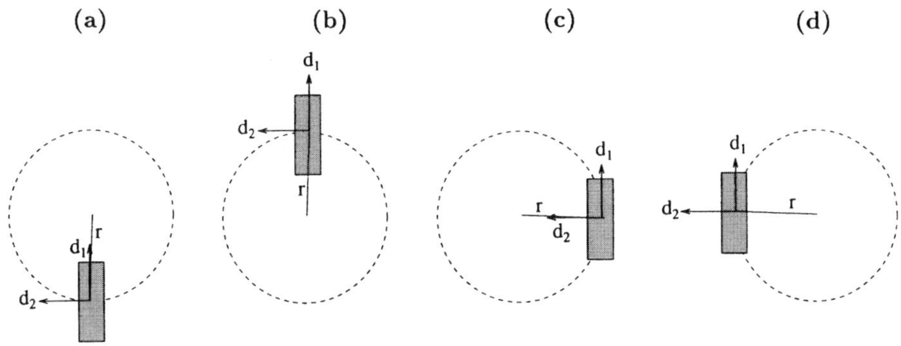

Fig. 8. Sketch of the rod's cross-sectional orientation in the symmetric section (i.e., at $t = 0$) for the primary buckling modes P_1 (a), P_2 (b), P_3 (c) and P_4 (d).

P_1, P_2, P_3 and P_4 onto the x–y plane are as depicted schematically in Fig. 8. In interpreting the cross-sectional orientations in this figure, recall from the discussion after (4) our convention that $a < b$, implying that the rod is elongated in the $\boldsymbol{d}_1$ direction.

Moving away from $t = 0$, for $\rho \leq \rho_c$ the two co-ordinate frames in (78) will rotate with respect to each other. For small ρ the relative speed (the internal twist $\dot{f}$) will be approximately ω. For larger ρ the internal twist becomes gradually more 'intermittent' along the length of the rod. To illustrate this, in Fig. 6 we have superimposed the angle f on a side view of the rod for the buckling modes P_1 and P_3. The angle f remains constant for long intervals and quickly makes a 180° turn at the pinched areas along the rod. Hence, the cross-section is mostly in the energetically favourable 'lying' position of Figs. 8(c) or (d), and adopts the unfavourable 'standing' position of Fig. 8(a) or (b) only during the small interval where the amplitude of the modulated helix is minimal.

Away from the buckling line and the linear regime, the relative phases of the four modes gradually change, although pairs 180° out of phase remain so by virtue of symmetry.

For $\rho > \rho_c$ only P_3 and P_4 exist. An analysis similar to the one leading to (79) yields

$$P_3: \quad \tilde{\kappa}_1(t) = -\tfrac{1}{2}r(1+\nu)(2+\rho), \quad \tilde{\kappa}_2(t) = 0, \quad \text{for all } t \in \mathbb{R} \ (r > 0), \tag{81}$$

where we have chosen $x = r$ positive in (72) to assure continuity of the solution P_3 across the critical point ρ_c (cf. (59)). It follows that $\boldsymbol{d}_1 = -\boldsymbol{b}$, $\boldsymbol{d}_2 = \boldsymbol{n}$, and $f \equiv 0$. For P_4 (with $x = -r$) we find $\boldsymbol{d}_1 = \boldsymbol{b}$, $\boldsymbol{d}_2 = -\boldsymbol{n}$ and $f \equiv \pi$. For both modes $\dot{f} = 0$, i.e., there is no internal twist.

Thus we see that of the four buckling modes with 90° phase differences existing in the mildly anisotropic regime only the two energetically favourable 'flat' modes survive into the strongly anisotropic regime.

7. Discussion

We have seen that in breaking the rotational symmetry, by considering rods with different bending stiffnesses in two orthogonal directions, the circle of homoclinic orbits present in the symmetric case breaks up and we are left with four isolated primary homoclinic orbits. For mildly anisotropic rods these orbits are born at buckling with their cross-sections roughly at right angles to each other. Due to internal twist the cross-sections rotate along the length of the rod. In the strongly anisotropic regime the rod buckles into a one-twist-per-wave helical mode with

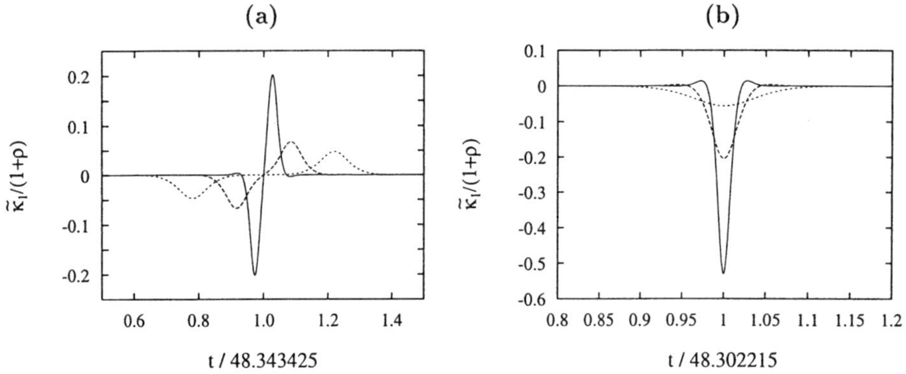

Fig. 9. Behaviour of the primary solutions P_1 (a) and P_3 (b) under an increase of m while $\rho > \rho_c$. The mode P_1 is shown at $m = 0.63$ (solid curve), $m = 0.6337$ (dashed curve) and $m = 0.633894$ (dotted curve), and dies at $m_{c_1} = 0.633918$; the mode P_3 is shown at $m = 0.6$ (solid curve), $m = 0.63$ (dashed curve) and $m = 0.633818$ (dotted curve), and dies at $m_{c_3} = 0.634072$ on the buckling line. ($\nu = 0.2$, $\rho = 10.0$.)

no internal twist. This means that the rod behaves as a true tape already at finite values of its cross-sectional aspect ratio. However, now only the energetically favourable pair of flat tape configurations occurs.

This raises the question what happens to the four localised post-buckling solutions existing for $\rho < \rho_c$ if we increase ρ beyond the critical point. To be sure, all four solutions can be continued into the $\rho > \rho_c$ region of Fig. 1(a) provided that m is sufficiently low, as was demonstrated in [13]. However, we have seen that the rod cannot actually buckle into a P_1 or P_2 mode if $\rho > \rho_c$. So, the P_1 and P_2-type homoclinic orbits must somehow bifurcate when we vary parameters such as to approach the line m_{c_3}. This is indeed observed numerically as illustrated in Fig. 9 contrasting the behaviour of the mode P_1 with that of the mode P_3. Whereas P_3 simply relaxes to the zero solution as a single-hump solution, P_1 splits into two anti-symmetric humps that move off in different directions as m is increased.

From the work in [7,13] we know that besides primary homoclinic orbits there exist also (countably infinite families of) multi-modal homoclinic orbits roughly consisting of multiple copies of primaries. The existence of these solutions is in agreement with the analysis in [4]. They should generically be expected to exist in the entire region of the ρ–m parameter plane where we have a complex quadruple of eigenvalues, provided the invariant manifolds of the fixed point intersect transversally. This latter property is proved (at least asymptotically for small ρ) in [23]. We have not considered these multi-modals in this paper because they do not bifurcate at the buckling line. Fig. 9, however, shows a connection between primary and bi-modal homoclinic orbits. The bi-modal, evolving from the P_1 primary, dies at the line m_{c_1} (for $\nu = 0.2$, $\rho = 10.0$ we have $m_{c_1} = 0.633918$). Interestingly, close to m_{c_1} this solution shows flat tape behaviour, with the rod's cross-sections at the two humps 180° rotated with respect to each other, as dictated by symmetry.

There is an interesting analogy between the problem of localised buckling of twisted anisotropic rods and the problem of steady water waves in the presence of gravity and surface tension [1,15,19]. In the latter problem the two-parameter bifurcation plane of Bond versus Froude number for the trivial equilibrium is qualitatively similar to Fig. 1(a), including the codimension-two point at $\rho = \rho_c$ (corresponding to Bond number $\frac{1}{3}$ and Froude number 1). The difference between the two problems is that the twisted rod problem has an additional reversibility. Thus the unique (up to translation) solitary wave solution of depression found by Amick and Kirchgässner [1] corresponds in our problem to the symmetric pair of Duffing-type homoclinic orbits (54) bifurcating from the origin upon crossing the m_{c_3}-part of the buckling line (and essentially depicted in Fig. 9(b)).

We would like to end by pointing out that our results on the existence of a transition to lock-on behaviour are valid for a much wider class of objects than the solid elliptical rods considered in the preceding sections. Indeed, solid rods form just one example of a class of structures with parametrised bending and torsional stiffnesses (the parameters being ρ and ν as defined in (9)). By tracing back our condition for criticality (20) in terms of the B_1, B_2 and C we have a criterion for lock-on valid for all structures with well-defined stiffnesses about the axes of the local co-ordinate frame $\{\boldsymbol{d}_1, \boldsymbol{d}_2, \boldsymbol{d}_3\}$.

A particularly nice and compact geometric representation of the potential for a lock-on transition in a given class of structures can be obtained as follows. Because we are interested in the ratios $B_1 : B_2 : C$ rather than the actual values of the individual stiffnesses we may impose the condition

$$B_1 + B_2 + C = 1. \tag{82}$$

Now this situation can be conveniently depicted by using so-called trilinear co-ordinates with respect to an equilateral triangle (see Fig. 10, top left). In the plane of such a triangle B_1B_2C of given side the trilinear co-ordinates of an arbitrary point P are defined to be the (perpendicular) distances b_1, b_2, c from P to the three sides B_2C, B_1C, B_1B_2, regarded as positive when P is inside the triangle (since in our case B_1, B_2 and C are non-negative we can restrict ourselves to points inside the triangle).

The above construction relies on the property

$$b_1 + b_2 + c = h, \tag{83}$$

where h is the altitude of the equilateral triangle. This property is easily proved by considering the areas of triangular regions defined by the point P and the vertices B_1, B_2 and C.

So we have that each point inside the triangle corresponds to a set of structures with certain stiffness ratios, and vice versa. The top right of Fig. 10 shows the stiffness triangle again with two shaded areas bounded by curves labelled L. The bounding curves correspond to the critical lock-on condition (20). There are two such curves because of reflectional symmetry of the diagram with respect to the altitude line from C. This is related to the freedom to choose either B_1 or B_2 to be the larger of the two bending stiffnesses. Our convention in this paper has been $B_2 \geq B_1$, but we have included the curve in the left part of the diagram for purposes of generality. Structures with stiffness ratios landing in the shaded regions of the triangle will show lock-on behaviour at buckling.

Our parametrised family of solid elliptical rods is represented in the stiffness diagram by a set of curves running from the B_1 to the B_2 vertex (corresponding to $-1 \leq \rho < \infty$). We have drawn three examples in the bottom left triangle of Fig. 10 corresponding to the critical case $\nu = \frac{1}{3}$ and the two extremes $\nu = 0$ and $\nu = \frac{1}{2}$, the latter two serving to indicate the region of the triangle occupied by solid elliptical rods. The line for $\nu = \frac{1}{3}$ is exactly tangent to the critical lines L at B_1 and B_2. For $\nu < \frac{1}{3}$ the solid rod lines eventually enter the critical region, implying that the corresponding buckling will be tape-like. (Lines for rectangular rods were found to look qualitatively the same.)

The shaded triangular region B_2ED in the bottom right of Fig. 10 is not entered by the solid elliptical (or rectangular) rod lines, and hence the associated behaviour is not physical for such rods. This behaviour is displayed, however, in the (hypothetical) family of structures considered in [13] (giving rise to straight lines in the stiffness triangle, one of which is indicated in Fig. 10) and it would in our case if we were to take $\nu < 0$. What happens upon entering the region B_2ED is that by varying m we can no longer get double real eigenvalues (as on the branches m_{c_1} in Fig. 1). The mechanism responsible for this is a fold occurring along the line m_{c_1} integrating the two separate branches in Fig. 1(a) into one single curve. Recalling that infinities of multi-modal buckling solutions exist in the region of the ρ–m parameter plane where we have a complex quadruple of eigenvalues [13], this means that in the B_2ED region we have no spatial chaos whereas outside of this region we do. Obviously, the fold causing this

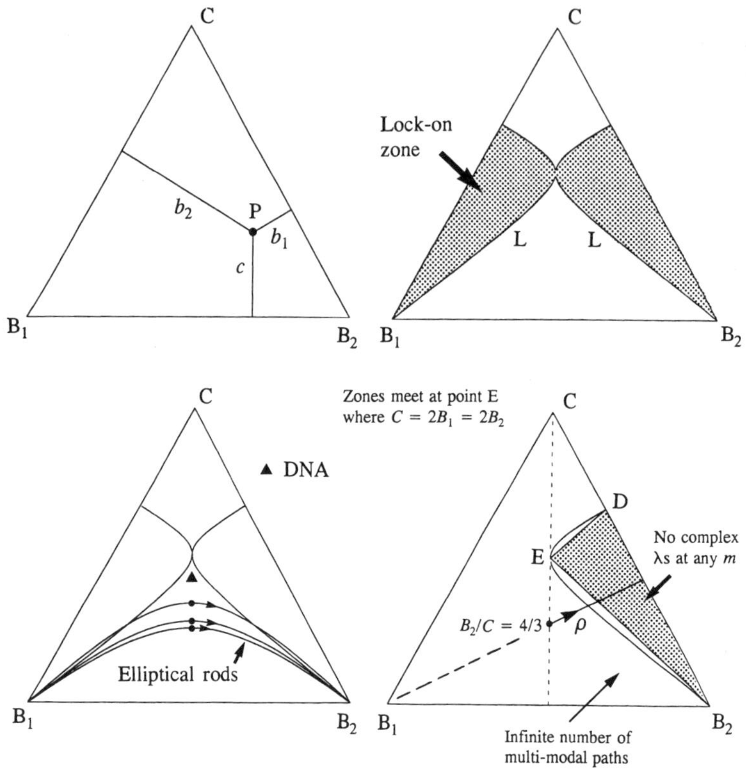

Fig. 10. The stiffness triangle as a means of classifying buckling behaviour in arbitrary structures. The three solid elliptical rod lines are for (from top to bottom) $\nu = 0$, $\frac{1}{3}$ and $\frac{1}{2}$. See main text for further explanation.

change of behaviour can only occur if a ρ_c is also present, as is nicely confirmed in Fig. 10 where the triangular region B_2ED lies entirely within the lock-on zone.

Lines for parametrised families of structures other than solid rods can be drawn in Fig. 10 and may occupy completely different regions of the triangle. Of course, one is not restricted to parametrised families of objects. Individual values of B_1, B_2 and C obtained from measurements on a certain object can also be directly included. In all these cases the stiffness triangle offers a quick way of ascertaining whether a given object is likely to show lock-on behaviour when subjected to twisting moments. For instance, the bottom left triangle in Fig. 10 shows a data point corresponding to the relations $C = 1.5B_1 = 1.5B_2$, which were used by Manning et al. [22] in a continuum rod model for 150-base-pair DNA sequences. Their results were found to compare well with results obtained from experiments. (In [24] solid circular rod theory was applied to DNA using a Poisson's ratio of 0, implying $C = B_1 = B_2$.)

Given the analogy between twisted rod statics and rigid body dynamics the stiffness triangle may also be useful in classifying the behaviour of spinning tops. In [21] a linearisation about certain simple solutions of the top (steady-state rotations about the vertical) yielded a parameter plane with bifurcation curves similar to those in our Fig. 1. (Note, though, that for tops curves other than those forming our buckling line will be physically relevant.) These curves can be drawn in the triangle B_1B_2C in Fig. 10, where B_1, B_2 and C now represent the principal moments of inertia.

A final remark concerns the intriguing codimension-three point E, defined by $C = 2B_1 = 2B_2$, where the two symmetric lock-on zones in Fig. 10 touch. Clearly, this point, which acts as the central singularity in our three-parameter family of reversible vector fields, is not physical for elliptical rods. In the parametrisation (9) it has $\nu = -\frac{1}{2}$, $\rho_c = 0$, $m_c = 2$, and would correspond to having the critical lock-on point for the integrable case of a symmetric cross-section. Also note that for these parameter values all four Hamiltonian–Hopf normal form coefficients (44) vanish. Since, in addition, $\alpha = \beta_1 = \beta_2 = 0$, the linearised vector field A is already in normal form on the standard basis:

$$A = \begin{pmatrix} 0 & 1 & 0 & 0 \\ 0 & 0 & 0 & 0 \\ 0 & 0 & 0 & 1 \\ 0 & 0 & 0 & 0 \end{pmatrix}. \tag{84}$$

In the complex variables $z_1 = x + \mathrm{i}\, y$, $z_2 = \dot{x} + \mathrm{i}\, \dot{y}$ the vector field F_r has the form (up to cubic terms)

$$\dot{z}_1 = z_2, \qquad \dot{z}_2 = -\tfrac{1}{8} z_1 |z_1|^2 + \tfrac{1}{2} z_2 (\mathrm{i}|z_1|^2 - z_1\bar{z}_2 - \bar{z}_1 z_2), \tag{85}$$

the other two equations just being the complex conjugates. The system (85) has O_2-symmetry, as should be expected in view of the fact that $\rho_c = 0$. Normal forms for vector fields with linear part given by (84) can be found in [10] (or [9]) for the general case without any symmetries, and in [16] for the O_2-symmetric case. It is not clear whether there exist parametrised families of structures in which the codimension-three bifurcation actually occurs.

Acknowledgements

The first author was supported by a European Union Human Capital and Mobility grant under no. ERBCHBGCT93-0427.

References

[1] C.J. Amick, K. Kirchgässner, A theory of solitary water-waves in the presence of surface tension, Arch. Rat. Mech. Anal. 105 (1989) 1–49.
[2] S.S. Antman, Nonlinear Problems of Elasticity, Springer, Berlin, 1995.
[3] S.S. Antman, K.B. Jordan, Qualitative aspects of the spatial deformation of non-linearly elastic rods, Proc. Roy. Soc. Edinburgh A 73 (5) (1974) 85–105.
[4] L.A. Belyakov, L.P. Šil'nikov, Homoclinic curves and complex solitary waves, Selecta Math. Soviet. 9 (1991) 219–228.
[5] E. Buzano, Secondary bifurcations of a thin rod under axial compression, SIAM J. Math. Anal. 17 (2) (1986) 312–321.
[6] E. Buzano, G. Geymonat, T. Poston, Post-buckling behaviour of a non-linearly hyperelastic thin rod with cross-section invariant under the dihedral group D_n, Arch. Rat. Mech. Anal. 89 (1985) 307–388.
[7] A.R. Champneys, J.M.T. Thompson, A multiplicity of localised buckling modes for twisted rod equations, Proc. Roy. Soc. London A 452 (1996) 2467–2491.
[8] A.R. Champneys, G.H.M. van der Heijden, J.M.T. Thompson, Spatially complex localisation after one-twist-per-wave equilibria in twisted circular rods with initial curvature, Phil. Trans. Roy. Soc. London A 355 (1997) 2151–2174.
[9] R. Cushman, J. Sanders, Nilpotent normal forms and representation theory of sl (2, $\mathbb{R}$), in: M. Golubitsky, J.M. Guckenheimer (Eds.), Multiparameter bifurcation theory, Contemporary Mathematics Vol. 56, American Mathematical Society, Providence, RI, 1986.
[10] C. Elphick, E. Tirapegui, M.E. Brachet, P. Coullet, G. Iooss, A simple global characterization for normal forms of singular vector fields, Physica D 29 (1987) 95–127.

[11] J.M. Gere, S.P. Timoshenko, Mechanics of Materials, 3rd SI ed., Chapman and Hall, London, 1991.

[12] J. Guckenheimer, P. Holmes, Nonlinear Oscillations, Dynamical Systems and Bifurcations of Vector Fields, Springer, New York, 1983.

[13] G.H.M. van der Heijden, A.R. Champneys, J.M.T. Thompson, The spatial complexity of localised buckling in rods with non-circular cross-section, SIAM J. Appl. Math. (1997), to appear.

[14] G. Iooss, A codimension 2 bifurcation for reversible vector fields, Fields Institute Commun. 4 (1995) 201–217.

[15] G. Iooss, K. Kirchgässner, Water waves for small surface tension: an approach via normal form, Proc. Roy. Soc. Edinburgh A 122 (1992) 267–299.

[16] G. Iooss, A. Mielke, Time-periodic Ginzburg–Landau equations for one-dimensional patterns with large wavelength, ZAMP 43 (1992) 125–138.

[17] G. Iooss, A. Mielke, Y. Demay, Theory of steady Ginzburg–Landau equation in hydrodynamic stability problems, Eur. J. Mech. B 8 (1989) 229–268.

[18] G. Iooss, M.C. Pérouème, Perturbed homoclinic solutions in reversible 1 : 1 resonance vector fields, J. Differential Equations 102 (1993) 62–88.

[19] K. Kirchgässner, Nonlinearly resonant surface waves and homoclinic bifurcation, Adv. in Appl. Mech. 26 (1988) 135–181.

[20] A.E.H. Love, A Treatise on the Mathematical Theory of Elasticity, 4th ed., Cambridge University Press, Cambridge, 1927.

[21] J.H. Maddocks, On the stability of relative equilibria, IMA J. Appl. Math. 46 (1991) 71–99.

[22] R.S. Manning, J.H. Maddocks, J.D. Kahn, A continuum rod model of sequence-dependent DNA structure, J. Chem. Phys. 105 (1996) 5626–5646.

[23] A. Mielke, P. Holmes, Spatially complex equilibria of buckled rods, Arch. Rat. Mech. Anal. 101 (1988) 319–348.

[24] Y. Shi, J.E. Hearst, The Kirchhoff elastic rod, the nonlinear Schrödinger equation, and DNA supercoiling, J. Chem. Phys. 101 (1994) 5186–5200.

[25] J.M.T. Thompson, A.R. Champneys, From helix to localized writhing in the torsional post-buckling of elastic rods, Proc. Roy. Soc. London A 452 (1996) 117–138.

[26] S.P. Timoshenko, J.M. Gere, Theory of Elastic Stability, 2nd ed., McGraw-Hill, New York, 1961.

[27] S.P. Timoshenko, J.N. Goodier, Theory of Elasticity, 2nd ed., McGraw-Hill, New York, 1951.

Spatially complex localization after one-twist-per-wave equilibria in twisted circular rods with initial curvature

BY A. R. CHAMPNEYS[1], G. H. M. VAN DER HEIJDEN[2] AND J. M. T. THOMPSON[2]

[1]*Department of Engineering Mathematics, University of Bristol, University Walk, Bristol BS8 1TR, UK*

[2]*Centre for Nonlinear Dynamics, University College London, Gower Street, London WC1E 6BT, UK*

In experiments on long rubber rods subject to end tension and moment, a one-twist-per-wave deformation is often observed on the fundamental path prior to the onset of localized buckling. An analysis is undertaken here to account for this observed behaviour. First we derive general equilibrium equations using the Cosserat theory, incorporating the effects of non-symmetric cross-section, shear deformation, gravity and a uniform initial curvature of the unstressed rod. Each of these effects in turn can be expressed as a perturbation of the classical completely integrable Kirchhoff–Love differential equations which are equivalent to those describing a spinning symmetric top. Non-symmetric cross-section was dealt with in earlier papers. Here, after demonstrating that shear deformation alone makes little qualitative difference, the case of initial curvature is examined in some detail.

It is shown that the straight configuration of the rod is replaced by a one-twist-per-wave equilibrium whose amplitude varies with pre-buckling load. Superimposed on this equilibrium is a localized buckling mode, which can be described as a homoclinic orbit to the new fundamental path. The dependence is measured of the pre-buckled state and critical buckling load on the amount of initial curvature. Numerical techniques are used to explore the multiplicity of localized buckling modes, given that non-zero initial curvature breaks the complete integrability of the differential equations, and also one of a pair of reversibilities.

Finally, the physical implications of the results are assessed and are shown to match qualitatively what is observed in an experiment.

1. Introduction

The torsional buckling of long rods subject to end moment and tension is a fundamental problem, with applications to DNA supercoiling and pipeline, cable and optical fibre kinking (see Thompson & Champneys 1996 and references therein). The mathematical modelling of such phenomena has a long history, going at least as far back as Kirchhoff (1859). See Antmann (1995, ch. VIII) and references therein for a modern account. In earlier work (Thompson & Champneys 1996; Champneys &

Phil. Trans. R. Soc. Lond. A (1997) **355**, 2151–2174

Thompson 1996; van der Heijden *et al.* 1997; van der Heijden & Thompson 1997) we considered the nature of buckling of a long initially straight rod with uniform cross-section subject to end moment M and tension T. Experiments and energy analysis suggested that a localized form of deformation was the preferred mode of buckling at the critical buckling load. This buckling load, for a perfect inextensible rod with symmetric cross-section, is given by the simple dimensionless condition

$$m = m_c = 2, \quad \text{where } m := M/\sqrt{BT}, \tag{1.1}$$

and B is the bending stiffness (Timoshenko & Gere 1961, equation 2–74).

For moderately large deflection (so that the buckled rod does not have a non-trivial contact with itself) the spatial equilibrium of a rod is described (with one caveat, see below) by the Kirchhoff–Love, or more generally the Cosserat, theory of rods (see § 2 below for the details). Here the rod is described by arclength along its centre line and three directors which are aligned along the rod's principal axes of inertia. This leads to a system of ordinary differential equations for the contact forces and moments in the arclength-dependent coordinate basis described by the directors. It was Kirchhoff (1859) who noticed that the equilibrium equations for infinitely long symmetric rods have identical form to the classical completely integrable system describing the dynamics of a symmetric spinning top. At the buckling load, the differential equations undergo a Hamiltonian–Hopf bifurcation (van der Meer 1985; Iooss & Pérouème 1993), included in the normal form of which is a homoclinic solution. This solution represents a localized helical deformation that tends asymptotically to the flat unbuckled equilibrium state of the rod. That the two end states must be physically aligned was not made clear in our previous work, but is a subtle property of the structure of the underlying equations (see Kehrbaum & Maddocks, this volume).

The situation is more complex, however, for rods without circular cross-section—more precisely for rods whose principal bending stiffness about two orthogonal planes are not equal—because the complete integrability of the differential equations is broken. For mildly anisotropic rods, this leads to spatial chaos in the differential equations and, more importantly to the problem of interest here, to an infinite multiplicity of multi-modal localized buckling paths (Mielke & Holmes 1988; Champneys & Thompson 1996; van der Heijden *et al.* 1997) existing right up to the buckling state. For strongly anisotropic rods with certain physical properties (sufficiently low Poisson's ratio, for example), a transition occurs to different buckling behaviour, with localized buckling being 'locked-on to tape-like behaviour' (van der Heijden & Thompson 1997), and multi-modal solutions only existing in a region of parameter space bounded away from the buckling state.

Qualitative experiments on circular and square cross-section rubber rods reported in Champneys & Thompson (1996) agree well with the theory, but with one caveat (see Thompson & Champneys (1996), figures 4 and 6 and also figures 12 and 13 in § 5 below). Prior to buckling a marked spatially periodic helical deformation is observed, on top of which at the critical load the localized buckling mode is superimposed. The characteristics of the observed pre-buckled and post-buckled helical deformations are quite distinct. The pre-buckled state has one twist per wave whereas the buckling mode has approximately three twists per wave.

The aim of this paper is to analyse an effect that is responsible for such a pre-buckled one-twist-per-wave deformation. Effects missing from our earlier analysis were: the effect of gravity (most experiments were carried out with the specimen held horizontally, often seen to hang in a shallow catenary); shear deformation (the rod was assumed to be inextensible and transversely unshearable in the analysis—inextensibility is clearly violated for the observations made under rigid loading); and initial curvature of the rod in its unstressed state (a permanent deformation caused by the fact that most specimens are stored in loosely wound coils). It is interesting to note the observation mentioned briefly in Thompson & Champneys (1996, § 6) that the one-twist-per-wave mode was not observed in neutral-buoyancy under-water tests on rubber rods that had been cast in a rectangular trough and hence had no initial curvature.

The rest of the paper is outlined as follows. In § 2, we review the Cosserat theory of rods applied to the current problem and use it to derive non-dimensional equilibrium equations for infinite rods subject to shear deformation and gravity, and having initial curvature and not necessarily symmetric cross-section. Each of these four effects can be treated as perturbations of the underlying 'perfect' problem. Our programme of work is to treat each perturbation separately. Non-symmetric cross-section was treated in earlier work and gravity will be left for future work. The main thrust of the present work will be to treat initial curvature.

In § 3 we review the unperturbed problem and use it to explain the numerical methods which will be used in the following. Also in § 3, the introduction of shear deformation alone is shown to make little difference to what is observed.

Section 4 contains the analysis of the effects of initial curvature. First, it is shown that the fundamental flat state is replaced by a new spatially homogeneous state in the arclength-dependent coordinates. It is then shown explicitly that this state has one twist per wave when interpreted physically and the amplitude and end-shortening associated with it are computed as a function of the strength of initial curvature κ_0. Next, a linearization about this state reveals the new critical buckling load, replacing (1.1) as a function, $m(\kappa_0)$, of the initial curvature. Finally, numerical methods are used to compute localized buckling modes superimposed on the one-twist-per-wave equilibria. Initial curvature, like non-symmetric cross-section, is again demonstrated to imply an infinity of post-buckling modes.

Section 5 draws conclusions and considers some physical implications of the results. We also present some experimental results. These confirm the finding of this paper, that a circular rod with initial curvature, when twisted, does indeed form one-twist-per-wave pre-buckling equilibria in which bending is in the direction of the unstressed curvature.

2. Equilibrium equations

(a) *The Cosserat theory*

Consider the formulation of equilibrium equations for a rod that can undergo flexure, tension, extension and shear as elucidated in Antman (1995, ch. VIII). We are interested in the case of an infinite rod, so we begin by choosing an origin for an arclength coordinate s and define $\boldsymbol{r}(s)$ to be the position vector of the centre line of the rod with respect to a fixed Cartesian coordinate system with basis vectors $\{\boldsymbol{i}, \boldsymbol{j}, \boldsymbol{k}\}$. The configuration of the rod in a deformed state is then defined by $\boldsymbol{r}(s)$, and two orthonormal vectors $\boldsymbol{d}_1(s)$ and $\boldsymbol{d}_2(s)$ which define the position of two orthogonal lines in the cross-section of the rod at s. In the case where the rod is straight and prismatic when unstrained, with its centre line aligned along the z-axis, we may regard $\boldsymbol{d}_{1,2}(s)$ as defining the deformed position of the material originally aligned along the x- and y-axes, respectively. Upon setting

$$\boldsymbol{d}_3 := \boldsymbol{d}_1 \times \boldsymbol{d}_2,$$

the triple $\{\boldsymbol{d}_1(s), \boldsymbol{d}_2(s), \boldsymbol{d}_3(s)\}$ defines a moving rod-centred coordinate system, called the *directors* of the rod. Note that in the presence of shear, in general $\boldsymbol{d}_3(s) \neq \boldsymbol{r}'(s)$, where the prime denotes differentiation with respect to arclength. Henceforth all vectors are expressed with respect to the moving basis $\{\boldsymbol{d}_1, \boldsymbol{d}_2, \boldsymbol{d}_3\}$, unless otherwise stated.

The *stress* in the rod is defined by

$$\boldsymbol{n}(s) = n_1\boldsymbol{d}_1 + n_2\boldsymbol{d}_2 + n_3\boldsymbol{d}_3 \quad \text{and} \quad \boldsymbol{m}(s) = m_1\boldsymbol{d}_1 + m_2\boldsymbol{d}_2 + m_3\boldsymbol{d}_3,$$

which are, respectively, the contact force and couple exerted by the material parametrized by $\tilde{s} < s$ on that parametrized by $\tilde{s} > s$. Here $n_{1,2}$ are shear forces, n_3 is tension, $m_{1,2}$ are bending moments about the axes $\boldsymbol{d}_{1,2}$ and m_3 is the twisting moment about $\boldsymbol{d}_3$.

The *strain* of the rod is defined by two vectors

$$\boldsymbol{u}(s) = u_1\boldsymbol{d}_1 + u_2\boldsymbol{d}_2 + u_3\boldsymbol{d}_3 \quad \text{and} \quad \boldsymbol{v}(s) = v_1\boldsymbol{d}_1 + v_2\boldsymbol{d}_2 + v_3\boldsymbol{d}_3,$$

defined by

$$\boldsymbol{d}_i' = \boldsymbol{u} \times \boldsymbol{d}_i, \quad i = 1, 2, 3, \quad \text{and} \quad \boldsymbol{r}' = \boldsymbol{v}. \tag{2.1}$$

Here, u_1 and u_2 measure the flexure, u_3 measures torsion, v_1 and v_2 measure transverse shear and v_3 measures dilation or axial extension. Note that if the rod were unshearable and inextensible, then $\boldsymbol{v} = \boldsymbol{d}_3$. It is thus reasonable to redefine the shear vector as

$$\boldsymbol{v} = \boldsymbol{d}_3 + \boldsymbol{y}$$

(cf. Antman 1995, equation VIII.15.13), where $\boldsymbol{y}$ is small in a sense to be made precise in the non-dimensionalization to follow.

Balancing forces and moments leads to the following equilibrium equations

$$\frac{\mathrm{d}\boldsymbol{n}}{\mathrm{d}s} + \boldsymbol{f} = 0, \tag{2.2}$$

$$\frac{\mathrm{d}\boldsymbol{m}}{\mathrm{d}s} + \boldsymbol{r}' \times \boldsymbol{n} + \boldsymbol{l} = 0, \tag{2.3}$$

where $\boldsymbol{f}$ and $\boldsymbol{l}$ are distributed external forces and couples acting on the rod along its length. The only body force or couple we shall be interested in is gravity. Therefore, we take in (2.2) and (2.3)

$$\boldsymbol{l} = \boldsymbol{0} \quad \text{and} \quad \boldsymbol{f} = -Dg\boldsymbol{k}, \tag{2.4}$$

where D is the density (per unit original arclength) of the cross-section and g is the acceleration due to gravity. Note that (2.4) assumes that the $\boldsymbol{k}$ axis points vertically upwards.

The differentials on the left-hand side of (2.2) and (2.3) represent the total derivative with respect to arclength which is related to the derivative with respect to s in the moving coordinate frame (which is denoted by a prime) with the addition of terms which describe the derivative of the coordinate frame with respect to s. For example,

$$\begin{aligned}\frac{\mathrm{d}\boldsymbol{n}}{\mathrm{d}s} &= \frac{\mathrm{d}}{\mathrm{d}s}(n_1\boldsymbol{d}_1 + n_2\boldsymbol{d}_2 + n_3\boldsymbol{d}_3)\\ &= (n_1'\boldsymbol{d}_1 + n_2'\boldsymbol{d}_2 + n_3'\boldsymbol{d}_3) + (n_1\boldsymbol{d}_1' + n_2\boldsymbol{d}_2' + n_3\boldsymbol{d}_3')\\ &= \boldsymbol{n}' + (n_1\boldsymbol{u}\times\boldsymbol{d}_1 + n_2\boldsymbol{u}\times\boldsymbol{d}_2 + n_3\boldsymbol{u}\times\boldsymbol{d}_3), \qquad \text{(by (2.1))}\\ &= \boldsymbol{n}' + \boldsymbol{u}\times\boldsymbol{n}.\end{aligned}$$

Hence, using (2.1) and (2.4), (2.2) and (2.3) become

$$\boldsymbol{n}' = \boldsymbol{n}\times\boldsymbol{u} + Dg\boldsymbol{k}, \tag{2.5}$$

$$\boldsymbol{m}' = \boldsymbol{m}\times\boldsymbol{u} + \boldsymbol{n}\times\boldsymbol{v}. \tag{2.6}$$

(*b*) *Constitutive relations and boundary conditions*

It remains to specify the curvatures $\boldsymbol{u}$ and shears $\boldsymbol{v}$ in terms of the contact forces $\boldsymbol{n}$ and couples $\boldsymbol{m}$. These constitutive relations may take a quite general form (Antman 1995, equation (2.2)), but here we shall take the simplest linear constitutive relations allowing for the effects of initial curvature. There is evidence to suggest (Antman & Jordan 1974; Champneys & Thompson 1996) that physically realistic nonlinear constitutive laws make little qualitative difference to localized buckling. Specifically, we assume relations of the form

$$\left.\begin{aligned} u_1(s) = u_0 + m_1(s)/A, \quad u_2(s) = m_2(s)/B, \quad u_3(s) = m_3(s)/C,\\ y_1(s) = n_1(s)/H, \quad y_2(s) = n_2(s)/J, \quad y_3(s) = n_3(s)/K. \end{aligned}\right\} \tag{2.7}$$

Here, A and B are the principal bending stiffnesses about $\boldsymbol{d}_1$ and $\boldsymbol{d}_2$, C is the torsional stiffness of the rod and u_0 is an assumed initial curvature of the rod, which without loss of generality we assume to occur about the $\boldsymbol{d}_1$ axis. H and J are the transverse shear stiffnesses, and K is the axial stiffness.

Using (2.7) it is now possible to express (2.5) and (2.6) as a six-dimensional system of ordinary differential equations for $\boldsymbol{m}(s)$ and $\boldsymbol{n}(s)$. In order to solve this system, we must discuss its boundary conditions. In this work it is assumed that the rod is held with boundary conditions at infinity which supply a constant twisting moment M and tension T. For the time being, we consider these loads as fixed (either pre-fixed or control) parameters. In interpretation of solutions, however, one should allow for the possibility of rigid loading in which either or both of M or T are regarded as free parameters under controlled end rotation or end shortening. The boundary conditions for $\boldsymbol{n}$ and $\boldsymbol{m}$ must then satisfy

$$\sqrt{\boldsymbol{n}\cdots\boldsymbol{n}} \to T, \quad \text{as } s\to\pm\infty, \quad \sqrt{\boldsymbol{m}\cdots\boldsymbol{m}} \to M, \quad \text{as } s\to\pm\infty. \tag{2.8}$$

Note that we do not put any constraint on the directions in which $\boldsymbol{n}$ and $\boldsymbol{m}$ are pointing. These will follow from the analysis.

(*c*) *Non-dimensionalization and dynamical system formulation*

Equations (2.5) and (2.6) can be made dimensionless by setting

$$\left.\begin{aligned} x_1 = n_1/T, \quad x_2 = n_2/T, \quad x_3 = n_3/T,\\ x_4 = m_1/M, \quad x_5 = m_2/M, \quad x_6 = m_3/M, \end{aligned}\right\} \tag{2.9}$$

rescaling the axial length

$$t = (M/B)s, \tag{2.10}$$

and introducing the dimensionless parameters

$$\left.\begin{aligned} m = M/\sqrt{BT}, \quad \rho = (B/A) - 1, \quad \nu = (B/C) - 1, \quad \kappa_0 = Au_0/M, \\ \sigma = (J/H) - 1, \quad \gamma = (J/K) - 1, \quad \epsilon = T/J, \quad \delta = DgB/(MT). \end{aligned}\right\} \tag{2.11}$$

The parameter m is a single dimensionless load parameter, showing the equivalence (under dead loading) of increasing moment and reducing tension. ρ may be regarded as measuring the non-circularity of the cross-section of the rod. More precisely, $\rho = 0$ corresponds to the bending stiffnesses about the two principal axes being equal, which could also represent a square cross-section, for example. The parameter $\nu + 1$ measures the ratio of torsional stiffness to the greater principal bending stiffness (here we assume $B \geqslant A$ and hence $\rho \geqslant 0$). For solid rods with circular cross-section ν is equal to Poisson's ratio, with a typical physical value being $\nu = \frac{1}{3}$. For rods with anisotropic cross-section, however, it should be pointed out that ν no longer equals Poisson's ratio which is given by a certain combination of ν and ρ, depending on the geometry (van der Heijden & Thompson 1997). κ_0 is the dimensionless initial curvature, and σ and γ are the analogues of ρ and ν for the shear and extension coefficients. ϵ measures the relative importance of shear compared with bending and δ measures the relative strength of gravity compared with the applied loads.

Typically we think of ϵ, δ and κ_0 as small whereas $\rho + 1$, $\nu + 1$, $\sigma + 1$ and $\gamma + 1$ are all $\mathcal{O}(1)$. It is also reasonable to set $\rho = \sigma$, with both parameters measuring the degree of non-circularity of the cross-section.

Under the non-dimensionalization (2.9), (2.10) and (2.11), the equilibrium equations (2.5) and (2.6) become

$$\left.\begin{aligned} \dot{x}_1 &= (1+\nu)x_2x_6 - x_3x_5 + \delta k_1(t), \\ \dot{x}_2 &= (1+\rho)x_3(x_4+\kappa_0) - (1+\nu)x_1x_6 + \delta k_2(t), \\ \dot{x}_3 &= x_1x_5 - (1+\rho)x_2(x_4+\kappa_0) + \delta k_3(t), \\ \dot{x}_4 &= \nu x_5x_6 + (1/m^2)[x_2 + \epsilon\gamma x_2x_3], \\ \dot{x}_5 &= (1+\rho)(x_4+\kappa_0)x_6 - (1+\nu)x_4x_6 + (1/m^2)[\epsilon(\sigma-\gamma)x_1x_3 - x_1], \\ \dot{x}_6 &= x_4x_5 - (1+\rho)(x_4+\kappa_0)x_5 - (\epsilon/m^2)\sigma x_1x_2, \end{aligned}\right\} \tag{2.12}$$

where (k_1, k_2, k_3) are the components of $\boldsymbol{k}$ with respect to $\{\boldsymbol{d}_1, \boldsymbol{d}_2, \boldsymbol{d}_3\}$, and a dot denotes differentiation with respect to the scaled arclength t.

In the gravity-free case, $\delta = 0$ and (2.12), when regarded as an initial-value problem, represents an autonomous six-dimensional dynamical system

$$\dot{\boldsymbol{x}} = \boldsymbol{f}(\boldsymbol{x}, \lambda), \quad \boldsymbol{x} \in \mathbb{R}^6, \quad \lambda \in \mathbb{R}^p,$$

with p the number of parameters considered to be controlled. It can be shown to be Hamiltonian with two independent integrals of the motion (Mielke & Holmes 1988). We are interested in solutions which are localized to some portion of the axial length. Hence we apply the homoclinic boundary conditions

$$\boldsymbol{x}(t) \to \boldsymbol{x}^*, \quad \text{as } t \to \pm\infty, \tag{2.13}$$

where $\boldsymbol{x}^*$ is a spatially homogeneous equilibrium position of the rod, that is $\boldsymbol{f}(\boldsymbol{x}^*, \lambda) = 0$, which is subject to the constraint imposed by the dimensionless version of (2.8),

$$\sqrt{x_1^{*2} + x_2^{*2} + x_3^{*2}} = 1, \quad \sqrt{x_4^{*2} + x_5^{*2} + x_6^{*2}} = 1. \tag{2.14}$$

In order to interpret solutions to (2.12), (2.13) as configurations of the rod in physical coordinates, the twelve-dimensional system (2.1) must be solved for the $\boldsymbol{i}$, $\boldsymbol{j}$ and $\boldsymbol{k}$ components of each $\boldsymbol{d}_i$ and $\boldsymbol{r}$. The first nine of these equations are the Frenet–Serret equations of differential geometry. The dimensionless versions of (2.1) are

$$\dot{\boldsymbol{d}}_{\mathrm{i}} = \tilde{\boldsymbol{u}} \times \boldsymbol{d}_{\mathrm{i}}, \quad i = 1, 2, 3, \quad \text{and} \quad \dot{\tilde{\boldsymbol{r}}} = \tilde{\boldsymbol{v}}, \tag{2.15}$$

in which

$$\tilde{\boldsymbol{u}} = (1+\rho)(x_4 + \kappa_0)\boldsymbol{d}_1 + x_5\boldsymbol{d}_2 + (1+\nu)x_6\boldsymbol{d}_3,$$
$$\tilde{\boldsymbol{v}} = \epsilon(1+\sigma)x_1\boldsymbol{d}_1 + \epsilon x_2\boldsymbol{d}_2 + (1+\epsilon(1+\gamma)x_3)\boldsymbol{d}_3, \quad \tilde{\boldsymbol{r}} = (M/B)\boldsymbol{r}.$$

These equations can be solved after a solution to (2.12) has been found. $\boldsymbol{r}$ and $\boldsymbol{d}_{\mathrm{i}}$ are typically subject to initial conditions specifying the position and orientation of the rod at its left-hand endpoint, $t = -\infty$.

If gravity effects are included, then $\delta \neq 0$ and (2.12) can no longer be regarded in isolation, but must be solved in tandem with the equations for $\boldsymbol{d}_{\mathrm{i}}$ in (2.15) in order to calculate $\boldsymbol{k}(t)$. Notice that $\boldsymbol{k}$ is obtained from the solution of (2.15) by forming the matrix

$$B = (\boldsymbol{d}_1|\boldsymbol{d}_2|\boldsymbol{d}_3)$$

and taking

$$\boldsymbol{k} = B^{-1}(0,0,1)^T.$$

We may consider each of the parameters ρ $(=\sigma)$, ϵ, κ_0, δ as measuring the strength of perturbations to the 'perfect' symmetric unshearable initially straight rod in the absence of gravity. The case of non-symmetric rods, with $\epsilon = \delta = \kappa_0 = 0$, $\rho \neq 0$, was analysed in depth in Champneys & Thompson (1996), van der Heijden *et al.* (1997) and van der Heijden & Thompson (1997) to which we refer the reader for the details. For the rest of this paper we shall consider only rods with symmetric cross-section, $\rho = \sigma = 0$. Also, we shall not, here, analyse the effects of gravity, that is we set $\delta = 0$. The following section reviews the unperturbed case and then considers the effect of axial extensibility and transverse shear ($\epsilon \neq 0$). The bulk of our analysis, in § 4, will be for a rod with curvature when unstrained ($\kappa_0 \neq 0$) with the other perturbations fixed at zero.

3. The unperturbed case

Consider (2.12) for $\rho = \epsilon = \kappa_0 = \delta = 0$. First, observe that the unperturbed system (and, indeed, the wider class of systems with $\rho \neq 0$) is invariant under the following two involutions:

$$R_1: \quad (x_1, x_2, x_3, x_4, x_5, x_6) \to (-x_1, x_2, x_3, -x_4, x_5, x_6), \quad t \to -t, \tag{3.1}$$
$$R_2: \quad (x_1, x_2, x_3, x_4, x_5, x_6) \to (x_1, -x_2, x_3, x_4, -x_5, x_6), \quad t \to -t. \tag{3.2}$$

This will have implications for the multiplicity of homoclinic solutions. Next, the unperturbed system (2.12) is completely integrable with the following three independent 'constants of the motion':

$$I_1 := x_1^2 + x_2^2 + x_3^2 = \text{const.}, \tag{3.3}$$
$$I_2 := x_1x_4 + x_2x_5 + x_3x_6 = \text{const.}, \tag{3.4}$$
$$I_3 := x_6 = \text{const.} \tag{3.5}$$

The integrals (3.3) and (3.4) correspond to the conservation along the axial length of the rod of the magnitude of force and the component of torque about the loading axis, respectively. (3.5) describes conservation of torque about the body (i.e. symmetry) axis of the rod. (Actually, the twisted rod equations are completely integrable for a much wider class of constitutive relations (see Antman 1995).)

Note that for given m, the only spatially homogeneous equilibrium solutions of (2.12) that satisfy the constraint (2.14) are $\boldsymbol{x} = (0, 0, \pm 1, 0, 0, \pm 1)$. We can ignore the solutions with minus signs because we are considering tension rather than compression and by convention of the direction in which we apply end moment. Therefore, we set

$$\boldsymbol{x}^* = (0, 0, 1, 0, 0, 1). \tag{3.6}$$

Linearizing about this equilibrium one finds that there are two zero eigenvalues corresponding to translations in the directions of $\boldsymbol{m}$ and $\boldsymbol{n}$. Imposing (2.14), in effect, fixes this translation and, through (3.6), the values of the integrals (3.3), (3.4), (3.5) for the spatially homogeneous solution and its localized deformations. Examining the other four eigenvalues, the trivial equilibrium loses physical stability (four complex eigenvalues coalesce on the imaginary axis) as m is increased through the critical value $m = 2$. Here a Hamiltonian–Hopf bifurcation occurs (van der Meer 1985; Iooss & Pérouème 1993).

As a result of complete integrability, at the Hamiltonian–Hopf point we have the subcritical bifurcation of a manifold of localized buckling modes. Each solution in the manifold consists of a rigid rotation of a primary localized helix, that is a homoclinic solution of (2.12). The physical stability of these localized buckling modes under rigid loading was investigated in Thompson & Champneys (1996).

(a) *Measuring twist*

Before proceeding, let us first demonstrate how the Frenet–Serret and centre line differential equations (2.15) may be used to give the degree of twist about the central axis for a particular physical configuration. Specifically we demonstrate that when interpreted in dimensional coordinates, the trivial equilibrium (3.6) corresponds to a straight rod with the tension T and moment M being carried along the centre line of the rod. Taking $\boldsymbol{x} = \boldsymbol{x}^*$ given by (3.6), we find that (2.15) gives

$$\dot{\boldsymbol{d}}_3 = 0, \quad \ddot{\boldsymbol{d}}_1 = -(1+\nu)^2 \boldsymbol{d}_1, \quad \dot{\tilde{\boldsymbol{r}}} = \text{const}. \tag{3.7}$$

Choosing initial conditions so that $\boldsymbol{d}_3 = \boldsymbol{i}$ is horizontal, and hence

$$\tilde{\boldsymbol{r}} = t\boldsymbol{i}, \tag{3.8}$$

the second equation of (3.7) gives a rotation of the material of the rod about the horizontal with angular frequency (fundamental twist) $\tau = 1 + \nu$.

In contrast, the wavelength of the helical deformation at the critical buckling load can be obtained by linearizing about the the trivial solution at $m = 2$ (van der Heijden & Thompson 1997). This analysis uses a description of the rod in terms of Euler angles $\theta(s)$, $\psi(s)$ and $\phi(s)$ (see Love (1927), art. 253, for the definition of the angles defined using this notation). Along the trivial solution we have $\dot{\phi} = \frac{1}{2} + \nu$ while at buckling (i.e. $\theta = 0$), $\dot{\phi} + \dot{\psi} = \tau = 1 + \nu$. Hence, $\dot{\psi} = \frac{1}{2}$. Now, ψ is the azimuthal angle determining the wavelength of the helix. The number of twists per wave (tpw) can therefore be obtained as

$$\text{tpw} = \frac{\tau}{\dot{\psi}} = 2(1+\nu). \tag{3.9}$$

For rubber rods ($\nu \approx 0.5$) this gives the three twists per wave observed in experiments mentioned in § 1 (see also § 5 below).

(b) Numerical methods

Let us briefly outline how the numerical methods used in Champneys & Thompson (1996) and van der Heijden *et al.* (1997) can be adapted for the present setting. For homoclinic orbits to a fixed point we use the shooting method developed in Champneys & Spence (1993). Load–deflection-type bifurcation diagrams are subsequently obtained by continuation of these orbits as parameters are varied using the code AUTO (Doedel *et al.* 1991) for continuation of solutions to boundary-value problems.

This combined use of shooting and continuation is robust and was described in detail in earlier papers (see, for example, van der Heijden *et al.* 1997) to which the reader is referred for the details. We remark, however, that these methods typically truncate the problem to an approximation of half the rod ($t \in [0, \mathcal{T}]$, $\mathcal{T}$ large), and only find solutions which are invariant under the reversibilities R_1 or R_2. Non-symmetric homoclinic orbits may also be found with more effort, by shooting over an approximation to the full length of the rod ($t \in [-\mathcal{T}, \mathcal{T}]$).

In order to compute load–deflection diagrams for localized buckling modes it is necessary to also measure the end displacement D and end rotation R from the spatially homogeneous reference position. Once a well-posed solution to (2.12) has been found, we solve (2.15) subject to the initial conditions

$$\tilde{\boldsymbol{r}}(0) = (0,0,0), \quad \boldsymbol{d}_1(0) = (1,0,0), \quad \boldsymbol{d}_2(0) = (0,1,0), \quad \boldsymbol{d}_3(0) = (0,0,1). \tag{3.10}$$

The end displacement and end rotation can then be obtained as

$$D = \mathcal{T} - \tilde{r}_3(\mathcal{T}), \quad \cos R = \langle \boldsymbol{d}_1(\mathcal{T}), (1,0,0)\rangle, \quad \sin R = \langle \boldsymbol{d}_1(\mathcal{T}), (0,1,0)\rangle, \tag{3.11}$$

provided $\boldsymbol{d}_3(\mathcal{T})$ is sufficiently aligned with the initial direction $\boldsymbol{d}_3(0) = (0,0,1)$. Although this alignment is not assured at the symmetric section, it is, for localized solutions, at the end points of the rod. Therefore, in applying (3.11) $\mathcal{T}$ is taken to be the full length of the rod. (The case that end vectors do not line up because the fundamental configuration of the rod is not a straight one, as happens when the rod has initial curvature, can be easily accommodated; see § 4*c*.) In (3.10) and (3.11) vectors are expressed with respect to the fixed frame $\{\boldsymbol{i}, \boldsymbol{j}, \boldsymbol{k}\}$. Note that the angle R is only defined modulo 2π in (3.11). Continuity with respect to the arclength parameter t, however, will allow us to keep track of the number of full turns made by the end point.

(c) The effect of shear and extension

In this section we demonstrate that including the effect of shear alone makes little qualitative difference. To that end we take $\epsilon > 0$ in (2.12) while keeping $\rho = \sigma = \delta = \kappa_0 = 0$. First note that, as for the unperturbed equation, the only spatially homogeneous equilibrium that satisfies the constraint (2.14) is the trivial solution (3.6). Setting $\boldsymbol{x} = \boldsymbol{x}^* + \boldsymbol{z}$, one obtains

$$\left.\begin{aligned}
\dot{z}_1 &= (1+\nu)z_2(1+z_6) - (1+z_3)z_5,\\
\dot{z}_2 &= (1+z_3)z_4 - (1+\nu)z_1(1+z_6),\\
\dot{z}_3 &= z_1 z_5 - z_2 z_4,\\
\dot{z}_4 &= \nu z_5(1+z_6) + (1/m^2)[z_2 + \epsilon\gamma z_2(1+z_3)],\\
\dot{z}_5 &= -\nu z_4(1+z_6) + (1/m^2)[-\epsilon\gamma z_1(1+z_3) - z_1],\\
\dot{z}_6 &= 0.
\end{aligned}\right\} \tag{3.12}$$

Here we note that the linear terms of (3.12) are identical to those when $\epsilon = 0$ provided $1/m^2$ is replaced by $(1+\epsilon\gamma)/m^2$. Hence we conclude that the local bifurcation

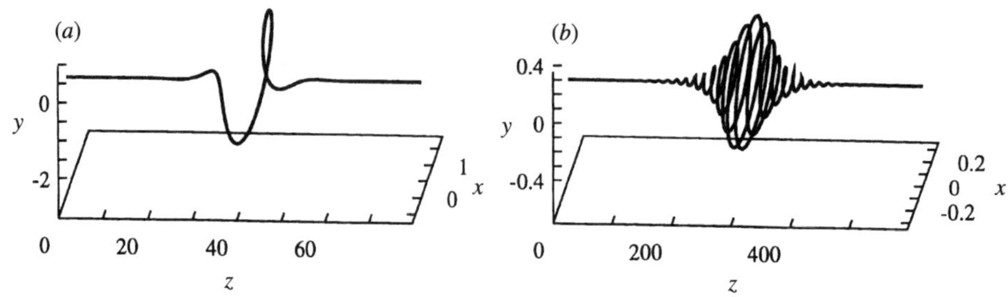

Figure 1. The primary buckling mode at two different loads: (a) $m = 1.7$; (b) $m = 2.09$ ($\nu = \frac{1}{3}$, $\epsilon = 0.1$, $\gamma = 1$, implying $m_c = 2.097618$).

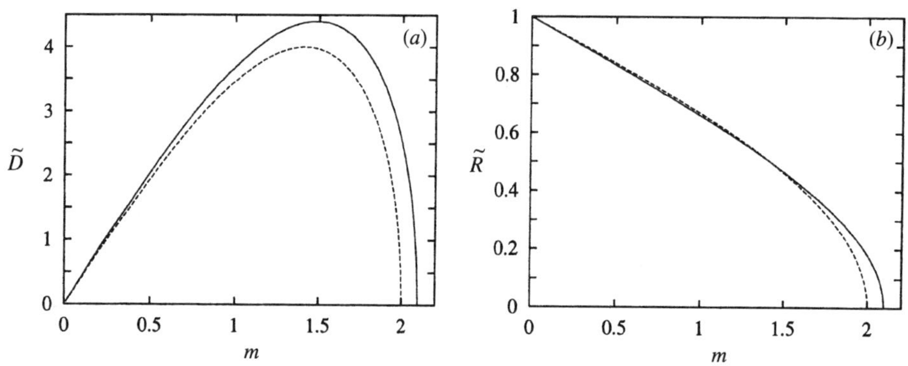

Figure 2. (a) End shortening $\tilde{D}$ and (b) end rotation $\tilde{R}$ versus load m for the primary homoclinic orbit ($\nu = \frac{1}{3}$, $\epsilon = 0.1$, $\gamma = 1$, implying $m_c = 2.097618$). Data for the unperturbed case have been included in dashed lines for comparison.

behaviour is qualitatively identical to that in the shear-free case, with the critical buckling condition now given by

$$m_c = 2\sqrt{1 + \gamma\epsilon}. \tag{3.13}$$

Using (2.11) to write this in terms of the original parameters we get

$$\frac{M}{\sqrt{BT}} = 2\sqrt{1 + T(K^{-1} - H^{-1})}, \tag{3.14}$$

where H^{-1} ($= J^{-1}$) and K^{-1} are the shear and axial flexibility, respectively. We conclude that compared with the unperturbed case ($H^{-1} = 0$, $K^{-1} = 0$), shear flexibility tends to lower the critical buckling load m_c, while axial flexibility tends to raise m_c.

Figure 1 shows examples of three-dimensional rod shapes for two different values of the load m. The solutions were computed on a truncated interval using the shooting method explained in § 4, followed by an integration of the centre line equations in (2.15).

Load–deflection characteristics of the localized buckling mode are given in figure 2. The diagrams were computed by monitoring end displacement, D, and end rotation, R, while numerically continuing the homoclinic orbit for varying m. Actually plotted are the corrected quantities $\tilde{D}$ and $\tilde{R}$ obtained from D and R by subtracting the trivial contributions, proportional to the length of the rod, which the straight unbuckled rod would have, and dividing the angle by 2π:

$$\tilde{D} = (1+\epsilon(1+\gamma))\mathcal{T} - \tilde{r}_3(\mathcal{T}), \quad \tilde{R} = \frac{R-(1+\nu)\mathcal{T}}{2\pi}, \tag{3.15}$$

where $\mathcal{T}$ denotes the full length of the rod. Thus, $\tilde{D}$ and $\tilde{R}$ represent the true buckling contributions to end displacement and end rotation.

For comparison we have also included in figure 2 the deflections in the unperturbed case which can be computed exactly:

$$\tilde{D} = 2m\sqrt{4-m^2}, \quad \tilde{R} = 4\arccos\tfrac{1}{2}m \tag{3.16}$$

(see Thompson & Champneys 1996, equations (5.3) and (5.4)). We observe that shear causes an increase in the maximum end-shortening.

4. Initial curvature

Consider now a rod that is naturally curved, $\kappa_0 \neq 0$, while ignoring the effects of shear, gravity and non-symmetric cross-section, $\rho = \sigma = \epsilon = \delta = 0$.

(*a*) *One-twist-per-wave equilibria*

Spatially homogeneous equilibria of (2.12), $\boldsymbol{x}(t) \equiv \boldsymbol{x}^*$, are determined by setting the right-hand sides of (2.12) to zero:

$$0 = (1+\nu)x_2^*x_6^* - x_3^*x_5^*, \tag{4.1}$$

$$0 = x_3^*(x_4^*+\kappa_0) - (1+\nu)x_1^*x_6^*, \tag{4.2}$$

$$0 = x_1^*x_5^* - x_2^*(x_4^*+\kappa_0), \tag{4.3}$$

$$0 = \nu x_5^*x_6^* + x_2^*/m^2, \tag{4.4}$$

$$0 = (x_4^*+\kappa_0)x_6^* - (1+\nu)x_4^*x_6^* - x_1^*/m^2, \tag{4.5}$$

$$0 = -\kappa_0 x_5^*. \tag{4.6}$$

In addition, an equilibrium must satisfy the boundary constraints (2.14). From (4.6) we obtain $x_5^* = 0$ and hence from (4.4) that $x_2^* = 0$. Solving (4.2) and (4.5) for x_1^* and x_4^* in terms of x_3^* and x_6^* yields

$$x_1^* = \frac{\kappa_0 m^2(1+\nu)x_3^*x_6^*}{x_3^* + x_6^{*2}\nu(1+\nu)m^2} \quad \text{and} \quad x_4^* = \frac{\kappa_0[m^2(1+\nu)x_6^{*2} - x_3^*]}{x_3^* + m^2\nu(1+\nu)x_6^{*2}}. \tag{4.7}$$

Substitution into (4.7) of x_3^* and x_6^* from (2.14) (assuming them to be positive), leads to two coupled equations for x_1^* and x_4^*. A straightforward application of the Implicit Function Theorem shows that there is a unique solution for small κ_0 tending to $x_1^* = x_4^* = 0$ as $\kappa_0 \to 0$. This solution can be expressed as a series expansion for small κ_0:

$$x_1^* = \frac{m^2(1+\nu)}{\nu m^2(1+\nu)+1}\kappa_0 - \frac{m^2(1+\nu)[m^2(1+\nu)(m^2(2\nu^2+3\nu+1)-(\nu+2))+1]}{2(\nu m^2(1+\nu)+1)^4}\kappa_0^3 + \mathcal{O}(\kappa_0^5),$$

$$x_2^* = 0,$$

$$x_3^* = 1 - \frac{m^4(1+\nu)^2}{2(\nu m^2(1+\nu)+1)^2}\kappa_0^2 + \mathcal{O}(\kappa_0^4),$$

$$x_4^* = \frac{m^2(1+\nu)-1}{\nu m^2(1+\nu)+1}\kappa_0 - \frac{m^2(1+\nu)^2[m^4(1+\nu)^2+4m^2(1-\nu)+2]}{2(\nu m^2(1+\nu)+1)^4}\kappa_0^3 + \mathcal{O}(\kappa_0^5),$$

$$x_5^* = 0,$$

$$x_6^* = 1 - \frac{(m^2(1+\nu)-1)^2}{2(\nu m^2(1+\nu)+1)^2}\kappa_0^2 + \mathcal{O}(\kappa_0^4). \tag{4.8}$$

Without using this expansion we shall now explicitly demonstrate that this equilibrium, when substituted into the Frenet–Serret and centre line differential equations (2.15), results in a one-twist-per-wave solution. Now,

$$\tilde{\boldsymbol{u}} = (x_4^* + \kappa_0)\boldsymbol{d}_1 + (1+\nu)x_6^*\boldsymbol{d}_3 =: J(\kappa_0,\nu,m)\boldsymbol{d}_1 + H(\kappa_0,\nu,m)\boldsymbol{d}_3, \tag{4.9}$$

where J and H are constants determined from the right-hand sides of (4.8). Hence equations (2.15) become

$$\dot{\boldsymbol{d}}_i = J(\boldsymbol{d}_1 \times \boldsymbol{d}_i) + H(\boldsymbol{d}_3 \times \boldsymbol{d}_i), \quad i = 1,2,3 \quad \text{and} \quad \dot{\tilde{\boldsymbol{r}}} = \boldsymbol{d}_3, \tag{4.10}$$

with the last equation arising because $\tilde{\boldsymbol{v}} = \boldsymbol{d}_3$ in the absence of shear. Note that for κ_0 small

$$J = \mathcal{O}(\kappa_0) \quad \text{and} \quad H = \mathcal{O}(1), \quad \text{as } \kappa_0 \to 0. \tag{4.11}$$

The first three equations in (4.10) are

$$\dot{\boldsymbol{d}}_1 = H\boldsymbol{d}_2, \quad \dot{\boldsymbol{d}}_2 = J\boldsymbol{d}_3 - H\boldsymbol{d}_1, \quad \dot{\boldsymbol{d}}_3 = -J\boldsymbol{d}_2,$$

from which we can eliminate $\boldsymbol{d}_2$ to give

$$\begin{pmatrix} \ddot{\boldsymbol{d}}_1 \\ \ddot{\boldsymbol{d}}_3 \end{pmatrix} = \begin{bmatrix} -H^2 & HJ \\ HJ & -J^2 \end{bmatrix} \begin{pmatrix} \boldsymbol{d}_1 \\ \boldsymbol{d}_3 \end{pmatrix}. \tag{4.12}$$

Upon defining new orthonormal coordinate axes

$$\boldsymbol{e}_1 = \frac{1}{\sqrt{H^2+J^2}}(H\boldsymbol{d}_1 - J\boldsymbol{d}_3), \quad \boldsymbol{e}_3 = \frac{1}{\sqrt{H^2+J^2}}(J\boldsymbol{d}_1 + H\boldsymbol{d}_3), \quad \boldsymbol{e}_2 = \boldsymbol{e}_3 \times \boldsymbol{e}_1 = \boldsymbol{d}_2, \tag{4.13}$$

one finds from (4.12) that

$$\ddot{\boldsymbol{e}}_1 = -\Omega^2\boldsymbol{e}_1 \quad \text{and} \quad \ddot{\boldsymbol{e}}_3 = 0, \tag{4.14}$$

where $\Omega = \sqrt{H^2+J^2}$. The solution to these equations is

$$\boldsymbol{e}_1(t) = \boldsymbol{A}\cos\Omega t + \boldsymbol{B}\sin\Omega t, \quad \boldsymbol{e}_3(t) = \boldsymbol{C}, \tag{4.15}$$

where $\boldsymbol{A}$, $\boldsymbol{B}$ and $\boldsymbol{C}$ are three constant orthonormal vectors determined by the initial conditions. Since there is no gravity, we are free to choose an origin for arclength and an orientation such that

$$\boldsymbol{A} = \boldsymbol{i}, \quad \boldsymbol{B} = \boldsymbol{j}, \quad \boldsymbol{C} = \boldsymbol{k}.$$

Note that the coefficient of t in the solution for $\boldsymbol{e}_3$ from (4.14) is chosen to be zero in order to avoid terms which grow as $t \to \pm\infty$. Inverting the transformation (4.13), we obtain an expression for the centre line of the rod via

$$\dot{\tilde{\boldsymbol{r}}} = \boldsymbol{d}_3 = (1/\Omega)(H\boldsymbol{e}_3 - J\boldsymbol{e}_1),$$

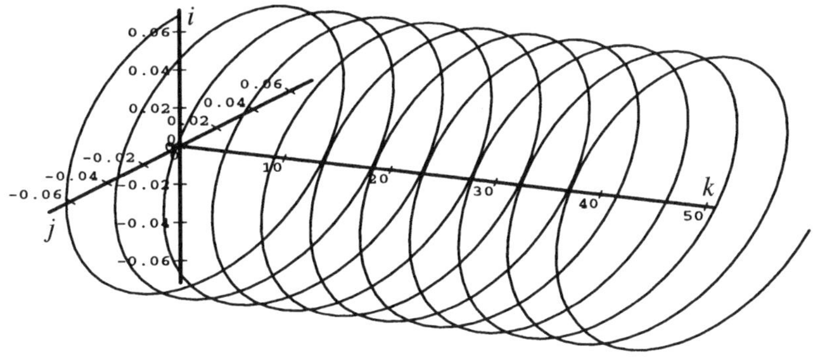

Figure 3. The dimensionless spatial configuration of the equilibrium solution (4.16) for $\nu = \frac{1}{3}$, $\kappa_0 = 0.1$ and $m = 1$.

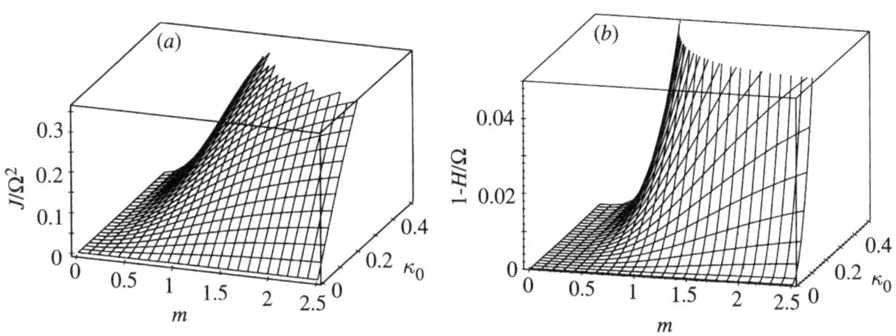

Figure 4. Three-dimensional plots of the amplitude (*a*) and end-shortening (*b*) of the helix (4.16) for $\nu = \frac{1}{3}$, $0 \leqslant \kappa_0 \leqslant 0.5$ and $0 < m \leqslant 2.5$.

to give

$$\tilde{\boldsymbol{r}}(t) = (H/\Omega)t\boldsymbol{k} + (J/\Omega^2)(\cos\Omega t\boldsymbol{j} - \sin\Omega t\boldsymbol{i}). \tag{4.16}$$

Equation (4.16) describes a helix with amplitude J/Ω^2 and pitch angle $\frac{1}{2}\pi - \theta$, where $\theta = \arctan(J/H)$, the angle of rotation in the transformation (4.13). The central axis is directed along $\boldsymbol{k}$ and the angular frequency Ω of the helix about its central axis is identical to that of the director $\boldsymbol{d}_1$, say, about that axis. *Hence (4.16) is a one-twist-per-wave spatially homogeneous equilibrium configuration of the rod.*

Figure 3 shows a solution (4.16) for some typical values of κ_0, m and ν, where H and J were determined by numerically solving the equations (4.7) and (2.14). Comparison with (3.8) gives that the dimensionless end-shortening per unit length is

$$\tilde{E} = 1 - (H/\Omega).$$

Recall the relative sizes (4.11) of H and J for small κ_0. Hence note that $\tilde{E} \to 0$ as $\kappa_0 \to 0$, as does the amplitude of the helix J/Ω^2. Figure 4 shows how the amplitude and end-shortening vary with m and κ_0.

(*b*) *Linearization*

Linearization around the equilibrium $\boldsymbol{x}^*$ shows that (2.12) always has two zero eigenvalues, as in the unperturbed case, which correspond to the rotational symmetry

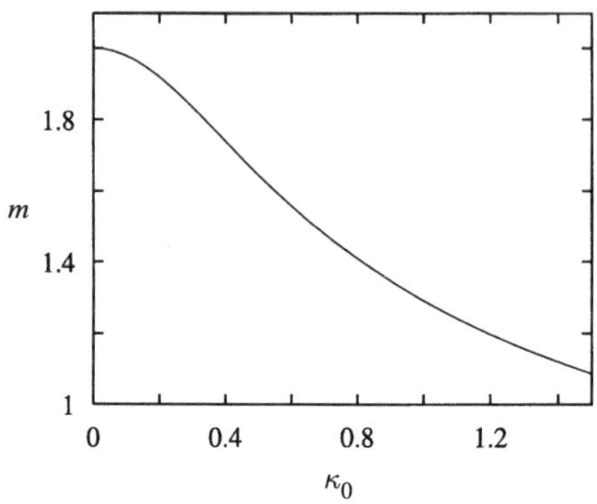

Figure 5. Locus in the (m, κ_0)-plane of Hamiltonian–Hopf bifurcation points.

in the problem. The other four eigenvalues satisfy the characteristic equation

$$\begin{aligned}\lambda^4 &+ (-2x_3^* m^2 + \tfrac{5}{3} m^4 \kappa_0 x_4^* + \tfrac{17}{9} m^4 x_6^{*2} + 2m^4 \kappa_0^2 + m^4 x_4^{*2})\lambda^2 m^{-4} \\ &+(-m^2 x_3^* x_4^{*2} - \tfrac{10}{27} m^4 x_6^{*2} \kappa_0 x_4^* + \tfrac{16}{81} m^4 x_6^{*4} + \tfrac{8}{9} x_6^{*2} m^2 x_3^* + x_3^{*2} \\ &+\tfrac{5}{3} x_6^* m^2 x_4^* x_1^* + \tfrac{35}{9} x_6^* m^2 \kappa_0 x_1^* - 2m^2 x_3^* \kappa_0^2 - \tfrac{1}{3} m^4 x_4^{*3} \kappa_0 + \tfrac{1}{3} m^4 x_4^{*2} \kappa_0^2 \tfrac{17}{9} m^4 x_6^{*2} \kappa_0^2 \\ &+ + m^4 \kappa_0^4 + \tfrac{5}{3} m^4 x_4^* \kappa_0^3 + \tfrac{1}{9} m^4 x_6^{*2} x_4^{*2} - \tfrac{5}{3} m^2 x_3^* \kappa_0 x_4^*) m^{-4} = 0, \end{aligned} \tag{4.17}$$

where, for simplicity, it is assumed that Poisson's ratio takes the numerical value $\nu = \frac{1}{3}$. This value will be taken in all numerical simulations which follow.

Analysis of (4.17) using Maple (solving for x_1, x_3, x_4 and x_6 using (4.7) and (2.14)) reveals that the situation is qualitatively the same as the unperturbed case with $\rho = 0$, $\kappa_0 = 0$. Namely that for $m < m_c(\kappa_0)$ there are four complex eigenvalues $\pm\lambda \pm i\omega$, implying physical stability of the spatially homogeneous equilibrium, and that at $m = m_c$ there is a Hamiltonian–Hopf bifurcation, with physical instability (four imaginary eigenvalues) occurring for $m > m_c$. A plot of $m_c(\kappa_0)$ is given in figure 5, which was determined numerically by solving for double roots of (4.17). Note that the value of m required for the buckling bifurcation to occur is initially flat and then decreases with increasing initial curvature κ_0.

(c) *Localized buckling solutions*

First note that adding initial curvature destroys part of the symmetry of the unperturbed system. Specifically, of the two symmetries R_1 and R_2 only R_2 remains (R_1 would have remained had we taken initial curvature about $\boldsymbol{d}_2$ instead of $\boldsymbol{d}_1$). Consequently, we should expect the existence of two rather than four reversible primary buckling modes. Our shooting method indeed gives us the two homoclinic orbits shown in figure 6. These solutions were computed for $\nu = \frac{1}{3}$, $\kappa_0 = 0.02$ and $m = 1.7$, while in the shooting method we used $\epsilon = 10^{-5}$ and

$$\begin{aligned}\boldsymbol{v}_1 &= (0.582029 \times 10^{-4}, 0.609510, -0.196407 \times 10^{-5}, \\ &\qquad 0.188564, 0.304719, -0.00636259), \\ \boldsymbol{v}_2 &= (-0.609205, 0, 0.0205577, -0.304777, 0.188629, 0.00494804).\end{aligned}$$

See van der Heijden (1997) for the meaning of ϵ, δ and $\boldsymbol{v}_i$.

Three-dimensional rod shapes, obtained by solving the centre line equation in

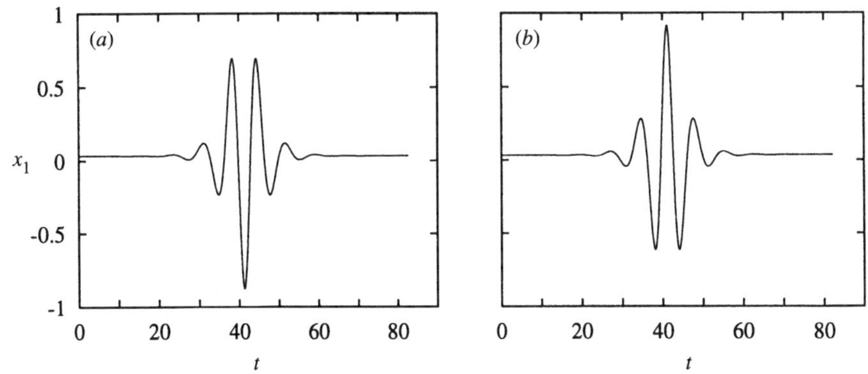

Figure 6. The two R_2-reversible primary buckling modes: (*a*) P_1 ($\delta = 5.156917$, $\mathcal{T} = 41.299427$); (*b*) P_2 ($\delta = 1.807504$, $\mathcal{T} = 41.097204$). ($\nu = \frac{1}{3}$, $\kappa_0 = 0.02$, $m = 1.7$.)

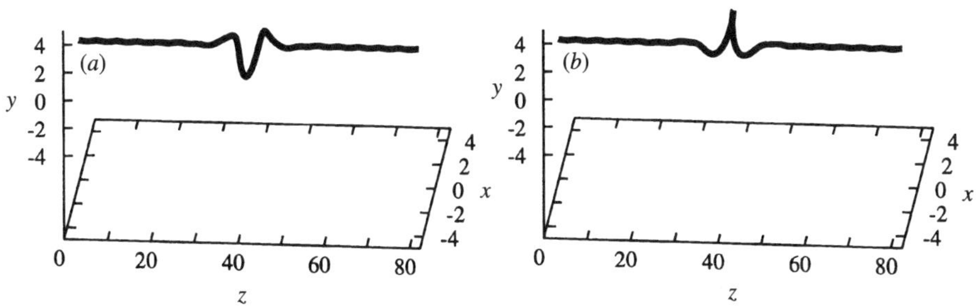

Figure 7. The two primary buckling modes P_1 (*a*) and P_2 (*b*) ($\nu = \frac{1}{3}$, $\kappa_0 = 0.02$, $m = 1.7$).

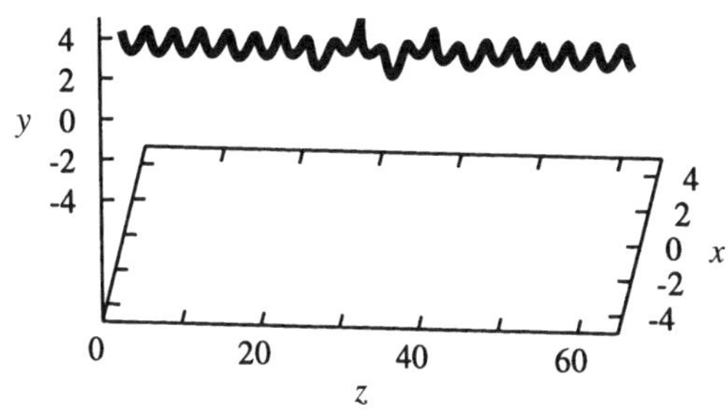

Figure 8. The primary buckling mode P_1 for large initial curvature ($\nu = \frac{1}{3}$, $\kappa_0 = 0.4$, $m = 1.7$).

(2.15), are depicted in figure 7. The figure nicely shows the small-amplitude helical solution of the rod away from the large deformation. Figure 8 shows one primary mode for large initial curvature $\kappa_0 = 0.4$. Here the distinction between the pre-buckled one-twist-per-wave equilibrium and the large deformation is somewhat blurred.

Because asymptotically towards infinity the spatially homogeneous solution is a helical rather than a straight rod, the $\boldsymbol{d}_3$ director of the local coordinate frames at the beginning and end point do not line up. We can not, therefore, use (3.11) directly to obtain the end deflections. End shortening and end rotation are now naturally computed with respect to the central axis $\boldsymbol{e}_3 = \boldsymbol{k}$ of the helix. Thus, to get the

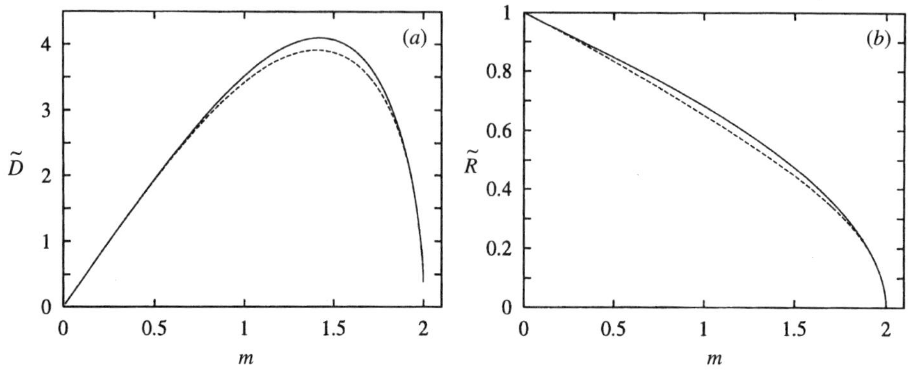

Figure 9. (a) End shortening $\tilde{D}$ and (b) end rotation $\tilde{R}$ versus load m for the primary homoclinic orbits P_1 (solid curves) and P_2 (dashed curves) ($\nu = \frac{1}{3}$, $\kappa_0 = 0.02$, implying $m_c = 1.999116$).

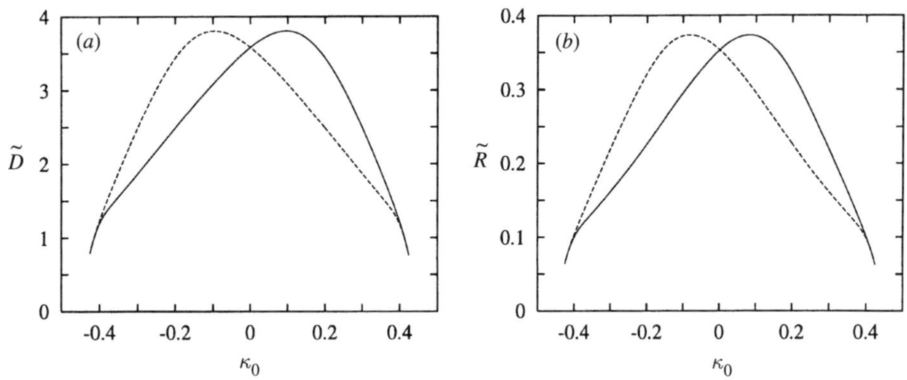

Figure 10. (a) End shortening $\tilde{D}$ and (b) end rotation $\tilde{R}$ versus initial curvature κ_0 for the primary homoclinic orbits P_1 (solid curves) and P_2 (dashed curves) ($\nu = \frac{1}{3}$, $m = 1.7$, implying the critical buckling value $\kappa_0 = 0.439763$).

end rotation we determine the projections $\hat{\boldsymbol{d}}_1$, $\hat{\boldsymbol{d}}_2$ of the vectors $\boldsymbol{d}_1$, $\boldsymbol{d}_2$ of the local coordinate frames at both ends of the rod onto the plane $(\boldsymbol{e}_1, \boldsymbol{e}_2)$ orthogonal to the axis $\boldsymbol{k}$. D and R are then computed from

$$D = \mathcal{T} - \boldsymbol{e}_3(\mathcal{T}), \quad \cos R = \langle \hat{\boldsymbol{d}}_1(\mathcal{T}), \hat{\boldsymbol{d}}_1(0) \rangle, \quad \sin R = \langle \hat{\boldsymbol{d}}_1(\mathcal{T}), \hat{\boldsymbol{d}}_2(0) \rangle, \tag{4.18}$$

where $\mathcal{T}$ represents the full length of the rod. Note that, in view of (4.13), $\hat{\boldsymbol{d}}_2 = \boldsymbol{d}_2$, and hence $\hat{\boldsymbol{d}}_2(0) = (0, 1, 0)$. The clean buckling contributions to the end deflections are again obtained from D and R by subtracting the contributions from the trivial solution, this time a helix:

$$\tilde{D} = (1 - \tilde{E})\mathcal{T} - \boldsymbol{e}_3(\mathcal{T}), \quad \tilde{R} = \frac{R - \Omega\mathcal{T}}{2\pi}. \tag{4.19}$$

Load-deflection diagrams are given in figure 9. Figure 10 shows how end shortening and end rotation change as the initial curvature κ_0 is varied. Note that the diagrams of figure 10 are symmetric about $\kappa_0 = 0$ (with a change of role of P_1 and P_2), as expected. At $\kappa_0 = 0$ the symmetric modes P_1 and P_2 are part of a full circle of primary homoclinic orbits, all with the same load–deflection characteristics. Indeed,

Table 1. *Sequence of R_2-reversible bi-modal homoclinic orbits with labels (P_1, n, P_1) (left) and (P_2, n, P_2) (right) for rods with initial curvature ($\nu = \frac{1}{3}$, $m = 1.7$, $\kappa_0 = 0.02$)*

n	δ_n	$\mathcal{T}_n$	$\mathcal{T}_n - \mathcal{T}_{n-1}$	n	δ_n	$\mathcal{T}_n$	$\mathcal{T}_n - \mathcal{T}_{n-1}$
1	5.527777	52.129804		1	2.520175	50.544884	
2	5.008232	53.422228	1.292424	2			
3	5.199237	55.519861	2.097633	3	1.913620	53.569436	
4	5.143442	57.342668	1.822808	4	1.769202	55.268184	1.698748
5	5.161089	59.246022	1.903353	5	1.818802	57.217025	1.948841
6	5.155616	61.124430	1.878408	6	1.803924	59.081629	1.864604
7	5.157323	63.010459	1.886029	7	1.808612	60.972140	1.890511
8	5.156791	64.894152	1.883693	8	1.807158	62.854410	1.882270
9	5.156957	66.778561	1.884408	9	1.807611	64.739273	1.884863
10	5.156905	68.662750	1.884189	10	1.807470	66.623317	1.884044
11	5.156921	70.547006	1.884256	11	1.807514	68.507620	1.884303
12	5.156916	72.431244	1.884237	12	1.807500	70.391841	1.884221
13	5.156918	74.315485	1.884241	13	1.807505	72.276088	1.884247
14	5.156917	76.199725	1.884240	14	1.807503	74.160327	1.884239
15	5.156917	78.083965	1.884240	15	1.807504	76.044568	1.884241
P_1	5.156917	41.299427	$(\pi/2\omega)$ $= 1.884241$	P_2	1.807504	41.097204	$(\pi/2\omega)$ $= 1.884241$

the corresponding $\tilde{D}$ and $\tilde{R}$ values are those of the unperturbed case included in dashed lines in figure 2.

The continuation code with fixed $\mathcal{T} = 82.5$ (full-length rod) has problems in the steep regions close to buckling, which is why the load–deflection curves do not reach right down to the κ_0-axis. A numerical check with larger values of the truncation interval $\mathcal{T}$ shows that the curves can be followed further down as $\mathcal{T}$ is increased. Hence, we conclude that we are dealing with an effect due to finite $\mathcal{T}$ and that P_1 and P_2 really bifurcate from the trivial solution at $\kappa_0 = 0.439763$.

Apart from breaking the symmetry of the unperturbed system, initial curvature also breaks complete integrability. Specifically, we no longer have conservation of torque about the body axis of the rod, i.e. $I_3 \neq$ const. As a result, we have a situation analogous to the case of an anisotropic rod studied extensively in van der Heijden *et al.* (1997), with a countable infinity of multi-modal homoclinic orbits amidst spatial chaos. These multi-modal solutions are located in the transverse intersection of the invariant manifolds of the fixed point and roughly consist of multiple copies of primary solutions separated by finite numbers of small-amplitude oscillations.

Table 1 gives data for the first 15 members of two families of R_2-reversible bimodals. The homoclinic orbits of one family are built up of two P_1-type humps and are accordingly labelled (P_1, n, P_1), where n is a measure of the number of small oscillations between the two large deformations (the choice of origin for n is somewhat arbitrary). Similarly, the orbits of the other family look like two copies of P_2 and are labelled (P_2, n, P_2). Data for $(P_2, 2, P_2)$ are missing because this orbit could not be found. It is not clear whether this solution somehow does not exist for the parameters

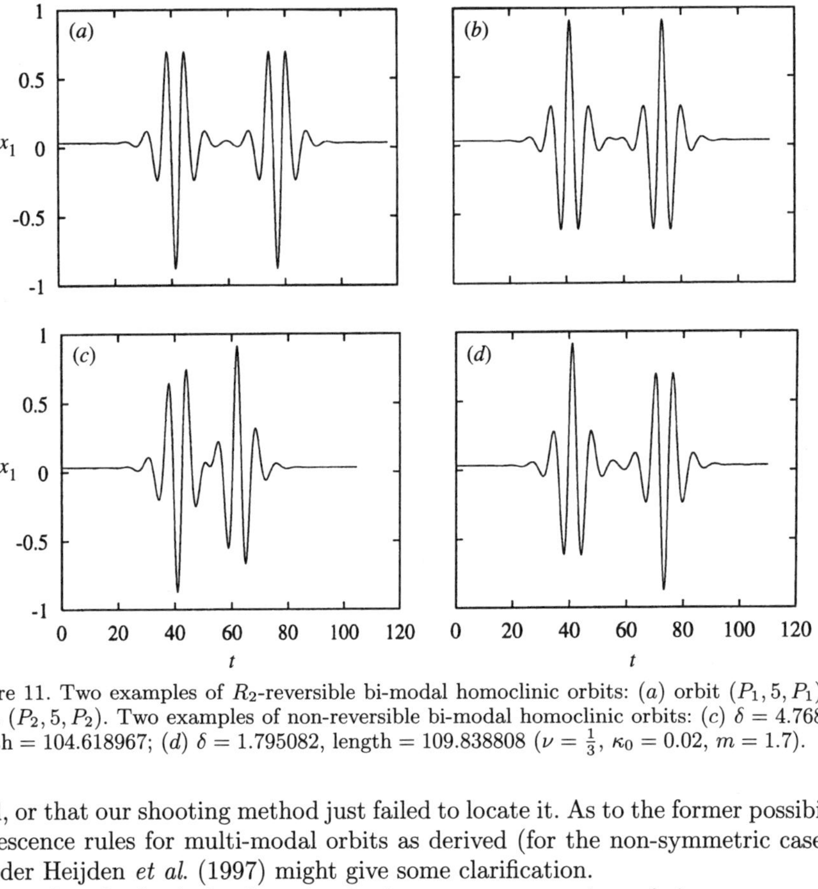

Figure 11. Two examples of R_2-reversible bi-modal homoclinic orbits: (*a*) orbit $(P_1, 5, P_1)$; (*b*) orbit $(P_2, 5, P_2)$. Two examples of non-reversible bi-modal homoclinic orbits: (*c*) $\delta = 4.768824$, length $= 104.618967$; (*d*) $\delta = 1.795082$, length $= 109.838808$ ($\nu = \frac{1}{3}$, $\kappa_0 = 0.02$, $m = 1.7$).

used, or that our shooting method just failed to locate it. As to the former possibility, coalescence rules for multi-modal orbits as derived (for the non-symmetric case) in van der Heijden *et al.* (1997) might give some clarification.

Note that for both families the δ-values converge to that of the corresponding primary orbit, while the difference between successive $\mathcal{T}$-values tends to $(\pi/2\omega)$, where ω is the (positive) imaginary part of the complex quadruple of eigenvalues at the fixed point $\boldsymbol{x}^*$. This limiting behaviour can be understood by imagining, in phase space, each successive homoclinic orbit in the sequence making an extra quarter turn around the fixed point immediately prior to hitting the symmetric section. Figures 11*a*, *b* show x_1–t plots for one member of each of the above two families of bi-modals.

The two infinite families (P_1, n, P_1) and (P_2, n, P_2) define all bi-modals which are themselves reversible. Non-reversible bi-modals with labels (P_1, n, P_2) and (P_2, n, P_1) do exist but are not captured by our shooting method in the present form. Two examples of non-reversible bi-modals, obtained by shooting over the full length of the rod, are given in figures 11*c*, *d*. As in the case of an anisotropic rod, reversible tri-modals, four-modals and higher-order multi-modals occur in similar families parametrized by multiple integers.

We should stress that our boundary conditions ensure that all the homoclinic orbits are computed for the same values of the two integrals of motion I_1 and I_2, namely those of the one-twist-per-wave equilibrium to which the localized solutions asymptote as $t \to \pm\infty$.

5. Physical interpretation

Let us finally make some physical interpretations of the results obtained. The main achievement of this paper is to demonstrate that a uniform initial curvature in the stress-free state does indeed give rise to one-twist-per-wave equilibria upon which localized buckling is superimposed at the critical load. Note from figure 5 that the critical buckling condition is not changed much by the introduction of small initial curvature. Neither is the post-buckling path of the primary buckling mode (compare figures 9 and 2).

We have also shown that (at least for the initially straight rod) allowing for shear/axial deformations makes little qualitative difference. However, when interpreting experiments under rigid loading it is important to note that 'shear' deformations, specifically axial extension, must occur if the rod is seen to buckle into a non-straight configuration. We must also take care not to make glib statements on experiments based on the so-far *dimensionless* analysis.

Note from the non-dimensionalization (2.11) that both $\tilde{\boldsymbol{r}}$ and t scale with M/B. Recalling that we are assuming a symmetric cross-section (i.e. $A = B$), the equilibrium solution (4.16) can then be written in dimensional coordinates as

$$\boldsymbol{r}(s) = \frac{H}{\Omega} s\boldsymbol{k} + \frac{AJ}{M\Omega^2}(\cos\omega s\boldsymbol{j} - \sin\omega s\boldsymbol{i}), \quad \text{where } \omega = \frac{M\Omega}{A}. \tag{5.1}$$

From (2.11), (4.8) and (4.9) it follows that for small initial curvature u_0 the dimensional helical characteristics are given by

$$\begin{aligned} \text{radius} &= \frac{AJ}{M\Omega^2} = \frac{AC^2}{M^2(A-C)+TC^2}u_0 + \mathcal{O}(u_0^3), \\ \text{frequency} = \omega &= \frac{M}{A}\Omega = \frac{M}{A}(1+\nu) + \mathcal{O}(u_0^2) = \frac{M}{C} + \mathcal{O}(u_0^2), \\ \tfrac{1}{2}\pi - \text{pitch angle} = \theta &= \arctan\left(\frac{J}{H}\right) = \frac{ACM}{M^2(A-C)+TC^2}u_0 + \mathcal{O}(u_0^3). \end{aligned} \tag{5.2}$$

We have explicitly computed the one-twist-per-wave helical equilibrium for the case where initial curvature is the only perturbation allowed in figure 13. It is easy to see, however, that the whole analysis goes through when anisotropy (ρ) and shear/extension (ϵ) are also taken into account. The crucial observation is that in the more general case we still have

$$x_2^* = x_5^* = 0. \tag{5.3}$$

Hence, although Ω will vary with ρ and ϵ we will still have a one-twist-per-wave solution. Note that (5.3) means that along the rod there is no force or moment in the direction $\boldsymbol{d}_2$; the helical rod is bent in the direction $\boldsymbol{d}_1$ of initial curvature, and forces and moments are in the tangent plane to the imaginary cylinder on which the helix winds. Indeed, it can be verified that the force, $\boldsymbol{n}$, is directed along the axis of the helix. The moment, $\boldsymbol{m}$, however, is not. In experiments one is likely to control the axial moment, $M_{\mathrm{ax}} := \boldsymbol{m}\cdot\boldsymbol{e}_3$, rather than the total moment M. In the Appendix, therefore, an analysis is given in terms of the axial moment. To first order in κ_0 (or u_0) taking M_{ax} or M does not make any difference: (5.2) is also valid with M replaced by M_{ax}.

If, in addition to initial curvature u_0, we also take M to be small, but such that $\kappa_0 = Au_0/M$ remains small (for instance by taking $u_0 = \mathcal{O}(\epsilon)$, $M = \mathcal{O}(\sqrt{\epsilon})$, in terms of an arbitrary small parameter ϵ), then to a good approximation (5.2) can be written as

$$\text{radius} = \frac{Au_0}{T}, \quad \text{frequency} = \frac{M}{C}, \quad \tfrac{1}{2}\pi - \text{pitch angle} = \frac{AMu_0}{TC}, \tag{5.4}$$

which is in agreement with the approximations (A 5) given in the Appendix.

We end by discussing some experiments serving to confirm the central result of this paper. The one-twist-per-wave pre-buckling deformations observed experimentally by Thompson & Champneys (1996) are shown in figure 12 for a circular rubber rod. Moving down through the sequence of pictures, as the loading is increased, we see a large-amplitude localized buckling mode with about three twists per wave superimposed on top of this pre-buckling response. The significance of the three is that it is approximately $2(1+\nu)$, as ν, Poisson's ratio of rubber, is approximately 0.5. This rod certainly had an initial curvature in its unstressed state, and the work of the present paper suggests very strongly indeed that this curvature was the cause of the one-twist-per-wave behaviour.

This conclusion is reinforced by the more recent experiments illustrated in figure 13 for a rod of square cross-section. This highly perfect rod, with no measurable initial curvature, was cast in silicone rubber. It was kindly supplied by C. R. Calladine of the Cambridge Engineering Laboratory. It was tested under water, under conditions of approximately neutral buoyancy. No pre-buckling deformation is observed, and the localized buckling mode, with about three twists per wave, arises directly from the trivial highly twisted straight configuration. Under the rigid loading of the test, a dynamic jump at constant end-shortening and constant end-rotation carries the rod from a buckled configuration (like that shown in the third picture) to self-contact in the writhing configuration of the fourth picture. The bottom picture shows the freely floating straight rod after the test. The high quality silicone rubber shows no sign of having suffered any permanent deformation during the test.

Appendix A. Physical analysis of one-twist-per-wave equilibria

It is instructive to make a direct physical study of the possible one-twist-per-wave equilibrium solutions of twisted rods. Consider a rod with principal bending stiffnesses A and B, which may or may not be equal, and torsional stiffness C. Let the rod have an initial curvature κ_0, which may or may not be zero, in the principal direction associated with A. We use as a measure of this curvature the applied bending moment that would just straighten the rod, namely $N = A\kappa_0$.

Inducing only (additional) bending in the A direction, consider a helical deformation in which the rod is wrapped with one twist per helical wave around an imaginary cylinder of radius r. Since there is no bending in the B direction, it follows that B will not appear in our analysis: it may or may not be equal to A. The assumed helical form is illustrated in figure 14 for a rod with a square cross-section. Here it should be supposed that in its unloaded state the initial curvature was such that the rod lay in a circle with the outer edge black.

Cutting the homogeneous helix at any point, the resultant moment vector, M, is the resultant of the bending moment $A(\kappa - \kappa_0)$ and the twisting moment $C\tau$, as illustrated. The bending moment is at right angles to the rod, the twisting moment is along the rod, and all three vectors lie in the tangent plane of the cylinder. Here, κ is the curvature of the helical space curve, given by the standard formula $\kappa = \sin^2\theta/r$,

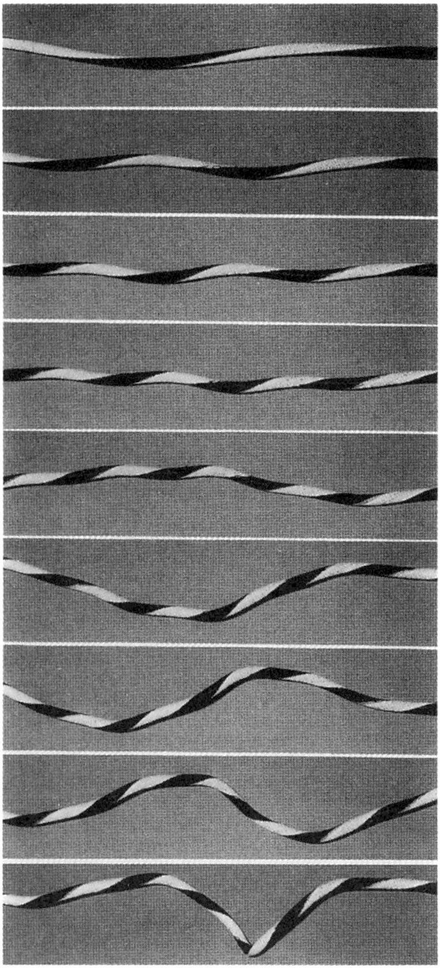

Figure 12. A photographic sequence during the test of an initially curved rod of circular cross-section. A one-twist-per-wave helical deformation is observed before the localized buckling.

where $\frac{1}{2}\pi - \theta$ is the pitch angle of the helix. With one twist per wave there is no internal twist ($\tau_{\mathrm{i}} = 0$), so the twist, τ, is equal to the torsion of the helical space curve, $\tau = \tau_{\mathrm{s}} = \sin\theta\cos\theta/r$.

For a homogeneous solution, the resultant force at the cut must be parallel to the axis, and we denote its magnitude by T. For the purpose of overall statics, we can transfer the line of action of T to the central axis of the cylinder with the introduction of the transfer moment of magnitude Tr lying in the tangent plane. Now the constant support reactions supplied at the remote left-hand end of the helix must be in balance with the above forces and moments at whatever phase, within a wave, the helix is cut. This is only possible if the vector sum of M and the transfer moment is parallel to the axis of the cylinder; notice that the vector of magnitude Tr drawn in the figure is the reversed transfer moment.

Under these conditions the helix is held in equilibrium by a wrench (compare Love

Figure 13. A photographic sequence during the under-water test of a silicone rubber rod of square cross-section. The rod had no initial curvature, and only a straight trivial state is observed before the localized buckling.

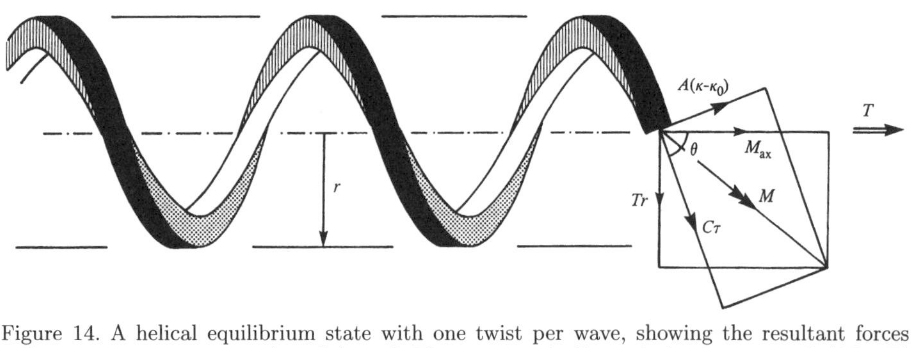

Figure 14. A helical equilibrium state with one twist per wave, showing the resultant forces and moments at a cross-section of the rod. These are statically equivalent to a wrench $(M_{\rm ax}, T)$ acting along the central chain-dotted axis.

1927, art. 270) comprising the axial tension, T, and the axial moment, $M_{\rm ax}$. Notice that $M_{\rm ax}$ is the axial component of the resultant moment M considered in the main body of this paper.

Setting the axial and circumferential components of M equal to $M_{\rm ax}$ and Tr, respectively, we have

$$\begin{aligned} M_{\mathrm{ax}} &= C\tau\cos\theta + (A/r)\sin^3\theta - N\sin\theta, \\ Tr &= C\tau\sin\theta - (A/r)\sin^2\theta\cos\theta + N\cos\theta. \end{aligned} \tag{A 1}$$

Applying the one-twist-per-wave requirement, $\tau = \tau_{\mathrm{s}} = \sin\theta\cos\theta/r$, gives

$$\begin{aligned} M_{\mathrm{ax}} &= (C/r)\sin\theta\cos^2\theta + (A/r)\sin^3\theta - N\sin\theta, \\ Tr &= (C/r)\sin^2\theta\cos\theta - (A/r)\sin^2\theta\cos\theta + N\cos\theta. \end{aligned} \tag{A 2}$$

Prescribing N, M_{ax} and T, these are two equations for the two helical variables θ and r. A first-order solution valid for small θ is given by

$$M_{\mathrm{ax}}/\theta = C/r - N, \tag{A 3}$$

$$Tr = (C - A)\theta^2/r + N. \tag{A 4}$$

(*a*) *Bending solution for the initially curved rod* ($N \neq 0$)

Let us suppose that under a fixed T we gradually increase M_{ax} from zero. With $M_{\mathrm{ax}} = 0$ we have the trivial physical solution with $\theta = 0$, $Tr = N$. This satisfies (A 4), and leaves (A 3) indeterminate. A first-order solution close to this trivial state, valid for small N, is given by

$$\theta = \frac{M_{\mathrm{ax}}N}{CT}, \quad r = \frac{N}{T}. \tag{A 5}$$

So to first order, for fixed T and N, the radius r is a constant, while the angle θ increases linearly with M_{ax}.

(*b*) *Buckling solution for a perfect* ($N = 0$) *anisotropic rod*

We next focus on the perfect ($N = 0$) rod with unequal principal bending stiffnesses. For such an anisotropic rod we would expect the one-twist-per-wave tape-like behaviour to involve bending about the weak axis. So we are led to assume $B > A$.

For a perfect rod, with $N = 0$, equation (A 3) gives

$$\frac{\theta}{r} = \frac{M_{\mathrm{ax}}}{C}, \tag{A 6}$$

and substituting into (A 4) gives

$$\frac{M_{\mathrm{ax}}^2}{T} = \frac{C^2}{C - A}. \tag{A 7}$$

Now the perfect rod (with fixed T, say) has the trivial solution $\theta = r = 0$, valid for all $M_{\mathrm{ax}}(= M)$. We can therefore identify (A 6), (A 7) (valid as $\theta \to 0$) as an eigenvalue problem with a non-trivial solution determined by the critical value of M_{ax}. Thus, (A 7) gives the critical buckling condition.

We see that for a real buckling solution we need C greater than the relevant bending stiffness (here, A). For a solid section, C is usually of the order of the lesser bending stiffness. So this analysis shows that there will usually be no possible equilibrium state for a tape-like rod standing, rather than lying, on the imaginary cylinder, a result also reached in van der Heijden & Thompson (1997).

The dimensionless version of (A 7), using (2.11), reads

$$m^2 = \frac{1 + \rho}{(1 + \nu)(\rho - \nu)}. \tag{A 8}$$

This is the equation for the curve m_{c_3} in the ρ–m parameter planes in Champneys & Thompson (1996) and van der Heijden *et al.* (1997). It is shown in van der Heijden & Thompson (1997) that under some conditions this curve is part of the buckling line, describing (lying) tape-like buckling.

References

Antman, S. 1995 *Nonlinear problems of elasticity.* Applied Mathematical Sciences 107. Berlin: Springer.

Antman, S. & Jordan, K. 1974 Qualitative aspects of the spatial deformation of non-linearly elastic rods. *Proc. R. Soc. Edinb.* A **73**, 85–105.

Champneys, A. R. & Spence, A. 1993 Hunting for homoclinic orbits in reversible systems: a shooting technique. *Adv. Comp. Math.* **1**, 81–108.

Champneys, A. R. & Thompson, J. M. T. 1996 A multiplicity of localized buckling modes for twisted rod equations. *Proc. R. Soc. Lond.* A **452**, 2467–2491.

Doedel, E., Keller, H. & Kernévez, J. 1991 Numerical analysis and control of bifurcation problems. I. Bifurcation in finite dimensions. *Int. J. Bifurcation and Chaos* **1**, 493–520, 745–772.

Iooss, G. & Pérouème, M. 1993 Perturbed homoclinic solutions in reversible 1:1 resonance vector fields. *J. Differ. Equat.* **102**, 62–88.

Kirchhoff, G. 1859 Über das Gleichgewicht und die Bewegung eines unendlich dünnen elastischen Stabes. *J. für Mathematik (Crelle)* **56**, 285–313.

Love, A. E. H. 1927 *A treatise on the mathematical theory of elasticity*, IVth edn. Cambridge University Press.

Mielke, A. & Holmes, P. 1988 Spatially complex equilibria of buckled rods. *Arch. Ration. Mech. Analysis* **101**, 319–348.

Thompson, J. M. T. & Champneys, A. R. 1996 From the helix to localized writhing in the torsional post buckling of elastic rods. *Proc. R. Soc. Lond.* A **452**, 117–138.

Timoshenko, S. P. & Gere, J. M. 1961 *Theory of elastic stability.* New York: McGraw-Hill.

Van der Heijden, G. & Thompson, J. M. T. 1997 Lock-on to tape-like behaviour in the torsional buckling of anisotropic rods. *Physica* D (In the press.)

Van der Heijden, G., Champneys, A. R. & Thompson, J. M. T. 1997 Load–deflection characteristics of localised torsional buckling modes in rods with non-circular cross-section. *SIAM J. Appl. Math.* (In the press.)

Van der Meer, J. C. 1985 *The Hamiltonian Hopf bifurcation (Lecture Notes in Mathematics 1160).* Berlin: Springer.

Elastic rods, rigid bodies, quaternions and the last quadrature

By S. Kehrbaum and J. H. Maddocks†

Institute for Physical Science and Technology and Department of Mathematics, University of Maryland, College Park, MD 20742, USA

The Kirchhoff kinetic analogy relates the governing equations for the statics of elastic rods and the dynamics of rigid bodies. We discuss the analogy in light of several different Hamiltonian formulations, including a non-canonical description of rod equilibria. We focus on the last three quadratures that are required to reconstruct the rod centreline from the frame variables, which form the complete configuration space in the rigid body interpretation. In particular, we demonstrate that if the frame evolution is formulated as a canonical Hamiltonian system involving quaternions (or Euler parameters), then the last quadratures can all be computed explicitly in terms of algebraic relations involving invariants (or integrals) of the evolution, independent of whether or not the entire system is completely integrable.

1. Introduction

Long elastic rods provide a class of examples within solid mechanics that can exhibit localization effects both within equilibrium (e.g. Champneys & Thompson 1996*a*,*b*; van der Heijden *et al.* 1997) and dynamic (e.g. Dichmann *et al.* 1996; Coleman *et al.* 1993) theories. Within the special Cosserat description a rod is taken to be a framed curve, i.e. a centreline $\boldsymbol{r}$ along with an orthonormal frame $\{\boldsymbol{d}_i\}_{i=1}^3$, where for equilibria (or indeed for travelling wave solutions of dynamics) the functions $\boldsymbol{r}$ and $\{\boldsymbol{d}_i\}$ are each parametrized by a single variable s.

The three observations that follow in the main part of this article apply to quite general *hyperelastic* Cosserat rod models encompassing effects of shear and extension. (A rod is hyperelastic if there is an associated scalar strain energy function that generates all associated constitutive relations.) However, for the sake of brevity we shall here restrict ourselves to the case of an inextensible, unshearable rod, for which the pointwise constraint

$$\boldsymbol{r}' = \boldsymbol{d}_3 \tag{1.1}$$

is satisfied everywhere (more general cases can be found in Kehrbaum (1997)). Therefore, the parameter s can be interpreted as arclength in any configuration and $\boldsymbol{d}_3$ coincides with the unit tangent to the centreline. In the absence of any external loads acting on the rod, such as gravity, the centreline $\boldsymbol{r}(s)$ does not explicitly appear in the potential energy of the rod, and as a consequence the force $\boldsymbol{n}(s)$ acting across

† Present address: Départment de Mathematiques, Ecole Polytechnique Fédérale de Lausanne, CH-1015 Lausanne, Switzerland.

Phil. Trans. R. Soc. Lond. A (1997) **355**, 2117–2136

each cross-section is in fact a constant $\boldsymbol{n}$ independent of s. Accordingly, the system of equations governing the equilibria of a rod can be decoupled into a part independent of $\boldsymbol{r}$ governing the frame $\{\boldsymbol{d}_i\}$ and once these equations have been solved (in some sense), the centreline $\boldsymbol{r}(s)$ can be recovered by carrying out the three quadratures inherent in the vector constraint (1.1). This is the observation at the heart of the *Kirchhoff kinetic analogy*, which, in its simplest form, states that for a linearly elastic rod, the reduced equations governing the rod frame $\{\boldsymbol{d}_i(s)\}$, parametrized by arclength s, are of a form identical with those governing the motion of a frame $\{\boldsymbol{d}_i(t)\}$, parametrized by time t, that is embedded in a heavy rigid body tumbling about a fixed point.

The Kirchhoff kinetic analogy is extremely useful for rod mechanics. Results from the vast literature concerning rigid body dynamics can be immediately translated to yield information for linearly elastic rods. In particular, the classic integrable case of the Lagrange top carries over to yield the integrable case of an isotropic, uniform rod, along with associated closed form solutions for the directors in terms of Euler angles (or other parametrizations of $SO(3)$). Recently, there has been renewed interest in tumbling rigid body dynamics as an example of a Hamiltonian system possessing a non-canonical Poisson bracket (cf. for example, Novikov 1982; Holmes & Marsden 1983) and, by invocation of the kinetic analogy, that non-canonical formulation can be exploited in the context of rods, as has been done by Mielke & Holmes (1988).

However, as is well known but sometimes ignored, the Kirchhoff kinetic analogy should not be pushed too far. For example, in the tumbling rigid body problem, the most natural side conditions involve an initial value problem, so that all the variables are known at $t = 0$, say. (Initial value problems combine well with closed form solutions known in completely integrable cases, because all the constants of integration arising in the associated quadratures are then known explicitly.) In contrast, many problems in rod mechanics involve two-point boundary value problems, so that unless the problem is statically determinate, the constant value for the force $\boldsymbol{n}$ is not known explicitly and must instead be determined as part of the solution satisfying prescribed boundary conditions. That issue has no simple analogy in rigid body dynamics. (It would correspond to motion in a gravity field whose direction and magnitude are unknown.)

The analogy is further strained by the fact that the centreline $\boldsymbol{r}(s)$ has no counterpart in the tumbling rigid body problem and the additional quadratures necessary to determine $\boldsymbol{r}(s)$ from (1.1) do not arise. Recently, it has been shown (from several rather different perspectives, e.g. Langer & Singer (1996); Shi & Hearst (1994); Ilyukhin (1979); Li (1986)) that the additional quadratures (1.1) can be carried out explicitly when the reduced problem corresponds to the integrable case of the Lagrange top. These closed form solutions provide much interesting information; for example, they imply that the centreline $\boldsymbol{r}(s) \in \mathbb{R}^3$ of all closed configurations of an isotropic, uniform rod lie on the surface of a torus of revolution. However, these closed form solutions again do not represent truly explicit solutions of many two-point boundary value problems for rods, because of the presence of undetermined constants. Tan & Witz (1995) discuss certain rod boundary value problems that can be solved explicitly, but these cases can in some sense be regarded as exceptional. In particular, while $\boldsymbol{r}(s)$ does not usually appear in the rod equilibrium equations, both $\boldsymbol{r}(0)$ and $\boldsymbol{r}(1)$ often do appear explicitly in the boundary conditions, and it is in such cases that the Kirchhoff kinetic analogy, the associated reduction to the tumbling rigid body problem and the last quadratures are particularly problematic.

In this article we wish to make three observations. First we show that the unreduced rod problem can be written as a twelve-dimensional non-canonical Hamiltonian system in which the associated Poisson bracket is an extension of the rigid body Poisson bracket (cf. Novikov 1982; Holmes & Marsden 1983; Mielke & Holmes 1988). (To our knowledge this twelve-dimensional extended bracket has not been described previously, although it is closely related to a nine-dimensional bracket that was introduced and exploited within the context of satellite dynamics (see Wang *et al.* 1991, 1992).) The non-canonical Hamiltonian structure persists in the presence of various external potentials such as an external uniform gravity field. The existence and commutation relations of the integrals that arise in the extended system yield yet another perspective on the quadratures (1.1) and complete integrability of the rod problem.

Second, considerable interesting information is lost if the reduction inherent to the kinetic analogy is invoked too soon. In particular, we show that two of the three quadratures inherent in (1.1) can always be eliminated (or performed explicitly, depending on your point of view), independent of whether or not the problem is integrable. In particular, once a solution to the three degree-of-freedom tumbling rigid body problem has been found, a single additional scalar quadrature is required to reconstruct the centreline of the rod. The analysis we follow is a variant of that given in Ilyukhin (1979) (with a special case appearing in Landau & Lifshitz (1970)).

Third and finally, we observe that if the reduction inherent to the kinetic analogy is performed on a seven degree-of-freedom canonical Hamiltonian description of the rod equilibrium equations that exploits a quaternion (or Euler parameter) parametrization of the director frame (as introduced by Li & Maddocks (1997)), then all three of the quadratures (1.1) can be explicitly performed in terms of the variables appearing in the resulting four degree-of-freedom system. In particular, the additional degree of freedom introduced in the four parameter, singularity-free description of the locally three-dimensional group $SO(3)$ provides an additional integral that allows the last quadrature in (1.1) to be carried out explicitly. As a consequence, any two-point boundary value problem for a rod, including ones with boundary conditions explicitly involving both $\boldsymbol{r}(0)$ and $\boldsymbol{r}(1)$, can be rewritten as a two-point boundary value problem for the reduced four degree-of-freedom system that describes the frame evolution in terms of quaternions.

The paper is structured as follows. Section 2 contains a brief synopsis of necessary background material, namely a description of the equilibrium theory of Cosserat rods in § 2 *a*, a description of the equations of a tumbling rigid body and the Kirchhoff kinetic analogy in § 2 *b* and certain relevant issues concerning Hamiltonian systems in § 2 *c*. Then § 3 deals with different Hamiltonian formulations of the equations of tumbling rigid body dynamics and the equilibria of elastic rods. Finally in § 4 we discuss how to reduce the quadratures (1.1) to a single, scalar (*last*) *quadrature*; the latter can be resolved by exploiting the canonical quaternion Hamiltonian description of elastic rods.

2. Background material

(*a*) *Cosserat theory of elastic rods*

In this section we review the director theory of elastic, inextensible and unshearable (Cosserat) rods (cf. Antman 1995, ch. VIII, IX). The configuration of a rod is specified by a smooth curve $\boldsymbol{r}(s)$, that can be interpreted as the centreline of the rod, and an

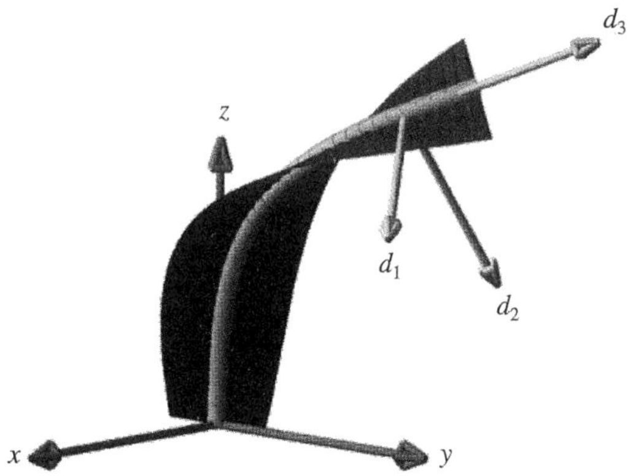

Figure 1. A typical configuration of a rod. The tube represents the centreline and the director $\boldsymbol{d}_1$ traces the ribbon, representing twisting.

orthonormal frame $\{\boldsymbol{d}_i\}_{i=1}^3$, called the *directors*, that serves to locate the orientation of a cross-section of the rod (cf. figure 1). We assume that the rod is effectively unshearable and inextensible. In mathematical terms, inextensibility is expressed as the condition $|\boldsymbol{r}'| = 1$ and unshearability is $\boldsymbol{r}' \cdot \boldsymbol{d}_1 = \boldsymbol{r}' \cdot \boldsymbol{d}_2 = 0$. Consequently, we may write $\boldsymbol{r}' = \boldsymbol{d}_3$, so that the parameter s can be interpreted as non-dimensionalized arclength in any configuration of the rod. It will be natural to express components of vectors with respect to the director frame; for any vector $\boldsymbol{p}$, the triple of components $p_i = \boldsymbol{p} \cdot \boldsymbol{d}_i$ will be denoted with the sans-serif symbol p. The triple p will be referred to as the body components of $\boldsymbol{p}$.

Orthonormality of the directors $\{\boldsymbol{d}_i\}$ implies the existence of a *strain vector* $\boldsymbol{u}$ satisfying

$$\boldsymbol{d}_i' = \boldsymbol{u} \times \boldsymbol{d}_i, \quad i = 1, 2, 3. \tag{2.1}$$

The components $u_i = \boldsymbol{u} \cdot \boldsymbol{d}_i$ with respect to the frame $\{\boldsymbol{d}_i\}$ are the bending ($i = 1, 2$) and twisting ($i = 3$) strains.

In rod mechanics, the forces and moments acting across a material cross-section are averaged to yield a net force $\boldsymbol{n}(s)$ and a net moment $\boldsymbol{m}(s)$. Then balance of moments and forces imply the coordinate-free equilibrium equations

$$\boldsymbol{m}' + \boldsymbol{r}' \times \boldsymbol{n} = \boldsymbol{0}, \quad \boldsymbol{n}' = \boldsymbol{0}. \tag{2.2}$$

The components $m_i = \boldsymbol{m} \cdot \boldsymbol{d}_i$ are the bending and twisting moments in the rod. The moments m_i must be related to the strains u_i via suitable constitutive relations. Here we make the assumption of hyperelasticity: for $\mathsf{v} = (v_1, v_2, v_3)$, there exists a convex, coercive strain energy density function $W(\mathsf{v}, s)$ such that $W_{\mathsf{v}}(\boldsymbol{0}, s) = \boldsymbol{0}$. The function W is *coercive*, if

$$\lim_{|\mathsf{v}| \to \infty} \frac{W(\mathsf{v}, s)}{|\mathsf{v}|} = \infty$$

(cf. Antman 1995, ch. VIII). The constitutive relations are taken to be of the form

$$\mathsf{m} = W_{\mathsf{v}}(\mathsf{u} - \hat{\mathsf{u}}, s), \tag{2.3}$$

where $\hat{\mathsf{u}}(s) = (\hat{u}_1(s), \hat{u}_2(s), \hat{u}_3(s))$ are the components of strain in the minimum energy unstressed configuration. A standard approximation in rod theory is to assume the rod to be linearly elastic (i.e. the strain energy is a quadratic function of the strains); in particular, one often adopts the diagonal quadratic form

$$W(\mathsf{v}, s) = \frac{1}{2}\sum_{i=1}^{3} K_i(s)v_i^2. \tag{2.4}$$

This example of the Cosserat theory is often called the Kirchhoff rod.

We will call the constitutive relation *isotropic* if the strain energy density function $W(\mathsf{v}, s)$ is invariant under rotations about the $\boldsymbol{d}_3$-axis; equivalently,

$$\frac{\partial}{\partial\alpha} W(\kappa\cos\alpha, \kappa\sin\alpha, v_3, s) = 0,$$

for all $0 \leqslant \alpha < 2\pi$, $\kappa > 0$ and $v_3 \in \mathbb{R}$ (cf. Maddocks 1984). The constitutive relation (2.4) is isotropic if $K_1(s) \equiv K_2(s)$. We will further describe the *rod* as isotropic if it has no preferred material direction for bending; provided that the constitutive relation is isotropic, the rod is isotropic if, in addition, $\hat{u}_1(s) \equiv \hat{u}_2(s) \equiv 0$. The constitutive relation is *uniform* if the strain energy function W does not explicitly depend on arclength s. The rod is uniform if, in addition, the components of unstressed strain $\hat{\mathsf{u}}$ are constant.

(*b*) *Kirchhoff's kinetic analogy and the dynamics of tumbling rigid bodies*

According to Love (1944), the classic form of the Kirchhoff kinetic analogy is based on a rod theory with linear constitutive relations of the form

$$\mathsf{m} = \mathsf{K}(s)(\mathsf{u} - \hat{\mathsf{u}}(s)), \tag{2.5}$$

where K is a positive definite matrix. If the matrix K is constant and $\hat{\mathsf{u}} \equiv \mathbf{0}$, equations (2.2) are equivalent to the classic equations of motion of a heavy rigid body in a uniform gravitational field tumbling about a fixed point; in this case, the frame $\{\boldsymbol{d}_i\}$ is fixed in the body (and referred to as the body frame) in such a way that the centre of mass lies on the $\boldsymbol{d}_3$-axis, the matrix K is the inertia tensor of the rigid body about the fixed point referred to the $\{\boldsymbol{d}_i\}$ frame and equations (2.2) are typically rewritten as

$$\boldsymbol{m}' + \boldsymbol{d}_3 \times \boldsymbol{n} = \mathbf{0}, \quad \boldsymbol{n}' = \mathbf{0}, \tag{2.6}$$

with the independent parameter s now interpreted as time (and usually written t). We shall refer to this problem as the tumbling rigid body. It is also often assumed that the body frame coincides with the principal axes of inertia, so that K is a diagonal matrix with entries $K_1 \leqslant K_2 \leqslant K_3$, which is a restriction on the possible locations of the centre of mass. Nevertheless, in this special case, the equations describing the dynamics of a tumbling rigid body are identical in form to those describing the equilibria of a Kirchhoff rod with diagonal strain energy density (2.4).

Two important special cases in which the problem is completely integrable (in the classic sense) have been studied extensively. A tumbling rigid body that is free to move with no external forces acting on it (or equivalently the fixed point is at the centre of mass) is often described as the free rigid body. In the context of rods the free rigid body corresponds to special problems where the end-loadings are pure moments, so that $\boldsymbol{n} = \mathbf{0}$. The second case is that of a Lagrange top, which is a tumbling rigid body with a diagonalized inertia tensor in which K_1 and K_2 are equal. The Lagrange top corresponds to an isotropic, uniform Kirchhoff rod.

A less widely known fact is that the Kirchhoff kinetic analogy persists for non-zero $\hat{u}$. For example, Wittenburg (1977) considers a gyrostat comprising a carrier rigid body in which is mounted a spinning sub-body (often called a momentum wheel). The function $\hat{u}$ can then be interpreted as the angular velocity of the momentum wheel relative to the carrier and the analogy applies if one assumes that these relative velocities are prescribed functions of time. (The analogy can also be made for more general systems; for example, for gyrostats with multiple momentum wheels.)

(c) *Hamiltonian systems*

We will require some basic concepts and terminology from finite-dimensional Hamiltonian systems (cf. for example, Olver 1993, ch. 6). Let $\boldsymbol{X}$ be a smooth, m-dimensional manifold (often $\boldsymbol{X} = \mathbb{R}^{2n}$). A Poisson bracket is an operator that assigns a smooth real-valued function $\{F, G\}$ to each pair of real-valued, smooth functions F and G defined on $\boldsymbol{X}$. The operation is bilinear, skew-symmetric and satisfies the Jacobi identity and the Leibniz rule. In local coordinates $\boldsymbol{z} = (z_1, \dots, z_m)$, any Poisson bracket can be written

$$\{F, G\}(\boldsymbol{z}) = \nabla F(\boldsymbol{z}) \cdot \mathcal{J}(\boldsymbol{z}) \nabla G(\boldsymbol{z}),$$

where $\mathcal{J}(\boldsymbol{z}) = (\mathcal{J}_{ij}(\boldsymbol{z}))$ is a skew-symmetric *structure matrix* defined by $\mathcal{J}_{ij}(\boldsymbol{z}) = \{z_i, z_j\}$, $1 \leqslant i, j, \leqslant m$, and ∇ denotes the gradient with respect to the phase variables $\boldsymbol{z}$. Given a Hamiltonian function $\mathcal{H} : \boldsymbol{X} \times \boldsymbol{R} \to \mathbb{R}$, a Hamiltonian system is the m-dimensional system of ordinary differential equations

$$\boldsymbol{z}' = \mathcal{J}(\boldsymbol{z}) \nabla \mathcal{H}(\boldsymbol{z}, t), \tag{2.7}$$

so that for any solution $\boldsymbol{z}(t)$, and any smooth function $F(\boldsymbol{z})$,

$$\frac{\mathrm{d}}{\mathrm{d}t} F(\boldsymbol{z}) = \{F, H\}(\boldsymbol{z}).$$

The coordinates $\boldsymbol{z}$ on an even-dimensional manifold $\boldsymbol{X} = \mathbb{R}^{2n}$ will be called canonical if the associated structure matrix is the standard constant matrix

$$\boldsymbol{J}_{2n} = \begin{pmatrix} \boldsymbol{0} & \boldsymbol{I} \\ -\boldsymbol{I} & \boldsymbol{0} \end{pmatrix}, \tag{2.8}$$

where $\boldsymbol{I}$ and $\boldsymbol{0}$ are the identity and null matrices of dimension n.

A function $I : \boldsymbol{X} \to \mathbb{R}$ is an integral of (2.7) if it is constant along trajectories. Equivalently, I is an integral if

$$\{I, \mathcal{H}\}(\boldsymbol{z}) = \nabla I(\boldsymbol{z}) \cdot \mathcal{J}(\boldsymbol{z}) \nabla \mathcal{H}(\boldsymbol{z}, t) = 0.$$

In particular, there exist integrals called Casimirs that exist for any Hamiltonian. The function $C : \boldsymbol{X} \to \mathbb{R}$ is a Casimir if $\{C, F\} = 0$ for all smooth functions F. Equivalently, C is a Casimir if ∇C is in the null-space of $\mathcal{J}$ (cf. Olver 1993). In a canonical formulation there are no Casimirs since the operator $\boldsymbol{J}_{2n}$ is non-singular.

The Hamiltonian system (2.7) is called completely integrable if its solution can be reduced to quadratures, so that (at least in principle) a solution to the initial value problem can be found in closed form. Two integrals I_1 and I_2 are said to be in involution (or to commute) if I_1 is an integral with respect to the flow generated by I_2, i.e.

$$\{I_1, I_2\}(\boldsymbol{z}) = \nabla I_1(\boldsymbol{z}) \cdot \mathcal{J}(\boldsymbol{z}) \nabla I_2(\boldsymbol{z}) = 0.$$

Suppose the Hamiltonian system (2.7) possesses k Casimirs and an additional l integrals in involution whose gradients are linearly independent. Then the system is completely integrable if

$$2l + k = m.$$

Notice that there may well be more than $\frac{1}{2}(m-k)$ integrals with linearly independent gradients, without there being a sufficient number of commuting integrals for the system to be completely integrable. This circumstance arises in the Hamiltonian systems governing rod equilibria that we describe below.

3. Hamiltonian formulations

In this section we derive several Hamiltonian formulations governing the equilibria of elastic rods and the motion of a tumbling rigid body. In § 3 *a* we review a non-canonical Hamiltonian formulation for the dynamics of a tumbling rigid body (cf. Novikov 1982; Holmes & Marsden 1983). This Hamiltonian rigid body formulation has been used by Mielke & Holmes (1988) in the context of rod mechanics to determine the orientation of the director frame, and they then recover the spatial components of the variable $\boldsymbol{r}$ by solving the additional differential equations (1.1) and (2.1), but from a non-Hamiltonian perspective. In § 3 *b* we present an extended non-canonical Hamiltonian formulation that includes the body components of both the centreline $\boldsymbol{r}$ and an additional fixed vector $\boldsymbol{i}$ in the phase space. The integrals of this extended Hamiltonian system provide a different perspective on the final quadratures (1.1). The spatial components of the centreline $\boldsymbol{r}$ can be obtained by simple algebraic manipulations from the solution in the body coordinate phase space. A more classical approach is to obtain a canonical Hamiltonian formulation by combining the spatial components of $\boldsymbol{r}$ with some parametrization of the rotation group $SO(3)$. In § 3 *c* we describe a seven degree-of-freedom canonical Hamiltonian system that arises in this way when a parametrization of $SO(3)$ in terms of quaternions is adopted. Finally, in § 3 *d* we discuss the various manifestations of integrals, commutativity properties and completely integrable cases that arise in the different Hamiltonian formulations.

(*a*) *Non-canonical formulation for the dynamics of a tumbling rigid body*

In this formulation we rewrite the balance equations (2.6) in body coordinates

$$\mathsf{m}' = \mathsf{m} \times \mathsf{u} + \mathsf{n} \times \mathsf{e}_3, \tag{3.1}$$

$$\mathsf{n}' = \mathsf{n} \times \mathsf{u}, \tag{3.2}$$

where $\mathsf{e}_3 = (0,0,1)^{\mathrm{T}}$ and u and m are related via (2.5). The Hamiltonian structure of this system has been investigated by Novikov (1982) and Holmes & Marsden (1983), for example. Specifically, a Poisson bracket can be defined in local coordinates by the structure matrix

$$\boldsymbol{J}(\mathsf{m},\mathsf{n}) = \begin{pmatrix} \mathsf{m}\times & \mathsf{n}\times \\ \mathsf{n}\times & \mathbf{0} \end{pmatrix}, \tag{3.3}$$

where for any triple $\mathsf{p} = (p_1, p_2, p_3)$ we use the notation $\mathsf{p}\times$ to denote the matrix

$$\mathsf{p}\times = \begin{pmatrix} 0 & -p_3 & p_2 \\ p_3 & 0 & -p_1 \\ -p_2 & p_1 & 0 \end{pmatrix},$$

corresponding to the vector product. Then equations (3.1) and (3.2) are equivalent to the Hamiltonian system

$$\begin{pmatrix} m \\ n \end{pmatrix}' = \boldsymbol{J}(m, n)\nabla H(m, n, s), \tag{3.4}$$

with Hamiltonian

$$H(m, n, s) = \tfrac{1}{2} m \cdot K^{-1} m + \hat{u}(s) \cdot m + n \cdot e_3.$$

Equations (3.4) are the well-known equations of Euler and Poisson for the motion of a tumbling rigid body.

(b) *Non-canonical formulation for equilibria of rods*

We now rewrite the equations governing rod equilibria entirely in terms of body variables to extend the formulation of the previous section to a twelve-dimensional non-canonical Hamiltonian system. Assume $\boldsymbol{n} \neq \boldsymbol{0}$, let $\boldsymbol{k} = \boldsymbol{n}/|\boldsymbol{n}|$, $\boldsymbol{i}$ be a fixed unit vector and define $\boldsymbol{j} = \boldsymbol{k} \times \boldsymbol{i}$. Then $\{\boldsymbol{i}, \boldsymbol{j}, \boldsymbol{k}\}$ define a basis of $\boldsymbol{R}^3$, provided that $\boldsymbol{i}$ and $\boldsymbol{n}$ are non-collinear.

We rewrite the differential equations (1.1) and (2.2) in terms of body coordinates to obtain

$$r' = r \times u + e_3, \tag{3.5}$$

$$m' = m \times u + n \times e_3, \tag{3.6}$$

$$n' = n \times u. \tag{3.7}$$

To complete the description we also rewrite the trivial equation $\boldsymbol{i}' = \boldsymbol{0}$ in body coordinates, i.e.

$$i' = i \times u, \tag{3.8}$$

to obtain an additional Poisson equation. The constitutive relation (2.3) relates u and m, which can be inverted to yield

$$u = \frac{\partial W^*(m, s)}{\partial m} + \hat{u},$$

where $W^*(m, s)$ is the Legendre transform of $W(u, s)$, i.e.

$$W^*(m, s) = \sup_{v \in \mathbb{R}^3} \{m \cdot v - W(v, s)\} \tag{3.9}$$

(cf. Li & Maddocks 1997; Kehrbaum & Maddocks 1997). Equations (3.5)–(3.8) can then be seen to possess a Hamiltonian structure. Define the Hamiltonian

$$H(i, r, m, n, s) = W^*(m, s) + m \cdot \hat{u} + n \cdot e_3. \tag{3.10}$$

Then Hamilton's equations are of the form

$$\begin{pmatrix} i \\ r \\ m \\ n \end{pmatrix}' = \boldsymbol{J}(i, r, m, n)\nabla H(i, r, m, n, s), \tag{3.11}$$

where the skew-symmetric operator $\boldsymbol{J}$ is defined by

$$\boldsymbol{J}(i,r,m,n)=\begin{pmatrix} \mathbf{0} & \mathbf{0} & i\times & \mathbf{0} \\ \mathbf{0} & \mathbf{0} & r\times & \boldsymbol{I} \\ i\times & r\times & m\times & n\times \\ \mathbf{0} & -\boldsymbol{I} & n\times & \mathbf{0} \end{pmatrix}. \tag{3.12}$$

In order that $\boldsymbol{J}$ actually generate a Poisson bracket, the Jacobi identity must be verified (cf. Olver 1993, proposition 6.8). The necessary calculations are computationally tedious, but have been directly checked using a symbolic manipulation package (Mathematica).

For the non-canonical formulation to be meaningful it is important that the fixed frame components of all variables be determined from the phase variables i, r, m and n. We now show that this is indeed the case. For convenience we assume that the fixed frame $\{\boldsymbol{i},\boldsymbol{j},\boldsymbol{k}\}$ is orthonormal. The rotation matrix can then be constructed in terms of the nine direction cosines relating the fixed frame $\{\boldsymbol{i},\boldsymbol{j},\boldsymbol{k}\}$ and the director frame $\{\boldsymbol{d}_i\}_{i=1}^3$. We write the direction cosines in the form

$$\boldsymbol{R}=\begin{pmatrix} \boldsymbol{d}_1\cdot\boldsymbol{i} & \boldsymbol{d}_1\cdot\boldsymbol{j} & \boldsymbol{d}_1\cdot\boldsymbol{k} \\ \boldsymbol{d}_2\cdot\boldsymbol{i} & \boldsymbol{d}_2\cdot\boldsymbol{j} & \boldsymbol{d}_2\cdot\boldsymbol{k} \\ \boldsymbol{d}_3\cdot\boldsymbol{i} & \boldsymbol{d}_3\cdot\boldsymbol{j} & \boldsymbol{d}_3\cdot\boldsymbol{k} \end{pmatrix}=(i,j,k)=\left(i,\frac{n}{|n|}\times i,\frac{n}{|n|}\right).$$

Then the triple p of body components of any vector is related to the triple $\underline{p}$ of components in the fixed frame $\{\boldsymbol{i},\boldsymbol{j},\boldsymbol{k}\}$ through the relation $p=\boldsymbol{R}\underline{p}$, or equivalently $\underline{p}=\boldsymbol{R}^{\mathrm{T}}p$.

It is noteworthy that the Hamiltonian structure defined by (3.12) persists under the presence of a gravitational field acting in the direction $\boldsymbol{i}$, say. For such a heavy elastic rod the equilibrium equations are given by (3.11), with H replaced by the Hamiltonian

$$\tilde{H}(i,r,m,n,s)=W^*(m,s)+\hat{u}\cdot m+n\cdot e_3+\gamma r\cdot i,$$

with γ an appropriate parameter.

(c) *Canonical formulations for rod equilibria and dynamics of tumbling rigid bodies*

To obtain a canonical Hamiltonian description of the rod equilibrium equations that directly includes the spatial components of the variables $\boldsymbol{r}$ and $\boldsymbol{n}$ as phase variables, we introduce a parametrization for the orthonormal directors $\{\boldsymbol{d}_i\}$, i.e. a parametrization of the group $SO(3)$ of 3×3 proper rotation matrices. The classic choice of parametrization is Euler angles (cf. for example, Antman 1995, ch. VIII). Any appropriate choice of three angles provides a local parametrization of $SO(3)$, but a polar singularity of some form always arises, which can give rise to awkward analytical and numerical difficulties. In contrast we shall adopt the four parameter description of $SO(3)$ in terms of quaternions that provides a singularity-free, double covering.

In § 3 *c* (i) we discuss the derivation of a seven degree-of-freedom Hamiltonian system involving quaternions. For completeness, in § 3 *c* (ii) we state an analogous six degree-of-freedom system in terms of Euler angles that is valid away from the polar singularity. Then in § 3 *c* (iii) we show how a reduction leads to lower-dimensional systems describing the dynamics of tumbling rigid bodies.

(i) *Quaternion formulation*

We now outline a canonical Hamiltonian formulation of the rod equilibrium equations in terms of quaternions due to Li & Maddocks (1997), which should be consulted for additional details. The derivation of the Hamiltonian system is not entirely standard because the adoption of a four parameter description of $SO(3)$ means that the appropriate Lagrangian is not a strictly convex function of the generalized velocities.

A quaternion is defined by a vector $\boldsymbol{q} = (q_1, q_2, q_3, q_4) \in \mathbb{R}^4$ satisfying the norm constraint

$$|\boldsymbol{q}| = 1. \tag{3.13}$$

An orthonormal set of directors $\{\boldsymbol{d}_i\}$ can be explicitly written in terms of the quaternion $\boldsymbol{q}$, e.g.

$$\boldsymbol{d}_3 = \begin{pmatrix} 2(q_1q_3 + q_2q_4) \\ 2(-q_1q_4 + q_2q_3) \\ -q_1^2 - q_2^2 + q_3^2 + q_4^2 \end{pmatrix}.$$

With the notation (see, for example, Dichmann 1994)

$$\boldsymbol{B}_1 = \begin{pmatrix} 0 & 0 & 0 & 1 \\ 0 & 0 & 1 & 0 \\ 0 & -1 & 0 & 0 \\ -1 & 0 & 0 & 0 \end{pmatrix}, \quad \boldsymbol{B}_2 = \begin{pmatrix} 0 & 0 & -1 & 0 \\ 0 & 0 & 0 & 1 \\ 1 & 0 & 0 & 0 \\ 0 & -1 & 0 & 0 \end{pmatrix}, \quad \boldsymbol{B}_3 = \begin{pmatrix} 0 & 1 & 0 & 0 \\ -1 & 0 & 0 & 0 \\ 0 & 0 & 0 & 1 \\ 0 & 0 & -1 & 0 \end{pmatrix},$$

the strains u_i can be expressed in terms of the quaternion $\boldsymbol{q}$ and its derivative

$$u_i = \frac{2\boldsymbol{B}_i\boldsymbol{q} \cdot \boldsymbol{q}'}{|\boldsymbol{q}|^2}.$$

The rod equilibria are then the constrained critical points of the energy functional

$$\int_0^1 W(u(\boldsymbol{q}, \boldsymbol{q}') - \hat{u}, s)\,\mathrm{d}s,$$

subject to the constraints of inextensibility and unshearability (1.1), and the quaternion norm condition written in the form $\boldsymbol{q}(s) \cdot \boldsymbol{q}'(s) = 0$. Equivalently, one can introduce Lagrange multipliers $\boldsymbol{\lambda}(s) \in \mathbb{R}^3$ and $\nu \in \mathbb{R}$ such that the equilibria are unconstrained critical points of the Lagrangian

$$\mathcal{L}(\boldsymbol{r}, \boldsymbol{r}', \boldsymbol{q}, \boldsymbol{q}', s) = W(u(\boldsymbol{q}, \boldsymbol{q}') - \hat{u}, s) + \boldsymbol{\lambda} \cdot (\boldsymbol{r}' - \boldsymbol{d}_3) + \nu \boldsymbol{q} \cdot \boldsymbol{q}'.$$

The conjugate variables are introduced through the definitions

$$\boldsymbol{n} = \frac{\partial \mathcal{L}}{\partial \boldsymbol{r}'} = \boldsymbol{\lambda}, \quad \boldsymbol{\mu} = \frac{\partial \mathcal{L}}{\partial \boldsymbol{q}'} = \frac{2}{|\boldsymbol{q}|^2} \sum_i W_{u_i} \boldsymbol{B}_i \boldsymbol{q} + \nu \boldsymbol{q}.$$

The conjugate variable $\boldsymbol{n}$ is merely the net force introduced in § 2 *a*. A complete physical interpretation of the conjugate variable $\boldsymbol{\mu}$ is less obvious, but the moments $\boldsymbol{m} = (m_1, m_2, m_3)$ are given in terms of $\boldsymbol{q}$ and $\boldsymbol{\mu}$ by the expressions

$$m_i = \tfrac{1}{2}\boldsymbol{B}_i\boldsymbol{q} \cdot \boldsymbol{\mu}. \tag{3.14}$$

It can then be verified that the rod equilibrium conditions are equivalent to the seven degree-of-freedom Hamiltonian system

$$\begin{pmatrix} \boldsymbol{r} \\ \boldsymbol{q} \\ \boldsymbol{n} \\ \boldsymbol{\mu} \end{pmatrix}' = \boldsymbol{J}_{14} \nabla H^{\boldsymbol{q}}(\boldsymbol{r}, \boldsymbol{q}, \boldsymbol{n}, \boldsymbol{\mu}, s), \tag{3.15}$$

in which the operator $\boldsymbol{J}_{14}$ corresponds to the canonical Poisson bracket defined in (2.8) and the Hamiltonian is merely (3.10) expressed in terms of the new variables

$$H^{\boldsymbol{q}}(\boldsymbol{r}, \boldsymbol{q}, \boldsymbol{n}, \boldsymbol{\mu}, s) = W^*(\mathsf{m}, s) + \mathsf{m} \cdot \hat{\mathsf{u}} + \boldsymbol{n} \cdot \boldsymbol{d}_3,$$

using (3.14).

(ii) *Euler angle formulation*

With a specific sequence of Euler angles $\boldsymbol{\phi} = (\phi, \theta, \psi)$ (we adopt here the particular sequence of Euler angles that can be found, for example, in Dichmann (1994) or Maddocks (1984)), one can parametrize the directors as e.g.

$$\boldsymbol{d}_3 = \begin{pmatrix} \sin\theta\cos\phi \\ \sin\theta\sin\phi \\ \cos\theta \end{pmatrix}.$$

As above, one constructs a constrained variational principle and introduces conjugate variables $\boldsymbol{n}$ and $\boldsymbol{p}_\phi = (p_\phi, p_\theta, p_\psi)$ to obtain a six degree-of-freedom Hamiltonian system

$$\begin{pmatrix} \boldsymbol{r} \\ \boldsymbol{\phi} \\ \boldsymbol{n} \\ \boldsymbol{p}_\phi \end{pmatrix}' = \boldsymbol{J}_{12} \nabla H^{\phi}(\boldsymbol{r}, \boldsymbol{\phi}, \boldsymbol{n}, \boldsymbol{p}_\phi, s),$$

where the Hamiltonian is (3.10) rewritten as

$$H^{\phi}(\boldsymbol{r}, \boldsymbol{\phi}, \boldsymbol{n}, \boldsymbol{p}_\phi, s) = W^*(\mathsf{m}, s) + \hat{\mathsf{u}} \cdot \mathsf{m} + \boldsymbol{n} \cdot \boldsymbol{d}_3,$$

in which

$$m_1 = -p_\phi \cos\psi \csc\theta + p_\theta \sin\psi + p_\psi \cos\psi \cot\theta,$$
$$m_2 = p_\phi \csc\theta \sin\psi + p_\theta \cos\psi - p_\psi \cot\theta \sin\psi, \quad m_3 = p_\psi.$$

(Compare, for example, Holmes & Marsden 1983.) Notice that the polar singularity is manifested in the appearance of $\csc\theta$ and $\cot\theta$ terms.

(iii) *Formulations for tumbling rigid body dynamics*

In both of the above canonical formulations the centreline $\boldsymbol{r}$ represents three ignorable (or cyclic) variables, i.e. the variable $\boldsymbol{r}$ does not appear explicitly in the Hamiltonian. Accordingly, the conjugate variables $\boldsymbol{n}$ automatically provide three commuting integrals that may therefore be used to reduce the dimension of the Hamiltonian system by six. Thus, for prescribed values of $\boldsymbol{n}$, the reduction to the frame variables inherent to the Kirchhoff kinetic analogy is entirely standard from the Hamiltonian perspective. Notice, however, that with the quaternion representation the reduced system has four degrees of freedom, whereas a parametrization in terms of Euler angles leads to a reduced system with three degrees of freedom.

(d) *Integrals and complete integrability*

We now describe the integrals that arise in the various Hamiltonian formulations for the elastic rod and the tumbling rigid body, and discuss associated completely integrable systems.

Considering the balance laws (2.2) it is apparent that the functions

$$\boldsymbol{m} + \boldsymbol{r} \times \boldsymbol{n} \tag{3.16}$$

and

$$\boldsymbol{n} \tag{3.17}$$

must be integrals of any of the Hamiltonian systems describing the equilibria of elastic rods. We will show how these (and additional) integrals arise in each of the formulations.

(i) *Integrals for the non-canonical formulation*

Consider the non-canonical Hamiltonian system (3.11). We need to distinguish trivial integrals that arise due to degeneracies of the operator $\boldsymbol{J}$ (i.e. Casimirs) and integrals that depend on the specific Hamiltonian under study. In our case, the operator (3.12) has a two-dimensional null-space, spanned by the vectors

$$\nabla C_1 = \begin{pmatrix} m + r \times n \\ n \times i \\ i \\ i \times r \end{pmatrix}, \quad \nabla C_2 = \begin{pmatrix} i \\ 0 \\ 0 \\ 0 \end{pmatrix}.$$

The associated distinguished integrals are $C_1 = i \cdot (m + r \times n)$ and $C_2 = \frac{1}{2}|i|^2$; the integral C_2 arises since $\boldsymbol{i}$ is a fixed vector, while the integral C_1 is the component of (3.16) with respect to $\boldsymbol{i}$ written in body coordinates.

Consider a general Hamiltonian $\mathcal{H}(m, n, s)$, dependent only on the variables m and n (and s), but not r and i. Then it is easy to verify that the components of $\boldsymbol{m}+\boldsymbol{r}\times\boldsymbol{n}$ with respect to $\boldsymbol{n}$ and $\boldsymbol{j}$ are integrals with respect to $\mathcal{H}$; in body coordinates these integrals are $m \cdot n$ and $(n \times i) \cdot (m + r \times n)$. Furthermore, the components of the constant force vector $\boldsymbol{n}$ in the basis $\{\boldsymbol{i}, \boldsymbol{j}, \boldsymbol{k}\}$ are trivially integrals of $\mathcal{H}$; in body coordinates $|n|^2$ and $n \cdot i$ are the only independent integrals. In the absence of further symmetry properties of the constitutive function, and therefore of the Hamiltonian, the six quantities (3.16) and (3.17) constitute all known integrals. However, if the rod is isotropic, the twisting moment m_3 is also an integral, and if the rod is uniform, the Hamiltonian becomes autonomous and so itself provides an additional integral.

(ii) *Integrals for the heavy rod equations*

For the Hamiltonian description of a heavy elastic rod the two Casimir integrals persist, since they form the null-space of (3.12). While all remaining integrals arising for system (3.11) are individually destroyed, a linear combination of them persists, namely $|n \times i|^2 = |j|^2$; and, as above, the integrals m_3 and H arise if the rod is isotropic and uniform, respectively.

(iii) *Integrals for the quaternion formulation*

The integrals (3.16) and (3.17) are independent of the particular variables adopted; they must persist in the canonical Hamiltonian formulation (3.15) written in terms of quaternions (or indeed Euler angles) and merely need to be re-expressed in terms of the quaternion $\boldsymbol{q}$ and its conjugate momentum $\boldsymbol{\mu}$.

The point of interest is that the parametrization in terms of quaternions gives rise to *additional* integrals (see Li & Maddocks (1997) for more details and a discussion of the associated symmetries). First, the quaternion norm constraint (3.13) is manifested as the integral

$$\boldsymbol{q}\cdot\boldsymbol{q}, \tag{3.18}$$

associated with the gauge freedom generated by the absence of strict convexity of the corresponding Lagrangian with respect to the generalized velocities. Second, the function

$$\boldsymbol{\mu}\cdot\boldsymbol{q}+2\,\boldsymbol{r}\cdot\boldsymbol{n} \tag{3.19}$$

is an integral. Thus in the quaternion description (3.15) there are, in general, always eight integrals, namely (3.16)–(3.19), and isotropy and uniformity can yield two additional integrals.

(iv) *Integrals for the tumbling rigid body equations*

We now specialize to the equations of rigid body motion and consider the force vector $\boldsymbol{n}$ as a parameter. Furthermore, we assume a constitutive law of the particular form (2.5), as in the Kirchhoff kinetic analogy. In the reduced non-canonical formulation (3.4), the operator (3.3) has a two-dimensional null-space spanned by the vectors

$$\nabla C_1=\begin{pmatrix} \mathsf{n}\\ \mathsf{m}\end{pmatrix},\quad \nabla C_2=\begin{pmatrix} \mathbf{0}\\ \mathsf{n}\end{pmatrix}.$$

The associated Casimirs $C_1=\mathsf{m}\cdot\mathsf{n}$ and $C_2=\frac{1}{2}|\mathsf{n}|^2$ are the trivial integrals of (3.4), independent of the Hamiltonian. In the non-canonical and the Euler angle representation, C_1 and C_2 constitute all known integrals of the general system. In the quaternion formulation, the additional integral (3.18) remains (while the integral (3.19) does not persist because it involves the variable $\boldsymbol{r}$). In addition, for a symmetric tumbling rigid body (or Lagrange top), the moment m_3 is an integral. The free rigid body has the integral $|\mathsf{m}|$. And finally, the Hamiltonian is an integral, provided that the system is autonomous (as is typically the case).

(v) *Completely integrable systems*

Finally, we describe the various integrable cases that arise in the above Hamiltonian formulations. One needs to verify the appropriate count on integrals in involution to obtain complete integrability (see § 2 *c*). These calculations may be cumbersome, but can be efficiently implemented with a symbolic manipulation package (such as Mathematica). For the non-canonical Hamiltonian system (3.11), there always exist two Casimirs and four integrals, three of which are in involution. Isotropy implies the integral m_3, which is involution with all the others. Finally, uniformity implies that the Hamiltonian is an integral, yielding a total of two Casimirs and five integrals in involution. Thus a uniform and isotropic rod is integrable. A parametrization in terms of quaternions (respectively Euler angles) always yields eight (six) integrals,

Table 1. *Numbers of integrals for Hamiltonian formulations describing elastic rods*
(The table indicates both the numbers of integrals that always exist and, in parentheses, the numbers that exist in the presence of the additional symmetries of isotropy and uniformity. The cases in parentheses all correspond to complete integrability.)

	non-canonical	Euler angles	quaternions
no. of Casimirs	2	0	0
no. of integrals	4 (6)	6 (8)	8 (10)
no. in involution	3 (5)	4 (6)	5 (7)

Table 2. *Numbers of integrals for Hamiltonian formulations describing a tumbling body*
(The table indicates both the numbers of integrals that always exist and, in parentheses, the numbers that exist for the case of a Lagrange top. The cases in parentheses all correspond to complete integrability.)

	non-canonical	Euler angles	quaternions
no. of Casimirs	2	0	0
no. of integrals	0 (2)	2 (4)	3 (5)
no. in involution	0 (2)	2 (3)	2 (4)

five (four) of which are in involution. Isotropy and uniformity each yield an additional integral, which raises the total number of integrals in involution to seven (six) and implies complete integrability for the system. A similar count arises for the reduced Hamiltonian formulations describing the dynamics of a tumbling rigid body. The results are summarized in tables 1 and 2.

4. The last quadrature

In this section we discuss the steps necessary to recover the full set of variables pertinent to rod mechanics, in particular the spatial coordinates of the centreline $\boldsymbol{r}$, when it is assumed that a solution to the reduced problem corresponding to the frame evolution is known (in some sense). The centreline $\boldsymbol{r}$ is certainly always computed via the constraint (1.1), but we demonstrate that the use of various integrals (i.e. invariant quantities) allows the components of $\boldsymbol{r}$ in the fixed frame to be computed directly, without explicitly having to perform the three quadratures implicit in (1.1). We first review the manner in which two of the quadratures can always be eliminated. We then make the observation that the extra information, and extra integral, in the four degree-of-freedom quaternion Hamiltonian formulation of the frame evolution can be exploited to finesse the third and last quadrature.

We also remark that, despite the kinetic analogy, considerable information concerning the shape of the centreline $\boldsymbol{r}$ is lost if the constraint (1.1) is ignored. In particular, the quadratures in (1.1) often seem to average over the potentially very complicated motion of the tangent indicatrix (i.e. the director $\boldsymbol{d}_3$) to yield a rather regular behaviour of the centreline, at least on relatively long length scales. These striking differences are indicated in figures 2, 3 and 4, which depict the evolution

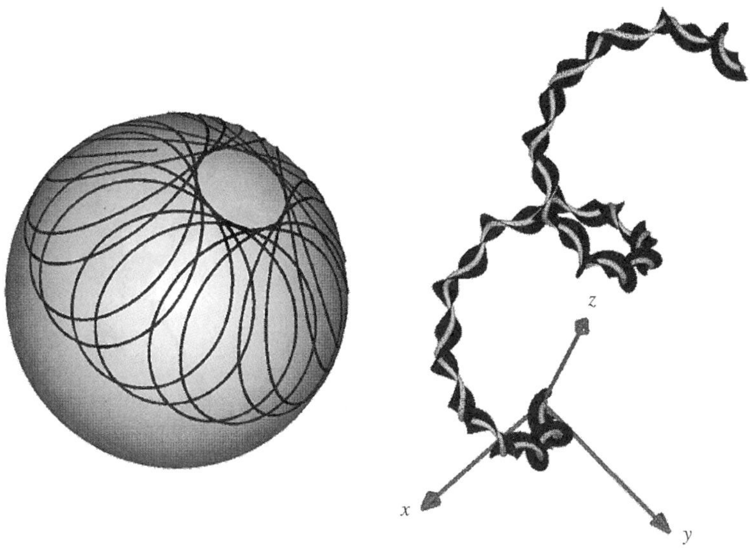

Figure 2. A configuration of an integrable rod and the corresponding tangent indicatrix, tracing the director $\boldsymbol{d}_3$ on the unit sphere. Both representations are highly structured.

of both the tangent indicatrix and the corresponding shape of the centreline $\boldsymbol{r}$ in various integrable and non-integrable cases. In figure 2 a uniform, isotropic rod is shown. The Hamiltonian system is integrable and exhibits a typical 'helix on helix' structure. The corresponding motion of the Lagrange top is indicated by the tangent indicatrix oscillating between two parallel circles on the unit sphere. For a general tumbling rigid body this regular behaviour of the tangent indicatrix is destroyed, as indicated in figures 3 and 4. The tangent indicatrix does not seem to follow any obvious structured pattern. However, the corresponding shape of the rod centreline does, despite a rather irregular local behaviour, average on a longer length scale to a helical structure. A desire to understand this behaviour motivated our consideration of the last quadratures.

(a) *Elimination of two quadratures*

The goal is to compute the components of the centreline in the fixed basis $\{\boldsymbol{i}, \boldsymbol{j}, \boldsymbol{k}\}$, which we assume for convenience to be orthonormal. The vector $\boldsymbol{k}$ is chosen to be parallel to the force vector $\boldsymbol{n}$. Ilyukhin (1979) shows that the $\boldsymbol{i}$ and $\boldsymbol{j}$ components of $\boldsymbol{r}$ are readily computed using the integral (3.16) arising from the balance equations (2.2). Assume that the value of the integral is $\boldsymbol{c}$ and rewrite

$$\boldsymbol{m} + \boldsymbol{r} \times \boldsymbol{n} = \boldsymbol{c},$$

as

$$\boldsymbol{m} + \left(\boldsymbol{r} - \frac{\boldsymbol{n}}{|\boldsymbol{n}|^2} \times \boldsymbol{c}\right) \times \boldsymbol{n} = \frac{\boldsymbol{c} \cdot \boldsymbol{n}}{|\boldsymbol{n}|^2}\boldsymbol{n} = \frac{\boldsymbol{m} \cdot \boldsymbol{n}}{|\boldsymbol{n}|^2}\boldsymbol{n},$$

using the fact that $\boldsymbol{m} \cdot \boldsymbol{n} = \boldsymbol{c} \cdot \boldsymbol{n}$. Without loss of generality we may assume that the fixed coordinate system is translated by a shift $\boldsymbol{r}_0 = \boldsymbol{n}/|\boldsymbol{n}|^2 \times \boldsymbol{c}$, so that

$$\boldsymbol{m} + \boldsymbol{r} \times \boldsymbol{n} = \frac{\boldsymbol{m} \cdot \boldsymbol{n}}{|\boldsymbol{n}|^2}\boldsymbol{n}.$$

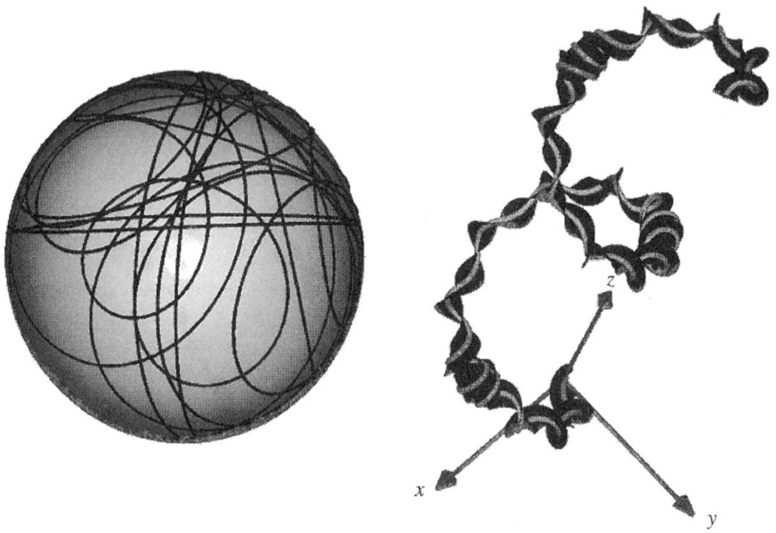

Figure 3. The full configuration of a non-integrable rod exhibits a fairly regular structure, while the tangent indicatrix does not.

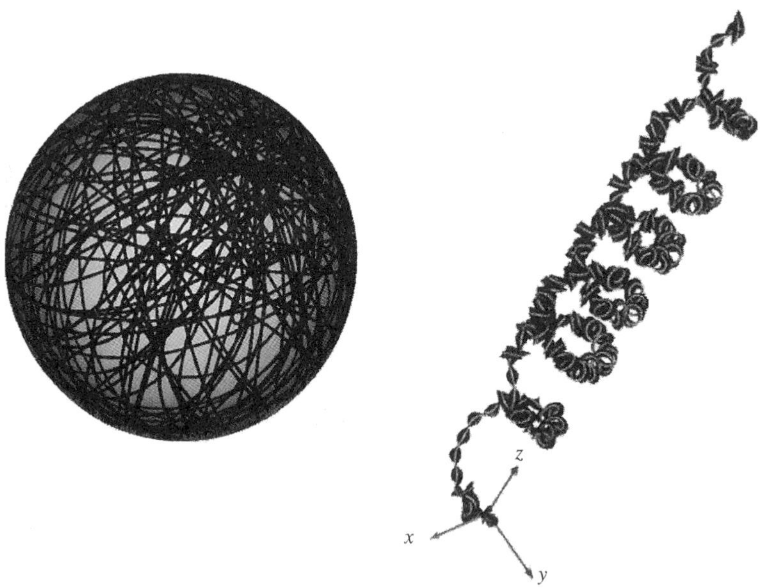

Figure 4. On a relatively long length scale, the rod exhibits a regular averaged structure, which is obscured in the tangent indicatrix.

The components of $\boldsymbol{r}$ with respect to $\boldsymbol{i}$ and $\boldsymbol{j}$ are now readily computed

$$\boldsymbol{r} \cdot \boldsymbol{i} = \frac{\boldsymbol{m} \cdot \boldsymbol{j}}{|\boldsymbol{n}|} = \frac{m \cdot j}{|n|}, \tag{4.1}$$

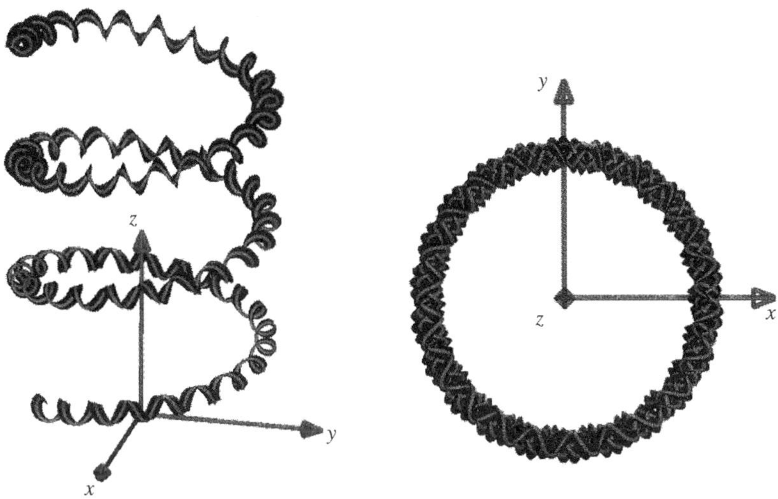

Figure 5. An integrable configuration. The centreline of the rod lies between two concentric circular cylinders (as indicated in the projection).

$$\boldsymbol{r} \cdot \boldsymbol{j} = -\frac{\boldsymbol{m} \cdot \boldsymbol{i}}{|\boldsymbol{n}|} = -\frac{m \cdot i}{|n|}. \tag{4.2}$$

As observed by several authors (cf. Shi & Hearst 1994; Ilyukhin 1979; Langer & Singer 1993), the above argument shows that cylindrical coordinates parallel to the constant force vector $\boldsymbol{n}$ are very special. In particular, the above manipulations allow us to compute the projection of the centreline $\boldsymbol{r}$ into the $\boldsymbol{ij}$-plane. From formulas (4.1) and (4.2) we see that this projection is simply a rotation and scaling of the projection of the moment vector $\boldsymbol{m}$ into the same plane, and two of the quadratures appearing in (1.1) are eliminated.

If the rod is uniform, the Hamiltonian is an integral and we see from (3.10) that $W^*(m) + m \cdot \hat{u}$ is bounded. The coercivity of W is equivalent to the coercivity of W^* (see, for example, Evans 1994, § 3.3.2), i.e. W^* grows super-linearly at infinity, and we conclude that m is bounded. Thus the projection of $\boldsymbol{r}$ in the $\boldsymbol{ij}$-plane is also bounded; we interpret this boundedness as the centreline of the rod evolving between two concentric cylinders. Figures 5–7 illustrate this conclusion for various integrable and non-integrable, linearly elastic, uniform rod configurations. In each figure we show a skew projection of the rod together with a projection into the $\boldsymbol{ij}$-plane. For the isotropic rod, depicted in figure 5, this conclusion was known, as it follows from the quasi-periodic behaviour of the solution. However, the more general, non-integrable case in figure 6 illustrates that, despite the generally complicated behaviour of the solution, the projection parallel to the $\boldsymbol{n}$-axis is bounded and in fact may resemble that of an isotropic rod.

It is also possible that the radius of the inner cylinder can vanish. In fact this happens exactly if $\boldsymbol{m}(s)$ is parallel to $\boldsymbol{n}(s)$ at some value of s. Vanishing of the inner radius is necessary for configurations to exhibit localization. However, there are also non-localized rod configurations for which the radius of the inner cylinder vanishes (cf. figure 7). Again, in this configuration the skew projection appears rather complicated, while the projection along the $\boldsymbol{n}$-axis has an almost periodic behaviour.

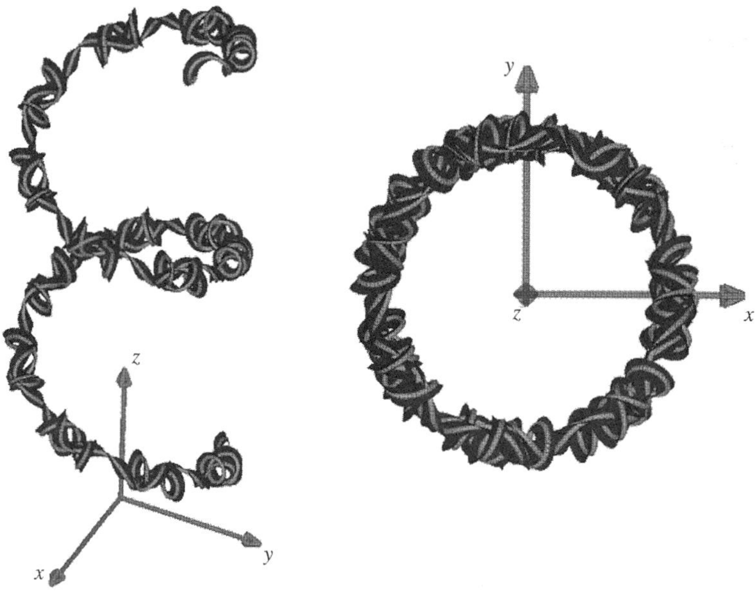

Figure 6. For a non-integrable rod, the overall structure becomes rather complicated. However, the existence of the two bounding cylinders persists.

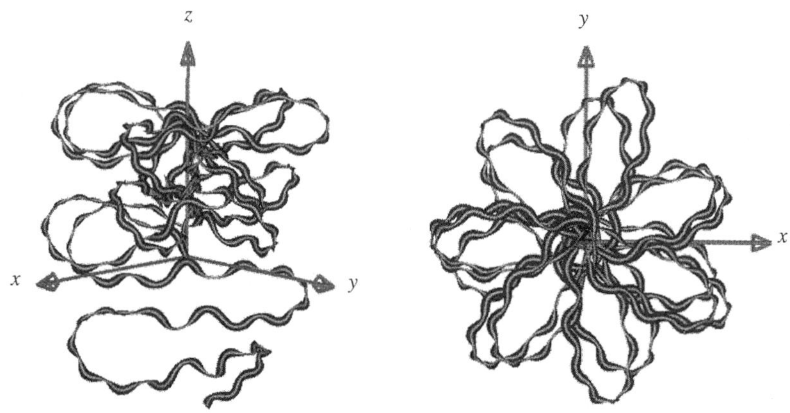

Figure 7. A configuration where the radius of the inner cylinder vanishes. The side view suggests a complicated structure, while the projection seems almost periodic.

(b) *The last quadrature resolved*

It remains to compute the $\boldsymbol{k}$ component of the centreline $\boldsymbol{r}$ via the *last quadrature*

$$\boldsymbol{r}' \cdot \boldsymbol{k} = \boldsymbol{d}_3 \cdot \boldsymbol{k} = \frac{n_3}{|n|}.$$

When an integrable rod is of interest it is possible to explicitly carry out this quadrature. Our observation is that if a solution to the reduced frame evolution is known in terms of quaternions, then, whether or not the rod is integrable, the last quadrature can be performed explicitly.

Consider the specific representation of the Hamiltonian system (3.15) in terms of quaternions. It is only here that the integral (3.19) arises. Assuming that the value of this integral is $\boldsymbol{\mu} \cdot \boldsymbol{q} + 2\boldsymbol{r} \cdot \boldsymbol{n} = c$, the last quadrature is

$$\boldsymbol{r} \cdot \boldsymbol{k} = \frac{1}{2|\boldsymbol{n}|}(\boldsymbol{\mu} \cdot \boldsymbol{q} - c). \tag{4.3}$$

The value of c merely represents a shift in origin of $\boldsymbol{r}$ parallel to $\boldsymbol{n}$.

We may therefore conclude that the extra degree of freedom in a quaternion Hamiltonian description of the equilibria of elastic rods contains additional information that provides a strengthened kinetic analogy. The relations (4.1), (4.2) and (4.3) allow a complete reduction of two-point boundary value problems for rods to a two-point boundary value problem for the lower-dimensional reduced quaternion Hamiltonian description of the frame evolution. This fact has been exploited for numerical simulations by Warner (1997), and he observes an increase in performance for the reduced problem by approximately a factor of three.

The authors thank both Professor P. J. Olver for discussions pertaining to Lie–Poisson brackets and the other members of the Laboratory for Computation and Visualization in Mechanics at the University of Maryland for their valuable contributions to this work. This research was supported by the AFOSR and ONR.

References

Antman, S. S. 1995 *Nonlinear problems of elasticity*, 1st edn. Berlin: Springer.

Champneys, A. R. & Thompson, J. M. T. 1996 A multiplicity of localized buckling modes for twisted rod equations. *Proc. R. Soc. Lond.* A **452**, 2467–2491.

Champneys, A. R. & Thompson, J. M. T. 1996 From helix to localized writhing in the torsional post-buckling of elastic rods. *Proc. R. Soc. Lond.* A **452**, 117–138.

Coleman, B. D., Dill, E. H., Lembo, M., Lu, Z. & Tobias, I. 1993 On the dynamics of rods in the theory of Kirchhoff and Clebsch. *Arch. Ration. Mech. Analysis* **121**, 339–359.

Dichmann, D. J. 1994 Hamiltonian dynamics of an elastica and the stability of solitary waves. Ph.D. thesis, University of Maryland.

Dichmann, D. J., Maddocks, J. H. & Pego, R. L. 1996 Hamiltonian dynamics of an elastica and the stability of solitary waves. *Arch. Ration. Mech. Analysis* **135**, 357–396.

Evans, L. C. 1994 *Partial differential equations.* Berkeley Mathematics Lecture Notes, vol. 3A. Berkeley: Center for Pure and Applied Mathematics, University of California.

Holmes, P. & Marsden, J. 1983 Horseshoes and Arnold diffusion for Hamiltonian systems on Lie groups. *Indiana Univ. Math. J.* **32**, 273–309.

Ilyukhin, A. A. 1979 *Spatial problems of nonlinear theory of elastic rods.* Kiev: Naukova Dumka. (In Russian.)

Kehrbaum, S. 1997 Hamiltonian formulations of the equilibrium conditions governing elastic rods: qualitative analysis and effective properties. Ph.D. thesis, University of Maryland.

Kehrbaum, S. & Maddocks, J. H. 1997 Effective properties of elastic rods with high intrinsic twist. Preprint.

Landau, L. D. & Lifshitz, E. M. 1970 *Theory of elasticity. Course of theoretical physics*, vol. 7. London: Pergamon.

Langer, J. & Singer, D. A. 1996 Lagrangian aspects of the Kirchhoff elastic rod. *SIAM Rev.* **38**, 605–618.

Li, H.-H. 1986 Geometrically constrained curves and the calculus of variations. Ph.D. thesis, Rice University.

Li, Y. & Maddocks, J. H. 1997 On the computation of equilibria of elastic rods. Part I. Integrals, symmetry and a Hamiltonian formulation. *J. Comp. Phys.* (In the press.)

Love, A. E. H. 1944 *A treatise on the mathematical theory of elasticity*, 4th edn, § 260. New York: Dover.

Maddocks, J. H. 1984 Stability of nonlinearly elastic rods. *Arch. Ration. Mech. Analysis* **85**, 180–198.

Mielke, A. & Holmes, P. 1988 Spatially complex equilibria of buckled rods. *Arch. Ration. Mech. Analysis* **101**, 318–348.

Novikov, S. P. 1982 The Hamiltonian formalism and a many-valued analogue of Morse theory. *Russ. Math. Surv.* **37**, 1–56.

Olver, P. J. 1993 *Applications of Lie groups to differential equations*, 2nd edn. Berlin: Springer.

Shi, Y. & Hearst, J. E. 1994 The Kirchhoff elastic rod, the nonlinear Schrödinger equation, and DNA supercoiling. *J. Chem. Phys.* **101**, 5186–5200.

Tan, Z. & Witz, J. A. 1995 On the deflected configuration of a slender elastic rod subject to parallel terminal forces and moments. *Proc. R. Soc. Lond.* A **449**, 337–349.

Van der Heijden, G. H. M, Champneys, A. R. & Thompson, J. M. T. 1997 Load-deflection characteristics of localised torsional buckling modes in rods with non-circular cross-section. *SIAM J. Appl. Math.* (Submitted.)

Wang, L. S., Krishnaprasad, P. K. & Maddocks, J. H. 1991 Hamiltonian dynamics of a rigid body in a central gravitational field. *Celest. Mech. Dyn. Astron.* **50**, 59–96.

Wang, L. S., Maddocks, J. H. & Krishnaprasad, P. K. 1992 Steady rigid-body motions in a central gravitational field. *J. Astronaut. Sci.* **40**, 449–478.

Warner, J. A. 1997 Numerical simulations of elastic rods. Ph.D. thesis, University of Maryland.

Wittenburg, J. 1977 *Dynamics of systems of rigid bodies*, § 4.6. Stuttgart: Teubner.

The Royal Society

The writhing of circular cross-section rods: undersea cables to DNA supercoils

By D. M. Stump[1], W. B. Fraser[2] and K. E. Gates[1]

[1]*Department of Mathematics, The University of Queensland, St. Lucia, QLD 4072, Australia,*

[2]*School of Mathematics and Statistics, The University of Sydney, NSW 2006, Australia*

Received 2 July 1997; accepted 27 October 1997

The large deflection theory of circular cross-section elastic rods is used to consider the writhing of long straight rods subjected to tension and torque, such as undersea cables, and to closed loops with inserted twist, such as DNA supercoils.

The writhed shape of the long straight rod under tension and torque is easily generated by twisting a piece of string with the fingers and consists of three separate parts: a balanced-ply region, a free end loop, and two tail regions. The solution for the rod shape in each of the regions is found. The results are then joined together to ensure continuity of the position and tangent vectors of the strand centreline through the introduction of point forces and moments at the points where the strands enter and exit the balanced ply. The results of the model are consistent with simple experiments on long braided rope.

The writhed shape of the closed loop with twist inserted between the ends prior to closure is modelled as a balanced ply joined to two end loops. The analysis combines the mechanics solution with the conservation of topological *link* to provide a simple formula which *quantitatively* predicts the approximate shape and helix angle of the supercoil. The results are in good agreement with simple experiments on rope and with available data on DNA supercoils.

Keywords: interwound supercoils; link, twist, writhe; tortuousity and writhe; large deflection rods; linear elastic rods

1. Introduction

The writhing of a twisted rod is a process familiar to almost everyone and is easily demonstrated with a piece of string that is held under tension and twisted with the fingers. As the ends are slowly displaced towards each other, the rod 'writhes' into the shape shown in figure 1. Greenhill (1883) showed that a 'long' straight rod loaded under a remote tension T_∞ and torsional moment Q_∞ first buckles into a helical shape when

$$T_\infty = \frac{Q_\infty^2}{4B},$$

where the bending rigidity $B = EI$, E is Young's modulus and I is the second area moment of the cross-section. The helix quickly localizes into a shape which is similar to a solitary wave (Coyne 1990), and eventually the rod contacts itself to form a

TEX Paper

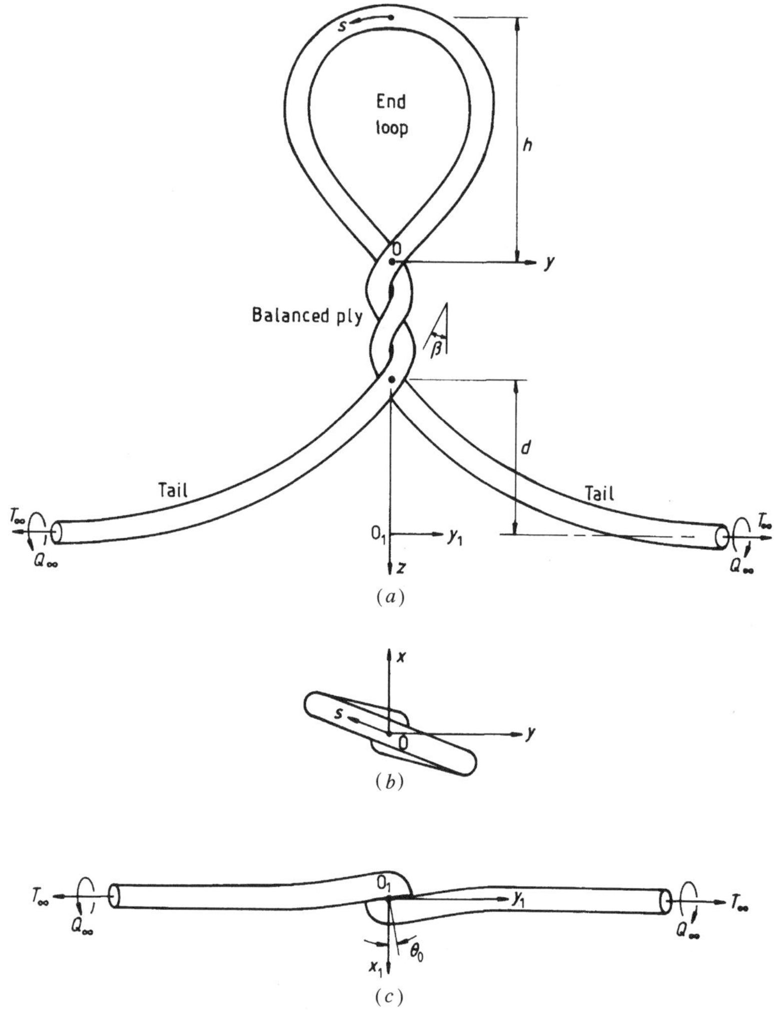

Figure 1. General view of the writhed rod: (*a*) side elevation; (*b*) plan view from above the end loop with the tails omitted; (*c*) plan view from beneath the tails with the end loop omitted.

small loop. Further inward displacements of the remote ends produce the writhed structure shown in figure 1 (see also figure 6 of Thompson & Champneys 1996), which consists of three parts:

(1) the *balanced-ply* region where the strands are in contact and exert forces on each other;

(2) the *end loop* region which is attached to the two strands of the balanced ply and is free from any lateral forces or moments; and

(3) the *tails* which are attached to the bottom of the balanced ply and extend to the remote region of constant applied T_∞ and Q_∞.

The objective of this paper is to use the large deflection theory of elastic rods of circular cross-section to find the shape, forces and moments in the three parts of the writhed rod. The range of application of the final solution is extensive since writhing occurs in a large number of applications, from undersea cables (Coyne 1990) to the interwound and toroidal supercoiling of DNA (Calladine & Drew 1992; Bauer *et al.* 1980).

The study is organized as follows. In § 2, the mathematical formulation for the shape of the rod is presented along with other model assumptions. In § 3 the solution for the balanced-ply region is discussed. Sections 4 and 5 develop the equations and jump conditions for the end loop and tails, respectively. Section 6 details the numerical solution for the end loop, while results of the calculations for the tails and end loop are presented in § 7. In § 8 these results are applied to the long writhed rod in figure 1 and to the writhing of a closed loop. This section also contains a more detailed discussion of the important topological constraint that influences writhing and compares the model predictions with results on DNA supercoiling and some simple experiments with ropes. Some final remarks are included in § 9.

The paper is written in distinct sections so that readers who are less interested in the mathematical details of the solution can proceed directly to §§ 7 and 8 which present the results of the calculations and the applications. The remainder of the introduction is devoted to a discussion of some background literature in mechanics and DNA. This study is quite lengthy and complete discussions on all of the relevant work (particularly on DNA supercoils) are not appropriate in this paper. Rather, the intent is to place this work within the larger context of literature and research on the subject.

(*a*) *The mechanics background*

The theory of bent and twisted rods of arbitrary cross-section has a long history and is discussed in ch. 18 of the treatise by Love (1927) where a theory for linear elastic rods that neglects cross-sectional warping is developed. Antman (1995) has summarized many aspects of the general theory in his recent treatise, with particular emphasis on nonlinear constitutive laws and more advanced models such as Cosserat theory.

Thompson & Champneys (1996), Champneys & Thompson (1996) and Champneys *et al.* (1997) have studied the buckling and post-buckling of straight elastic rods subject to equal and opposite torque and tension end loads but with no lateral forces. They have primarily investigated the chaotic solutions which develop for non-circular cross-sections and for nonlinear materials. While they have obtained experimental pictures of writhed rods in lathe fixtures, they have not actually obtained a mathematical description of the shape. It is their experimental results which have partially motivated this study.

A number of other investigators (Coleman & Dill, 1992; Coleman *et al.* 1993; Dichmann & Maddocks 1994) have studied the mathematical structure of the dynamical equations of elastic rods free of lateral forces and moments. These results contain a rigorous theory formulated in terms of a set of principal cross-sectional inertia axes, which is applicable to rods of arbitrary cross-sectional shape. Since this study considers only circular cross-sections, a simpler theory based upon the Frenet basis vectors is sufficient for the mathematical description of the rod.

The analysis of this paper is based upon the work of Fraser & Stump (1998*a*–*c*) and Stump & Fraser (1998), who have applied the theory of bent and twisted rods to the static and dynamic analysis of textile yarn-plying processes in which straight yarn segments are plyed into a helical structure. At the convergence point where the straight strands come together with the helical portions, the introduction of concentrated forces and moments between the strands is necessary in order to maintain overall structural equilibrium and continuity. This formulation is applied at the two convergence points in the writhed elastic rod where the strands enter and exit the balanced-ply region and is essential to the development of a solution for the writhed shape.

(*b*) *The DNA background*

The writhing of DNA strands into larger coiled structures is a process known as *supercoiling* and has attracted intense interest due to the importance of supercoils during the transcription and replication processes. Bauer *et al.* (1980), Calladine & Drew (1992) and Wang (1982) have presented non-specialist discussions of the process. Supercoils are formed when DNA loops of the order of μm in size are operated on by a variety of enzymes (e.g. gyrase and topoisomerases) which remove or insert turns of twist relative to the relaxed double-strand structure of approximately 10.2–10.4 base pairs (BP hereafter) per helical turn of the phosphate backbones. The removal or insertion of twist strains the strand, which responds by writhing into a structure that is described approximately by a balanced ply and two end loops (see figure 7). One of the objectives of this paper is to develop a simple analytic formula (without the need for electron micrographs) for predicting the gross features of the shape of the writhed molecule given the degree of under- or over-twisting of the DNA loop.

The methods by which supercoiling catalyses interactions with various enzymes are beyond the scope of this study. The articles by Wang (1982, 1995) and Wang & Liu (1990) describe the role of supercoiling in detail and present various mechanistic models. It is hoped the results of this study will provide further illumination to the interaction of supercoils and topoisomerases.

The mathematical study of DNA supercoiling has primarily used a topological approach to categorize the various supercoiled shapes that preserve the total link inserted into a looped strand of the molecule. The work of White (1969), Fuller (1971, 1978) and Crick (1976) proved instrumental to the development of the understanding that link is conserved and divided into writhe and twist during the coiling process. Much of the history of development of the topological understanding of DNA supercoiling is summarized in the review article by Cozzarelli *et al.* (1990). The related subject of knot theory has also contributed to the understanding of splicing reactions and is discussed by Murasugi (1996).

A variety of mechanics studies of supercoiling ranging from molecular dynamics to finite element continuum models have been undertaken. Recent review articles by Schlick (1995) and Olson (1996) summarize many of the investigations. The article by Schlick shows results (figure 3, p. 250) for supercoiled shapes computed by molecular dynamics techniques that are remarkably similar to the continuum model developed in this study. The advantage of the molecular approach is the ability to incorporate many different effects such as electochemical forces and thermal agitation, which are absent from the continuum approach. However, this comes at the expense of any simple formula for relating the gross features of the supercoiled shape to the degree of undertwisting and other parameters. Many of these studies are concerned with DNA behaviour on much shorter length scales than is appropriate for the strictly continuum rod model of supercoiling; although Marko (1997*a*) has recently adapted free-energy minimization techniques to consider the plectonemic (interwound) supercoiling of μm size molecules from a thermodynamic standpoint.

The bibliographies of Schlick (1995) and Olson (1996) indicate a paucity of continuum analyses of interwound supercoiling relative to the molecular approaches. Le Bret (1979) conducted a highly simplified analysis to consider the stability of an elastic ring with trapped twist. Calladine (1980, 1992) has considered toroidal supercoiling and the implications of DNA sequence on the elastic behaviour from a heuristic standpoint using the principles of solid mechanics. Tobias & Olson (1993) and Manning *et al.* (1996) have considered smaller-scale effects of BP variations on the local intrinsic curvature of a relaxed strand of DNA and the consequences that these variations will have on the distribution of twist and writhe along the strand. These latter effects are on a length scale much smaller than is considered here. This analysis is a continuum model applicable to the supercoiling of a long (i.e. several thousand BP) closed loop of DNA, such as those shown in electron micrographs in ch. 6 of Calladine & Drew (1992) and in Wang (1995).

The finite-element method has also been used to predict DNA supercoiling. Yang *et al.* (1993) have used Kirchhoff beam elements to model the writhing of a loop that has differential twist inserted between the two ends. Their results include a variety of realistic loop shapes, but their analysis provides no way of predicting the gross features of the supercoiled molecule shape (such as the supercoil helix angle or the length of the balanced-ply segments) for specified levels of link. Bauer *et al.* (1993) have also used the finite-element method to study the effects that local regions of initial curvature have on the supercoiling of a circular ring.

Finally, it is important to note that molecular dynamics techniques (see, for example, Marko & Siggia 1994, 1995; Marko 1997*b*) have been used in conjunction with experiments on single strands of DNA (see, for example, Smith *et al.* 1996; Wang *et al.* 1997; Strick *et al.* 1996) to predict and measure the elastic behaviour of DNA molecules. As summarized in Austin *et al.* (1997), the uncoiling of the DNA helical strand structure only becomes significant for forces in excess of 10 pN, which are well above the force levels predicted in this analysis. For the purposes of modelling a large supercoil (i.e. μm in size), the assumption of inextensibility proves quite reasonable (as shown in § 8).

2. The mathematical formulation

(*a*) *The model assumptions*

The rod is modelled as a strand of uniform circular cross-section and inextensible centreline. Linear bending moment–curvature and torque–torsion constitutive laws are assumed. Since the cross-section is circular, the problem can be formulated entirely in terms of the Frenet derivatives of the centreline.

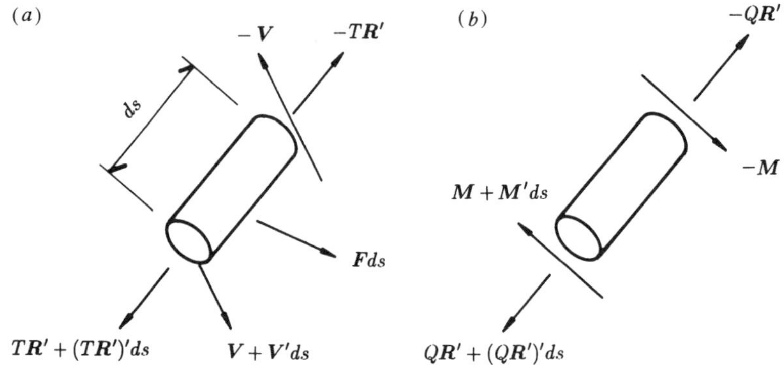

Figure 2. (a) The forces acting on a rod element: the tension T in the tangential direction and a perpendicular shear force $\boldsymbol{V}$. (b) The moments acting on a rod element: the torque Q in the tangential direction and the bending moment $\boldsymbol{M}$ in the binormal direction $\boldsymbol{b}$.

The equations for the bending and torsional stiffnesses, B and K, of a circular shaft of radius a are

$$K = \tfrac{1}{2}GAa^2, \quad B = \tfrac{1}{4}EAa^2, \tag{2.1}$$

where E is Young's modulus, G is the shear modulus and $A = \pi a^2$ is the cross-sectional area. If ν is Poisson's ratio, then $G = E/[2(1+\nu)]$, and

$$\kappa = K/B = 1/(1+\nu). \tag{2.2}$$

The important point is that the ratio K/B depends only on ν and is $O(1)$.

Let $\boldsymbol{R}(s)$ be the position vector of a material point of the strand axis with respect to the origin O of the Cartesian coordinate system $Oxyz$ in figure 1. The arc-length coordinate s is measured from the midpoint of the end loop. Since the shape is symmetric about the $\boldsymbol{k}$-axis running between the strands, it is only necessary to consider the region $s \geqslant 0$; the shape for $s \leqslant 0$ is obtained by a rotation of 180° about the $\boldsymbol{k}$-axis.

The centreline of the inextensible rod forms the Frenet basis vector system $\boldsymbol{R}'$, $\boldsymbol{n} = \boldsymbol{R}''/|\boldsymbol{R}''|$, and $\boldsymbol{b} = \boldsymbol{R}' \wedge \boldsymbol{n}$ which are, respectively, the tangent, principle normal and binormal of the path, where $(\ )' = \partial/\partial s$. The forces and moments acting on an element of the rod are shown in figures $2a, b$ and consist of: the tension force T and the torque Q in the direction of the tangent $\boldsymbol{R}'$; the bending moment $\boldsymbol{M}$ in the direction of the binormal $\boldsymbol{b}$; and the shear force $\boldsymbol{V}$ lying in the plane formed by the principle normal $\boldsymbol{n}$ and the binormal $\boldsymbol{b}$. $\boldsymbol{F}$ is the applied force exerted on the element.

Within the balanced-ply region the strands are in contact and exert forces upon each other. The model allows for both frictional and pressure contact forces on the strand surfaces. However, this study makes the simplifying assumption that the centreline of each strand within the balanced ply is described by the shape of a uniform helix of constant winding angle. The translational invariance of the helical shape along the ply axis allows only an effective pressure contact force which is perpendicular to both the ply axis and the strand axes. This is certainly not the only solution for the strand shape within the balanced ply, but the assumption of no friction within the ply due to the constant helix angle provides the most tractable approach for a first analysis.

(*b*) *The governing equations*

The formulation of the equations follows that in Fraser & Stump (1998*a*). In terms of dimensionless quantities defined in equation (2.7), the equations of force and moment equilibrium for the strand $s > 0$ are

$$\left.\begin{aligned} (T\boldsymbol{R}')' + \boldsymbol{V}' - p\boldsymbol{n}[H(s-L) - H(s-l)] + f_{\rm h}\boldsymbol{n}\delta(s-L) + f_{\rm t}\boldsymbol{n}\delta(s-l) = \boldsymbol{0}, \\ (Q\boldsymbol{R}')' + \boldsymbol{M}' + \boldsymbol{R}' \wedge \boldsymbol{V} + m_{\rm h}\boldsymbol{n}\delta(s-L) + m_{\rm t}\boldsymbol{n}\delta(s-l) = \boldsymbol{0}, \end{aligned}\right\} \tag{2.3}$$

where $H(s)$ is the Heaviside unit step function, $\delta(s)$ is Dirac's delta function, $-p\boldsymbol{n}$ is the pressure per unit strand length exerted by one strand on the other within the balanced ply, and L and l are, respectively, the locations where the strand enters the balanced ply from the end loop and exits the ply into the tails.

The concentrated forces $f_{\rm h}$ and $f_{\rm t}$, and moments $m_{\rm h}$ and $m_{\rm t}$, point in the principle normal direction $\boldsymbol{n}$ which lies within the xy-plane of rotational symmetry and represent the resultant forces and moments that the two strands exert on each other at the convergence points at the ends of the balanced ply (shown by the heavy dots in figure 1*a*). As discussed in Fraser & Stump (1998*a*), the deformation of the strands at the convergence points is highly three dimensional and therefore only approximately represented by the one-dimensional rod theory. Since the pressure and frictional stresses between the strands vary rapidly in the vicinity of the convergence point over a distance that is less than the strand diameter, only the resultant forces and moments associated with the stress distributions can be taken into account in the context of a one-dimensional rod theory. These are represented as a point force and moment acting in the normal direction between the strands at the convergence point.

The equations of equilibrium are supplemented with the inextensibility constraint and the condition of normality between shear force $\boldsymbol{V}$ and the axial tension $T\boldsymbol{R}'$:

$$\boldsymbol{R}' \cdot \boldsymbol{R}' = 1, \quad \boldsymbol{V} \cdot \boldsymbol{R}' = 0. \tag{2.4}$$

As in Fraser & Stump (1998*a*, *b*), N' is the rate of tangential angular rotation of an orthogonal material coordinate system ($\boldsymbol{e}_1 = \boldsymbol{R}', \boldsymbol{e}_2, \boldsymbol{e}_3$) embedded in the cross-section and is obtained from the torsion vector

$$\frac{\partial N}{\partial s}\boldsymbol{R}' = (\phi' - \boldsymbol{n} \cdot \boldsymbol{b}')\boldsymbol{R}'. \tag{2.5}$$

The torsion can be divided into two components:

(1) ϕ', which is the rate of change in angle between the material axes ($\boldsymbol{e}_1 = \boldsymbol{R}', \boldsymbol{e}_2, \boldsymbol{e}_3$) and the Frenet basis vectors ($\boldsymbol{R}', \boldsymbol{n}, \boldsymbol{b}$); and

(2) the tortuosity $-\boldsymbol{b}' \cdot \boldsymbol{n}$, which results from the osculation of the plane of curvature.

The linear bending moment–curvature and the torque–torsion constitutive laws are

$$\boldsymbol{M} = \boldsymbol{R}' \wedge \boldsymbol{R}'', \quad Q = \kappa\frac{\partial N}{\partial s}, \tag{2.6}$$

where the ratio $\kappa = K/B$. This completes the set of equations for the static system.

The dimensionless variables used in the equations are related to the physical variables (barred) by the following definitions:

$$\left.\begin{aligned} \boldsymbol{R} = \frac{\bar{\boldsymbol{R}}}{a} = \frac{\bar{x}}{a}\boldsymbol{i} + \frac{\bar{y}}{a}\boldsymbol{j} + \frac{\bar{z}}{a}\boldsymbol{k}, \quad s = \frac{\bar{s}}{a}, \quad p = \frac{\bar{p}a^3}{B}, \\ T = \frac{\bar{T}a^2}{B}, \quad \boldsymbol{V} = \frac{\bar{\boldsymbol{V}}a^2}{B}, \quad Q = \frac{\bar{Q}a}{B}, \quad \boldsymbol{M} = \frac{\bar{\boldsymbol{M}}a}{B}. \end{aligned}\right\} \tag{2.7}$$

(i) *The shear vector*

Formation of the vector product of $\boldsymbol{R}'$ with $(2.3)_2$ and the use of the triple vector product yields the formula

$$\boldsymbol{V} = Q(\boldsymbol{R}' \wedge \boldsymbol{R}'') + \boldsymbol{R}' \wedge \boldsymbol{M}' + m_{\mathrm{h}}\boldsymbol{b}\delta(s-L) + m_{\mathrm{t}}\boldsymbol{b}\delta(s-l) \tag{2.8}$$

within all three regions of the writhed rod. Once the rod shape $\boldsymbol{R}$ is known, $\boldsymbol{M}$ is computed by $(2.6)_1$, and the additional specification of Q gives the shear $\boldsymbol{V}$ from (2.8).

(ii) *Two first integrals*

Formation of the scalar product of $\boldsymbol{R}'$ with $(2.3)_2$ gives $Q' = 0$ so that Q is a constant throughout the entire rod. In accordance with Miekle & Holmes (1988), the use of (2.8) in $(2.3)_1$ and the formation of the scalar product with $\boldsymbol{R}'$, along with the result $\boldsymbol{V}' \cdot \boldsymbol{R}' = -\boldsymbol{V} \cdot \boldsymbol{R}''$ which follows from $(2.4)_2$, provides

$$T' = -\boldsymbol{R}'' \cdot \boldsymbol{R}'''. \tag{2.9}$$

This equation is integrated to obtain

$$T = T_{\mathrm{g}} - \tfrac{1}{2}|\boldsymbol{R}''|^2, \tag{2.10}$$

where T_{g} is a constant that differs in each part of the rod.

(iii) *The elastic strain energy*

The elastic strain energy of the bent and twisted rod is stored in the deformations associated with the torque Q and bending moment $\boldsymbol{M}$ along the length. The formula for U', the dimensionless strain energy per unit arc-length, is

$$U' = \frac{Q^2}{2\kappa} + \tfrac{1}{2}|\boldsymbol{M}|^2. \tag{2.11}$$

3. The balanced-ply region

The forces and moments within the balanced-ply region have been found as a specialization of a larger study of helically plied yarn structures by Fraser & Stump (1998*a*, *c*), and aspects of that solution are presented here for completeness. Within the balanced-ply region, the centreline path of strand 1 ($L \leqslant s \leqslant l$) is described by the equation of a right-hand helix:

$$\boldsymbol{R}(s) = \boldsymbol{i}\cos\theta + \boldsymbol{j}\sin\theta + \boldsymbol{k}(s-L)\cos\beta, \tag{3.1}$$

where $\theta = (s - L)\sin\beta$. The equation for strand 2 ($-l \leqslant s \leqslant -L$) is obtained by replacing θ with $\theta + \pi$. The Frenet basis vectors of strand 1 are

$$\left.\begin{aligned} \boldsymbol{R}' &= (-\boldsymbol{i}\sin\theta + \boldsymbol{j}\cos\theta)\sin\beta + \boldsymbol{k}\cos\beta, \\ \boldsymbol{n} &= -\boldsymbol{i}\cos\theta - \boldsymbol{j}\sin\theta, \\ \boldsymbol{b} &= (\boldsymbol{i}\sin\theta - \boldsymbol{j}\cos\theta)\cos\beta + \boldsymbol{k}\sin\beta. \end{aligned}\right\} \tag{3.2}$$

Substitution of the position vector (3.1) in $(2.6)_1$, (2.8), and $(2.3)_1$ determines $\boldsymbol{M}$, $\boldsymbol{V}$ and p in terms of the unknown T and torque Q in each of the helically plied strands:

$$\left.\begin{aligned} \boldsymbol{M} &= \sin^2\beta\boldsymbol{b}, \\ \boldsymbol{V} &= \{Q\sin^2\beta - \cos\beta\sin^3\beta\}\boldsymbol{b}, \\ p &= T\sin^2\beta - Q\cos\beta\sin^3\beta + \sin^4\beta\cos^2\beta. \end{aligned}\right\} \tag{3.3}$$

The overall tension T_0 and torque Q_0 acting in the $\boldsymbol{k}$-direction through the cross-sectional centre of a helical ply structure are expressed in terms of the two sets of strand variables (denoted by subscript 1 and 2) by

$$\left.\begin{aligned} T_0\boldsymbol{k} &= T(\boldsymbol{R}_1' + \boldsymbol{R}_2') + \boldsymbol{V}_1 + \boldsymbol{V}_2 \\ &= \{2T\cos\beta + 2Q\sin^3\beta - 2\cos\beta\sin^4\beta\}\boldsymbol{k}, \\ Q_0\boldsymbol{k} &= \boldsymbol{M}_1 + \boldsymbol{M}_2 + Q(\boldsymbol{R}_1' + \boldsymbol{R}_2') \\ &\quad + T(\boldsymbol{R}_1 \wedge \boldsymbol{R}_1' + \boldsymbol{R}_2 \wedge \boldsymbol{R}_2') + \boldsymbol{R}_1 \wedge \boldsymbol{V}_1 + \boldsymbol{R}_2 \wedge \boldsymbol{V}_2 \\ &= \{2T\sin\beta + 2Q\cos^3\beta + 2\sin^3\beta(1 + \cos^2\beta)\}\boldsymbol{k}. \end{aligned}\right\} \tag{3.4}$$

Since the balanced ply is attached to the end loop which is free from external forces and moments, the overall cross-sectional force and moment $T_0 = Q_0 = 0$. The solution of (3.4) for the unknown strand quantities T and Q and the use of these in (3.3) gives the formulae

$$\left.\begin{aligned} T = p = \frac{\sin^4\beta}{\cos 2\beta}, \quad Q = -\frac{2\cos\beta\sin^3\beta}{\cos 2\beta}, \\ \boldsymbol{V} = -\frac{\cos\beta\sin^3\beta}{\cos 2\beta}\boldsymbol{b}, \quad \boldsymbol{M} = \sin^2\beta\boldsymbol{b}, \end{aligned}\right\} \tag{3.5}$$

for the forces and moments in the balanced-ply strands. For later reference, note that the constant elastic energy storage per unit arc-length in each of the balanced-ply strands obtained from (2.11) and (3.5) is

$$U' = \frac{1}{2\kappa}\sin^4\beta(\tan^2 2\beta + \kappa). \tag{3.6}$$

Two points about the balanced-ply solution deserve mention. First, the solution is completely specified by the helix angle β so that the attached end loop and tails are also expected to be completely described by the angle β. Second, the formulae in (3.5) show that various quantities become unbounded as β approaches 45°. It does not appear possible that β in a balanced ply can be made to approach the critical 'lock-up' value of 57.52° for an incompressible strand material.

4. The end-loop region

In this section the equations for the shape of the end loop are formulated along with a set of jump conditions that must be satisfied between the end loop and ply at the convergence point $s = L$.

(*a*) *The governing equations for end-loaded rods*

In the loop and tail regions, the rod is subject only to end loads so that the equilibrium equations (2.3) reduce to

$$\left.\begin{aligned} (T\boldsymbol{R}')' + \boldsymbol{V}' = \boldsymbol{0}, \\ (Q\boldsymbol{R}')' + \boldsymbol{M}' + \boldsymbol{R}' \wedge \boldsymbol{V} = \boldsymbol{0}. \end{aligned}\right\} \tag{4.1}$$

The integral of $(4.1)_1$ gives

$$T\boldsymbol{R}' + \boldsymbol{V} = \boldsymbol{C}, \tag{4.2}$$

where $\boldsymbol{C}$ is a constant vector which represents the constant total force on any cross-section of the end-loaded rod. This equation is solved for $\boldsymbol{V}$ and inserted into $(4.1)_2$, which is then integrated to obtain

$$Q\boldsymbol{R}' + \boldsymbol{M} = \boldsymbol{D} - \boldsymbol{R} \wedge \boldsymbol{C}, \tag{4.3}$$

where $\boldsymbol{D}$ is a constant moment vector. This equation also has a simple physical meaning since the left side is the total moment on any cross-section of the end-loaded rod.

The differential equation for the end-loaded rod is obtained by replacing $\boldsymbol{M}$ with $(2.6)_1$, then forming the cross product of $\boldsymbol{R}'$ with (4.3), and finally using the triple vector product identity to derive

$$\boldsymbol{R}'' = \boldsymbol{R}' \wedge (\boldsymbol{R} \wedge \boldsymbol{C} - \boldsymbol{D}). \tag{4.4}$$

This equation must be integrated with a suitable set of boundary conditions for prescribed $\boldsymbol{C}$ and $\boldsymbol{D}$. Through a linear transformation of $\boldsymbol{R}$, equation (4.4) can be changed into a coordinate system in which the $\boldsymbol{C}$ and $\boldsymbol{D}$ are parallel and each have only a single component. The integration can then be accomplished in closed form and yields a general solution in terms of elliptic functions, which are transformed back to the original coordinate system. However, since the values of $\boldsymbol{C}$ and $\boldsymbol{D}$ are unknown, the necessary coordinate transformation cannot be specified *a priori*, and it is simpler to integrate numerically the equation system in the original coordinates.

Once $\boldsymbol{R}$ has been found, the forces and moments in the loop are found as follows. The scalar product of $\boldsymbol{R}'$ with (4.2) is formed, and the result is back-substituted in (4.2) to obtain

$$T = \boldsymbol{C} \cdot \boldsymbol{R}', \quad \boldsymbol{V} = \boldsymbol{C} - \boldsymbol{R}'(\boldsymbol{C} \cdot \boldsymbol{R}'). \tag{4.5}$$

Similarly, the scalar product of $\boldsymbol{R}'$ with (4.3) yields

$$Q = \boldsymbol{R}' \cdot (\boldsymbol{D} - \boldsymbol{R} \wedge \boldsymbol{C}), \tag{4.6}$$

which is back-substituted into (4.3) to obtain $\boldsymbol{M}$. Alternatively, $\boldsymbol{M}$ is available directly from $(2.6)_1$.

Expression $(4.5)_2$ for $\boldsymbol{V}$ is completely consistent with (2.8) as the use of the force and moment expressions for the end-loaded rod in (2.8) produces an identity. In addition, since (2.10) and $(4.5)_1$ are equivalent, this gives the formula

$$T_{\mathrm{g}} = \boldsymbol{C} \cdot \boldsymbol{R}' + \tfrac{1}{2}|\boldsymbol{R}''|^2 \tag{4.7}$$

for the constant T_{g} within the end-loaded segment.

An expression for the elastic energy storage per unit arc-length in the end-loaded rod is found by substitution of $(2.6)_1$ and (4.7) into (2.11) to obtain

$$U' = \frac{Q^2}{2\kappa} + \tfrac{1}{2}|\boldsymbol{R}''|^2 = \frac{Q^2}{2\kappa} + T_{\mathrm{g}} - \boldsymbol{R}' \cdot \boldsymbol{C}. \tag{4.8}$$

(b) The jump conditions

The Cartesian coordinate system used to describe the shape of the end loop coincides with the convergence point of the balanced ply and end-loop regions (see figure 1). The loop extends for an unknown height h in the $-\boldsymbol{k}$-direction, where the origin of the coordinate s is placed, and occupies the domain $-L \leqslant s \leqslant L$. Since the end loop is symmetric about the $\boldsymbol{k}$-axis, the vectors $\boldsymbol{C}$ and $\boldsymbol{D}$ must lie within the xy-plane so that the axial force and moment components C_z and D_z are zero.

The geometric boundary conditions on the loop are given by the continuity of the position and tangent vectors with those of the balanced ply at the two ends. These are obtained from (3.1) and $(3.2)_1$ as

$$\left.\begin{aligned} \boldsymbol{R}(-L) &= -\boldsymbol{i}, & \boldsymbol{R}'(-L) &= \boldsymbol{j}\sin\beta - \boldsymbol{k}\cos\beta, \\ \boldsymbol{R}(L) &= \boldsymbol{i}, & \boldsymbol{R}'(L) &= \boldsymbol{j}\sin\beta + \boldsymbol{k}\cos\beta. \end{aligned}\right\} \tag{4.9}$$

The integration of the equilibrium equations (2.3) across the convergence point from L^- to L^+ gives the jump equations

$$[\boldsymbol{V}] + f_{\mathrm{h}}\boldsymbol{n} = \boldsymbol{0}, \quad [\boldsymbol{M}] + m_{\mathrm{h}}\boldsymbol{n} = \boldsymbol{0}, \tag{4.10}$$

where the jump notation $[\boldsymbol{V}]$ denotes $\boldsymbol{V}(L^+) - \boldsymbol{V}(L^-)$ and the continuity of T and Q has been used. The form of the jump equations (4.10) shows that forces and moments at the convergence point are continuous in the tangential and binormal directions and discontinuous in the normal direction where the contact force and moment f_{h} and m_{h} act to pry the end loop apart.

Next, the balanced-ply quantities (3.5) and the end loop quantities obtained from (4.2) and (4.3) are substituted into (4.10) to get

$$\left.\begin{aligned} -\frac{\cos\beta\sin^3\beta}{\cos 2\beta}\boldsymbol{b} - C_x\boldsymbol{i} - C_y\boldsymbol{j} + \frac{\sin^4\beta}{\cos 2\beta}\boldsymbol{R}' + f_{\mathrm{h}}\boldsymbol{n} = \boldsymbol{0}, \\ \sin^2\beta\boldsymbol{b} - D_x\boldsymbol{i} - D_y\boldsymbol{j} + C_y\boldsymbol{k} - \frac{2\cos\beta\sin^3\beta}{\cos 2\beta}\boldsymbol{R}' + m_{\mathrm{h}}\boldsymbol{n} = \boldsymbol{0}, \end{aligned}\right\} \tag{4.11}$$

where the basis vectors are evaluated at the convergence point $s = L$. Equations (3.1) and (3.2) allow (4.11) to be separated into Cartesian components and yield the four independent jump conditions

$$\left.\begin{aligned} C_x &= -f_{\mathrm{h}}, & C_y &= \frac{\sin^3\beta}{\cos 2\beta}, \\ D_x &= -m_{\mathrm{h}}, & D_y &= -\frac{\cos\beta\sin^2\beta}{\cos 2\beta}, \end{aligned}\right\} \tag{4.12}$$

so that only the two constants C_x and D_x (or equivalently the point force $f_{\rm h}$ and point moment $m_{\rm h}$) are unknown. The final problem for the free loop consists of the integration of the vector differential equations (4.4) subject to the boundary conditions (4.9), which is solved with a shooting method to determine C_x, D_x and L for given β. The details of the numerical solution are discussed in § 6.

5. The tail regions

The coordinate system used to describe the tail region is shown in figure 1. The origin O_1 of the axes system $O_1x_1y_1z$ lies a distance d along the $\boldsymbol{k}$-axis beneath the convergence point where the tails join the balanced ply. The y_1-axis of the system is orientated to coincide with the direction of the remote applied tension and torque for the strand with $s > l$. All vector quantities in this section refer to this coordinate system, and the subscript notation is dropped hereafter. In this region the analytic solution first derived by Coyne (1990) can be used as is now shown.

(*a*) *Integration of the governing equations*

Equations (4.2) and (4.3) for the end-loaded rods derived in § 4 *a* also apply to the tail regions. However, in the tail region it is possible to determine the constants of integration $\boldsymbol{C}$ and $\boldsymbol{D}$ as follows: first note that in the limit $s \to \infty$, the position, tangent and curvature vectors approach the asymptotes $\boldsymbol{R} = s\boldsymbol{j}$, $\boldsymbol{R}' = \boldsymbol{j}$ and $\boldsymbol{R}'' = \boldsymbol{0}$, respectively, and the forces and moments approach the remote values

$$T = T_\infty, \quad Q = Q_\infty, \quad \boldsymbol{V} = \boldsymbol{M} = \boldsymbol{0}. \tag{5.1}$$

When these results are substituted into equations (4.2) and (4.3) the constants of integration are found to be

$$\boldsymbol{C} = T_\infty \boldsymbol{j}, \quad \boldsymbol{D} = Q_\infty \boldsymbol{j}, \tag{5.2}$$

so that equation (4.4) becomes

$$\boldsymbol{R}'' = \boldsymbol{R}' \wedge (\boldsymbol{R} \wedge T_\infty \boldsymbol{j} - Q_\infty \boldsymbol{j}). \tag{5.3}$$

The integration of this equation is most easily accomplished in terms of the cylindrical-polar coordinates (ρ, α, y) which are related to the $Oxyz$ coordinates shown in figure 1 by the transformation equations

$$x = \rho \sin\alpha, \quad y = y, \quad z = \rho\cos\alpha, \tag{5.4}$$

and

$$\boldsymbol{R}(s) = \rho(s)\boldsymbol{e}_\rho[\alpha(s)] + y(s)\boldsymbol{j}, \tag{5.5}$$

where $(\boldsymbol{e}_\rho, \boldsymbol{e}_\alpha, \boldsymbol{j})$ are the unit basis vectors of the cylindrical-polar coordinate system. The three components of equation (5.3) corresponding to these coordinates are

$$\rho'' - \rho\alpha'^2 = T_\infty \rho y' - Q_\infty \rho\alpha', \quad \rho\alpha'' + 2\rho'\alpha' = Q_\infty \rho', \quad y'' = -T_\infty \rho\rho'. \tag{5.6}$$

Equation $(5.6)_2$ can be integrated if it is first multiplied by ρ to yield

$$(\rho^2\alpha')' = (\tfrac{1}{2}\rho^2)' Q_\infty. \tag{5.7}$$

This equation is now integrated twice to give

$$\alpha = \tfrac{1}{2} Q_\infty s + \alpha_0, \tag{5.8}$$

where the constant from the first integration is shown to be zero because in the limit $s \to \infty$, $\rho \to 0$ and α' must be bounded; α_0 is the second constant of integration.

Equation $(5.6)_3$ is integrated once to give

$$y' = 1 - \tfrac{1}{2}T_\infty \rho^2, \tag{5.9}$$

where the constant of integration has been determined from the limit $s \to \infty$, $y' \to 1$.

When (5.8) and (5.9) are substituted back into equation $(5.6)_1$ a second-order differential equation for ρ is obtained:

$$\rho'' - \tfrac{1}{4}\gamma^2 \rho + \tfrac{1}{2}{T_\infty}^2 \rho^3 = 0, \tag{5.10}$$

where $\gamma^2 = 4T_\infty - {Q_\infty}^2$ must be non-negative for the validity of the solution derived below. When this equation is multiplied through by ρ' it can be integrated to give

$$\rho'^2 - \tfrac{1}{4}\gamma^2 \rho^2 + \tfrac{1}{4}{T_\infty}^2 \rho^4 = C, \tag{5.11}$$

where C is a constant of integration, which is seen to be zero when it is noted that in the limit $s \to \infty$, $\rho \to 0$ and $\rho' \to 0$. On rearrangement (5.11) becomes

$$\rho' = -\tfrac{1}{2}T_\infty \rho \sqrt{(\gamma^2/T_\infty^2) - \rho^2}, \tag{5.12}$$

where the negative sign has been taken for the square root because ρ must be a decreasing function of s in the tail region. Finally this first-order differential equation can be integrated by standard techniques to give the result

$$\rho = \frac{\gamma}{T_\infty} \operatorname{sech}[\tfrac{1}{2}\gamma(s - s_0)], \tag{5.13}$$

where s_0 is a constant of integration. This result is now substituted into equation (5.9), which on a further integration yields

$$y = s - \frac{\gamma}{T_\infty} \tanh[\tfrac{1}{2}\gamma(s - s_0)] + y_0, \tag{5.14}$$

where y_0 is the final constant of integration.

Solutions (5.8), (5.13) and (5.14) must now be joined smoothly to the balanced-ply region at the convergence point $s = l$ so that the position and tangent vectors $\boldsymbol{R}$ and $\boldsymbol{R}'$ are continuous functions of s, and the jump conditions derived below are satisfied.

(b) *The jump conditions*

Referred to the coordinate system shown in figure 1, the shape of the balanced-ply helix (3.1) becomes

$$\boldsymbol{R} = \boldsymbol{i}\cos\theta + \boldsymbol{j}\sin\theta + \boldsymbol{k}[(s - l)\cos\beta - d], \tag{5.15}$$

where $\theta = (s - l)\sin\beta + \theta_0$, and θ_0 is an unknown offset angle (see figure 1*c*). The system of Frenet basis vectors at $s = l$ is

$$\left.\begin{aligned} \boldsymbol{R}' &= -\boldsymbol{i}\sin\beta\sin\theta_0 + \boldsymbol{j}\sin\beta\cos\theta_0 + \boldsymbol{k}\cos\beta, \\ \boldsymbol{n} &= -\boldsymbol{i}\cos\theta_0 - \boldsymbol{j}\sin\theta_0, \\ \boldsymbol{b} &= \boldsymbol{i}\cos\beta\sin\theta_0 - \boldsymbol{j}\cos\beta\cos\theta_0 + \boldsymbol{k}\sin\beta. \end{aligned}\right\} \tag{5.16}$$

The position vector of the tails is expressed in the Cartesian coordinate system as

$$\boldsymbol{R} = \boldsymbol{i}\rho\sin\alpha + \boldsymbol{j}y + \boldsymbol{k}\rho\cos\alpha. \tag{5.17}$$

At the convergence point $s = l$, the continuity of the position and tangent vectors can be rearranged to give the five independent equations (only two components of the tangent are independent due to inextensibility)

$$\left.\begin{aligned} \cos^2\theta_0 + d^2 = \rho^2, \quad \sin\theta_0 = y, \quad \tan\alpha = -\frac{\cos\theta_0}{d}, \\ \sin\beta\cos\theta_0 = 1 - \tfrac{1}{2}T_\infty\rho^2, \quad \sin\beta\sin\theta_0\cos\theta_0 + d\cos\beta = -\rho\rho', \end{aligned}\right\} \tag{5.18}$$

where α, ρ and y are given by equations (5.8), (5.13) and (5.14), respectively, evaluated at $s = l$.

The force and moment jump conditions at the convergence point are obtained by integration of equations (2.3) from $s = l^-$ to $s = l^+$, which yields

$$[\boldsymbol{V}] + f_t\boldsymbol{n} = \boldsymbol{0}, \quad [\boldsymbol{M}] + m_t\boldsymbol{n} = \boldsymbol{0}, \tag{5.19}$$

where the jump notation introduced in § 4 is used. These equations are analogous to (4.10) and imply the continuity of all forces and moments in the tangential and binormal directions at the convergence point.

The values of the various quantities at $s = l^-$ are given by the balanced-ply solution (3.5) and the basis vectors (5.16). The quantities in the tails at $s = l^+$ are obtained from equations (4.2) and (4.3) with $\boldsymbol{C}$ and $\boldsymbol{D}$ given by equations (5.2) and $\boldsymbol{R}$ given by (5.15). With these substitutions, the jump conditions (5.19) become

$$\left.\begin{aligned} T_\infty\boldsymbol{j} - \frac{\sin^4\beta}{\cos 2\beta}\boldsymbol{R}' + \frac{\cos\beta\sin^3\beta}{\cos 2\beta}\boldsymbol{b} + f_t\boldsymbol{n} = \boldsymbol{0}, \\ Q_\infty\boldsymbol{j} - \boldsymbol{R}\wedge T_\infty\boldsymbol{j} + \frac{2\cos\beta\sin^3\beta}{\cos 2\beta}\boldsymbol{R}' - \sin^2\beta\boldsymbol{b} + m_t\boldsymbol{n} = \boldsymbol{0}. \end{aligned}\right\} \tag{5.20}$$

The separation of these equations into Cartesian components leads to the four independent conditions

$$\left.\begin{aligned} T_\infty\cos\theta_0 = \frac{\sin^3\beta}{\cos 2\beta}, \quad T_\infty d\sin\theta_0 + Q_\infty\cos\theta_0 = -\frac{\cos\beta\sin^2\beta}{\cos 2\beta}, \\ T_\infty\sin\theta_0 = f_t, \quad Q_\infty\sin\theta_0 - T_\infty d\cos\theta_0 = m_t. \end{aligned}\right\} \tag{5.21}$$

The continuity and jump equations (5.18) and (5.21) comprise an overdetermined set of nine algebraic equations in eight unknowns $(\alpha_0, y_0, l - s_0, \theta_0, T_\infty, d, f_t, m_t)$. However, it is straightforward to reduce the system to a single transcendental equation. First, note that $(5.21)_1$, $(5.21)_2$ and $Q_\infty = Q$, where Q is given by $(3.5)_2$, are combined to eliminate T_∞ and yield

$$d = \cot\beta\cot\theta_0(2\sin\beta\cos\theta_0 - 1). \tag{5.22}$$

Next, $(5.18)_1$ and $(5.21)_1$ are substituted into $(5.18)_4$ to obtain

$$\sin\beta\cos\theta_0 = 1 - \frac{\sin^3\beta}{2\cos\theta_0\cos 2\beta}[d^2 + \cos^2\theta_0]. \tag{5.23}$$

Equation (5.22) is used to eliminate d from (5.23), which provides a single transcendental equation for θ_0 in terms of β. The solution is found with Newton's method, and the other quantities are obtained by back-substitution into the various expressions in (5.18), (5.21) and (5.22).

6. Numerical solution for the end loop

The numerical solution of the end-loop equations is discussed in this section. In order to simplify the numerical solution and the asymptotic analysis of Appendix A, the various quantities in the problem are rescaled against the dimensionless end-loop half-length L as the characteristic length scale, which makes all the quantities $O(1)$ in the rescaled system. The rescaled coordinates, forces and moments are

$$\tilde{s} = \frac{s}{L}, \quad \tilde{\boldsymbol{R}} = \frac{\boldsymbol{R}}{L}, \quad \tilde{Q} = LQ, \quad \tilde{\boldsymbol{D}} = L\boldsymbol{D}, \quad \tilde{\boldsymbol{C}} = L^2\boldsymbol{C}. \tag{6.1}$$

The rescaled differential equation (4.4) has the form

$$\tilde{\boldsymbol{R}}'' = \tilde{\boldsymbol{R}}' \wedge (\tilde{\boldsymbol{R}} \wedge \tilde{\boldsymbol{C}} - \tilde{\boldsymbol{D}}), \tag{6.2}$$

where the rescaled constant vectors of § 4 a are

$$\tilde{\boldsymbol{C}} = \boldsymbol{i}\tilde{C}_x + \boldsymbol{j}\frac{L^2 \sin^3\beta}{\cos 2\beta}, \quad \tilde{\boldsymbol{D}} = \boldsymbol{i}\tilde{D}_x - \boldsymbol{j}\frac{L\cos\beta\sin^2\beta}{\cos 2\beta}. \tag{6.3}$$

The rescaled torque constraint is obtained from (4.6) and $(3.5)_2$:

$$-\frac{2L\cos\beta\sin^3\beta}{\cos 2\beta} = \tilde{\boldsymbol{R}}' \cdot (\tilde{\boldsymbol{D}} - \tilde{\boldsymbol{R}} \wedge \tilde{\boldsymbol{C}}), \tag{6.4}$$

which must be enforced at a single location along the end loop. The transformed boundary conditions (4.9) are

$$\left.\begin{aligned} \tilde{\boldsymbol{R}}(-1) &= -\boldsymbol{i}/L, & \tilde{\boldsymbol{R}}'(-1) &= \boldsymbol{j}\sin\beta - \boldsymbol{k}\cos\beta, \\ \tilde{\boldsymbol{R}}(1) &= \boldsymbol{i}/L, & \tilde{\boldsymbol{R}}'(1) &= \boldsymbol{j}\sin\beta + \boldsymbol{k}\cos\beta. \end{aligned}\right\} \tag{6.5}$$

For a given β, the system of equations (6.2) and boundary conditions (6.4) and (6.5) comprise a two-point boundary-value problem involving the iterates L, $\tilde{C}_x$, and $\tilde{D}_x$. The system is integrated with a fourth-order Runge–Kutta scheme using $(6.5)_1$ as initial conditions, and with (6.4) and $(6.5)_2$ serving as constraint equations in a shooting method for the determination of L, $\tilde{C}_x$ and $\tilde{D}_x$. A high degree of accuracy is required in the integration of (6.2) in order for the Newton–Raphson iteration to converge.

7. Results for the end loop and tails

The results for the calculations of the end loop and tail regions are presented in this section.

(*a*) *The end loop*

The end-loop arc-length L is plotted in figure 3 as a function of balanced-ply helix angle β. There are two families of solutions: the upper branch, which consists of *large* loops that lie primarily in the yz-plane (see insert); and the lower branch, which is comprised of *small* loops that lie primarily in the xz-plane (see insert). (Details of these different types of solutions are discussed below.) The curve is a solid line where the contact pressure force at the convergence point is positive and dashed where it is negative. Along the upper branch of figure 3, L is very sensitive to β, while on

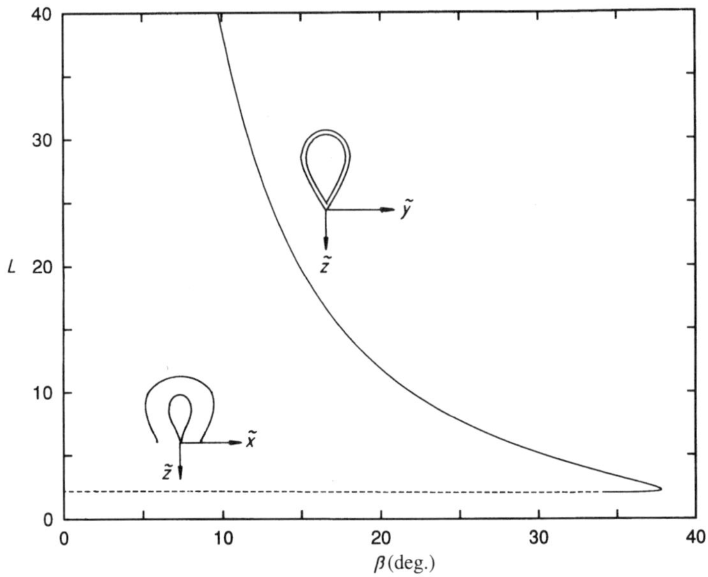

Figure 3. A plot of dimensionless end-loop half-length L versus the helix angle β of the balanced ply.

the lower branch L is almost a constant value independent of β. Interestingly, the diagram shows a turning point at the value $\beta = 37.792°$ beyond which no solution is possible. In the region $34.1° \leqslant \beta \leqslant 37.792°$, there are two solutions consistent with positive contact pressure.

(i) *Large loops*

The upper branch of figure 3 is associated with loops that become unbounded in size as $\beta \to 0$. The asymptotic analysis in Appendix A shows that on the upper branch $L \sim 3.14/\beta^{3/2}$ (β in radians) as $\beta \to 0$, while the loop shape normalized by L remains well behaved and lies primarily in the yz-plane. Figure 4*a* shows the $\tilde{y}\tilde{z}$-plane projection of large loops corresponding to the values $\beta = 0°$, $10°$, $20°$ and $30°$. The asymptotic shape for $\beta \to 0$ is expressed in terms of elliptic functions in Appendix A.

Since large loops are associated exclusively with the physically realistic case of positive contact pressure where the ends of the loop act to pry each other apart, these are the shapes that would normally be observed in experimental models. The strong sensitivity of L to β is noteworthy since this implies that considerable care will need to be taken in the experimental verification of this curve.

(ii) *Small loops*

The lower branch of figure 3 is associated with the domain $2.04 \leqslant L \leqslant 2.24$ and has a finite limit as $\beta \to 0$. Most of the lower branch of figure 3 is shown as a dashed line representing a contact tension, which is physically unrealistic since it implies

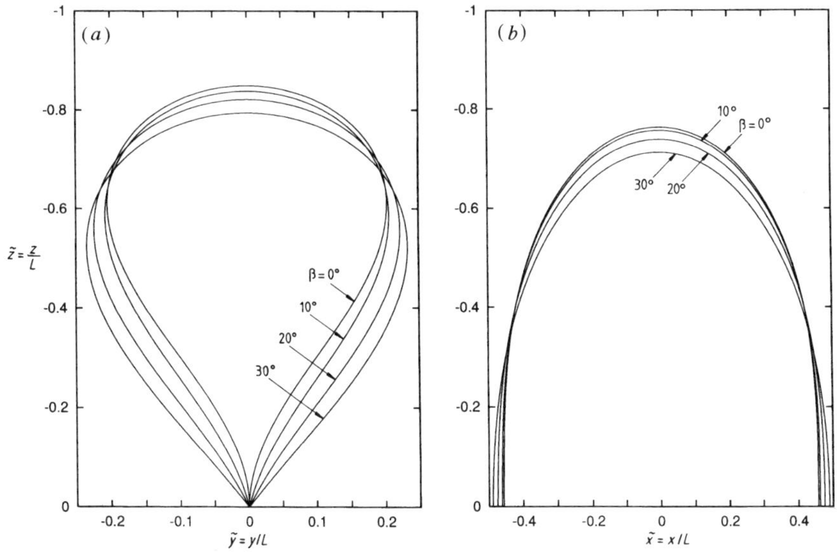

Figure 4. (a) The yz-plane projection of large loops rescaled by the loop length L for various values of β. (b) The xz-plane projection of small loops rescaled by the loop length L for various values of β.

that the ends of the loop are pulling each other together. Figure 4b shows the $\tilde{x}\tilde{z}$-plane projection of the centreline for the small loops associated with $\beta = 0°$, $10°$, $20°$ and $30°$. The shape for $\beta = 0°$ is also expressed in terms of elliptic functions in Appendix A. It should be noted that the small size of these loops means that most of the space inside the curves of figure 4b is occupied by the rod itself.

(iii) *The dimensions*

The introduction of the rescaled quantities in § 6 has the effect of making all dimensions $O(1)$ in the $\tilde{\boldsymbol{R}}$ coordinates so that the essential physical size-scaling behaviour in β is contained in figure 3. However, the relative height and width of the loops are also of interest. (Throughout the remaining plots in this section, solid lines and dashed lines are used to indicate positive and negative contact pressures, respectively.) The loop height $\tilde{h} = h/L = -z(0)/L$ is plotted as a function of β in figure 5a. The height as a proportion of the length L varies between 0.70 and 0.85 for both large and small loops. The asymptotic limit for the large loops is $\tilde{h} = 0.8486$, while the limit for small loops is $\tilde{h} = 0.7628$. The value $\tilde{h} \sim 0.83$ for large loops. It is somewhat surprising that $\tilde{h}$ is relatively insensitive to β.

The maximum projected widths $\tilde{w}_x = w_x/L$ and $\tilde{w}_y = w_y/L$ in the x and y directions are shown as functions of β in figure 5b. Since the loop shape changes from lying primarily in the yz-plane on the upper branch of figure 3 to lying mainly in the xz-plane on the lower branch of figure 3, the widths $\tilde{w}_x$ and $\tilde{w}_y$ approach zero for large and small loops, respectively. The non-zero asymptotes are given by $\tilde{w}_x \sim 0.9139$ and $\tilde{w}_y \sim 0.4084$, which are found from the two asymptotic shapes

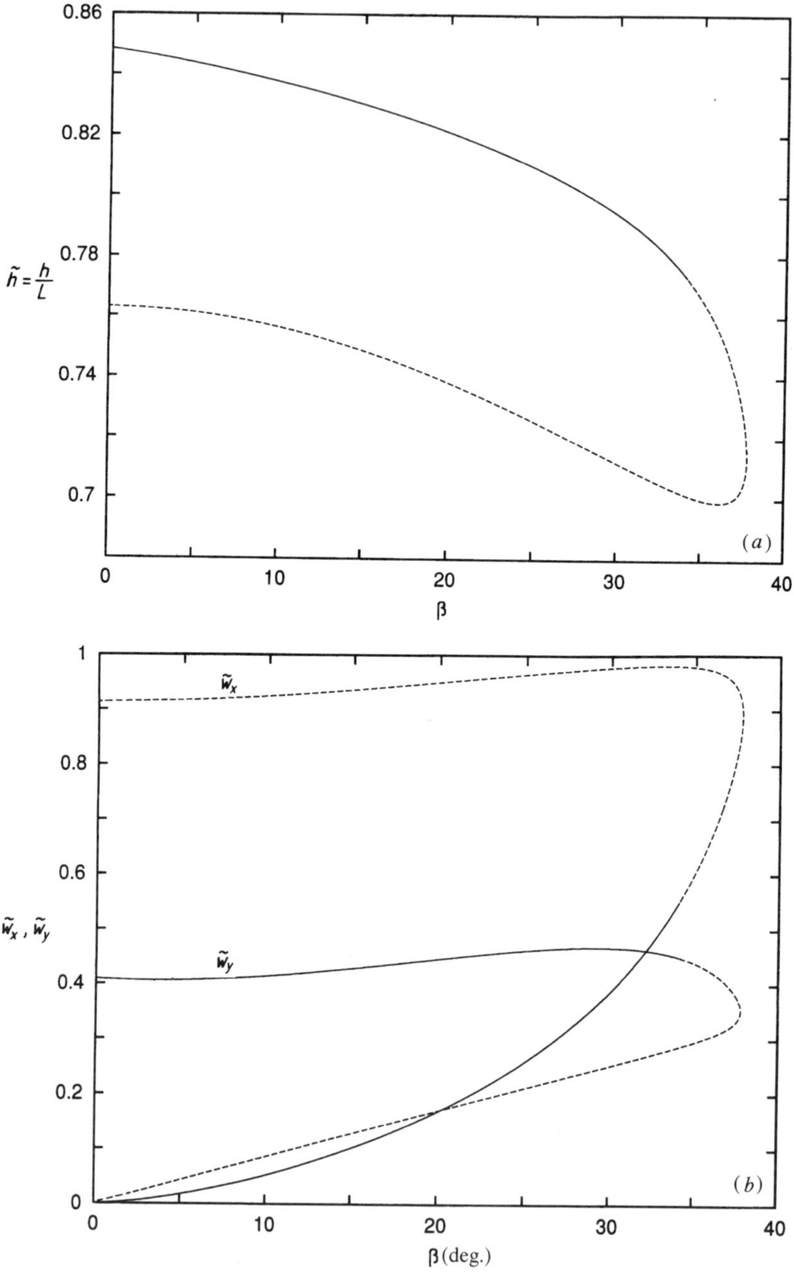

Figure 5. (*a*) The rescaled loop height $\tilde{h}$ versus β. (*b*) The rescaled maximum x- and y-direction widths $\tilde{w}_x$ and $\tilde{w}_y$ versus β.

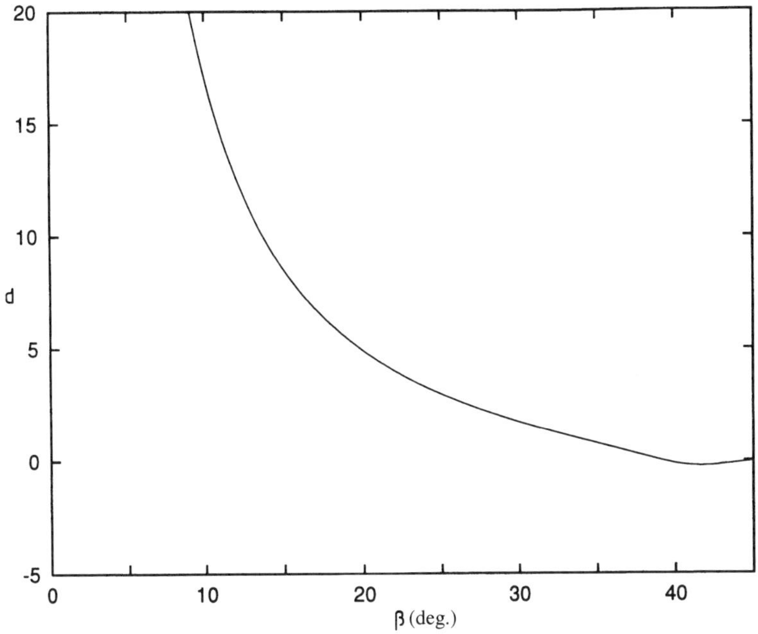

Figure 6. A plot of d, the height of the convergence point above the the tails, versus β.

shown in figure 4. The values of $\tilde{w}_x$ and $\tilde{w}_y$ are remarkably constant over the lower and upper branch, respectively, of figure 3. This behaviour in combination with the height behaviour of figure 5*a* indicates that the loop shape does not vary much along the two branches of figure 3, even though the arc length L of large loops is very sensitive to β.

(*b*) *The tail regions*

The solution for the rod shape in the tail regions developed in § 5 is closed form and is joined to the balanced-ply solution by a set of continuity and jump conditions. The three quantities of physical interest from the solution are the remote tension T_∞, the torque $Q_\infty = Q$ and the height d of the convergence point above the line of action of the remote tension and torque (see figure 1). These are given respectively by $(5.21)_1$, $(3.5)_2$ and (5.22).

The height of the tails is shown in figure 6 where d has been plotted against β. As $\beta \to 0$, the height d diverges while both T_∞ and Q approach zero. The value of d drops to only a few rod radii for helix angles of $20^\circ \leqslant \beta \leqslant 35^\circ$, and even becomes slightly negative in the region $42^\circ \leqslant \beta \leqslant 45^\circ$ before approaching zero as $\beta \to 45^\circ$.

8. Application of the results

The results of the elastic rod model are applied to the writhing of closed loops and long elastic rods.

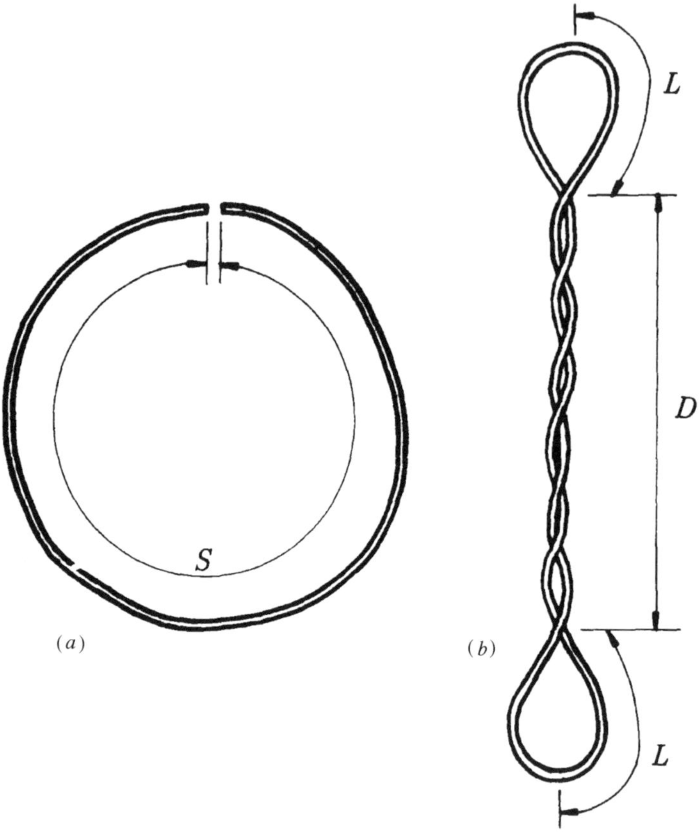

Figure 7. (*a*) The circular loop of DNA with n turns of twisted inserted between the end. (*b*) The same strand in a writhed configuration with ply length D and end-loop half-length L.

(*a*) *The closed loop*

Consider the planar closed rod shown in figure 7*a*. Start with the rod in a straight unstressed reference configuration and draw a straight black line along one side of the cylinder. Next, bend the rod into a circle and twist the two ends relative to one another. After $-n$ integral turns (left-hand rotation is negative in the coordinate system used here), connect the two ends so that the black line forms a continuous curve on the surface of a torus.

When a sufficient number of turns are inserted into the planar loop, it becomes unstable and must be restrained on the table by applied forces. Once these forces are removed, the loop writhes into a shape which consists (at least approximately) of a series of balanced plies and end loops. Figure 7*b* shows the case of a single balanced ply connected to two end loops and is used as a basis for calculations throughout this section unless stated otherwise.

A model consisting of balanced-ply and end-loop segments developed in the preceding sections is applied to predict the writhed shape of figure 7*b* for a given number of inserted turns n and specified arc length S. This is an important problem in the study of DNA supercoiling, and the restriction that arises from the trapping of the inserted turns is discussed first.

(i) *Link, twist and writhe*

The writhing of closed loops of underwound DNA due to the trapped turns has been considered by a number of investigators (White 1969; Fuller 1971; Crick 1976; Bauer *et al.* 1980; Pohl 1980). An important conservation equation which applies to the closed loop of figure 7*b* is reviewed here. More complete discussions and derivations can be found in the references.

The *link* Lk is the number of differential turns inserted between the two loop ends prior to their closure. For figure 7*b*, $Lk = -n$. The link is invariant throughout the writhing process. The formal proof of this assertion (see, for example, Fuller 1971) is based upon topological properties of the centreline and the reference line drawn on the surface of the rod.

The *topological twist* Tw is related to the rotation of the surface reference line about the centreline of the strand and can be obtained from the integration of the rotation rate N' between the two joined ends of the closed loop. For a writhed state with a constant Q along the strand, which includes writhed loops constructed from the balanced-ply and end-loop segments developed here, the integration of $(2.6)_2$ gives the formula

$$Tw = \frac{N(S) - N(0)}{2\pi} = \frac{QS}{2\pi\kappa}. \tag{8.1}$$

The topological twist Tw is not constant as the closed loop writhes. This can be verified by a simple experiment with a piece of string held between the fingers under fixed end rotation. As the fingers are displaced towards one another and the string writhes into a shape similar to figure 1, the torque, hence Tw, decreases noticeably. In the examples discussed here, the link is given an integral value. However, the topological twist Tw as computed from (8.1) need not have an integral value in a writhed state even though the ends of the reference line remain joined together. Finally, it should be noted that the terminology *topological twist* Tw is different from the textile twist $\phi'/2\pi$ which has been discussed in Fraser & Stump (1998*a–c*) and Stump & Fraser (1998).

The *writhe* Wr is defined to be the difference between Lk and Tw:

$$Wr = Lk - Tw, \tag{8.2}$$

and, as explained by Fuller (1971), it is a property of just the strand centreline. The writhe Wr is a measure of the overall 'curviness' of the path when viewed from the embedding space. Pohl (1980) defines the writhe number Wr of any closed curve by the formula

$$Wr = \frac{1}{4\pi} \oint_{s_1} \oint_{s_2} \boldsymbol{R}'(s_1) \wedge \boldsymbol{R}'(s_2) \cdot \frac{\boldsymbol{R}(s_1) - \boldsymbol{R}(s_2)}{|\boldsymbol{R}(s_1) - \boldsymbol{R}(s_2)|^3} \, \mathrm{d}s_2 \, \mathrm{d}s_1, \tag{8.3}$$

which involves a double integral over the path of the centreline.

Fuller (1978) has given an interpretation of this formula for 'simple' closed paths (i.e. non-knotted) by translating the instantaneous tangent vector to the centre of a unit sphere and then calculating the spherical surface area within the image curve traced by the tangent as the closed loop is traversed. The formula obtained from applying Fuller's approach to this particular problem is

$$Wr + 1 = \frac{A}{2\pi}, \tag{8.4}$$

where A is the total spherical surface area (which may be positive or negative) enclosed within the image. (It is straightforward to extend this formula to other paths, see Fuller (1978).) This formula is used in Appendix B to determine the writhe of figure 7*b*.

(ii) *The constraint equations*

The two fundamental restrictions relating the circular and writhed states of figure 7 are the constancy of arc length and the constancy of link. The requirement of fixed arc length is given by the equation

$$\frac{2D}{\cos\beta} + 4L = S, \tag{8.5}$$

where D is the axial length of the balanced-ply segment and L is the half arc-length of each end loop. L depends only on β and can be predicted with good accuracy by using the asymptotic series developed in Appendix A in combination with a numerical scheme to fit the upper branch of figure 3 with a polynomial in powers of $\beta^{k/2}$ where $k = (-3, -2, -1, \dots)$. The four-term series (β in radians)

$$L = \frac{3.1489}{\beta^{3/2}} - \frac{0.1365}{\beta} - \frac{1.041}{\beta^{1/2}} - 1.375 \tag{8.6}$$

is quite accurate for $\beta \leqslant 0.55$ (31.5°). (It should be remembered that all lengths are dimensionless with respect to the rod radius in accordance with the formulae of (2.7).)

The second fundmental equation is obtained from (8.2) by setting $Lk = -n$. For writhed loops constructed from the balanced-ply and end-loop solutions developed in the previous sections, the value of Q is given by $(3.5)_2$ so that the twist formula (8.1) becomes

$$Tw = -\frac{S\cos\beta\sin^3\beta}{\pi\kappa\cos 2\beta}. \tag{8.7}$$

This is a new expression because it contains information from the continuum mechanics analysis of the writhed shape, which has not been considered in previous studies. In Appendix B, (8.4) is used as a basis for calculating the writhe expression

$$Wr = -\frac{D}{\pi}\sin\beta + F_{\text{loop}} - 1. \tag{8.8}$$

The factor $F_{\text{loop}} = A_{\text{loop}}/2\pi$, where A_{loop} is the surface area on the the unit sphere swept out by the tangent during the traversal of the two end loop segments, is plotted as a function of β in figure 8. The solid line shows the writhe associated with large loops on the upper branch of figure 3, and using the asymptotic analysis of Appendix A it is well approximated by the five-term series (β in radians and $\beta \leqslant 0.55$)

$$F_{\text{loop}} = -0.4355\beta^{1/2} - 10.87\beta^{3/2} + 56.11\beta^2 - 89.27\beta^{5/2} + 49.22\beta^3. \tag{8.9}$$

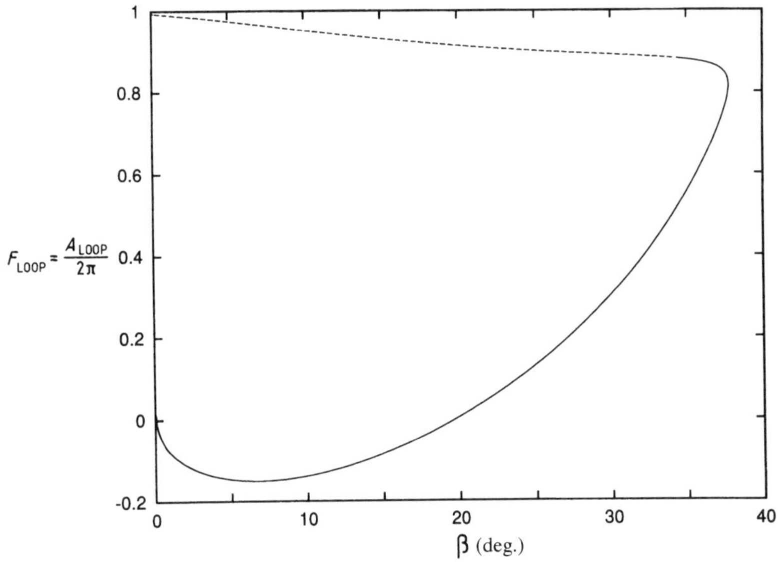

Figure 8. A plot of the spherical area factor $F_{\mathrm{loop}} = A_{\mathrm{loop}}/2\pi$ for a set of end loops versus β.

With these expressions for twist and writhe, the equation for constancy of link is

$$-n = -\frac{S\cos\beta\sin^3\beta}{\pi\kappa\cos 2\beta} - \frac{D}{\pi}\sin\beta + F_{\mathrm{loop}} - 1. \tag{8.10}$$

The use of (8.5) to eliminate D provides the single transcendental equation for the conservation of link in terms of β:

$$n = \frac{S\cos\beta\sin^3\beta}{\pi\kappa\cos 2\beta} + \frac{(S-4L)}{2\pi}\cos\beta\sin\beta - F_{\mathrm{loop}} + 1. \tag{8.11}$$

Once β is determined, the length of the balanced ply, the twist and the writhe are computed from (8.5), (8.7) and (8.8). The formula (8.11) is now applied to several examples.

(iii) *Applications of the formula*

A 110 cm length of braided 5 mm radius rope was given 25 turns of left-hand twist and then closed. When the external forces were released, the rope writhed into a shape similar to figure 7*b* with 12 crossovers in the balanced-ply segment, a helix angle of approximately 25° and end-loop half-lengths of 4.5 cm. The formula (8.11) is used to predict β using $n = 25$, $S = 110/0.5 = 220$, $\kappa = \frac{2}{3}$ and gives $\beta = 26°$. Back-susbstitution yields $Tw = -12.3$, $Wr = -12.7$, $D = 86$ and $L = 7.3$. (The dimensional values of the two lengths are $\bar{D} = 43$ cm and $\bar{L} = 3.7$ cm.) The ratio of the balanced-ply axis length to the overall loop length is $D/S = 0.39$, and the number of strand crossovers in the ply is given by $2D/\lambda = D\sin\beta/\pi = 12$, which agree well with the rope data. The model predictions are in quite reasonable agreement with simple experiments considering the high helix angle and the assumption of linear constitutive laws. The difficulty in defining where the end loop begins and ends makes a direct comparison with the calculations uncertain. More rigorous experiments using Nitinal shape memory alloy are planned.

As a second example, the DNA loop on p. 126 of Calladine & Drew (1992) is analysed. They quote a strand length of 7000 BP and approximately 40 turns of underwinding. The circular loop contains about 660 turns so that the removal of 40 turns represents an underwinding of 6%. The assumptions of a strand radius $a = 10$ Å and the axial separation distance 3.2 Å/BP gives a dimensionless strand length $S = 2240$. The formula (8.11) is adjusted to account for the micrograph shape which shows four end loops rather than the two assumed in the derivation and yields $\beta \sim 7.5°$ for the value $\kappa = \frac{2}{3}$. Back-substitution gives $Tw = -2.5$ $Wr = -37.5$, $D = 869$ and $L = 61$. The lengths correspond to the physical dimensions $\bar{D} = 8690$ Å and $\bar{L} = 610$ Å. In terms of BPs, the axial length of the balanced ply is 2700 BP and the *total* length of each end loop is 380 BP. The ratio of the balanced-ply axis length to the overall loop length is $D/S = 0.39$, which is the same as for the simple experiment with the twisted rope. Finally, the number of strand crossovers in the balanced ply is given by $2D/\lambda = D \sin\beta/\pi = 36$, which corresponds to approximately 150 BP for each wavelength of the balanced-ply helix.

These results are comparible with those of Calladine & Drew (1992) who estimate $Wr = -36$ by counting the number crossovers and infer $Tw = -4$ from the constancy of link. Given that the actual link of the strand is unknown and the relatively low forces and moments from (3.5) opposing the electrostatic repulsion of the phosphate backbones along the strand, the model results are in reasonable agreement with the experimental inferences.

Cozzarelli *et al.* (1990) have presented a model of supercoiling based upon experimental measurements. They assert that the ratio of $D/S = 0.41$ (their l/L) remains unchanged over a wide range of relative undertwisting (2–12%). This certainly agrees with the two examples. However, their model is more complicated since it accounts also for the electrostatic repulsion of the strands, and they present evidence (figure 16, p. 171) that the strands in the balanced ply do not actually touch but are separated from each other by a distance that depends upon the degree of specific linking difference (i.e. percentage underwinding). They also assert that the pitch of the superhelix (their γ which is the complement of β) is constant at $\gamma = 54°$ over the same range of undertwisting. This angle corresponds to a $\beta = 36°$ in our system, which is significantly greater than our prediction based upon contacting strands. However, these results can be understood with this simpler model as follows.

First, the model developed here is the simplest approach and does not include any electrostatic repulsion due to the phosphate groups along the backbone so that the strands within the ply are in direct physical contact. The pressure p between the strands is given in terms of β by $(3.5)_1$, and results from direct contact. This model can be extended to include a pressure term due to electrostatic repusion, which would predict a stable structure with a stand-off distance related to the helix angle β.

Second, Cozzarelli *et al.* (1990) report that the stand-off distance is the only variable that changes significantly with increased underwinding; the angle γ and the superhelix wavelength λ remain essentially constant (see their figure 15, p. 169 for a sketch of several shapes.) If these experimental facts are accepted then the model developed here can be adapted to describe the stand-off effect by drawing two right triangles with λ as the hypotenuse. One triangle relates the model parameters λ,

β, and $2\pi a$ for strands in contact, while the other relates λ, γ, and $2\pi r$, for the non-contact model. The constancy of λ between the two models means that

$$\frac{2\pi a}{\sin\beta} = \frac{2\pi r}{\cos\gamma}.$$

This expression along with figure 16 of Cozzarelli *et al.* (1990) allows for conversion between models by pulling the strands apart and changing the apparent helix angle while keeping the other features of the model the same. Consider the value of $\gamma = 54°$ for the example of 6% underwinding described above, for which Cozzarelli *et al.* (1990) predict a stand-off distance of $r = 50$ Å. For the value $a = 10$ Å, the conversion formula predicts $\beta = 6.7°$, which is consistent with the contact model value of $\beta = 7.5°$. Essentially, the electrical repulsion pushes the strands apart and increases the pitch to compensate without altering the other features of the model. (The end-loop length will also change, but this is a much smaller adjustment.)

Several points deserve further mention. First, with the energy density expressions U' developed for the end loop and balanced ply, it is straightforward to show that the energy stored in the writhed shape is lower than the circular configuration. Second, the issue of whether a particular loop writhes into a single balanced ply connected to two end loops or into a more complicated arrangement with a larger number of end loops and balanced plies could be answered in part by comparing the overall elastic energy contained in the various possible writhed shapes constructed from this model. However, this would only yield an approximate qualitative picture due to the limitations of the uniform elastic rod model of DNA. It is highly likely that any regions of local weakness in the strand (such as ATAT rich segments) also contribute to nucleating end loops. Third, the strand tension $\bar{T}$ computed for the DNA example above can be calculated from (2.7), the value $E \sim O(10^8)$ N m^{-2} (Austin *et al.* 1997), $\beta = 7.5°$, and $a = 10^{-9}$ m and is predicted to be $\bar{T} \sim 0.03$ pN, which is well below the 10 pN level needed to extend the BP separation distance within the double-stranded helical structure. This verifies the assumption of inextensibility.

(*b*) *The long elastic rod*

The results are now applied to describe the long straight rod loaded by remote tension and torque. Since the tail solution depends upon $s \to \infty$, examine an infinitely long straight rod subjected to a fixed remote applied torque Q_∞ and to an adjustable remote tension T_∞, as is shown figure 9*a*. Consider how this structure develops into the writhed rod, and note that there are an infinite number of turns in the rod.

When the tension T_∞ is reduced to $\frac{1}{4}Q_\infty^2$ the rod buckles, and under increasing T_∞ the Coyne (1990) solution develops, which is given here by (5.8), (5.13) and (5.14) with the constants of integration all set equal to zero. An interesting aspect of the Coyne solution is that the inward displacement of each point at infinity,

$$\Delta_T = \lim_{s\to\infty} \tfrac{1}{2}[y(s) - y(-s)] = -\frac{2}{T_\infty}\sqrt{4T_\infty - Q_\infty^2}, \tag{8.12}$$

increases in opposition to the increasing tension in the range $\frac{1}{4}Q_\infty^2 \leqslant T_\infty \leqslant \frac{1}{2}Q_\infty^2$ so that the rod does work against the applied tension load (figure 9*b*). At $T_\infty = \frac{1}{2}Q_\infty^2$, the inward displacement of the Coyne solution reaches its maximum value (figure 9*c*). Further inward movement of the remote points causes the end loop to flip around dynamically and to contact itself. The precise details of this shape have not been

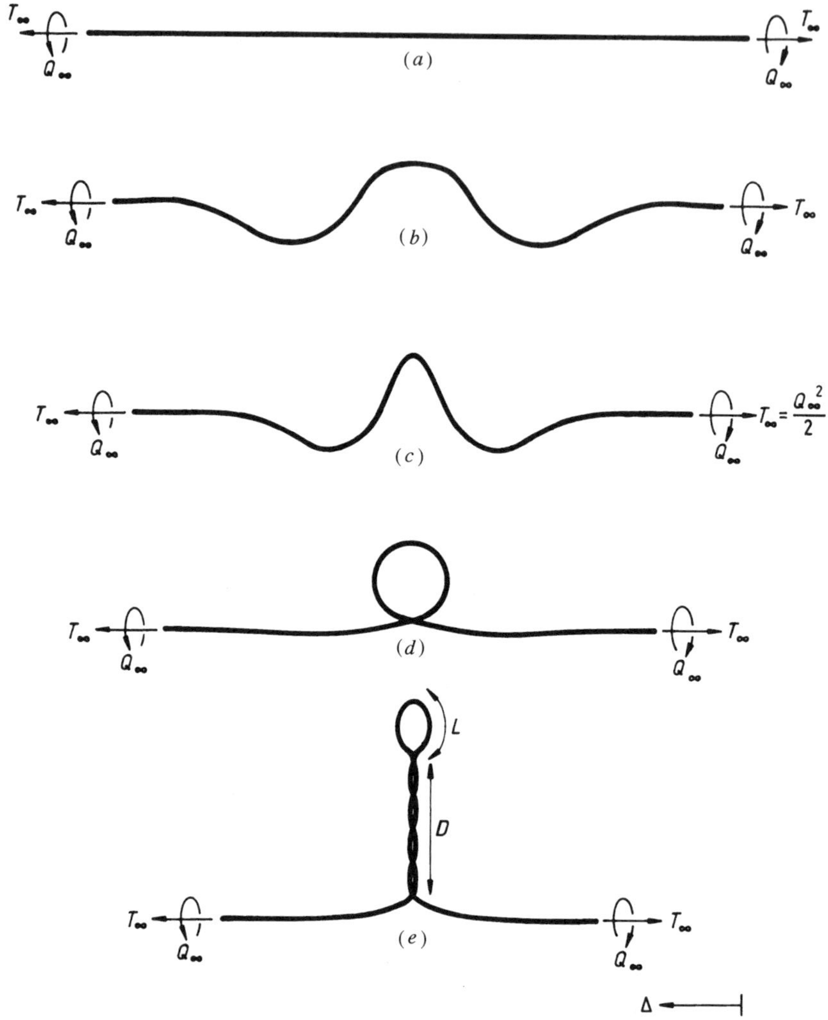

Figure 9. (a) The straight rod under remote tension T_∞ and torque Q_∞. (b) The rod after buckling and while the Coyne solution applies. (c) The maximum inward displacement of the Coyne solution. (d) The rod just after the contacting loop has formed. (e) The writhed shape that develops under significant inward displacement of the remote ends.

calculated but this can be done in principle by combining the equations of § 4 *a* with the tail analysis of § 5 *a* and then formulating an appropriate set of continuity and jump conditions. The shape of the end loop at first contact is not the same as the end loop attached to the balanced-ply helix so that as the remote points are displaced further inward, there appears to be some sliding of the two strands against each other during the winding up of the helix region. This is a complicated transient process which we have not attempted to model (see figure 9*d*).

After a sufficient amount of remote displacement, the writhed structure shown in figure 9*e*, and more closely in figure 1, is obtained. The end loop and tails have reached the proper shape and the balanced-ply region has developed. The inward displacement of each remote loading point Δ is absorbed completely by the arc length of the end loop and the balanced ply, which gives the formula

$$\Delta = L + \frac{D}{\cos\beta} \tag{8.13}$$

for D the balanced-ply length. The formulation of the problem has been based on a set of simple experiments with flexible ropes and only limited instrumentation. Verification of the writhing process requires a much more carefully conducted experiment where the remote displacement, rotation, tension and torque can be controlled and measured.

The long-rod analysis has so far considered the case when Q_∞ is held constant throughout the writhing process, which requires the rotation of the remote ends to input extra energy due to the bending of the rod in the writhed state. It is also possible to allow a finite-length rod to writhe after a fixed number of rotations have been inserted between the ends of the straight rod (for example, by locking the rod into a lathe fixture with fixed rotation at the ends). In this case, a form of the link restriction (8.2) applies during the course of writhing. A simple experiment with a piece of rope, say 2 m long, which is made to writhe under fixed end rotations shows that the helix angle of the balanced ply changes noticeably during the fixed-link writhing process (see also figure 6 of Thompson & Champneys (1996)). This requires a more complicated analysis of the balanced-ply region and will be considered in subsequent work.

9. Concluding remarks

The large deflection theory of elastic rods has been used to study two problems of interest: the writhing of closed loops, and the writhing of long elastic rods. The prescription of a balanced-ply region that is a uniform helix means that the shapes of both structures are completely parametrized by the helix angle β.

The model predictions for the writhed shape of closed loops are in approximate quantitative agreement with data obtained from electron microscopy of DNA and simple experiments with ropes. The writhing of the long elastic rod has not been compared with experimental data, but the analysis appears to predict the important features of the final rod shape. Experimental confirmation of both applications remains a priority.

In applying this model to the description of interwound DNA supercoiling, it should be kept in mind that the purpose of (8.11) is to provide a simple predictive tool that captures the gross features of the supercoiled shape rather than to predict the precise strand shape in all circumstances and on all length scales. The model developed here constitutes a theoretical advance since it brings together the continuum mechanics and topological approaches to provide a better understanding of the interaction of the two effects and provides a means for making *quantitative* estimates of supercoil shape.

This work has been supported by an Australian Research Council Large Grant to D.M.S. and W.B.F., and by a grant to W.B.F. from the US National Textile Centre at Clemson University,

South Carolina. Discussions with Charlie Macaskill, Colin Please, Tony Watts, Larry Forbes, John Maddocks, Gert van der Heijden, Mike Thompson, Alan Champneys and Louis Kauffman have been most helpful and appreciated. Mr Stuart Lee assisted with some of the calculations and in the preparation of various figures.

Appendix A. Asymptotic end-loop analysis

The two families of loops associated with $\beta \to 0$ are considered with perturbation series, but first the complete Cartesian components of the differential equation (6.2) and the torque constraint (6.4) are presented.

The Cartesian components of (6.2) are

$$\left.\begin{aligned} \tilde{x}'' &= \tilde{y}'(\tilde{x}\tilde{C}_y - \tilde{y}\tilde{C}_x) - \tilde{z}'(\tilde{z}\tilde{C}_x - \tilde{D}_y), \\ \tilde{y}'' &= -\tilde{z}'(\tilde{z}\tilde{C}_y + \tilde{D}_x) - \tilde{x}'(\tilde{x}\tilde{C}_y - \tilde{y}\tilde{C}_x), \\ \tilde{z}'' &= \tilde{x}'(\tilde{z}\tilde{C}_x - \tilde{D}_y) + \tilde{y}'(\tilde{z}\tilde{C}_y + \tilde{D}_x), \end{aligned}\right\} \tag{A 1}$$

where $\tilde{C}_y$ and $\tilde{D}_y$ are given by (6.3). The torque equation (6.4) is

$$-\frac{2L\cos\beta\sin^3\beta}{\cos 2\beta\cdot} = \tilde{x}'(\tilde{D}_x + \tilde{z}\tilde{C}_y) + \tilde{y}'(\tilde{D}_y - \tilde{z}\tilde{C}_x) + \tilde{z}'(\tilde{y}\tilde{C}_x - \tilde{x}\tilde{C}_y). \tag{A 2}$$

(*a*) *The large loop*

The loop shape, forces and moments associated with the upper branch of the L versus β curve shown in figure 3 are obtained by introducing the following perturbation expansions in powers of $\beta^{1/2}$:

$$\left.\begin{aligned} L &= L_{-3/2}\beta^{-3/2} + L_{-1}\beta^{-1} + \dots, \\ \tilde{C}_x &= \tilde{C}_{1/2}\beta^{1/2} + \tilde{C}_1\beta + \dots, \\ \tilde{D}_x &= \tilde{D}_0 + \tilde{D}_{1/2}\beta^{1/2} + \dots, \\ \tilde{x}(\tilde{s}) &= \tilde{x}_{3/2}(\tilde{s})\beta^{3/2} + \tilde{x}_2(\tilde{s})\beta^2 + \dots, \\ \tilde{y}(\tilde{s}) &= \tilde{y}_0(\tilde{s}) + \tilde{y}_{1/2}(\tilde{s})\beta^{1/2} + \dots, \\ \tilde{z}(\tilde{s}) &= \tilde{z}_0(\tilde{s}) + \tilde{z}_{1/2}(\tilde{s})\beta^{1/2} + \dots. \end{aligned}\right\} \tag{A 3}$$

These expansions are substituted into the rescaled equations for the end loop (A 1), the torque equation (A 2), the inextensibility equation $(2.4)_1$ and the boundary conditions (6.5). The resulting system of equations is then separated into a series of boundary-value problems for different $O(\beta^{n/2})$ with $n = (0, 1, 2, \dots)$. The leading $O(1)$ system of equations is

$$\tilde{y}_0'' = -L_{-3/2}^2\tilde{z}_0\tilde{z}_0' - \tilde{D}_0\tilde{z}_0', \quad \tilde{z}_0'' = \tilde{y}_0'\tilde{z}_0L_{-3/2}^2 + \tilde{D}_0\tilde{y}_0', \quad (\tilde{y}_0')^2 + (\tilde{z}_0')^2 = 1. \tag{A 4}$$

The associated $O(1)$ boundary conditions are

$$\left.\begin{aligned} \tilde{y}_0(-1) = \tilde{z}_0(-1) = \tilde{y}_0'(-1) = 0, &\qquad \tilde{z}_0'(-1) = -1, \\ \tilde{y}_0(1) = \tilde{z}_0(1) = \tilde{y}_0'(1) = 0, &\qquad \tilde{z}_0'(1) = 1. \end{aligned}\right\} \tag{A 5}$$

This system can be integrated in closed form as follows. The integral of $(\mathrm{A}\,4)_1$ in combination with the boundary conditions (A 5) gives

$$\tilde{y}_0' = -\tfrac{1}{2}L_{-3/2}^2\tilde{z}_0^2 - \tilde{D}_0\tilde{z}_0. \tag{A 6}$$

The insertion of (A 6) into $(\mathrm{A}\,4)_3$ directly or in combination with the integration of $(\mathrm{A}\,4)_2$ gives the equation

$$\tilde{z}_0' = \sqrt{1-\tilde{z}_0^2[\tfrac{1}{2}L_{-3/2}^2\tilde{z}_0+\tilde{D}_0]^2}, \tag{A 7}$$

which is valid in the region $0 \leqslant \tilde{s} \leqslant 1$ since $\tilde{z}_0$ is an even function of $\tilde{s}$. With the variable change $\tilde{z}_0 = -u - \tilde{D}_0/L_{-3/2}^2$ the integral of this equation can be put in the form

$$\int_u^b \frac{\mathrm{d}u}{\sqrt{(c^2+u^2)(b^2-u^2)}} = \tfrac{1}{2}\tilde{s}L_{-3/2}^2, \tag{A 8}$$

where

$$b^2 = \frac{2}{L_{-3/2}^2} + \frac{\tilde{D}_0^2}{L_{-3/2}^4}, \quad c^2 = \frac{2}{L_{-3/2}^2} - \frac{\tilde{D}_0^2}{L_{-3/2}^4}. \tag{A 9}$$

This integral has a closed-form representation in terms of Jacobi elliptic functions (equation 17.4.49 of Abramowitz & Stegun (1965)), and the coordinate $\tilde{z}_0$ is given by

$$\tilde{z}_0 = -b\,\mathrm{cn}[\tilde{s}L_{-3/2};m] - \frac{\tilde{D}_0}{L_{-3/2}^2}, \tag{A 10}$$

where the index $m = b^2/(b^2+c^2)$. This expression is used in (A 6) which is then integrated (equation 16.26.2 of Abramowitz & Stegun (1965)) to obtain

$$\tilde{y}_0 = \frac{\tilde{s}\tilde{D}_0^2}{2L_{-3/2}^2} - \frac{b^2L_{-3/2}}{2m}\{\mathrm{E}[\tilde{s}L_{-3/2};m] - m_1\tilde{s}L_{-3/2}\}, \tag{A 11}$$

where $m_1 = 1-m$ and E is the incomplete elliptic integral of the second kind.

The assertion of the boundary conditions $\tilde{z}_0(1) = 1$ and $\tilde{y}_0(1) = 0$ gives a set of transcendental equations for $L_{-3/2}$ and $\tilde{D}_0$ which are solved by a Newton–Raphson procedure to obtain $L_{-3/2} = 3.1484$ and $\tilde{D}_0 = 3.0276$.

(*b*) *The small loop*

The lower branch of the L versus β plot in figure 3 is associated with a family of end-loop shapes that remain bounded in size as $\beta \to 0$. This family of solutions is described by introducing the perturbation series expansions in powers of β:

$$\left.\begin{aligned} L &= L_0 + L_1\beta + \dots, \\ \tilde{C}_x &= \tilde{C}_0 + \tilde{C}_1\beta + \dots, \\ \tilde{D}_x &= \tilde{D}_1\beta + \tilde{D}_2\beta^2 + \dots, \\ \tilde{x}(\tilde{s}) &= \tilde{x}_0(\tilde{s}) + \tilde{x}_1(\tilde{s})\beta + \dots, \\ \tilde{y}(\tilde{s}) &= \tilde{y}_1(\tilde{s})\beta + \dots, \\ \tilde{z}(\tilde{s}) &= \tilde{z}_0(\tilde{s}) + \tilde{z}_1(\tilde{s})\beta + \dots. \end{aligned}\right\} \tag{A 12}$$

These expansions are substituted into the rescaled equations for the end loop (A 1), the torque equation (A 2), the inextensibility condition $(2.4)_1$ and the boundary conditions (6.5) to obtain the leading $O(1)$ system of equations

$$\tilde{x}_0'' = -\tilde{z}_0\tilde{z}_0'\tilde{C}_0, \quad \tilde{z}_0'' = \tilde{z}_0\tilde{x}_0'\tilde{C}_0, \quad (\tilde{x}_0')^2 + (\tilde{z}_0')^2 = 1 \tag{A 13}$$

and boundary conditions

$$\left.\begin{aligned} \tilde{z}_0(-1) = \tilde{x}_0'(-1) = 0, \qquad \tilde{z}_0'(-1) = -1, \qquad \tilde{x}_0(-1) = -1/L_0, \\ \tilde{z}_0(1) = \tilde{x}_0'(1) = 0, \qquad \tilde{z}_0'(1) = 1, \qquad \tilde{x}_0(1) = 1/L_0. \end{aligned}\right\} \tag{A 14}$$

The equations (A 13) are integrated as follows. The integral of $(\text{A } 13)_1$ and the boundary conditions (A 14) provide

$$\tilde{x}_0' = -\tfrac{1}{2}\tilde{C}_0\tilde{z}_0^2, \tag{A 15}$$

which is then used in $(\text{A } 13)_3$ to obtain

$$\tilde{z}_0' = \sqrt{1 - \tfrac{1}{4}\tilde{C}_0^2\tilde{z}_0^4} \tag{A 16}$$

for the region $0 \leqslant \tilde{s} \leqslant 1$. With the variable change

$$u = -\tilde{z}_0\sqrt{\tfrac{1}{2}|\tilde{C}_0|},$$

the integral of this equation is rewritten

$$\int_u^1 \frac{\mathrm{d}t}{\sqrt{1-t^4}} = \tilde{s}\sqrt{\tfrac{1}{2}|\tilde{C}_0|}, \tag{A 17}$$

and is expressible in terms of elliptic functions (see equation 17.4.52 of Abramowitz & Stegun (1965)) as

$$\tilde{z}_0 = -\sqrt{\frac{2}{|\tilde{C}_0|}}\,\mathrm{cn}[\tilde{s}\sqrt{|\tilde{C}_0|}; \tfrac{1}{2}]. \tag{A 18}$$

This expression is inserted into (A 15), which is then integrated to yield

$$\tilde{x}_0 = \frac{2}{\sqrt{|\tilde{C}_0|}}\mathrm{E}[\tilde{s}\sqrt{|\tilde{C}_0|}; \tfrac{1}{2}] - \tilde{s}. \tag{A 19}$$

The assertion of the boundary condition $\tilde{z}_0(1) = 0$ gives $\tilde{C}_0 = -3.4375$, and the boundary condition $\tilde{x}_0(1) = 1/L_0$ yields $L_0 = 2.1884$.

Appendix B. The writhe of the closed loop

In this appendix the method of Fuller (1978), given by (8.4), is used to calculate the writhe number of a closed curve based upon the spherical surface area traced out by the spherical image (Struik 1961) of the tangent vector. Let the tangent vector $\boldsymbol{t} = \boldsymbol{R}'$ have the Cartesian components (t_1, t_2, t_3) and let (r, ω, η) be a coincident spherical coordinate system where ω is the circumferential angle measured from the x-axis and η is the polar angle measured from the z-axis. The spherical basis vectors are

$$\left.\begin{aligned} \boldsymbol{e}_r &= \boldsymbol{i}\cos\omega\sin\eta + \boldsymbol{j}\sin\omega\sin\eta + \boldsymbol{k}\cos\eta, \\ \boldsymbol{e}_\omega &= -\boldsymbol{i}\sin\omega + \boldsymbol{j}\cos\omega, \\ \boldsymbol{e}_\eta &= \boldsymbol{i}\cos\omega\cos\eta + \boldsymbol{j}\sin\omega\cos\eta - \boldsymbol{k}\sin\eta. \end{aligned}\right\} \tag{B 1}$$

Let $\boldsymbol{t} = \boldsymbol{e}_r$ so that the angles (ω, η) can be treated as functions of s as the close curve is traced out on the surface of the unit sphere. The spherical basis vectors can also be expressed in terms of the components of $\boldsymbol{t}$:

$$\left.\begin{aligned} \boldsymbol{e}_r &= \boldsymbol{i}t_1 + \boldsymbol{j}t_2 + \boldsymbol{k}t_3, \\ \boldsymbol{e}_\omega &= \frac{-\boldsymbol{i}t_2 + \boldsymbol{j}t_1}{\sqrt{t_1^2 + t_2^2}}, \\ \boldsymbol{e}_\eta &= \frac{\boldsymbol{i}t_1 t_3 + \boldsymbol{j}t_3 t_2}{\sqrt{t_1^2 + t_2^2}} - \boldsymbol{k}\sqrt{1 - t_3^2}. \end{aligned}\right\} \tag{B 2}$$

The element of surface area on the unit sphere is $\mathrm{d}A = \sin\eta\, \mathrm{d}\eta\, \mathrm{d}\omega$, and the area A of a closed curve as measured from the polar axis is

$$A = \oint \mathrm{d}\omega \int_0^{\eta(\omega)} \sin\eta\, \mathrm{d}\eta = \oint [1 - \cos\eta]\, \mathrm{d}\omega, \tag{B 3}$$

where the angle η has been taken as an implicit function of ω. With the parametrization $0 \leqslant s \leqslant s_0$ for the closed curve, (B 3) becomes

$$A = \int_0^{s_0} (1 - \cos\eta)\omega'\, \mathrm{d}s, \tag{B 4}$$

which can be expressed in terms of the components (t_1, t_2, t_3) and their derivatives. From $\boldsymbol{t} = \boldsymbol{e}_r$ and $\boldsymbol{t} \cdot \boldsymbol{t} = 1$, it is straightforward to obtain

$$\boldsymbol{t}' = \boldsymbol{e}_\omega \omega' \sin\eta + \boldsymbol{e}_\eta \eta', \tag{B 5}$$

so that

$$\cos\eta = \boldsymbol{t} \cdot \boldsymbol{k}, \quad \omega' \sin\eta = \boldsymbol{t}' \cdot \boldsymbol{e}_\omega. \tag{B 6}$$

With (B 5) and (B 6), the area integral (B 4) is given in terms of vector quantities by

$$A = \int_0^{s_0} \sqrt{\frac{1 - \boldsymbol{t} \cdot \boldsymbol{k}}{1 + \boldsymbol{t} \cdot \boldsymbol{k}}}\, \boldsymbol{t}' \cdot \boldsymbol{e}_\omega\, \mathrm{d}s. \tag{B 7}$$

With (B 2), this can be written out in terms of the Cartesian components

$$A = \int_0^{s_0} \frac{t_2' t_1 - t_1' t_2}{1 + t_3}\, \mathrm{d}s. \tag{B 8}$$

Alternate formulations using $\boldsymbol{i}$ or $\boldsymbol{j}$ as the polar axis are obtained by cyclicly permuting the indices $(1, 2, 3)$.

(*a*) *The balanced ply*

Within each wavelength $\lambda = 2\pi/\tan\beta$ along the axis of the balanced ply, the tangents of the two strands trace out a counter-clockwise circle at the colatitude β from the polar axis $\boldsymbol{k}$ and a clockwise circle with a colatitude β measured from -$\boldsymbol{k}$. The arc length of each strand in an axial wavelength is $s_0 = 2\pi/\sin\beta$. From (3.2) and (B 2), the quantities in (B 6) become

$$\boldsymbol{t} \cdot \boldsymbol{k} = \cos\beta, \quad \boldsymbol{t}' \cdot \boldsymbol{e}_\omega = \sin^2\beta. \tag{B 9}$$

A_1, the spherical surface area associated with strand 1 which traces out the counter-clockwise circle about the $\boldsymbol{k}$-axis, is obtained from using (B 9) in (B 8) to provide

$$A_1 = \int_0^{2\pi/\sin\beta} (1-\cos\beta)\sin\beta\,\mathrm{d}s = 2\pi(1-\cos\beta), \tag{B 10}$$

which agrees with Fuller (1978). A similar analysis for strand 2 taking account that the circle traces a clockwise circle around the south pole of the sphere gives the formula

$$A_2 = -2\pi(1+\cos\beta), \tag{B 11}$$

which is the negative of the supplementary area to (B 10). The *net* spherical surface area A_λ per axial wavelength of balanced ply is

$$A_\lambda = A_1 + A_2 = -4\pi\cos\beta, \tag{B 12}$$

which agrees with Tanaka & Takahashi (1985). For a ply length D, the number of turns is D/λ so that total spherical surface area is

$$A_{\mathrm{ply}} = -\frac{D4\pi}{\lambda}\cos\beta = -2D\sin\beta. \tag{B 13}$$

(b) The end loops

The spherical surface area traced out by each set of ends loops (such as those shown in figure 7*b*) can be calculated by joining the two loops together to form a continuous figure. Since the surface area is independent of the physical coordinate system, it is useful to employ the rescaling of § 6 so that $\boldsymbol{t} = \tilde{\boldsymbol{R}}'(\tilde{s})$. The coordinates $\tilde{\boldsymbol{R}}(\tilde{s})$ of the *upper* loop are denoted by a subscript 'u' and parametrized by $-1 \leqslant \tilde{s} \leqslant 1$. The coordinates of the *lower* loop (denoted by a subscript 'l') can be parametrized in the same way and are related to the upper loop coordinates by

$$\tilde{x}_{\mathrm{l}}(\tilde{s}) = -\tilde{x}_{\mathrm{u}}(\tilde{s}), \quad \tilde{y}_{\mathrm{l}}(\tilde{s}) = \tilde{y}_{\mathrm{u}}(\tilde{s}), \quad \tilde{z}_{\mathrm{l}}(\tilde{s}) = -\tilde{z}_{\mathrm{u}}(\tilde{s}). \tag{B 14}$$

With (B 14), the closed-loop integral (B 8) for the spherical area of the loop A_{loop} becomes

$$A_{\mathrm{loop}} = \int_{-1}^{1} \frac{t_2't_1 - t_1't_2}{1+t_3} + \frac{t_1't_2 - t_2't_1}{1-t_3}\,\mathrm{d}\tilde{s} = \int_{-1}^{1} \frac{2t_3(t_1't_2 - t_2't_1)}{1-t_3^2}\,\mathrm{d}\tilde{s}. \tag{B 15}$$

For numerical reasons it is preferable to permute the indices 1 and 3 so that $\boldsymbol{i}$ is the polar axis. The subsequent use of the notation of § 6 provides the formula

$$A_{\mathrm{loop}} = \int_{-1}^{1} \frac{2\tilde{x}'}{1-(\tilde{x}')^2}\{\tilde{y}''\tilde{z}' - \tilde{z}''\tilde{y}'\}\,\mathrm{d}\tilde{s}, \tag{B 16}$$

and the use of (A 1) and (A 2) reduces this to

$$A_{\mathrm{loop}} = \int_{-1}^{1} \frac{2\tilde{x}'}{1-(\tilde{x}')^2}\{\tilde{x}'\tilde{Q} - (\tilde{z}\tilde{C}_y + \tilde{D}_x)\}\,\mathrm{d}\tilde{s}. \tag{B 17}$$

Similar formulae using the $\boldsymbol{j}$- or $\boldsymbol{k}$-axes as poles can be developed from (B 15).

The normalized factor $F_{\mathrm{loop}} = A_{\mathrm{loop}}/2\pi$ is plotted in figure 8. In the limit $\beta \to 0$ for the large-loop solution, the figure lies in the $\tilde{y}\tilde{z}$-plane with a single crossover so that the loop writhe Wr_{loop} computed from (8.4) becomes $Wr_{\mathrm{loop}} = -1$. In the limit

$\beta \to 0$ for the small-loop solutions, the figure forms an open loop in the $\tilde{x}\tilde{z}$-plane so that $Wr_{\text{loop}} = 0$.

The *total* writhe number of the closed loop is calculated by using the total spherical surface area $A = A_{\text{ply}} + A_{\text{loop}}$ in (8.4) to obtain

$$Wr = -\frac{D}{\pi}\sin\beta + F_{\text{loop}} - 1. \tag{B 18}$$

References

Abramowitz, M. & Stegun, I. 1965 *Handbook of mathematical functions.* New York: Dover.

Antman, S. S. 1995 *Nonlinear problems of elasticity.* Berlin: Springer.

Austin, R. H., Brody, J. P., Cox, E. C., Duke, T. & Volkmuth, W. 1997 Stretch genes. *Phys. Today* **50**, 32–38.

Bauer, W. R., Crick, F. H. C. & White, J. H. 1980 Supercoiled DNA. *Sci. Am.* **243**, 100–113.

Bauer, W. R., Lund, R. A. & White, J. H. 1993 Twist and writhe of a DNA loop containing intrinsic bends. *Proc. Natn. Acad. Sci. USA* **90**, 833–837.

Calladine, C. R. 1980 Toroidal elastic supercoiling of DNA. *Biopolymers* **19**, 1705–1713.

Calladine, C. R. 1992 Application of structural mechanics to biological systems. *Theoretical and applied mechanics* (ed. S. R. Bodner *et al.*), pp. 205–219. New York: Elsevier.

Calladine, C. R. & Drew, H. R. 1992 *Understanding DNA, the molecule and how it works*, ch. 6, pp. 114–137. New York: Academic.

Champneys, A. R. & Thompson, J. M. T. 1996 A multiplicity of localized buckling modes. *Proc. R. Soc. Lond.* A **452**, 2467–2491.

Champneys, A. R., van der Heijden, G. H. M. & Thompson, J. M. T. 1997 Spatially complex localization after one-twist-per-wave equilibria in twisted circular rods with initial curvature. *Phil. Trans. R. Soc. Lond.* A **355**, 2151–2174.

Coleman, B. D. & Dill, E. H. 1992 Flexure waves in elastic rods. *J. Acoust. Soc. Am.* **91**, 2663–2673.

Coleman, B. D., Dill, E. H., Lembo, M., Lu, Z. & Tobias, I. 1993 On the dynamics of rods in the theory of Kirchhoff and Clebsch. *Arch. Ration. Mech. Analysis* **121**, 339–359.

Coyne, J. 1990 Analysis of the formation and elimination of loops in twisted cable. *IEEE J. Oceanic Engng* **15**, 72–83.

Cozzarelli, N. R., Boles, T. R. & White, J. H. 1990 Primer on the topology and geometry of DNA supercoiling. *DNA topology and its biological Effects*, pp. 139–184. Cold Spring Harbor Laboratory Press.

Crick, F. H. C. 1976 Linking numbers and nucleosomes. *Proc. Natn. Acad. Sci. USA* **73**, 2639–2643.

Dichmann, D. J. & Maddocks, J. H. 1994 Hamiltonian dynamics of an elastica and stability of solitary waves. *J. Elasticity* **34**, 83–96.

Fraser, W. B. & Stump, D. M. 1998*a* The equilibrium of the convergence point in two-strand yarn plying. *Int. J. Solids Struct.* **35**, 285–298.

Fraser, W. B. & Stump, D. M. 1998*b* Yarn twist in the ring spinning balloon. *Proc. R. Soc. Lond.* A **454**, 707–723.

Fraser, W. B. & Stump, D. M. 1998*c* Twist in balanced-ply structures. *J. Text. Inst.* (In the press.)

Fuller, F. B. 1971 The writhing number of a space curve. *Proc. Natn. Acad. Sci. USA* **68**, 815–819.

Fuller, F. B. 1978 Decomposition of the linking number of a closed ribbon: a problem from molecular biology. *Proc. Natn. Acad. Sci. USA* **75**, 3557–3561.

Greenhill, A. G. 1883 On the strength of shafting when exposed both to torsion and end thrust. *Inst. Mech. Eng. Proc.* **1883**, 182–209.

Le Bret, M. 1979 Catastrophic variation of twist and writhing of circular DNAs with constraint. *Biopolymers* **18**, 1709–1725.

Love, A. E. H. 1927 *A treatise on the mathematical theory of elasticity*, 4th edn, ch. 18, pp. 381–398. Cambridge University Press.

Manning, R. S., Maddacks, J. H. & Kahn, J. D. 1996 A continuum model of sequence dependent DNA structure. *J. Chem. Phys.* **105**, 5626–5646.

Marko, J. F. 1997*a* Supercoiled and braided DNA under tension. *Phys. Rev.* E **55**, 1758–1772.

Marko, J. F. 1997*b* Stretching must twist DNA. *Europhys. Lett.* **38**, 183–188.

Marko, J. F. & Siggia, E. D. 1994 Bending and twisting elasticity of DNA. *Macromolecules* **27**, 981–988.

Marko, J. F. & Siggia, E. D. 1995 Stretching DNA. *Macromolecules* **28**, 8759–8770.

Miekle, A. & Holmes, P. 1988 Spatially complex equilibria of buckled rodes. *Arch. Ration. Mech. Analysis* **101**, 319–348.

Murasugi, K. 1996 *Knot theory and its applications*, ch. 13, pp. 267–283. Boston, MA: Birkhauser.

Olson, W. K. 1996 Simulating DNA at low resolution. *Curr. Opin. Struct. Biol.* **6**, 242–256.

Pohl, W. F. 1980 DNA and differential geometry. *Math. Intelligencer* **3**, 20–27.

Schlick, T. 1995 Modelling superhelical DNA: recent analytical and dynamic approaches. *Curr. Opin. Struct. Biol.* **5**, 245–262.

Smith, S. B., Cui, Y. & Bustamante, C. 1996 Overstretching B-DNA: the elastic response of individual double-strand and single strand DNA molecules. *Science* **271**, 795–798.

Strick, T. R., Allemand, J. -F., Bensimon, D., Bensimon, A. & Croqutte, V. 1996 The elasticity of a single supercoiled DNA molecule. *Science* **271**, 1835–1837.

Struik, D. J. 1961 *Lectures on classical differential geometry*, ch. 1, pp. 1–54. London: Addison-Wesley.

Stump, D. M. & Fraser, W. B. 1998 On the dynamical theory of twist in yarn plying. *Math. Eng. Ind.* (In the press.)

Tanaka, F. & Takahashi, H. 1985 Elastic theory of supercoiled DNA. *J. Chem. Phys.* **83**, 6017–6026.

Thompson, J. M. T. & Champneys, A. R. 1996 From helix to localized writhing in the torsional post-buckling of elastic rods. *Proc. R. Soc. Lond.* A **452**, 117–138.

Tobias, I. & Olson, W. K. 1993 The effect of intrinsic curvature on supercoiling: predictions of elasticity theory. *Biopolymers* **33**, 639–646.

Wang, J. C. 1982 Topoisomerases. *Sci. Am.* **247**, 94–109.

Wang, J. C. 1995 Enzymes that supercoil DNA. *DNA-protein structural interactions* (ed. D. Lilly), pp. 76–87. Oxford University Press.

Wang, J. C. & Liu, L. F. 1990 DNA replication: topological aspects and the role of DNA topoisomerases. *DNA topology and its biological effects*, pp. 321–340. Cold Spring Harbor Laboratory Press.

Wang, M. D., Yin, H., Landick, R., Gelles, J. & Block, S. M. 1997 Stretching DNA with optical tweezers. *Biophys. J.* **72**, 1335–1346.

White, J. H. 1969 Self-linking and the Gauss integral in higher dimensions. *Am. J. Math.* **91**, 693–728.

Yang, Y., Tobias, I. & Olson, W. 1993 Finite element models of DNA supercoiling. *J. Chem. Phys.* **98**, 1673–1686.

THE ROYAL SOCIETY

Nonlinear dynamics of filaments. IV Spontaneous looping of twisted elastic rods

BY ALAIN GORIELY[1,2] AND MICHAEL TABOR[1]

[1] *University of Arizona, Program in Applied Mathematics and Department of Mathematics, Building 89, Tucson, AZ 85721, USA*

[2] *Université Libre de Bruxelles, Département de Mathématique, CP218/1, 1050 Brussels, Belgium*

Received 23 July 1997; revised 20 November 1997; accepted 27 January 1998

Everyday experience shows that twisted elastic filaments spontaneously form loops. We model the dynamics of this looping process as a sequence of bifurcations of the solutions to the Kirchhoff equation describing the evolution of thin elastic filaments. The control parameter is taken to be the initial twist density in a straight rod. The first bifurcation occurs when the twisted straight rod deforms into a helix. This helix is an exact solution of the Kirchhoff equations, whose stability can be studied. The secondary bifurcation is reached when the helix itself becomes unstable and the localization of the post-bifurcation modes is demonstrated for these solutions. Finally, the tertiary bifurcation takes place when a loop forms at the middle of the rod and the looping becomes ineluctable. Emphasis is put on the dynamical character of the phenomena by studying the dispersion relation and deriving amplitude equations for the different configurations.

Keywords: Kirchhoff equations; filament dynamics; spontaneous looping; writhing instability

1. Introduction

Everybody has been faced at one point or another with the impossible tangles formed by coiled telephone chords. The mechanism by which a telephone chord entangles with itself is a generic phenomenon encountered in many contexts. It is one of the most basic forms of instability encountered with elastic filaments and is usually referred to as the *writhing instability*, i.e. a change in spatial configuration of the filament to reduce the overall twist of the unstable structure. Consider the following experiment: a straight elastic filament is held between one's fingers and twist is injected by rotating one end while holding the other one fixed. After a small amount of twist has been added (small rotation), the filament is no longer straight but assumes a helical form (with very small radius). As the twist is increased, the deformation of the filament tends to localize in the middle of the rod and eventually a loop forms (see figure 1). Experimental studies of filament twisting (Thompson & Champneys 1996) have shown that this sequence of bifurcations is qualitatively correct. It is the purpose of this paper to study and describe this looping process as a sequence of dynamical instabilities within the framework of elasticity theory.

The starting point of almost all analyses of twisted filaments is the Kirchhoff equations. This set of partial differential equations describes the time and space

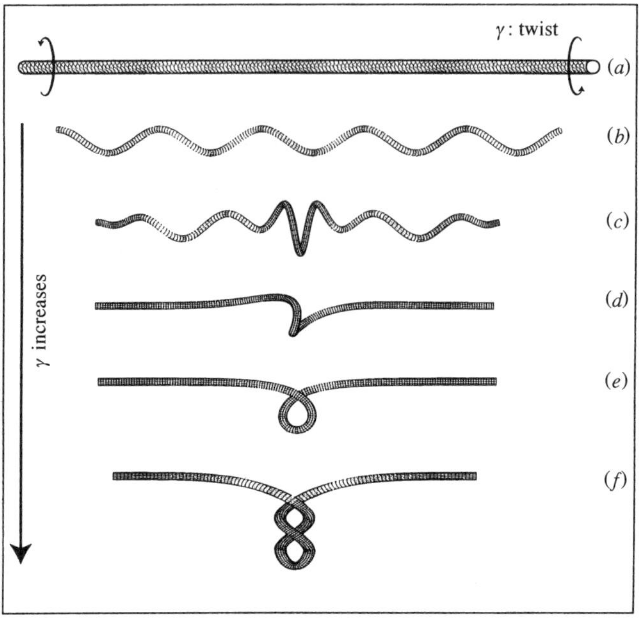

Figure 1. A schematic representation of the sequence of bifurcations of a looping process.

evolution of a filament subjected to external stresses (induced by applied forces and moments at the ends). The classical analysis (Love 1892; Timoshenko & Gere 1961) of the *stationary* equations suggests that a straight, twisted, elastic filament deforms into a helix. However, these analyses are limited in many respects. First, they do not include or explain the dynamical phenomena triggering such an instability. Second, while the period of the helix can be easily computed by linear theory, its radius cannot be obtained. Third, it is not known from this analysis whether the new helix is itself stable. Other analyses of the stationary equation (Coyne 1990) propose a completely different scenario for the looping by exhibiting a family of stationary solutions (valid only for a restricted class of boundary conditions) deforming continuously from a straight rod to a loop. We will come back to these solutions and their relationship with our model in the final discussion.

The analysis presented here departs radically from previous studies. We consider the *dynamical instabilities* as the main mechanism triggering the changes in filament configuration. The looping process is then explained as a sequence of bifurcations of the solutions to the Kirchhoff equations in which the twist density is taken as a control parameter. The first bifurcation occurs when the straight twisted filament becomes unstable. It deforms to a helix whose radius can be computed by a nonlinear analysis (Goriely & Tabor 1996, 1997*a*, *b*). It is then argued that this new helix is itself an exact solution to the Kirchhoff model and that its stability can in turn be studied by recently developed methods (Goriely & Tabor 1997*c*). The new physical parameters associated with this helix (which are required for the associated stability analysis), such as the twist density, can be deduced from energetic considerations.

Another view of looping, which does not utilize Kirchhoff dynamics, has been proposed by Ricca (1995) and is based on a consideration of the elastic energy of various space curve configurations.

The secondary bifurcation occurs when the helix itself becomes unstable. The unstable mode of the helix tends to localize the deformation at one point on the rod and forms a loop. Mode localization is a well-known experimental and theoretical phenomena for plates or filaments under stress (Tvergaard & Needleman 1980; Pomeau 1981; Damil & Pottier-Ferry 1986; Champneys & Thompson 1996). Our analysis provides a simple dynamical model explaining the tendency of stressed filamentary structures to localize the deformations. Finally, the nonlinear analysis is used to compute the amplitude of the deformation, and the tertiary bifurcation is reached when the loop collapses onto itself.

As already mentioned, Thompson and Champneys have performed careful experiments on the buckling, localization and looping of straight twisted rods (Thompson & Champneys 1996). They show that under rigid loading conditions a helical deformation consisting of one twist per wave (the state H_1 in their terminology) can often appear as a result of continuous deformation, i.e. as a pre-bifurcation phenomenon. The three twists per wave helical solution (the H_3 state) predicted by the classical theory is apparently not observed; rather they see the localized helical structure, L_3, (which also has approximately three twists per wave). It is this state that then undergoes a writhing instability into the looped state. The main difference between their experimental observations and our theoretical prediction is that we do not observe the H_1 state. However, this is not too surprising since subsequent studies by Champneys *et al.* (1997) show that the appearance of this solution results from the influence of a small initial curvature in the rod—an effect not included in our model. Beyond that, the experimental observations and our model are consistent since one of the conclusions that we draw from our analysis is that for long enough rods, the first bifurcation (to the helical form H_3) and the secondary bifurcation (to the localized helix) are so close that they would, in all likelihood, be indistinguishable. After the secondary bifurcation, our theory also predicts a continuous localization until the tertiary bifurcation point is reached (equivalent to the 'dynamic jump' of Thompson and Champneys).

The paper is organized as follows. In § 2, we describe the general set-up, the Kirchhoff model, and briefly review the linear and nonlinear methods that will be used. Since our proposed model is quite elaborate and uses ideas from both linear and nonlinear stability theory we give a summary of the proposed three-step process in § 3 and then each of these steps is discussed in detail in the sections that follow. In § 4, we study the primary bifurcation of the straight rod and obtain the post-bifurcation helix. In § 5, the linear and nonlinear analysis of the helix is performed and we show the tendency to localization. In § 6, we find the tertiary bifurcation where the looping first occurs.

2. General set-up

The derivation of the Kirchhoff model has been explained in detail in earlier papers (Goriely & Tabor 1997*a*, *b*). Here, we review the most important aspects of the theory relevant to the problem. First, we explain the kinematics of space curves by introducing a director basis attached to a general space curve and then the dynamics of rods is described within the approximation of linear elasticity theory. The section ends with a brief review of the linear and nonlinear stability methods.

(a) Spin and twist

Let $\boldsymbol{X} = \boldsymbol{X}(s,t)$: $\mathbb{R} \times \mathbb{R} \to \mathbb{R}^3$ be a time-dependent space curve, parametrized by the arc length s. A director basis $\{\boldsymbol{d}_1, \boldsymbol{d}_2, \boldsymbol{d}_3\}$ can be attached to the curve as follows. The vector $\boldsymbol{d}_3(s,t) = \boldsymbol{X}'(s,t)$ is the tangent vector of $\boldsymbol{X}$ at s (the prime denotes the s derivative). The vectors $\{\boldsymbol{d}_1(s,t), \boldsymbol{d}_2(s,t)\}$ are chosen such that $\{\boldsymbol{d}_1, \boldsymbol{d}_2, \boldsymbol{d}_3\}$ forms a right-handed orthonormal triad ($\boldsymbol{d}_1 \times \boldsymbol{d}_2 = \boldsymbol{d}_3$, $\boldsymbol{d}_2 \times \boldsymbol{d}_3 = \boldsymbol{d}_1$). If $\boldsymbol{d}_1$ is along $\boldsymbol{d}_3'$, the director basis specializes to the well-known Frenet triad for which $\boldsymbol{d}_1$ is the normal vector and $\boldsymbol{d}_2$ the bi-normal vector. The curve $\boldsymbol{X} = \boldsymbol{X}(s,t)$ can be obtained by integrating the tangent vector: $\boldsymbol{X}(s,t) = \int^s \boldsymbol{d}_3(s,t)\,\mathrm{d}s$.

The space and time evolution of the director basis specifies the kinematics of the space curve $\boldsymbol{X}$:

$$\boldsymbol{d}_i' = \sum_{j=1}^{3} K_{ij}\boldsymbol{d}_j, \quad i = 1,2,3, \tag{2.1 a}$$

$$\dot{\boldsymbol{d}}_i = \sum_{j=1}^{3} W_{ij}\boldsymbol{d}_j, \quad i = 1,2,3, \tag{2.1 b}$$

where $(\dot{\ })$ stands for the time derivative. The antisymmetry of W and K:

$$K = \begin{pmatrix} 0 & \kappa_3 & -\kappa_2 \\ -\kappa_3 & 0 & \kappa_1 \\ \kappa_2 & -\kappa_1 & 0 \end{pmatrix}, \qquad W = \begin{pmatrix} 0 & \omega_3 & -\omega_2 \\ -\omega_3 & 0 & \omega_1 \\ \omega_2 & -\omega_1 & 0 \end{pmatrix}, \tag{2.2}$$

is a consequence of the orthonormality of the basis. The elements of K and W make up the components of the *twist and spin vectors*; namely

$$\boldsymbol{\kappa} = \sum_{i=1}^{3} \kappa_i \boldsymbol{d}_i \quad \text{and} \quad \boldsymbol{\omega} = \sum_{i=1}^{3} \omega_i \boldsymbol{d}_i.$$

(b) The Kirchhoff model

The Kirchhoff model describes the space and time evolution of thin filaments, i.e. filaments whose length is much greater than their radius and whose curvature is sufficiently large relative to the small length-scales in the problem. In this approximation all physically relevant quantities characterizing the three-dimensional elastic body are averaged over the cross-sections attached to the central axis of the filament. This results in a one-dimensional theory.

The total force, $\boldsymbol{F} = \boldsymbol{F}(s,t)$, and the total moment, $\boldsymbol{M} = \boldsymbol{M}(s,t)$, are expanded in terms of the director basis; namely,

$$\boldsymbol{F} = \sum_{i=1}^{3} f_i \boldsymbol{d}_i \quad \text{and} \quad \boldsymbol{M} = \sum_{i=1}^{3} M_i \boldsymbol{d}_i.$$

The Kirchhoff model expresses the conservation of linear and angular momentum together with the constitutive relationship of linear elasticity relating the moments to the strains as characterized by the twist vector. For a naturally straight rod (i.e. no intrinsic curvature or twist) with circular cross-section the scaled equations read (Dill 1992; Coleman *et al.* 1993)

$$\boldsymbol{F}'' = \ddot{\boldsymbol{d}}_3, \tag{2.3a}$$

$$\boldsymbol{M}' + \boldsymbol{d}_3 \times \boldsymbol{F} = \boldsymbol{d}_1 \times \ddot{\boldsymbol{d}}_1 + \boldsymbol{d}_2 \times \ddot{\boldsymbol{d}}_2, \tag{2.3b}$$

$$\boldsymbol{M} = \kappa_1 \boldsymbol{d}_1 + \kappa_2 \boldsymbol{d}_2 + \Gamma \kappa_3 \boldsymbol{d}_3, \tag{2.3c}$$

where $\Gamma = 1/(1+\sigma)$ (where σ is the Poisson ratio) measures the ratio between bending and twisting coefficients of the rod. These equations, together with (2.1) can be reduced to a set of nine equations, second order in space and time, for the nine unknowns $(\boldsymbol{\kappa}, \boldsymbol{\omega}, \boldsymbol{f})$.

A constant twist, γ, of an elastic rod about its axis can be conveniently introduced by using the extra degree of freedom provided by the director basis. The vectors $\boldsymbol{d}_1$ and $\boldsymbol{d}_2$ can be chosen such that they rotate around $\boldsymbol{d}_3$ in the normal plane at a constant rate (with respect to the arc length), γ, independent of the space curve torsion. This axial twist is conveniently visualized as the twist of a 'ribbon' about the rod axis, and, where appropriate, it is referred to as the *ribbon twist.* An equivalent way of introducing the twist, γ, is to define a new set of variables rotating with the basis. That is, rather than using the variables $\{\boldsymbol{\kappa}, \boldsymbol{\omega}, \boldsymbol{f}\}$, a new set of variables $\{\hat{\boldsymbol{\kappa}}, \hat{\boldsymbol{\omega}}, \hat{\boldsymbol{f}}\}$ is obtained by introducing a constant rotation along $\boldsymbol{d}_3$ and a translation of the vector $\boldsymbol{\kappa}$. The rotated vectors are now

$$\hat{\boldsymbol{\omega}} = \boldsymbol{R} \cdot \boldsymbol{\omega}, \quad \hat{\boldsymbol{f}} = \boldsymbol{R} \cdot \boldsymbol{f}, \tag{2.4a}$$

$$\hat{\kappa}_i = \sum_{j=1}^{3} (R_{ij} - \delta_{j3}\gamma)\kappa_j, \tag{2.4b}$$

$$\hat{\boldsymbol{d}}_i = \sum_{j=1}^{3} R_{ij} \cdot \boldsymbol{d}_j, \quad i = 1, 2, 3, \tag{2.4c}$$

where

$$\boldsymbol{R} = \begin{pmatrix} \cos(\gamma s) & -\sin(\gamma s) & 0 \\ \sin(\gamma s) & \cos(\gamma s) & 0 \\ 0 & 0 & 1 \end{pmatrix} \tag{2.5}$$

The force, spin and twist vector are form invariant with respect to the change of variables:

$$\hat{\boldsymbol{\kappa}} = \sum_{i=1}^{3} \hat{\kappa}_i \hat{\boldsymbol{d}}_i, \qquad \hat{\boldsymbol{\omega}} = \sum_{i=1}^{3} \hat{\omega}_i \hat{\boldsymbol{d}}_i, \qquad \hat{\boldsymbol{F}} = \sum_{i=1}^{3} \hat{f}_i \hat{\boldsymbol{d}}_i.$$

The transformed Kirchhoff equations read:

$$\hat{\boldsymbol{F}}'' = \ddot{\hat{\boldsymbol{d}}}_3, \tag{2.6a}$$

$$\hat{\boldsymbol{M}}' + \hat{\boldsymbol{d}}_3 \times \hat{\boldsymbol{F}} = \hat{\boldsymbol{d}}_1 \times \ddot{\hat{\boldsymbol{d}}}_1 + \hat{\boldsymbol{d}}_2 \times \ddot{\hat{\boldsymbol{d}}}_2, \tag{2.6b}$$

$$\hat{\boldsymbol{M}} = \hat{\kappa}_1 \hat{\boldsymbol{d}}_1 + \hat{\kappa}_2 \hat{\boldsymbol{d}}_2 + \Gamma(\hat{\kappa}_3 + \gamma)\hat{\boldsymbol{d}}_3. \tag{2.6c}$$

Note also that due to the translation on the last component of the curvature vector, we now have: $\hat{\boldsymbol{\kappa}} \cdot \hat{\boldsymbol{d}}_3 = \hat{\kappa}_3 + \gamma$.

The main advantage of this representation is that simple stationary filament configurations can be conveniently expressed as constant solutions of the Kirchhoff equations. For instance, the straight rod with constant twist γ, and axial force f_3, has

$$\boldsymbol{\omega} = (0,0,0), \quad \boldsymbol{\kappa} = (0,0,\gamma), \quad \boldsymbol{f} = (0,0,f_3), \tag{2.7}$$

which transform under (2.4) to:

$$\hat{\boldsymbol{\kappa}} = \hat{\boldsymbol{\omega}} = (0,0,0), \quad \hat{\boldsymbol{f}} = (0,0,f_3). \tag{2.8}$$

In the same way, a helical filament with twist $\gamma = \gamma_{\mathrm{H}}$ and Frenet curvature and torsion $\kappa_{\mathrm{F}}, \tau_{\mathrm{F}}$ which takes the form (Goriely & Tabor 1997*c*)

$$\boldsymbol{X}_{\mathrm{H}} = \left(\frac{\tau_{\mathrm{F}} s}{\delta}, \frac{\kappa_{\mathrm{F}}}{\delta}\cos(\delta s), \frac{\kappa_{\mathrm{F}}}{\delta}\sin(\delta s) \right), \quad \delta^2 = \kappa_{\mathrm{F}}^2 + \tau_{\mathrm{F}}^2, \tag{2.9}$$

with

$$\boldsymbol{\kappa} = (\kappa_{\mathrm{F}}\sin(\gamma s), \kappa_{\mathrm{F}}\cos(\gamma s), \tau_{\mathrm{F}} + \gamma), \tag{2.10 a}$$

$$\boldsymbol{\omega} = (0,0,0), \tag{2.10 b}$$

$$\boldsymbol{f} = (f_0\sin(\gamma s), f_0\cos(\gamma s), (\tau_{\mathrm{F}}/\kappa_{\mathrm{F}})f_0), \tag{2.10 c}$$

where $f_0 = -\tau_{\mathrm{F}}\kappa_{\mathrm{F}} + \Gamma\kappa_{\mathrm{F}}\tau_{\mathrm{F}}$, can also be written as a constant solution in terms of $(\hat{\boldsymbol{\kappa}}, \hat{\boldsymbol{f}})$, i.e.

$$\hat{\boldsymbol{\kappa}} = (0, \kappa_{\mathrm{F}}, \tau_{\mathrm{F}}), \quad \boldsymbol{\omega} = (0,0,0), \tag{2.11 a}$$

$$\hat{\boldsymbol{f}} = (0, \kappa_{\mathrm{F}}\gamma\Gamma + \kappa_{\mathrm{F}}\tau_{\mathrm{F}}(\Gamma - 1), \tau_{\mathrm{F}}(\tau_{\mathrm{F}}(\Gamma-1) + \Gamma\gamma)). \tag{2.11 b}$$

We now drop the hats and consider system (2.6) as the main system of equations studied here.

(*c*) *Perturbation expansions*

To study the stability of stationary solutions, we developed a perturbation method at the level of the director basis (Goriely & Tabor 1996, 1997*a*, *b*, *c*). The main idea is to expand all local quantities around the local stationary solutions and close the system to each order in the perturbation parameter by demanding that the basis remain orthonormal up to a given order,

$$\boldsymbol{d}_i = \boldsymbol{d}_i^{(0)} + \epsilon \boldsymbol{d}_i^{(1)} + \epsilon^2 \boldsymbol{d}_i^{(2)} + \cdots, \quad i = 1,2,3. \tag{2.12}$$

The orthonormality condition $\boldsymbol{d}_i.\boldsymbol{d}_j = \delta_{ij}$ gives rise to a system of constraints which can be solved to each order

$$\boldsymbol{d}_i^{(1)} = \sum_{j=1}^{3} A_{ij}^{(1)} \boldsymbol{d}_j^{(0)}, \tag{2.13 a}$$

$$\boldsymbol{d}_i^{(k)} = \sum_{j=1}^{3} (A_{ij}^{(k)} + S_{ij}^{(k)}) \boldsymbol{d}_j^{(0)}, \quad k > 1, \tag{2.13 b}$$

where $S^{(k)}$ is a *symmetric* matrix whose entries depend only on $\alpha_i^{(j)}$ with $j < k$, and $A^{(k)}$ is the *antisymmetric* matrix

$$A^{(k)} = \begin{pmatrix} 0 & \alpha_3^{(k)} & -\alpha_2^{(k)} \\ -\alpha_3^{(k)} & 0 & \alpha_1^{(k)} \\ \alpha_2^{(k)} & -\alpha_1^{(k)} & 0 \end{pmatrix}. \tag{2.14}$$

Once the vector $\boldsymbol{\alpha}^{(k)}$ is known, it is an easy matter to obtain the perturbed solution by integrating the tangent vector up to any given order,

$$\boldsymbol{X}(s,t) = \int^s \mathrm{d}s \sum_{i=1}^{3} \left(\delta_{3i} + \epsilon A_{3i} + \sum_{j=2}^{k} \epsilon^j (A_{3i}^{(j)} + S_{3i}^{(j)}) \right) \boldsymbol{d}_i{}^{(0)} + O(\epsilon^{k+1}). \tag{2.15}$$

Any local vector

$$\boldsymbol{V} = \sum_{i=1}^{3} v_i \boldsymbol{d}_i$$

can be expanded in terms of the perturbed basis; namely $\boldsymbol{V} = \boldsymbol{V}^{(0)} + \epsilon \boldsymbol{V}^{(1)} + \epsilon^2 \boldsymbol{V}^{(2)} + \cdots$:

$$\boldsymbol{V}^{(1)} = \sum_i (v_i^{(1)} + (A^{(1)} v^{(0)})_i) \boldsymbol{d}_i{}^{(0)}. \tag{2.16}$$

Hence we can express the first-order perturbation of the twist and spin matrix, i.e. $K = K^{(0)} + \epsilon K^{(1)} + \cdots$, $W = W^{(0)} + \epsilon W^{(1)} + \cdots$, where

$$K^{(1)} = \frac{\partial A^{(1)}}{\partial s} + [A^{(1)}, K^{(0)}], \tag{2.17 a}$$

$$W^{(1)} = \frac{\partial A^{(1)}}{\partial t} + [A^{(1)}, W^{(0)}], \tag{2.17 b}$$

where $[A, B] = A.B - B.A$. Higher-order perturbations can be obtained in terms of the lower-order terms.

Using these equations, one can write the kth order perturbation of Newton's equation (2.6 *a*) and the moment's equation (2.6 *b*) in terms of $\{\boldsymbol{\alpha}^{(k)}, \boldsymbol{f}^{(k)}\}$. The first of this system ($k = 1$), referred to as the *dynamical variational equations*, controls the stability of the stationary solutions with respect to linear time-dependent modes. Higher-order modes provide better approximations of the perturbed solutions and are used to perform a nonlinear analysis. To emphasize the linear character of these equations we rewrite them as a linear system of six equations for the six-dimensional vector $\boldsymbol{\mu}^{(0)} = \{\boldsymbol{\kappa}^{(0)}, \boldsymbol{f}^{(0)}\}$, $\boldsymbol{\mu}^{(k)} = \{\boldsymbol{\alpha}^{(k)}, \boldsymbol{f}^{(k)}\}$, $k > 0$:

$$L_E(\boldsymbol{\mu}^{(0)}) \cdot \boldsymbol{\mu}^{(1)} = 0, \tag{2.18 a}$$

$$L_E(\boldsymbol{\mu}^{(0)}) \cdot \boldsymbol{\mu}^{(k)} = H_k(\boldsymbol{\mu}^{(0)}, \boldsymbol{\mu}^{(1)}, \ldots, \boldsymbol{\mu}^{(k-1)}), \quad k > 1, \tag{2.18 b}$$

where L_E is a linear differential operator in s and t whose coefficients may depend on s through the unperturbed solution $\boldsymbol{\mu}^{(0)} = \{\boldsymbol{\kappa}^{(0)}, \boldsymbol{f}^{(0)}\}$ and H_k are vector-valued functions depending on the variables $(\boldsymbol{\mu}^{(0)}, \ldots, \boldsymbol{\mu}^{(k-1)})$.

(*d*) *Linear stability analysis*

The linear stability of stationary solutions is determined through the use of (2.18 *a*). It has a set of fundamental solutions which we label by the spatial mode number n, i.e.

$$\boldsymbol{\mu}_n^{(1)} = \boldsymbol{\xi}_n \mathrm{e}^{\sigma t + \mathrm{i}(ns/L)}, \tag{2.19}$$

where $\boldsymbol{\xi}_n \in \mathbb{C}^6$. The growth rate of this mode, $\sigma = \sigma(n)$, is determined by the dispersion relation: $\Delta(\sigma, n) = \det(L_E) = 0$, obtained by substituting (2.19) into (2.18 *a*).

Typically, Δ is a very complicated expression which is best derived by symbolic manipulation. The threshold of instability is heralded by a change in sign of (the real part of) σ and can be determined by examining the *neutral curves* corresponding to the parameter values, for given n, for which $\sigma = 0$. These curves are thus solutions of $\Delta(0, n) = 0$. In what follows, all our statements concerning critical values of twist (or tension) at which a given configuration becomes unstable, are based on their determination from the dispersion relations.

Of course the linear analysis can only identify the initial instabilities as a function of the parameters. As these instabilities grow (exponentially) in time the linear approximation breaks down and any further description of the bifurcation requires a nonlinear analysis.

(*e*) *Nonlinear analysis*

The techniques of nonlinear analysis enable one to derive equations for the amplitude of a solution close to bifurcation. The distance from the bifurcation point is considered to be of the order of the perturbation itself and this relationship is used to introduce new, longer, space- and time-scales over which the solution varies. For the problems considered here, the twist, γ, is taken as the control (or 'stress') parameter and one sets

$$\epsilon^2 = \gamma - \gamma_c, \tag{2.20}$$

where γ_c is the critical value at which the bifurcation occurs.

Stretched time- and space-scales appropriate to the problem at hand are introduced:

$$t_0 = t, \qquad s_0 = s, \tag{2.21}$$

$$t_1 = \epsilon t, \qquad s_1 = \epsilon s, \tag{2.22}$$

and taking into account the expansion in the bifurcation parameter and the new scales, one seeks solutions of the full system (cf. equations (2.18)) order by order in ϵ:

$$O(\epsilon^0): \quad E(\boldsymbol{\mu}^{(0)}; s_0, t_0) = 0, \tag{2.23 a}$$

$$O(\epsilon^1): \quad L_E^{(0)}(\boldsymbol{\mu}^{(0)}) \cdot \boldsymbol{\mu}^{(1)} = 0, \tag{2.23 b}$$

$$O(\epsilon^2): \quad L_E^{(0)}(\boldsymbol{\mu}^{(0)}) \cdot \boldsymbol{\mu}^{(2)} + L_E^{(1)}(\boldsymbol{\mu}^{(0)}) \cdot \boldsymbol{\mu}^{(1)} = H_2(\boldsymbol{\mu}^{(1)}), \tag{2.23 c}$$

$$O(\epsilon^3): \quad L_E^{(0)}(\boldsymbol{\mu}^{(0)}) \cdot \boldsymbol{\mu}^{(3)} + L_E^{(1)}(\boldsymbol{\mu}^{(0)}) \cdot \boldsymbol{\mu}^{(2)} + L_E^{(2)}(\boldsymbol{\mu}^{(0)}) \cdot \boldsymbol{\mu}^{(1)} = H_3(\boldsymbol{\mu}^{(1)}, \boldsymbol{\mu}^{(2)}), \tag{2.23 d}$$

$$\vdots$$

where now the $\boldsymbol{\mu}^{(i)}$ are functions of the stretched variables, i.e.

$$\boldsymbol{\mu}^{(i)} = \boldsymbol{\mu}^{(i)}(s_0, s_1, \dots, t_0, t_1, \dots)$$

and $L_E = \sum \epsilon^i L_E^{(i)}$ is the expansion of the linear operator in terms of the new variables (s_i, t_i). These linear solutions (cf. (2.19)) are written as a superposition of the neutral modes: for example, to $O(\epsilon)$, $\boldsymbol{\mu}^{(1)}$ will involve terms of the form $X_n(s_1, t_1)\boldsymbol{\xi}_n \mathrm{e}^{\mathrm{i}(n_c s_0/L)}$ where $X_n(s_1, t_1)$ represents the slowly varying amplitude of the unstable mode $n = n_c$. A nonlinear amplitude equation for this slowly varying amplitude arises as a condition (the Fredholm alternative) that the solutions $\boldsymbol{\mu}^{(i)}$ remain bounded. Typically one finds this condition at $O(\epsilon^3)$ and its derivation involves, at least in the cases considered here, massive symbolic manipulation.

In the case of a straight rod, discussed in the next section, the nonlinear amplitude equation takes the form of a (system of) nonlinear partial differential equations and the amplitude(s) can be thought of as the envelope of a small packet of unstable wave numbers centred around the unstable mode. In other cases, such as the helical problem considered later, we are only interested in the amplitude of an isolated mode and in this case, as will be shown explicitly, one derives a simpler nonlinear ordinary differential equation.

3. Summary of looping mechanism

The sequence of bifurcations that constitutes our model of spontaneous looping is, in summary, as follows.

1. *Primary bifurcation*

   (a) Starting with a straight rod, with arbitrary twist, γ_S, and tension ϕ^2, linear stability analysis is used to show that it becomes unstable (for a given ϕ) at a critical twist value $\gamma_S = \gamma_1$ and then bifurcates into a helix.
   (b) The precise geometrical form of the helix is identified by using nonlinear analysis to determine its amplitude.
   (c) Using energy considerations the redistribution of the straight rod twist, γ_S, into twist and torsion of the new helix is determined. The helix is then specified by its curvature κ_F, torsion τ_F and twist γ_H.

2. *Secondary bifurcation*

   (a) A linear stability analysis of the helix obtained in step 1(a) is carried out and the critical twist value, $\gamma_H = \gamma_2$, at which it becomes unstable is determined.
   (b) The (linear) post-bifurcation solutions are constructed and a one-loop solution, of arbitrary amplitude B, is identified.
   (c) A nonlinear analysis of the unstable helical mode found in step 2(b) is used to determine the amplitude of the loop.

3. *Tertiary bifurcation*

   (a) A simple criterion is developed to determine the critical B value, B_c, at which the loop will flip (i.e. the onset of looping).
   (b) The twist value, γ_3, at which this loop amplitude equals B_c is found.

The various special values of the twist, γ_H, γ_2, γ_3, can all be related back to the initial twist density, γ_S, injected into the straight rod. Thus, in what follows, the term 'control' parameter refers to the value of γ_S.

Table 1. *The different bifurcations and the new parameters introduced at each step*

process	parameters	definition of parameters
initial straight	γ_S	twist density in straight rod
twisted rod	ϕ^2	tension in straight rod
(figure 1*a*)	γ_1	value of γ_S where the rod is unstable
primary bifurcation	κ_F, τ_F	Frenet curvature and torsion of helix
straight rod → helix	γ_H	axial twist
(figure 1*b*)	A	amplitude of the helix
secondary bifurcation	γ_2	value of γ_S where the helix is unstable
helix becomes linearly unstable (figure 1*c*)	B	amplitude of the first mode of deformation of the helix
tertiary bifurcation	γ_3	value of γ_S where the loop collapses
looping occurs (figure 1*d*)	B_c	critical value of the amplitude for looping

4. Primary bifurcation: instability of a straight twisted rod

We consider a stationary rod of length L, simply supported, with tension $T = \phi^2$ along the x-axis and twist γ_S. Rather than considering general boundary conditions, we just consider the case in which the tangents at the ends can assume any values while the tension along the x-axis is kept constant. We assume that these boundary conditions are maintained throughout the sequence of bifurcation and that the unique control parameter is the twist γ_S. These are the only boundary conditions considered in this paper, and it should of course be noted that both local and global behaviour can be very sensitive to different choices of conditions (Antman & Kenney 1981). The stationary solution is given by

$$\boldsymbol{\mu}^{(0)} = (0, 0, 0, 0, 0, \phi^2). \tag{4.1}$$

(*a*) *Linear analysis*

Using the techniques of linear stability analysis we are able to derive the dispersion relations for this case (Goriely & Tabor 1996, 1997*b*) and obtain the neutral curve from the relation $\Delta(0, n; \gamma_S) = 0$, where

$$\Delta(n; \gamma_S) = (L^2{\gamma_S}^2 - n^2)[(L^2{\gamma_S}^2(\Gamma - 1) - L^2\phi^2 - n^2)^2 - L^2{\gamma_S}^2(\Gamma - 2)^2 n^2]. \tag{4.2}$$

From this, one can deduce that the first instability occurs for the critical mode

$$n_c = \frac{\phi L(2 - \Gamma)}{\Gamma}, \tag{4.3}$$

with critical twist value

$$\gamma_1 = \frac{2\phi}{\Gamma}. \tag{4.4}$$

This condition for the first instability is the classical buckling condition that can be found in Timoshenko & Gere (1961) but written in rescaled variables. For γ_S bigger than, but close to, γ_1, the straight rod bifurcates to the helix

$$\boldsymbol{X}_{\rm H}(s) = \left(s, \frac{A}{\phi}\cos\phi s, \frac{A}{\phi}\sin\phi s\right). \tag{4.5}$$

(In the notation of Champneys & Thompson (1996), this corresponds to the classic H_3 solution.)

(*b*) *Nonlinear analysis*

The linear analysis does not specify the amplitude A of the new solution. However, a nonlinear analysis can be performed (Goriely & Tabor 1997*b*) yielding an amplitude equation consisting of a system of nonlinear partial differential equations coupling the amplitude of the unstable mode to the filament twist density. The stationary solution of this amplitude equation gives the amplitude A as a function of $\gamma_{\rm S}$, namely,

$$A^2 = 2\frac{(\gamma_{\rm S} - \gamma_1)}{\phi}. \tag{4.6}$$

(*c*) *Resummation*

The remarkable feature of the first bifurcation is that the new solution (4.5) is an exact solution of the Kirchhoff equations. Indeed, as was shown explicitly in the first section, it is well known that the Kirchhoff system sustains helicoidal solutions. This property can also be seen at the level of the perturbation expansion. Thus, if we carry on the perturbation analysis to higher order in ϵ and solve the subsequent linear system (2.18), it can be observed that the effect of higher-order corrections $\boldsymbol{\mu}^{(2)}, \boldsymbol{\mu}^{(3)}, \dots,$ is to provide correction to the radius of the helix without changing the overall form of the solution (that is, higher orders do not introduce higher-order 'harmonics'). Since the second order in ϵ provides correction of the radius of order A^2 and, as shown below, the radius, A, is small at the second bifurcation, the first-order approximation of the helix is sufficient for our analysis. Therefore, to study the stability of the perturbed helicoidal solution, one can study the stability of the new exact stationary solutions by taking the zeroth-order solution to be the helix itself with parameters given by the nonlinear analysis of the straight rod.

In order to study the stability of the helix (4.5), we first rewrite it in the standard form (2.9):

$$\boldsymbol{X}_{\rm H} = \left(\frac{\tau_{\rm F}\tilde{s}}{\delta}, \frac{\kappa_{\rm F}}{\delta}\cos(\delta\tilde{s}), \frac{\kappa_{\rm F}}{\delta}\sin(\delta\tilde{s})\right), \quad \delta^2 = \kappa_{\rm F}^2 + \tau_{\rm F}^2, \tag{4.7}$$

where $\tilde{s}$ is the arc length along $\boldsymbol{X}_{\rm H}$ (as opposed to s which is the arc length along the original straight rod). The identification of (4.7) with (4.5) gives

$$\tau_{\rm F} = \frac{\phi}{1+A^2}, \qquad \kappa_{\rm F} = \frac{A\phi}{1+A^2}, \tag{4.8a}$$

$$\delta = \frac{\phi}{\sqrt{1+A^2}}, \qquad \tilde{s} = \frac{s}{\sqrt{1+A^2}}, \tag{4.8b}$$

The tangential force is determined from the relation $f_3^{(H)} = \phi^2\cos\theta$, where θ is the pitch angle of the helix ($\tan\theta = \kappa_{\rm F}/\tau_{\rm F}$). Hence

$$f_3^{(H)} = \frac{\phi^2}{\sqrt{1+A^2}}. \tag{4.9}$$

So far, we have computed the geometrical parameters of the new helical (space) curve. However, our elastic filament also has an imposed twist. In the deformation from straight rod to helical filament, the twist density changes. In order to find the exact new twist density corresponding to the helical filament, we compute the energy for both structures.

The total energy of the system is

$$\mathcal{H} = \int_0^L \mathrm{d}s\,[\kappa_1^2 + \kappa_2^2 + \Gamma(\kappa_3+\gamma)^2 + \phi^2\cos\theta], \tag{4.10}$$

where the first two terms in the integral represent the elastic energy due to curvature effects, the third one the elastic energy due to torsion and twist and the last one, the potential energy due to external constraints.

For the straight rod and the helices, we have, respectively,

$$\boldsymbol{\kappa}_\mathrm{S} = (0,0,0), \qquad \gamma = \gamma_\mathrm{S}, \tag{4.11 a}$$

$$\boldsymbol{\kappa}_\mathrm{H} = (0,\kappa_\mathrm{F},\tau_\mathrm{F}), \quad \gamma = \gamma_\mathrm{H}, \tag{4.11 b}$$

where γ_H is the unknown *ribbon twist* superimposed on the helical curve of Frenet torsion τ_F and curvature κ_F. The ribbon twist is the actual twist of the rod: it represents the rotation in the cross-section of the director basis with respect to the Frenet basis. The energies of the straight rod and helix are, respectively,

$$\mathcal{H}_\mathrm{S} = L(\Gamma\gamma_\mathrm{S}^2 + \phi^2), \tag{4.12 a}$$

$$\mathcal{H}_\mathrm{H} = L[\kappa_\mathrm{F}^2 + \Gamma(\tau_\mathrm{F}+\gamma_\mathrm{H})^2 + \phi\delta]. \tag{4.12 b}$$

Assuming conservation of energy and equating (4.12 *a*) and (4.12 *b*), the helical ribbon twist is found to be

$$\gamma_\mathrm{H} = -\tau_\mathrm{F} \pm \sqrt{\frac{\Gamma\gamma_\mathrm{S}^2 + \phi(\phi-\delta) - \kappa_\mathrm{F}^2}{\Gamma}}, \tag{4.13}$$

where the choice of sign is yet to be determined.

There is a degeneracy in the limit where the radius of a helix shrinks to zero. Indeed, in the limit $A \to 0$, both torsion τ_F and twist γ_H contribute to the total twist of the straight rod γ_S. Formally, a twisted straight rod can be written as a helix with zero radius, zero twist and finite torsion, or as a helix with zero radius, zero torsion and finite twist, or any combination of the two. Therefore, in order to relate the ribbon twist γ_H to the twist γ_S as $A \to 0$ we introduce the *pseudo-torsion*, τ_0, and the *residual twist*, γ_0, of a straight rod as the limits

$$\tau_0 = \lim_{A\to 0} \tau_\mathrm{F}, \tag{4.14 a}$$

$$\gamma_0 = \lim_{A\to 0} \gamma_\mathrm{H}. \tag{4.14 b}$$

The twist of the straight rod, viewed as the limit of the helix with vanishing radius, is simply $\gamma_\mathrm{S} = \tau_0 + \gamma_0$, and in general we have

$$\boldsymbol{\kappa}\cdot\boldsymbol{d}_3 = \tau_\mathrm{F} + \gamma_\mathrm{H}, \tag{4.15}$$

for all $A \geqslant 0$.

In the limit $A \to 0$, the twist density goes to γ_1 and the residual twist is

$$\gamma_0 = \gamma_1 - \tau_0 = \phi\frac{(2-\Gamma)}{\Gamma}.$$

The sign determination in (4.13) can now be found by demanding that $\gamma_{\rm H} \to \gamma_0$ as the radius vanishes. Thus

$$\gamma_{\rm H} = -\tau_{\rm F} + \sqrt{\frac{\Gamma\gamma_{\rm S}^2 + \phi(\phi-\delta) - \kappa_{\rm F}^2}{\Gamma}}. \tag{4.16}$$

We now have all the parameters of the helix as a function of the initial twist density $\gamma_{\rm S}$. Indeed, the radius A is a function of $\gamma_{\rm S}$ (see (4.6)), and the Frenet curvature and torsion are functions of A through (4.8 *a*). Finally the ribbon twist is expressed in terms of all the other parameters and $\gamma_{\rm S}$. Therefore, the parameters $(\kappa_{\rm F}, \tau_{\rm F}, \gamma_{\rm H})$ describing the helical filament, are known in terms of $\gamma_{\rm S}$ and the stability of the helix as a function of the control parameter $\gamma_{\rm S}$ and the number of helical turns $N = \delta L/2\pi$ can now be analysed.

5. Secondary bifurcation: instability of the helix

An extensive study of the stability of helices has been given in Goriely & Tabor (1997*c*). We base the following analysis on this work and specialize it to the family of helices, obtained in the previous section, with one free parameter (the control parameter $\gamma_{\rm S}$). The stationary solution is written as (cf. equations (2.11)):

$$\boldsymbol{\mu}^{(0)} = \{0, \kappa_{\rm F}, \tau_{\rm F}, 0, \kappa_{\rm F}\gamma_{\rm H}\Gamma + \kappa_{\rm F}\tau_{\rm F}(\Gamma-1), \tau_{\rm F}\gamma_{\rm H}\Gamma + \tau_{\rm F}^2(\Gamma-1)\}, \tag{5.1}$$

where the helical ribbon twist, $\gamma_{\rm H}$, the curvature and the torsion should be thought of in terms of its relationship to the control parameter $\gamma_{\rm S}$.

(*a*) *Linear analysis*

The fundamental mode solutions to the variational equations (2.18 *a*) are of the form

$$\boldsymbol{\mu}_n^{(1)} = \boldsymbol{\xi}_n \mathrm{e}^{\sigma_n t + \mathrm{i}\delta(ns/N)}, \tag{5.2}$$

where $\boldsymbol{\xi}_n \in \mathbb{C}^6$ and n denotes the mode number which, due to the choice of boundary conditions, is an integer between 1 and N. The explicit form of the linear operator L_E as a function of the curvature, torsion, twist and tension is given in Goriely & Tabor (1997*c*) and from this the dispersion relations, $\Delta(\sigma, n; \gamma_{\rm H}) = 0$, can be obtained in the usual way. A typical plot of them is shown in figure 2 for increasing values of $\gamma_{\rm H}$.

The neutral curve is determined from $\Delta(0, n; \gamma_{\rm H}) = 0$, where

$$\Delta(0, n; \gamma_{\rm H}) = (\Gamma-1)(\Gamma-2)\tau_{\rm F}^2 + 2\Gamma(\Gamma-2)\gamma_{\rm H}\tau_{\rm F} + \Gamma^2\gamma_{\rm H}^2 + \delta^2(1 - n^2/N^2). \tag{5.3}$$

For given n, one can then read off the value of $\gamma_{\rm H}$ at which the instability occurs. In general, different modes of n can become unstable; however, within the family of helices parametrized by $\gamma_{\rm S}$, the mode $n = 1$ is always the first unstable mode as the control parameter is increased. It is this case that corresponds to our 'secondary' bifurcation.

Let γ_2 be the critical value at which new solutions appear, i.e. $\Delta(0, 1; \gamma_2) = 0$. This last relation is transcendental in $\gamma_{\rm S}$; therefore, the exact value of γ_2 as a function of $\gamma_{\rm S}$ cannot be obtained. Nevertheless, by expanding all the parameters to first order in $(\gamma_{\rm S} - \gamma_1)$, a remarkably good approximation of γ_2 is found,

$$\gamma_2 = \gamma_1 + \frac{4\phi}{4N^2 + (3\Gamma + 16)N + 4}. \tag{5.4}$$

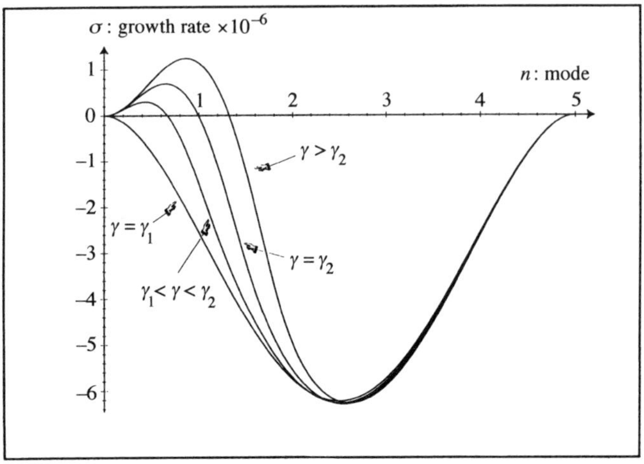

Figure 2. Solutions of the dispersion relation for σ^2 as a function of n with increasing values of $\gamma_S = 0.2267, 0.2675, 0.2687, 0.2695$. $\Gamma = \frac{3}{4}$, $\phi = \frac{1}{10}$, $N = 5$.

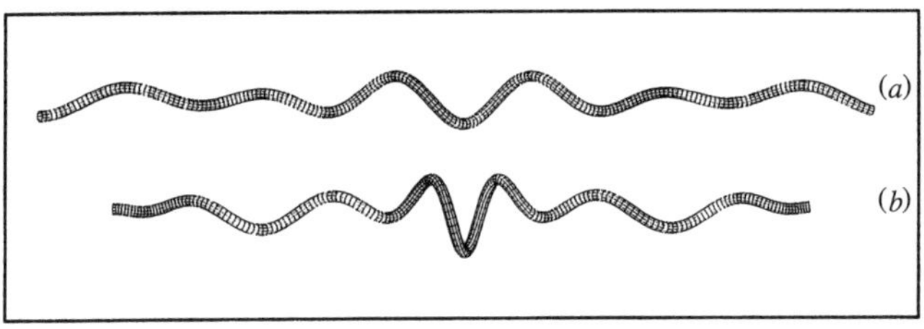

Figure 3. The deformation of a helix due to the unstable mode $n = 1$. (*a*) To first order; (*b*) to second order. The parameters are $\Gamma = \frac{3}{4}$, $\phi = \frac{1}{10}$, $N = 5$, $K = 20$ and s runs from 80 to $80 + 100\pi$.

The approximation obtained by expanding the parameters to second order in $(\gamma_S - \gamma_1)$ only slightly improves this last result. The difference $\gamma_2 - \gamma_1$ is always very small and decreases to zero as the number of loops N increases. In the limit of an infinite long rod, $\gamma_1 \to \gamma_2$. Therefore, the delay of the bifurcation is only due to the discretization of modes (induced by the boundary conditions).

At this point, it is of interest to reflect on the classical results of elasticity theory. The instability of the twisted, infinite (or finite), straight rod is a well-known result obtained by many authors (see, for example, Love 1892; Timoshenko & Gere 1961). However, the stability of the new helix obtained after bifurcation has, to the best of our knowledge, never been investigated (or even questioned). The condition (5.4) gives the instability threshold of the *twisted helix* obtained after the first bifurcation of a straight rod. It is different from the delayed bifurcation condition obtained for a finite *straight rod.* The derivation of this new condition stems from our determination of stability properties directly from the dynamical Kirchhoff equations. However, it is likely that in real, well-controlled, experiments on the bifurcation of rods (where the number of loops N may be large) the twisted rod could jump past the secondary bifurcation (see Thompson & Champneys (1996) for recent experimental data) and the helical solution might then not be observed.

Since the mode $n = 1$ is the first unstable mode, the effect of the instability is to localize the solution at one point. Indeed, to first order the explicit form of the bifurcated solution is

$$x_1(s,t) = P\delta s - \frac{2NKR\nu_1}{n\tau_F}\cos\left(\frac{n\delta s}{N}\right), \tag{5.5 a}$$

$$x_2(s,t) = R\cos(\delta s) - \frac{K}{\delta}\left[\frac{\nu_2-\nu_1}{n-N}\sin\left(\frac{n-N}{N}\delta s\right) + \frac{\nu_2+\nu_1}{n+N}\sin\left(\frac{n+N}{N}\delta s\right)\right], \tag{5.5 b}$$

$$x_3(s,t) = R\sin(\delta s) - \frac{K}{\delta}\left[\frac{\nu_2-\nu_1}{n-N}\cos\left(\frac{n-N}{N}\delta s\right) - \frac{\nu_2+\nu_1}{n+N}\cos\left(\frac{n+N}{N}\delta s\right)\right], \tag{5.5 c}$$

where $K = \epsilon B e^{\sigma t}$ and

$$\left.\begin{aligned} \nu_1 &= n^3\tau_F\kappa_F^2\delta^5\Gamma \\ &\quad\times[(n^2-N^2)^2\delta^4 n^2 + (n^2-N^2)^2N^2\delta^2(\sigma^2-2\kappa_F) + N^4\sigma^2(N^2+n^2)], \\ \nu_2 &= -Nn^4\delta^4\kappa_F^2\tau_F\Gamma[\delta^4(n^2-N^2)(2-\Gamma) + 2N^5n^4\sigma^2 s]. \end{aligned}\right\} \tag{5.6}$$

To second order, the approximate solution is much closer to the exact solution since the arc length is conserved to order $O(\epsilon^4)$. The shape of the bifurcated solution is shown on figure 3 to first and second order in ϵ.

(b) *Nonlinear analysis*

The unstable modes of the helix are discretized due to the boundary conditions. Therefore, a nonlinear analysis can only take into account the temporal evolution of these discrete modes. At the first instability, there is only one unstable mode, namely $n = 1$. Thus, we seek to derive an equation describing the temporal evolution of the mode amplitude B. The corresponding stationary solution of this amplitude equation will give us a relation between the amplitude and the control parameter. As discussed in § 2, the main idea behind a (weakly) nonlinear analysis is to look at the solution near the threshold and introduce new scales (in this case, a new time-scale) proportional to the distance to the bifurcation point and let the arbitrary amplitude, B, vary on this time-scale. If the system is close enough to the secondary bifurcation, the difference between the twist density and the critical twist γ_2, is proportional to the perturbation parameter itself, namely,

$$\epsilon^2 = \gamma_S - \gamma_2. \tag{5.7}$$

The new longer time-scale is $t_1 = \epsilon t$. The choice of this new scale can be justified by expanding the dispersion relation in powers of ϵ (see Goriely & Tabor 1997*b*). Taking into account the possibility of the solutions varying on these (independent) different scales, one can now solve the linear system (2.18). To first order one obtains the linear solution

$$\boldsymbol{\mu}^{(1)} = B(t_1)\boldsymbol{\xi}_1 e^{i(\delta s/N)} + \text{c.c.}, \tag{5.8}$$

where c.c. stands for the complex conjugate and $\boldsymbol{\xi}_1 \in \mathbb{C}^6$ is specified such that $\boldsymbol{\xi}_1 \cdot \boldsymbol{\xi}_1^* = 1$.

The second-order solution can be found in the same way by solving (2.18 *b*) with $k = 2$

$$\boldsymbol{\mu}^{(2)} = \boldsymbol{\chi}_0 + \boldsymbol{\chi}_1 e^{i(\delta s/N)} + \boldsymbol{\chi}_2 e^{2i(\delta s/N)} + \text{c.c.} \tag{5.9}$$

In order to find a condition on the amplitude $B(t_1)$ we demand that the solutions to the third-order system (2.18 *b*) with $k = 3$ remain bounded. Thus, we apply the Fredholm alternative to the system (2.18 *b*): here this consists of integrating H_3 against all neutral solutions of the adjoint operator $L_E^\dagger$,

$$\int_0^L \zeta_1 . H_3(\boldsymbol{\mu}^{(0)}, \boldsymbol{\mu}^{(1)}, \boldsymbol{\mu}^{(2)})\, \mathrm{d}s = 0, \tag{5.10}$$

where $\boldsymbol{\zeta}_1$ is the adjoint solution to $\boldsymbol{\mu}_1^{(1)}$ (i.e. $L_E^\dagger \cdot \boldsymbol{\zeta}_1 = 0$). This compatibility condition gives rise to a differential equation for B as a function of t_1,

$$\frac{\partial^2 B}{\partial t_1^2} = B(c_1 - c_3|B|^2), \tag{5.11}$$

where c_1 and c_3 are

$$\begin{aligned}
c_1 = {} & 2\Gamma[(\Gamma-1)^2\tau_F^2 + 2\Gamma\gamma_c(\Gamma-1)\tau_F + \Gamma^2\gamma_c^2][(2-\Gamma)\tau_F - \Gamma\gamma_c] \\
& \times [(\Gamma^2 - 4\Gamma + 3)\tau_F^2 + (-4\Gamma\gamma_c + 2\Gamma^2\gamma_c)\tau_F + \Gamma^2\gamma_c^2 + \delta^2] \\
& \times [(2\Gamma^4 + 5 - 12\Gamma^3 + 23\Gamma^2 - 18\Gamma)\tau_F^4 \\
& \quad + (46\Gamma^2\gamma_c + 8\Gamma^4\gamma_c - 36\gamma_c\Gamma^3 - 18\Gamma\gamma_c)\tau_F^3 \\
& + (3\delta^2\Gamma^2 - 8\Gamma + 23\Gamma^2\gamma_c^2 + 3\delta^2 + 12\Gamma^4\gamma_c^2 \\
& \quad - 36\Gamma^3\gamma_c^2 + 6 + 2\Gamma^2 - 6\Gamma\delta^2)\tau_F^2 \\
& + \Gamma\gamma_c(6\Gamma\delta^2 - 12\Gamma^2\gamma_c^2 + 8\Gamma^3\gamma_c^2 - 8 + 4\Gamma - 6\delta^2)\tau_F \\
& \quad + 2\Gamma^2\gamma_c^2 + 2\Gamma^4\gamma_c^4 + 3\Gamma^2\gamma_c^2\delta^2 + 2\delta^2]^{-1},
\end{aligned} \tag{5.12}$$

$$\begin{aligned}
c_3 = {} & [2(\Gamma^2 - 4\Gamma + 3)\tau_F^2 + 2(-4\Gamma\gamma_c + 2\Gamma^2\gamma_c)\tau_F + 2\Gamma^2\gamma_2^2 + 2\delta^2] \\
& \times [(2\Gamma^3 - 9 + 17\Gamma - 10\Gamma^2)\tau_F^2 + \Gamma\gamma_2(10 - 12\Gamma + 4\Gamma^2)\tau_F \\
& \quad - 3\delta^2 - 2\Gamma^2\gamma_c^2 + 2\Gamma^3\gamma_c^2 + \Gamma\delta^2] \\
& \times [(\Gamma-1)^3\tau_F^3 + 3\Gamma\gamma_c(\Gamma-1)^2\tau_F^2 + 3\Gamma^2\gamma_c^2(\Gamma-1)\tau_F + \Gamma^3\gamma_c^3] \\
& \quad \times [(\Gamma-2)\tau_F + \Gamma\gamma_c] \\
& \times [(2\Gamma^4 + 5 - 12\Gamma^3 + 23\Gamma^2 - 18\Gamma)\tau_F^4 \\
& \quad + (46\Gamma^2\gamma_c + 8\Gamma^4\gamma_c - 36\gamma_c\Gamma^3 - 18\Gamma\gamma_c)\tau_F^3 \\
& + (3\delta^2\Gamma^2 - 8\Gamma + 23\Gamma^2\gamma_c^2 + 3\delta^2 + 12\Gamma^4\gamma_c^2 - 36\Gamma^3\gamma_c^2 + 6 + 2\Gamma^2 - 6\Gamma\delta^2)\tau_F^2 \\
& + \Gamma\gamma_2(6\Gamma\delta^2 - 12\Gamma^2\gamma_c^2 + 8\Gamma^3\gamma_c^2 - 8 + 4\Gamma - 6\delta^2)\tau_F \\
& + 2\Gamma^2\gamma_c^2 + 2\Gamma^4\gamma_c^4 + 3\Gamma^2\gamma_c^2\delta^2 + 2\delta^2]^{-1},
\end{aligned} \tag{5.13}$$

where $\gamma_c = \gamma_H(\gamma_2)$, that is the critical value of the helical ribbon twist (as given by (4.16) with $\gamma_S = \gamma_2$) at which the helix becomes unstable with respect to the first unstable mode $n = 1$.

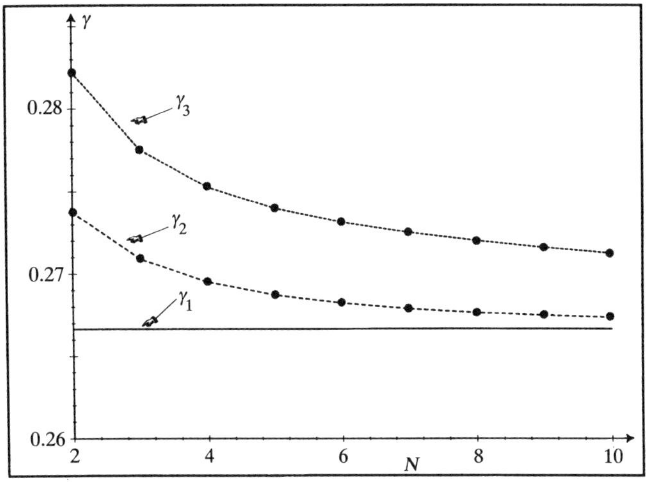

Figure 4. The primary, secondary and tertiary bifurcation as a function of the parameters for $N = 2$–10 with $\Gamma = \frac{3}{4}$, $\phi = \frac{1}{10}$.

All the parameters involved in the coefficients of the amplitude equation depend on the control parameter γ_S. The stationary solution of (5.11) $B^2_{\text{stat}} = c_1/c_3$ gives a relation between the amplitude and the initial twist density.

6. Tertiary bifurcation

The effect of the instability is to locate the perturbation at one point of the rod (chosen here to be the middle point). Eventually, as the amplitude B is varied, a critical point is reached where the projection of the solution in the x–y plane becomes multivalued. The situation is schematically depicted in figure 1*d*. Following Coyne, we take the critical point to be the one at which the loop becomes perpendicular to the x-axis (Coyne 1990). The critical value of the amplitude, B_c, at which the loop centred at $s = s_c$ will flip therefore corresponds to the condition

$$x'_1(s_c; B_c) = 0, \qquad x''_1(s_c; B_c) = 0. \tag{6.1}$$

The flipping condition (6.1) can be solved analytically using the second-order solution. However, the explicit form is too involved to be useful and is not given here.

It is now possible to find the value of the tertiary bifurcation by finding the value of $\gamma_S = \gamma_3$ such that $B_{\text{stat}} = B_c$. A plot of the bifurcation curves is shown in figure 4 for different values of N. Figure 5 shows the filament as γ_3 is varied.

7. Discussion and conclusions

(*a*) *The dynamical picture*

In the previous section, we studied the looping process as a sequence of bifurcations as the control parameter is varied from $\gamma_S < \gamma_1$ to $\gamma_S > \gamma_3$. This sequence of configurations can be obtained by considering that the system attains its stationary

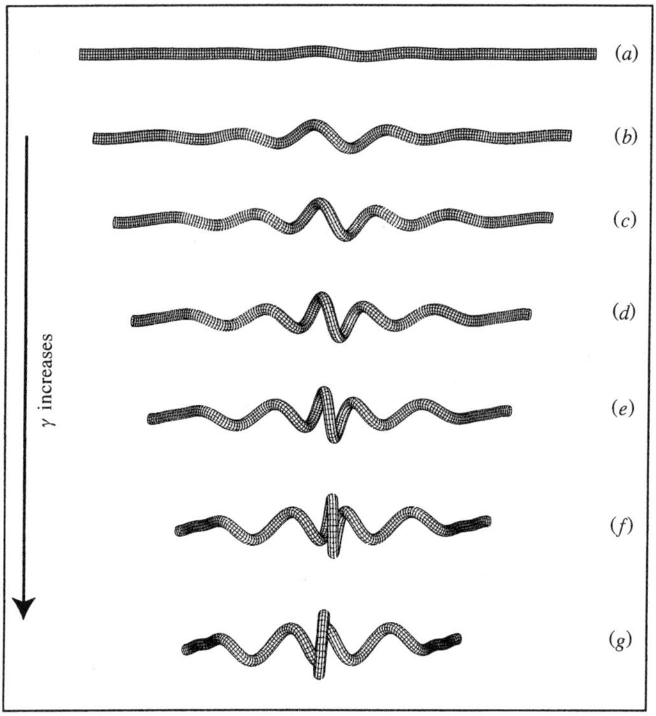

Figure 5. A sequence of helical filaments for varying values of γ_S. (*a*) $\gamma_S < \gamma_1$; (*b*) $\gamma_1 < \gamma_S < \gamma_2$; (*c*) $\gamma_2 < \gamma_S > \gamma_3$; (*d*) $\gamma_S = \gamma_3$; (*e*) $\gamma_S > \gamma_3$. The parameters are $\Gamma = \frac{3}{4}$, $\phi = \frac{1}{10}$, $N = 5$, $\gamma_1 = 0.2667$, $\gamma_2 = 0.2684$, $\gamma_3 = 0.27376$ and $\gamma_S = \gamma_2 + \mu$ with $\mu \times 10^3 = 0.1$, 1.5, 3, 4.9, 5.36, 5.4.

state for every small change of the control parameter. In order to obtain these configurations we rely on two assumptions: first, that in a real system, the presence of damping will allow the system to relax in time (otherwise, none of the stationary solution could be obtained as the system is conservative); second, that the change in the control parameter occurs on a much smaller time-scale than the relaxation time, thereby giving rise to a sequence of quasi-stationary configurations. However, these assumptions are not necessary for the basic phenomena of looping. Indeed, we chose to present the sequence of bifurcations as quasi-stationary for the sake of simplicity. The real problem of looping consists of bringing a system to a region of instability where looping takes place, that is (in our setting) by suddenly varying the twist density of a stable straight rod to a value $\gamma_S > \gamma_3$. Then, the dynamics of the system will directly take the system from a straight rod to a loop without passing through the intermediary steps (this situation is actually closer to the everyday experience of playing with telephone chords). The sequence of configurations shown in figure 5 can then be seen as a dynamical sequence.

In this analysis, we have used the twist density as a control parameter. This is only one of the possible choices. We could have used, equivalently, the tension (think of suddenly decreasing the tension at the end of a straight twisted rod to produce the same dynamics) or the intrinsic twist density. This last choice might be relevant to some problems occurring in biology where filaments may grow with no twist but a large intrinsic twist deficit, effectively creating a twist density resulting in the formation of a loop. (For a remarkable example of this biological phenomena in the growth of bacterial filaments, see Mendelson (1978, 1990).) In any cases, the dynamics obtained by using different parameters as control parameters are equivalent, and it is likely that in any real system changes in all parameters will actually occur simultaneously.

(*b*) *Why do loops forms at the middle of the rod?*

One commonly observes that the loops created by twisting the ends of a rod usually form in the middle of the rod. This is due to the choice of boundary conditions. Here for the sake of simplicity we have considered that the tangents at the ends are free. In doing so, all the points of the strings are actually equivalent, and the loop can form at any point. However, a simple argument can show why loops form at the middle for different boundary conditions. If the ends are held clamped (that is the tangents at the ends are constrained along the axis), the first bifurcation does not lead to a helix but rather (as shown in Goriely & Tabor 1997*b*) to a helix modulated by a envelope-like shape (so that the radius of the modulated helix goes to zero at the ends). The maximal amplitude of this solution is reached at the middle. We have shown here that the secondary instability is triggered by increasing the radius of the helix (controlled by the initial twist density). Therefore, the secondary instability for a twisted rod with clamped ends will be first triggered at the middle of the string where the amplitude of the deformation is maximal.

(*c*) *Conclusions*

We have now completed our picture of the looping as a dynamical process: a phenomenon triggered by a sequence of instabilities for different configurations of the filament. The first bifurcation occurs when a pulled, twisted, straight rod becomes unstable. The linear analysis predicts that it deforms to a helix. The radius of such a helix can be computed via a nonlinear analysis (Goriely & Tabor 1996, 1997*b*) while its twist density can be obtained by a consideration of energetics. The important feature of the bifurcated solution is that it is an exact solution of the model. Therefore, its stability can be easily studied by the same method used to study the stability of the straight rod. The linear stability of the helix shows that it rapidly becomes unstable, giving rise to a secondary bifurcation. Moreover, the instability tends to localize the deformation of the rod at one point. The amplitude of this localization can be obtained by performing a one-mode amplitude expansion. This nonlinear analysis provides a relationship between the amplitude of the deformation and the control parameters. The tertiary bifurcation is reached when a loop is formed at the middle and looping takes place.

The dynamical mechanism proposed here differs in many respects from previous analyses and exploits a series of results on the linear and nonlinear stability of filaments. These techniques are quite general and should be applicable to other problems involving dynamical changes in form.

This work was supported by NSF grant DMS-9704421 and Flinn Foundation's Biomathematics and Dynamics Initiative program (ID no. 048-1000206-94).

References

Antman, S. S. & Kenney, C. S. 1981 Large buckled states of nonlinearly elastic rods under torsion, thrust and gravity. *Arch. Ration. Mech. Analysis* **76**, 289–338.

Champneys, A. R. & Thompson, J. M. T. 1996 A multiplicity of localized buckling modes for twisted rod equations. *Proc. R. Soc. Lond.* A **452**, 2467–2491.

Coleman, B. D., Dill, E. H., Lembo, M., Lu, Z. & Tobias, I. 1993 On the dynamics of rods in the theory of Kirchhoff and Clebsch. *Arch. Ration. Mech. Analysis* **121**, 339–359.

Coyne, J. 1990 Analysis of the formation and elimination of loops in twisted cable. *IEE J. Oceanic Engng* **15**, 72–83.

Damil, N. & Pottier-Ferry, M. 1986 Wavelength selection in the postbuckling of a long rectangular plate. *Int. J. Solids Struct.* **22**, 511–526.

Dill, E. H. 1992 Kirchhoff's theory of rods. *Arch. Hist. Exact. Sci.* **44**, 2–23.

Goriely, A. & Tabor, M. 1996 New amplitude equations for thin elastic rods. *Phys. Rev. Lett.* **77**, 3537–3540.

Goriely, A. & Tabor, M. 1997*a* Nonlinear dynamics of filaments. I. Dynamical instabilities. *Physica* D **105**, 20–44.

Goriely, A. & Tabor, M. 1997*b* Nonlinear dynamics of filaments. II. Nonlinear analysis. *Physica* D **105**, 45–61.

Goriely, A. & Tabor, M. 1997*c* Nonlinear dynamics of filaments. III. Instabilities of helical rods. *Proc. R. Soc. Lond.* A **453**, 2583–2601.

Love, A. E. H. 1892 *A treatise on the mathematical theory of elasticity.* Cambridge University Press.

Mendelson, N. H. 1978 Helical *Bacillus subtilis* macrofibers: morphogenesis of a bacterial multicellular macroorganism. *Proc. Natn. Acad. Sci. USA* **75**, 2472–2482.

Mendelson, N. H. 1990 Bacterial macrofibers: the morphogenesis of complex multicellular bacterial forms. *Sci. Progress Oxford* **74**, 425–441.

Pomeau, Y. 1981 Nonlinear pattern selection in a problem of elasticity. *J. Physique Lett.* **42**, L1–L4.

Ricca, R. L. 1995 The energy spectrum of a twisted flexible string under elastic relaxation. *J. Phys.* A **28**, 2335–2352.

Thompson, J. M. T. & Champneys, A. R. 1996 From helix to localized writhing in the torsional post-buckling of elastic rods. *Proc. R. Soc. Lond.* A **452**, 117–138.

Timoshenko, S. P. & Gere, J. M. 1961 *Theory of elastic stability.* New York: McGraw-Hill.

Tvergaard, V. & Needleman, A. 1980 On the localization of buckling patterns. *J. Appl. Mech.* **47**, 613–619.

Chapter 5
Cylindrical shells

The buckling and post-buckling behaviour of long, elastic, axially-compressed cylindrical shells is a classic nonlinear problem of importance in many civil engineering applications. Traditionally the application of linear buckling techniques applied to such shell structures is not accurate, because of "snap-back" behaviour, where the initial bifurcation is markedly subcritical. Hence, under controlled end displacement, finite-amplitude buckle patterns appear instantaneously, with greatly reduced load-carrying capacity. Thus understanding the shape of the nonlinear postbuckled response is of paramount importance. With comparatively few exceptions, analytical approaches have assumed *a priori* that postbuckling modes are periodic both axially and circumferentially. However, this conflicts with experimental evidence which strongly suggests that, for all but very short shells, buckle patterns tend to be periodic circumferentially but localized to some portion of the axial length.

This chapter presents three articles that apply the methodology of this book to the von Kármán–Donnell thin-shell equations. The first article by HUNT & LUCENA NETO reviews some experimental evidence for localized buckling and compares the results with localized solutions of a low-order modal decomposition of the equations. The next article by LORD, CHAMPNEYS & HUNT considers a numerical approximation using spectral methods and specific boundary conditions for localized solutions. The final article considers a link between what is observed numerically, asymptotically and the so-called Maxwell load hypothesis for predicting the effective buckling load.

Localized Buckling in long axially loaded cylindrical shells. HUNT,

G.W. & Lucena Neto, E. *J. Mech. Phys. Solids* **39** (1991) 881-894

Computation of localized post buckling in long axially compressed cylindrical shells. Lord, G.J, Champneys, A.R. & Hunt, G.W. *Phil. Trans. Roy. Soc. A* **355** (1997) 2137-2150.

Homoclinic and heteroclinic orbits underlying the post-buckling of axially-compressed cylindrical shells. Hunt, G.W., Lord, G.L. & Champneys, A.R. *Comp. Methods Appl. Mech. Eng.* to appear (1999)

J. Mech. Phys. Solids Vol. 39, No. 7, pp. 881–894, 1991.
Printed in Great Britain.
0022–5096/91 \$3.00+0.00

LOCALIZED BUCKLING IN LONG AXIALLY-LOADED CYLINDRICAL SHELLS

G. W. HUNT and E. LUCENA NETO†‡

Department of Civil Engineering, Imperial College London, London SW7 2BU, U.K.
† Instituto Tecnológico de Aeronáutica, 12225 São José dos Campos-SP, Brazil

(*Received* 30 *April* 1990)

ABSTRACT

DOUBLE-SCALE perturbation analysis of a long elastic cylindrical shell under axial compression reveals a second-order non-linear differential equation for the buckle–pattern amplitude in slow-space. Numerical solution then suggests that the most easily-triggered failure mode is localized along the length. The method is extended to include mode interaction, giving three coupled second-order non-linear differential equations in slow-space. Localized solutions are again found, by combining features of the Lagrangian function with a systematic numerical search procedure. The predicted extent of the localization, about one-and-a-half axial wavelengths when fully developed, compares well with published experiments on long cylinders. Moreover, in contrast to the associated periodic solutions, "square" waves at the minimum critical load are denied; the predominant waveform turns out to be long axially, again as seen experimentally.

1. INTRODUCTION

THE buckling of a long elastic cylindrical shell under axial compression is a classic problem of structural mechanics. So much has been written that it is inappropriate to attempt a review, save to mention a few key contributions (DONNELL, 1934; VON KÁRMÁN and TSIEN, 1941; KOITER, 1945; HOFF *et al.*, 1966). It is the case, however, that periodicity in the axial direction is often either explicitly or implicitly assumed in the process of formulation (see, for example, HANSEN, 1975; HUNT *et al.*, 1986). This does not square with the experimental evidence, which indicates that, however long the cylinder, buckling occurs only over a range of about one-and-a-half wavelengths of the final diamond pattern (SENDELBECK *et al.*, 1967; YAMAKI, 1984).

The localization of buckling along a length is a known phenomenon for long structures, notably railway lines and pipelines, and carries significant analytical implications. Successful formulations must be both non-linear and capable of admitting multiple solutions, since the localized buckle can occur anywhere along the length. Periodic solutions are also part of the overall picture, which is most usefully seen in phase-space (HUNT *et al.*, 1989).

‡ Present address: Department of Civil Engineering, Imperial College London, London SW7 2BU, U.K.

With a complex problem like the long cylinder, it is often the periodic solution that first catches the eye. The potential to localize, however, introduces significant extra ingredients. Differential equations of amplitude modulation replace simple algebraic coupling equations, and their solution can result in localized modes in differing combinations from their periodic counterparts. Also, localized solutions approach the pre-buckled state asymptotically, and standard structural boundary conditions hence carry little practical significance once the extent of the localization is down to the specimen length or less; the length of localization is thus a key feature, useful for distinguishing between "long" and "short" cylinders, for example.

The most useful tool for the study of localized solutions is double-scale analysis, based on the underlying differential equation (SEGEL, 1969; NEWELL and WHITEHEAD, 1969). For the pair of fourth-order von Kármán–Donnell equations relevant to the axially-compressed cylinder, this supplies valuable new information. In particular, for a perfect cylinder, localized solutions are initiated for modes with wavelengths on the well-known Koiter circle, but not for "square" waves at the crown of the circle. This is a new effect, not found in periodic studies, but reflected in the experimental evidence. Indeed, comparisons with published experimental results of YAMAKI (1984) and SENDELBECK *et al.* (1967) are most encouraging.

2. LINEAR ANALYSIS

Based on the linear membrane equations (BRUSH and ALMROTH, 1975), prior to buckling the perfect axially-compressed cylinder is in a uniform membrane state

$$N_x = -P, \qquad N_y = N_{xy} = 0, \tag{1}$$

with an associated outward radial displacement

$$w = -\frac{vP}{\rho E t}, \tag{2}$$

where P is the applied compressive axial load per unit length; E and v are Young's modulus and Poisson's ratio; and ρ and t denote the shell curvature and thickness, respectively. The solution above can be considered as rigorous for infinitely long cylinders.

As a result of buckling, contributions arise to the in-plane stress resultants and to the radial displacement. Using Eqs (1) and (2) these contributions are related to each other by von Kármán–Donnell equations given as follows:

$$k^2\nabla^4 w + \lambda\frac{\partial^2 w}{\partial x^2} - \rho\frac{\partial^2 \phi}{\partial x^2} = \frac{\partial^2 w}{\partial x^2}\frac{\partial^2 \phi}{\partial y^2} + \frac{\partial^2 w}{\partial y^2}\frac{\partial^2 \phi}{\partial x^2} - 2\frac{\partial^2 w}{\partial x\,\partial y}\frac{\partial^2 \phi}{\partial x\,\partial y} \tag{3}$$

$$\nabla^4\phi + \rho\frac{\partial^2 w}{\partial x^2} = \left(\frac{\partial^2 w}{\partial x\,\partial y}\right)^2 - \frac{\partial^2 w}{\partial x^2}\frac{\partial^2 w}{\partial y^2}, \tag{4}$$

where x and y are the axial and circumferential coordinates; ∇^4 denotes the two-biharmonic operator; $k^2 = t^2/12(1-v^2)$; $\lambda - P/Et$; and the stress function ϕ is related to the in-plane stress resultant deviations from the uniform membrane state by

$$N_x = Et\frac{\partial^2\phi}{\partial y^2}, \qquad N_y = Et\frac{\partial^2\phi}{\partial x^2}, \qquad N_{xy} = -Et\frac{\partial^2\phi}{\partial x\,\partial y}. \tag{5}$$

The linear buckling equations

$$k^2\nabla^4 w + \lambda\frac{\partial^2 w}{\partial x^2} - \rho\frac{\partial^2\phi}{\partial x^2} = 0 \tag{6}$$

$$\nabla^4\phi + \rho\frac{\partial^2 w}{\partial x^2} = 0 \tag{7}$$

are obtained from the radial equilibrium equation (3) and the compatibility equation (4) by omitting the non-linear terms. If we assume

$$w = A\cos\gamma x\cos\beta\gamma y \tag{8}$$

Eq. (7) yields

$$\phi = a\cos\gamma x\cos\beta\gamma y, \tag{9}$$

where A is a constant; $a = \rho\alpha A/\gamma^2$; $\alpha = 1/(1+\beta^2)^2$; $\gamma = \rho n/\beta$; β is the modal aspect ratio (axial/circumferential wavelength); and n is the number of waves in the circumferential direction. Substitution of (8) and (9) into Eq. (6) leads to

$$\lambda^A = \frac{\gamma^2 k^2}{\alpha} + \frac{\alpha\rho^2}{\gamma^2} \tag{10}$$

as the condition of validity of the resulting equation for all values of x and y and non-zero values of A. λ^A is the buckling parameter load λ corresponding to the mode (8). Similarly, if we assume the axisymmetric mode

$$w = B\cos 2\gamma x \tag{11}$$

the corresponding buckling parameter load is

$$\lambda^B = 4\gamma^2 k^2 + \frac{\rho^2}{4\gamma^2}. \tag{12}$$

The analytical minimization of λ^A with respect to α gives

$$\lambda^A = 2\rho k, \qquad \alpha = \frac{\gamma^2 k}{\rho}. \tag{13}$$

$\lambda^A = 2\rho k$ corresponds to the classical solution and all the modes having this load are those on the well-known Koiter circle; $\alpha = \gamma^2 k/\rho$ is the required condition for the mode (8) to be on the Koiter circle.

3. DOUBLE SCALE ANALYSIS

Equations (3) and (4) can be converted into an infinite set of linear partial differential equations by expanding the radial displacement, the stress function and the load parameter into power series in terms of an arbitrary parameter s (THOMPSON and HUNT, 1973)

$$\begin{aligned} w(x,X,y) &= w_1(x,X,y)s + w_2(x,X,y)s^2 + w_3(x,X,y)s^3 + \cdots \\ \phi(x,X,y) &= \phi_1(x,X,y)s + \phi_2(x,X,y)s^2 + \phi_3(x,X,y)s^3 + \cdots \\ \lambda &= \lambda^A + \lambda_1 s + \lambda_2 s^2 + \lambda_3 s^3 + \cdots \end{aligned} \tag{14}$$

w, ϕ and λ are expanded about the critical point corresponding to the Koiter circle, as can easily be seen by substituting $s = 0$ into the above expanded forms. As an attempt to get not only periodic solutions but also localized ones, the slow-space variable X (LANGE and NEWELL, 1971; POTIER-FERRY, 1983; HUNT *et al.*, 1989) defined by

$$X = sx \tag{15}$$

is incorporated in w and ϕ, Eqs (14), by assuming that the amplitudes of the modes w_1, w_2, etc., depend on this variable.

Considering Eq. (15), derivatives with respect to x become

$$\begin{aligned} \frac{\partial}{\partial x} &= \frac{\partial}{\partial x} + \frac{\partial}{\partial X}s \\ \frac{\partial^2}{\partial x^2} &= \frac{\partial^2}{\partial x^2} + 2\frac{\partial^2}{\partial x\,\partial X}s + \frac{\partial^2}{\partial X^2}s^2, \end{aligned} \tag{16}$$

etc. Then, from the first series of Eqs (14), we can write

$$\begin{aligned} \frac{\partial w}{\partial x} &= \frac{\partial w_1}{\partial x}s + \left(\frac{\partial w_2}{\partial x} + w_1'\right)s^2 + \left(\frac{\partial w_3}{\partial x} + w_2'\right)s^3 + \cdots \\ \frac{\partial^2 w}{\partial x^2} &= \frac{\partial^2 w_1}{\partial x^2}s + \left(\frac{\partial^2 w_2}{\partial x^2} + 2\frac{\partial w_1'}{\partial x}\right)s^2 + \left(\frac{\partial^2 w_3}{\partial x^2} + 2\frac{\partial w_2'}{\partial x} + w_1''\right)s^3 + \cdots \\ \frac{\partial^3 w}{\partial x^3} &= \frac{\partial^3 w_1}{\partial x^3}s + \left(\frac{\partial^3 w_2}{\partial x^3} + 3\frac{\partial^2 w_1'}{\partial x^2}\right)s^2 + \left(\frac{\partial^3 w_3}{\partial x^3} + 3\frac{\partial^2 w_2'}{\partial x^2} + 3\frac{\partial w_1''}{\partial x}\right)s^3 + \cdots \\ \frac{\partial^4 w}{\partial x^4} &= \frac{\partial^4 w_1}{\partial x^4}s + \left(\frac{\partial^4 w_2}{\partial x^4} + 4\frac{\partial^3 w_1'}{\partial x^3}\right)s^2 + \left(\frac{\partial^4 w_3}{\partial x^4} + 4\frac{\partial^3 w_2'}{\partial x^3} + 6\frac{\partial^2 w_1''}{\partial x^2}\right)s^3 + \cdots \\ \nabla^4 w &= \nabla^4 w_1 s + \left(\nabla^4 w_2 + 4\frac{\partial^3 w_1'}{\partial x^3} + 4\frac{\partial^3 w_1'}{\partial x\,\partial y^2}\right)s^2 \\ &\quad + \left(\nabla^4 w_3 + 4\frac{\partial^3 w_2'}{\partial x^3} + 4\frac{\partial^3 w_2'}{\partial x\,\partial y^2} + 6\frac{\partial^2 w_1''}{\partial x^2} + 2\frac{\partial^2 w_1''}{\partial y^2}\right)s^3 + \cdots, \end{aligned} \tag{17}$$

where a prime denotes partial differentiation with respect to X. Similar expressions can be written for ϕ by simply replacing w with ϕ in above equations.

Since s was taken to be an arbitrary parameter, the stipulation that a power series in s vanishes requires that each coefficient of the power series vanishes as well. If Eqs (14) and (17) are substituted into Eqs (3) and (4), the requirement that each coefficient in the power series vanishes leads to the following linear equations, which are the first three of an infinite set:

Order s

$$k^2\nabla^4 w_1+\lambda^4\frac{\partial^2 w_1}{\partial x^2}-\rho\frac{\partial^2\phi_1}{\partial x^2}=0 \tag{18}$$

$$\nabla^4\phi_1+\rho\frac{\partial^2 w_1}{\partial x^2}=0. \tag{19}$$

Order s^2

$$\begin{aligned}k^2\nabla^4 w_2+\lambda^4\frac{\partial^2 w_2}{\partial x^2}-\rho\frac{\partial^2\phi_2}{\partial x^2}&=\frac{\partial^2 w_1}{\partial x^2}\frac{\partial^2\phi_1}{\partial y^2}+\frac{\partial^2 w_1}{\partial y^2}\frac{\partial^2\phi_1}{\partial x^2}-2\frac{\partial^2 w_1}{\partial x\,\partial y}\frac{\partial^2\phi_1}{\partial x\,\partial y}\\&\quad-\lambda_1\frac{\partial^2 w_1}{\partial x^2}-4k^2\left(\frac{\partial^3 w_1'}{\partial x^3}+\frac{\partial^3 w_1'}{\partial x\,\partial y^2}\right)-2\lambda^4\frac{\partial w_1'}{\partial x}+2\rho\frac{\partial\phi_1'}{\partial x}\end{aligned} \tag{20}$$

$$\nabla^4\phi_2+\rho\frac{\partial^2 w_2}{\partial x^2}=\left(\frac{\partial^2 w_1}{\partial x\,\partial y}\right)-\frac{\partial^2 w_1}{\partial x^2}\frac{\partial^2 w_1}{\partial y^2}-2\rho\frac{\partial w_1'}{\partial x}-4\left(\frac{\partial^3\phi_1'}{\partial x^3}+\frac{\partial^3\phi_1'}{\partial x\,\partial y^2}\right). \tag{21}$$

Order s^3

$$\begin{aligned}k^2\nabla^4 w_3+\lambda^4\frac{\partial^2 w_3}{\partial x^2}-\rho\frac{\partial^2\phi_3}{\partial x^2}&=\frac{\partial^2 w_1}{\partial x^2}\frac{\partial^2\phi_2}{\partial y^2}+\frac{\partial^2 w_1}{\partial y^2}\left(\frac{\partial^2\phi_2}{\partial x^2}+2\frac{\partial\phi_1'}{\partial x}\right)\\&\quad-2\frac{\partial^2 w_1}{\partial x\,\partial y}\left(\frac{\partial^2\phi_2}{\partial x\,\partial y}+\frac{\partial\phi_1'}{\partial y}\right)+\left(\frac{\partial^2 w_2}{\partial x^2}+2\frac{\partial w_1'}{\partial x}\right)\frac{\partial^2\phi_1}{\partial y^2}\\&\quad+\frac{\partial^2 w_2}{\partial y^2}\frac{\partial^2\phi_1}{\partial x^2}-2\left(\frac{\partial^2 w_2}{\partial x\,\partial y}+\frac{\partial w_1'}{\partial y}\right)\frac{\partial^2\phi_1}{\partial x\,\partial y}\\&\quad-\lambda_1\left(\frac{\partial^2 w_2}{\partial x^2}+2\frac{\partial w_1'}{\partial x}\right)-\lambda_2\frac{\partial^2 w_1}{\partial x^2}\\&\quad-2k^2\left(2\frac{\partial^3 w_2'}{\partial x^3}+2\frac{\partial^3 w_2'}{\partial x\,\partial y^2}+3\frac{\partial^2 w_1''}{\partial x^2}+\frac{\partial^2 w_1''}{\partial y^2}\right)\\&\quad-\lambda^4\left(2\frac{\partial w_2'}{\partial x}+w_1''\right)+\rho\left(2\frac{\partial\phi_2'}{\partial x}+\phi_1''\right)\end{aligned} \tag{22}$$

$$\nabla^4\phi_3+\rho\frac{\partial^2 w_3}{\partial x^2} = 2\frac{\partial^2 w_1}{\partial x\,\partial y}\left(\frac{\partial^2 w_2}{\partial x\,\partial y}+\frac{\partial w_1'}{\partial y}\right)-\frac{\partial^2 w_1}{\partial x^2}\frac{\partial^2 w_2}{\partial y^2}$$

$$-\frac{\partial^2 w_1}{\partial y^2}\left(\frac{\partial^2 w_2}{\partial x^2}+2\frac{\partial w_1'}{\partial x}\right)-\rho\left(2\frac{\partial w_2'}{\partial x}+w_1''\right)$$

$$-2\left(2\frac{\partial^3\phi_2'}{\partial x^3}+2\frac{\partial^3\phi_2'}{\partial x\,\partial y^2}+3\frac{\partial^2\phi_1''}{\partial x^2}+\frac{\partial^2\phi_1''}{\partial y^2}\right). \tag{23}$$

The order s^n of this sequence of equations can be written as

$$k^2\nabla^4 w_n+\lambda^A\frac{\partial^2 w_n}{\partial x^2}-\rho\frac{\partial^2\phi_n}{\partial x^2} = F_n(x,X,y)$$

$$\nabla^4\phi_n+\rho\frac{\partial^2 w_n}{\partial x^2} = G_n(x,X,y), \tag{24}$$

where $F_n(x, X, y)$ and $G_n(x, X, y)$ are determined from the solutions of the previous $(n-1)$ set of equations. Terms proportional to the solution of the homogeneous equations arising in $F_n(x, X, y)$ or $G_n(x, X, y)$ must be eliminated because they lead to unbounded particular solution (that is, secular).

The procedure employed of eliminating the secular terms is similar to that outlined by AMAZIGO and FRASER (1971) and is given as follows: suppose that the homogeneous solution of Eqs (24), that we will be concerned with, is

$$w_n = A_n\cos\gamma x\cos\beta\gamma y, \qquad \phi_n = a_n\cos\gamma x\cos\beta\gamma y, \tag{25}$$

with $a_n = kA_n$. Suppose that F_{n1} and G_{n1} are the coefficients of the terms proportional to $\cos\gamma x\cos\beta\gamma y$ arising in $F_n(x, X, y)$ and $G_n(x, X, y)$, respectively. Writing Eqs (24) with these terms only on the right-hand side, we have

$$k^2\nabla^4 w_n+\lambda^A\frac{\partial^2 w_n}{\partial x^2}-\rho\frac{\partial^2\phi_n}{\partial x^2} = F_{n1}\cos\gamma x\cos\beta\gamma y$$

$$\nabla^4\phi_n+\rho\frac{\partial^2 w_n}{\partial x^2} = G_{n1}\cos\gamma x\cos\beta\gamma y. \tag{26}$$

A solution to (26) in the form (25) gives

$$(\rho a_n-\rho kA_n)\cos\gamma x\cos\beta\gamma y = F_{n1}\cos\gamma x\cos\beta\gamma y \tag{27}$$

$$(\rho a_n-\rho kA_n)\cos\gamma x\cos\beta\gamma y = kG_{n1}\cos\gamma x\cos\beta\gamma y. \tag{28}$$

Dividing (27) by (28) gives

$$F_{n1}-kG_{n1} = 0. \tag{29}$$

This condition of suppressing secular terms arising from a nonhomogeneous expression in the form (25) holds for any secular terms arising from trigonometric expressions in the form of modes which are on the Koiter circle.

3.1. *First-order equations*

Equations (18) and (19) have solution similar to Eqs (6) and (7)

$$w_1 = A(X)\cos\gamma x\cos\beta\gamma y, \qquad \phi_1 = a(X)\cos\gamma x\cos\beta\gamma y, \tag{30}$$

where $a = kA$ and A is no longer a constant, but depends on X.

3.2. *Second-order equations*

After substitution of (30) and reduction of products of trigonometric functions to sums, Eqs (20) and (21) become

$$k^2\nabla^4 w_2 + \lambda^4\frac{\partial^2 w_2}{\partial x^2} - \rho\frac{\partial^2\phi_2}{\partial x^2} = \lambda_1\gamma^2 A\cos\gamma x\cos\beta\gamma y$$
$$+[2\gamma\rho k - 4\gamma^3 k^2(1+\beta^2)]A'\sin\gamma x\cos\beta\gamma y + \beta^2\gamma^4 kA^2(\cos 2\gamma x + \cos 2\beta\gamma y) \tag{31}$$

$$\nabla^4\phi_2 + \rho\frac{\partial^2 w_2}{\partial x^2} = \frac{2\gamma(\beta^2-1)\rho}{1+\beta^2}A'\sin\gamma x\cos\beta\gamma y - \frac{\beta^2\gamma^4}{2}A^2(\cos 2\gamma x + \cos 2\beta\gamma y). \tag{32}$$

Elimination of secular terms arising from $\cos\gamma x\cos\beta\gamma y$ by applying the condition (29) provides

$$\lambda_1 = 0. \tag{33}$$

The solution of (31) and (32) contains no secular terms arising from $\sin\gamma x\cos\beta\gamma y$ because the coefficients of these terms on the right-hand side of these equations satisfy condition (29) identically. If we assume the particular solution

$$w_2 = B(X)\cos 2\gamma x + C(X)\cos 2\beta\gamma y, \tag{34}$$

Eq. (32) provides

$$\phi_2 = b(X)\cos 2\gamma x + c(X)\cos 2\beta\gamma y + d(X)\sin\gamma x\cos\beta\gamma y, \tag{35}$$

where

$$b = \frac{\rho}{4\gamma^2}B - \frac{\beta^2}{32}A^2, \qquad c = -\frac{1}{32\beta^2}A^2, \qquad d = \frac{2(\beta^2-1)k}{\gamma(1+\beta^2)}A'. \tag{36}$$

Substituting Eqs (33)–(36) into (31) gives

$$B = \frac{\rho\beta^2(1+8\alpha)}{32\sigma}A^2, \qquad C = \frac{1}{16\beta^2 k}A^2, \tag{37}$$

with

$$\sigma = \lambda^B - \lambda^A. \tag{38}$$

3.3. *Secular terms in third-order equations*

The right-hand side of Eqs (22) and (23) can be determined by substituting previous results given by (30) and (33)–(37). Terms proportional to $\cos \gamma x \cos \beta\gamma y$, which arise in both equations, must be eliminated to avoid unbounded particular solutions. Let F_{31} and G_{31} be the coefficients of these terms in (22) and (23), respectively. Then

$$F_{31} = \gamma^2 k^2 (1+3\beta^4) A'' + \gamma^2 \lambda_2 A + \frac{\gamma^4}{16}\left[1 - \beta^4 + \frac{\rho^2 \beta^4}{4\gamma^2 \sigma}(1+4\alpha)(1+8\alpha)\right] A^3$$

$$G_{31} = -\gamma^2 k(\beta^4 - 8\beta^2 + 3) A'' - \frac{\gamma^4}{16}\left[\frac{2}{k} + \frac{\rho \beta^4}{\sigma}(1+8\alpha)\right] A^3. \tag{39}$$

Now, substituting (39) into (29) we have

$$4(1-\beta^2)^2 k^2 A'' + \lambda_2 A + \frac{1}{16}\left[\frac{\rho^2 \beta^4}{4\sigma}(1+8\alpha)^2 + \gamma^2(3-\beta^4)\right] A^3 = 0. \tag{40}$$

The periodic solution is recovered by setting $A'' = 0$. In this case, we get the same solution as that obtained by conventional post-buckling theory, with mode A represented by an active, and mode B a passive, coordinate (THOMPSON and HUNT, 1984). But localized solutions also exist. If we set the perturbation parameter s to $\sqrt{(\lambda^A - \lambda)}$, so that, on equating coefficients in (14), $\lambda_2 = -1$ for $\lambda < \lambda^A$, we retrieve the well-known amplitude modulation equation of NEWELL and WHITEHEAD (1969) and SEGEL (1969), with signs suitable for the localized form (POTIER-FERRY, 1983; HUNT *et al.*, 1989). As aspect ratio β is varied, the position of mode A moves around the Koiter circle, with the square waveform represented by $\beta = 1$ at the crown. Runge–Kutta solutions to Eq. (40), from the same starting state of $A_0 = 10^{-5}$, $A'_0 = 0$, are as shown in Fig. 1. For clarity, the curves have been drawn with different initial value X_0. We see that the curves flatten to the X-axis as $\beta \to 1$. Here critical loads for A and B coincide, so $\sigma = 0$ and the coefficient of A^3 is infinite while, simultaneously, that of A'' is zero; the latter forces periodicity, and the former suggests that the amplitude is zero.

We also note that solution of the third-order perturbation equations would involve terms of the form $\cos 3\gamma x \cos \beta\gamma y$ in w_3. These are of particular interest, since at $\beta = \sqrt{3}$ they also lie on the Koiter circle. Our single-mode view, which results from the inclusion just of mode A in the solution (30) for w_1, is thus called into question specifically at $\beta = 1$ and $\beta = \sqrt{3}$. The possibility of localization combined with mode interaction occurs for each, the first case involving two modes, A and B, and the second, three.

4. LOCALIZATION AND MODE INTERACTION

Let us take the new solution to the first-order perturbation equations to replace (30)

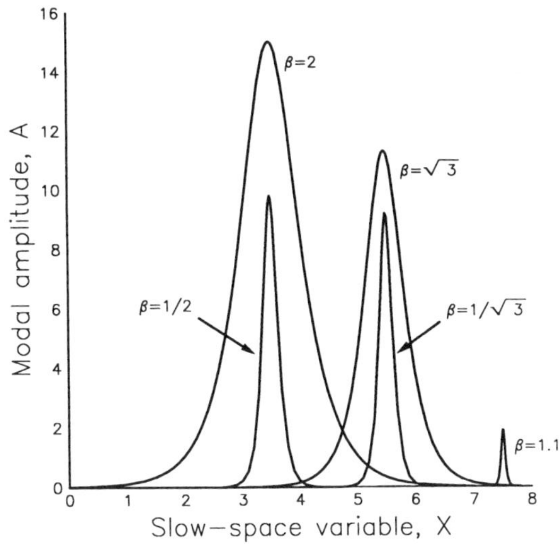

FIG. 1. Runge–Kutta solutions to Eq. (40), for representative modes on the Koiter circle.

$$w_1 = A(X)\cos\gamma x\cos\beta\gamma y + B(X)\cos(1+\beta^2)\gamma x + C(X)\cos\beta^2\gamma x\cos\beta\gamma y$$

$$\phi_1 = a(X)\cos\gamma x\cos\beta\gamma y + b(X)\cos(1+\beta^2)\gamma x + c(X)\cos\beta^2\gamma x\cos\beta\gamma y, \qquad (41)$$

where $a = kA$, $b = kB$, $c = kC$ and, as before, $k = \rho\alpha/\gamma^2$. All three modes appear on the Koiter circle, as seen in Fig. 2.

The same analytical sequence is now followed; this time, however, to account for the differing influence of non-linear terms, slow-space is defined by

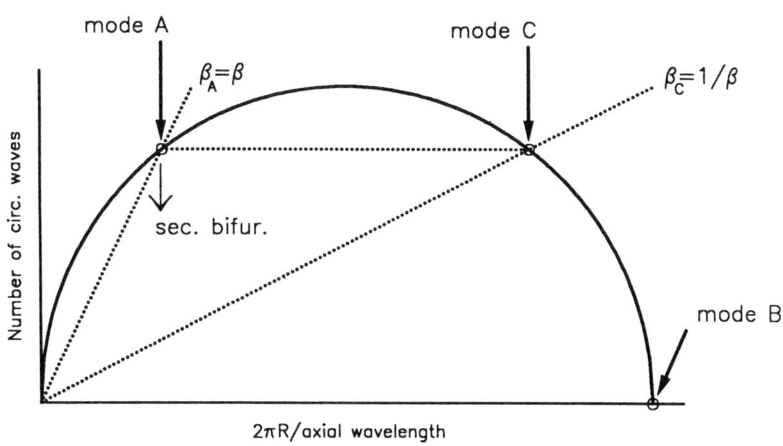

FIG. 2. Modal interactions and the Koiter circle. As the post-buckling develops, it is observed experimentally that there is a drift to within the circle, via a sequence of secondary bifurcations (see Fig. 4).

$$X = s^{1/2}x \tag{42}$$

to replace the linear relation (15). The remainder of the analysis proceeds much as above. No secular terms are found to arise in the solution of $s^{3/2}$ level equations. A particular solution is given by

$$w_{3/2} = 0$$

$$\phi_{3/2} = d(X) \sin \gamma x \cos \beta\gamma y + e(X) \sin (1+\beta^2)\gamma x + f(X) \sin \beta^2\gamma x \cos \beta\gamma y, \tag{43}$$

where

$$d = -\frac{2(1-\beta^2)k}{\gamma(1+\beta^2)} A', \qquad e = -\frac{2k}{\gamma(1+\beta^2)} B', \qquad f = \frac{2(1-\beta^2)k}{\gamma\beta^2(1+\beta^2)} C'. \tag{44}$$

Suppression of secular terms at the s^2 level then leads to the three coupled second-order differential equations

$$8(1-\beta^2)^2k^2A'' + 2\lambda_1 A + 3\rho\beta^2 BC = 0$$

$$16(1+\beta^2)^2k^2B'' + 4(1+\beta^2)^2\lambda_1 B + 3\rho\beta^2 AC = 0$$

$$8(1-\beta^2)^2k^2C'' + 2\beta^4\lambda_1 C + 3\rho\beta^2 AB = 0 \tag{45}$$

and we can set $s = \lambda^A - \lambda$, so that $\lambda_1 = -1$. We note that the coefficient of A'' again approaches zero as $\beta \to 1$. For the interaction at the crown of the Koiter circle, we thus conclude that no localized solutions exist. For the remaining values of β, the possibility of localized interactive solutions arises.

In the search for these, one useful technique is to borrow from the work of TOLAND (1986) and others, and take this as a system of equations associated with the Lagrangian

$$\mathscr{L} = T - V, \tag{46}$$

where we can identify a "kinetic energy" part

$$T = 4k^2[(1-\beta^2)^2A'^2 + 2(1+\beta^2)^2B'^2 + (1-\beta^2)^2C'^2] \tag{47}$$

and a "potential" part

$$V = -A^2 - 2(1+\beta^2)^2B^2 - \beta^4C^2 + 3\rho\beta^2 ABC. \tag{48}$$

We note that the quadratic form for T is positive-definite for $\beta \neq 1$. This being a Hamiltonian system, progress in X is along lines

$$\mathscr{H} = T + V = \text{constant}. \tag{49}$$

In the fundamental unbuckled state we can take $T = V = 0$, all amplitudes and first derivatives vanishing. The *symmetric section*, through which localized solutions must pass (HUNT *et al.*, 1989), is defined by $T = A' = B' = C' = 0$. The energy contour $V = 0$ then carries a particular significance; points on this locus in ABC space represent states on the symmetric section with just enough "energy" to arrive at the flat fundamental state with zero kinetic and potential parts. In the search for localized solutions, this is valuable extra information.

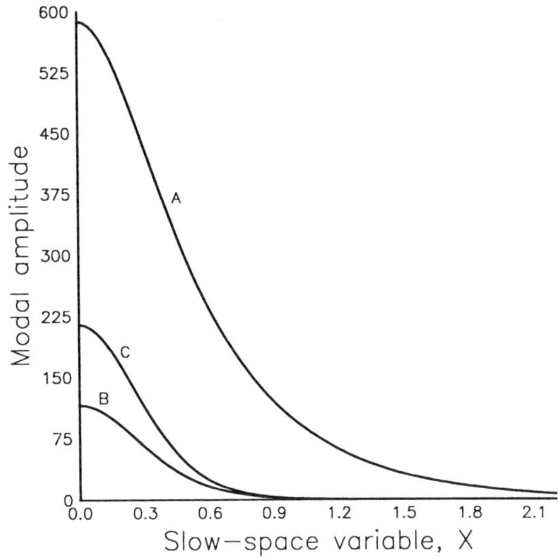

FIG. 3. Runge–Kutta solutions to Eqs (45), showing relative modulated amplitudes to the modes of (41), at $\beta = 2$.

Using a systematic numerical search in two dimensions, from starts both on the symmetric section and zero energy contour, apparent localizations at $\beta = \sqrt{3}$ and $\beta = 2$ have been found; solutions exist for which A, B and C all simultaneously decrease from finite amplitude and drop monotonically towards zero, as seen in Fig. 3. We note that the modal amplitudes found here appear in different proportions from the related periodic solutions and fully anticipate that such solutions exist for other values of $\beta \neq 1$. Further examination is left for later publication.

5. Comparison with Experiments

Initial buckling of the long thin axially-loaded cylinder is known to be violently unstable, with a sharp restabilization once the load has dropped to some fraction (perhaps one-fifth) of the classical critical value. Our first-order analyses pick up only the initial destabilization, not the subsequent restabilization. It is therefore possible to compare against published experiments only the wavelengths and predicted extent of the localization at about the correct load level, not the restabilizing load itself. It is hoped eventually to extend the analysis to include the restabilization, as has been the case with periodic solutions (Hunt *et al.*, 1986).

It is a further experimental observation (Yamaki, 1984) that the equilibrium path between destabilization and restabilization is punctuated by a sequence of secondary bifurcations into longer circumferential waves (see Fig. 4). Thus, although buckling starts on the Koiter circle of Fig. 2, post-buckling is marked by a vertical drop to within the circle, as shown. Because of the severity of the initial instability, even under

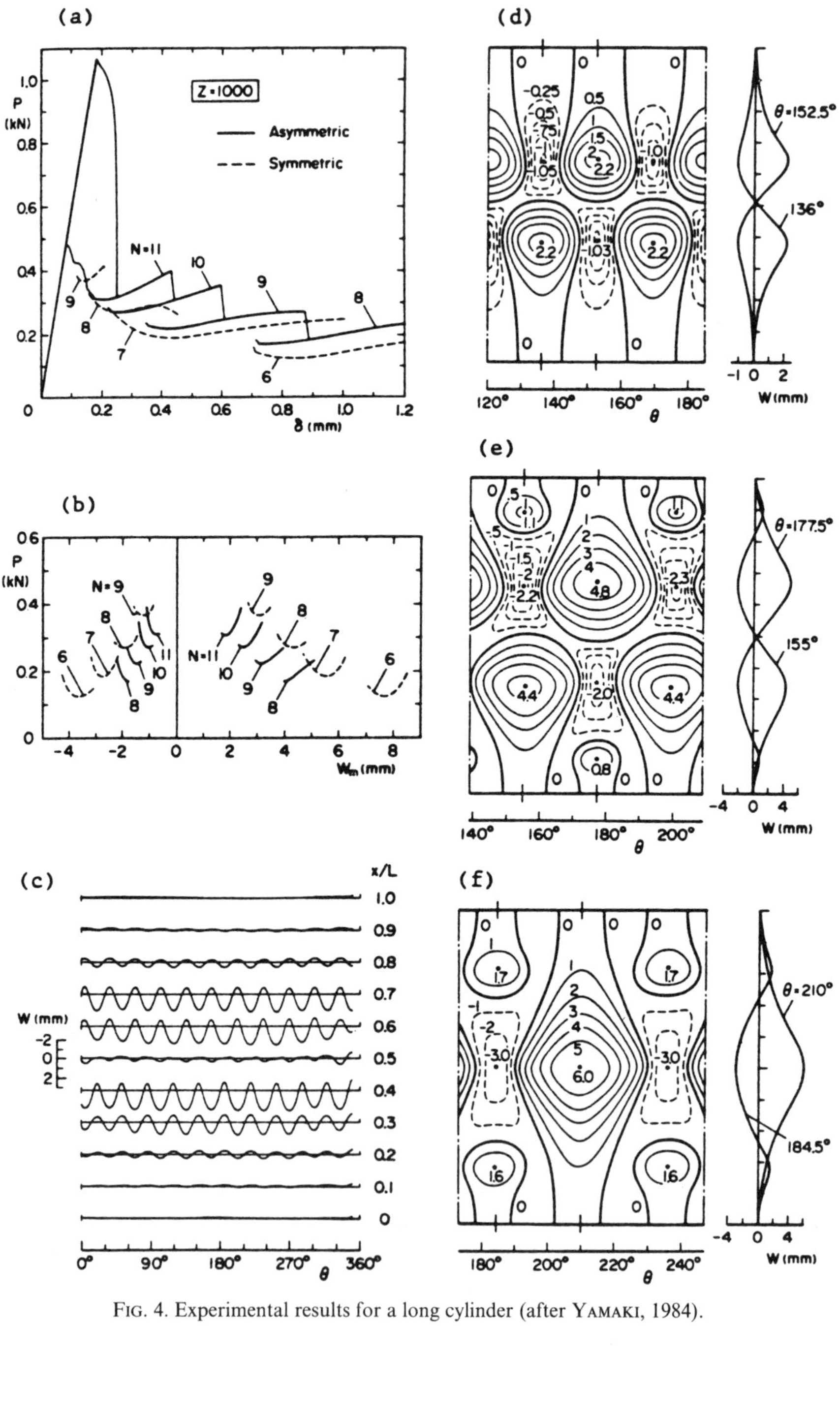

FIG. 4. Experimental results for a long cylinder (after YAMAKI, 1984).

rigid loading, the triggering bifurcation and its corresponding wavelengths are unlikely to be seen experimentally.

As an example, let us consider the longest of the experimental specimens of YAMAKI (1984), for which the shell radius $R = 100$ mm, $t = 0.247$ mm, $\nu = 0.3$, and the effective length L is chosen such that the Batdorf parameter $Z = \sqrt{(1-\nu^2)}L^2/Rt = 1000$. Substituting for the known values and rounding to the nearest integer then gives the typical circumferential wavenumbers for positions on the Koiter circle

$$n_{\beta=1} = 18, \qquad n_{\beta=\sqrt{3}} = n_{\beta=1/\sqrt{3}} = 16, \qquad n_{\beta=2} = n_{\beta=1/2} = 15. \tag{50}$$

Experimental results are given in Fig. 4, taken directly from Fig. 3.52e of YAMAKI (1984). Two types of deflection pattern are observed, symmetric and asymmetric; both are reproducible by the above analysis, the latter by replacing the cosine functions of Eq. (30) with equally valid sine functions. For the symmetric case, the highest circumferential wavenumber observed experimentally is 9. As the post-buckling develops, the sequence of secondary bifurcations is seen to bottom-out at $n = 6$; deflections at $n = 7$ are shown at the bottom-right of Fig. 4, and crude wavelength measurements indicate that here the diamond wavepattern is close to being of the "square" form $\beta = 1$. By projecting upwards from the appropriate point on Fig. 2, tracing in reverse the sequence of secondary bifurcations, the point of initial instability on the Koiter circle can be approximated; this suggests a triggering bifurcation at $n = 14$ and $\beta \approx 2.2$. If we assume a post-buckling load of about one-fifth of the classical critical load, the range of localization seen in Figs 1 and 3 is in each case over about one to one-and-a-half axial wavelengths; although no account is taken of the restabilization, nor the sequence of secondary bifurcations, it is encouraging to note that this is exactly as reported experimentally (SENDELBECK *et al.*, 1967; YAMAKI, 1984).

6. CONCLUDING REMARKS

We offer in this paper two complementary analyses for the initial post-buckling of long thin axially-loaded cylinders, the first assuming distinct critical behaviour (THOMPSON and HUNT, 1973, 1984), and the second allowing interaction between modes. The approach is distinguishable from earlier perturbation analyses by the fact that amplitudes are left free to modulate, which then opens the way for localized solutions. Some general conclusions can be drawn.

The most significant feature, which is reflected in published experiments on long cylinders (see Fig. 4), is that localization forces the critical modes away from the crown of the Koiter circle. It has been realized for some time that the initial triggering buckle pattern comprises long axial waves, but adequate explanations apparently have been lacking. The answer is of course found in the form of Eqs (40) and (45).

Extension of the analysis to find the minimum localized post-buckling load, via the sequence of secondary bifurcations, certainly seems possible. COWELL (1986) develops a perturbation approach for the interaction between the periodic crown mode and its axially-symmetric counterpart, which in one step includes both initial destabilization and subsequent restabilization of the post-buckling path. A similar approach, but allowing for modulating amplitudes and interactions away from the crown, could prove most rewarding.

REFERENCES

AMAZIGO, J. C. and FRASER, W. B. 1971 *Int. J. Solids Structures* **7,** 883.

BRUSH, D. O. and ALMROTH, B. O. 1975 *Buckling of Bars, Plates, and Shells.* McGraw-Hill, New York.

COWELL, R. G. 1986 *Dyn. Stab. Systems* **1,** 115.

DONNELL, L. H. 1934 *Trans ASME Aero. Engng* **AER-56-12,** 795.

HANSEN, J. S. 1975 *Int. J. Solids Structures* **11,** 1223.

HOFF, N. J., MADSEN, W. A. and MAYERS, J. 1966 *AIAA Jl* **4,** 126.

HUNT, G. W., WILLIAMS. K. A. J. and COWELL, R. G. 1986 *Int. J. Solids Structures* **22,** 1501.

HUNT, G. W., BOLT, H. M. and THOMPSON, J. M. T. 1989 *Proc. R. Soc. Lond.* **A425,** 245.

KOITER, W. T. 1945 Dissertation, Delft, The Netherlands.

LANGE, C. G. and NEWELL, A. C. 1971 *SIAM J. Appl. Math.* **21,** 605.

NEWELL, A. C. and WHITEHEAD, J. A. 1969 *J. Fluid Mech.* **38,** 279.

POTIER-FERRY, M. 1983 *Collapse: the Buckling of Structures in Theory and Practice* (IUTAM Symposium, London, 1982) (edited by J. M. T. THOMPSON and G. W. HUNT), p. 149. Cambridge University Press, Cambridge.

SEGEL, L. A. 1969 *J. Fluid Mech.* **38,** 203.

SENDELBECK, R. L., CARLSON, R. L. and HOFF, N. J. 1967 Report SUDAAR 318, Dept of Aeronautics and Astronautics. Stanford University.

THOMPSON, J. M. T. and HUNT, G. W. 1973 *A General Theory of Elastic Stability.* John Wiley, London.

THOMPSON, J. M. T. and HUNT, G. W. 1984 *Elastic Instability Phenomena.* John Wiley, Chichester.

TOLAND, J. F. 1986 *Am. Math. Soc., Proc. Symp. Pure Math.* **45,** Part 2, 447.

VON KÁRMAN, T. and TSIEN, H. S. 1941 *J. Aero. Sci.* **8,** 303.

YAMAKI, N. 1984 *Elastic Stability of Circular Cylindrical Shells.* Elsevier, Amsterdam.

Computation of localized post buckling in long axially compressed cylindrical shells

BY G. J. LORD[1], A. R. CHAMPNEYS[1] AND G. W. HUNT[2]
[1]*Department of Engineering Mathematics, University Walk, University of Bristol, Bristol BS8 1TR, UK*
[2]*Department of Mechanical Engineering, University of Bath, Bath BA2 7AY, UK*

Buckling is investigated of a long thin cylindrical shell under longitudinal compression as modelled by the von Kármán–Donnell equations. Evidence is reviewed for the buckling being localized to some portion of the axial length. In accordance with this observed behaviour the equations are first approximated circumferentially by a Galerkin procedure, whereupon cross-symmetric homoclinic solutions of the resulting system of ordinary differential equations are sought in the axial direction. Results are compared with experimental and other numerical data. Excellent agreement with experiments is achieved with fewer approximating modes than other methods.

1. Introduction

The buckling of a long thin circular cylindrical shell under axial compression is one of the classic problems of structural mechanics. If overall (strut or Euler) buckling is suppressed, experiments under rigid loading indicate a highly unstable snapback response, as described, for example, in Donnell (1934) or Yamaki (1984), in which a finite-amplitude buckle pattern appears instantaneously, accompanied by a dramatic drop in load (see figure 1). In addition, the system is notoriously imperfection sensitive (Koiter 1945). To quote Donnell & Wan (1950), referring to a particular set of experiments:

> Buckling comes suddenly, almost explosively, and usually occurs over only part of the wall.... The number of waves around the circumference is of the order of ten or so, and the wave shape ratio (ratio of wavelength in the circumferential to that in the axial direction) is always close to unity, with an average value of about 0.75.

This suggests that, while periodicity governs circumferentially, axially the buckle pattern may localize. A standard analytical approach adopted by Koiter (1945) and others is to assume that this pattern comprises a complex interaction of a number of linear *modes*, associated with zero eigenvalues at the critical load of the perfect system, that are periodic both axially and circumferentially. The governing equations do indeed admit such solutions, but it has emerged recently via double-scale asymptotic analysis that axially localized solutions are also admissible (Hunt & Lucena Neto 1991, 1993). Such behaviour has only been validated close to the theoretical critical load, yet real (imperfect) shells buckle at some fraction, perhaps 25%, of that load. This thoroughly unstable post-buckling characteristic led in the 1960s to

Phil. Trans. R. Soc. Lond. A (1997) **355**, 2137–2150
Printed in Great Britain

TEX Paper

a search for the so-called *minimum post-buckling load* that can be sustained in the buckled regime, a quest effectively ended by Hoff *et al.* (1966), who demonstrated for outright periodicity that it tends to fall to zero as shell thickness approaches zero. This would be expected from the periodic form of the *Yoshimura* or *diamond* pattern (Yoshimura 1930), that can be folded out of a flat sheet of paper and thus exist at zero load without membrane stretching.

For responses that localize axially, none of these arguments need apply. The need for non-zero membrane strain energy in a *localized* Yoshimura state means that some external loading is required, and the minimum localized post-bucking load is thus greater than zero. Hunt & Lucena Neto (1993) suggest, purely on asymptotic grounds, that the Maxwell critical load for periodic buckling (when energy levels are the same in the fundamental and restabilized post-buckled states) might represent a good approximation to the minimum post-buckling load for an associated localized response. Experimental correlation based on a small sample set (Yamaki 1984) is good, but until now numerical confirmation away from the asymptotic limit has been lacking.

The most widely used model for the buckling of thin cylinders rests with the von Kármán–Donnell equations, a coupled pair of nonlinear fourth-order partial differential equations (PDEs) governing radial displacement, w, and in-plane stress function, ϕ. We consider an infinitely long cylinder and discretize the von Kármán–Donnell equations circumferentially by the Galerkin spectral method. This yields a system of ordinary differential equations (ODEs) in the axial variable x, for which we seek homoclinic (axially localized) solutions, with the axial length taken to be infinite. We claim that this infinite-length approximation is valid when one seeks localized solutions. This is confirmed below by the agreement found with experimental results on only moderately long cylinders.

The key idea behind the direct numerical methods for computing homoclinic orbits to equilibria is to pose the homoclinic orbit as a boundary value problem and then use the linearization about the fundamental state to define boundary conditions with the correct asymptotic behaviour on a truncated interval (Beyn 1990; Friedman & Doedel 1994). The technique of reducing to a set of ODEs to find localized solutions in elliptic equations has used by other authors, such as Mielke (1991), Hagstrom & Keller (1987) and Kirchgässner (1982).

Here the approach is adapted to a Galerkin approximation of the von Kármán–Donnell equations which retains circumferential periodicity. We seek solutions axially that are *cross-symmetric* (symmetric in even-numbered waves and anti-symmetric in odd-numbered waves) about a section through the cylinder, as observed experimentally (Yamaki 1984); results on symmetric and other possible forms will be presented elsewhere. Buckling paths are investigated as the load parameter is varied by numerical continuation using the code AUTO (Doedel & Kernevez 1986). A detailed presentation of the numerical homotopies required to obtain an initial solution at fixed load is deferred to later work.

The remainder of the paper is outlined as follows. Section 2 introduces the von Kármán–Donnell equations and their approximation: properties of low-order approximations are examined, and their numerical computation and suitability discussed. In § 3 we present numerical results for approximations of increasing order. Section 4 discusses the relevance of these results in the context of other work on cylinder buckling, and draws some general conclusions.

2. The von Kármán–Donnell equations and their approximation

Consider an infinitely long cylinder of radius R and shell thickness t. The classical equilibrium equations for the in-plane stress function ϕ and outward radial displacement displacement w in the post-buckling regime of the cylinder are given by the von Kármán–Donnell equations:

$$\kappa^2\nabla^4 w + \lambda w_{xx} - \rho\phi_{xx} = w_{xx}\phi_{yy} + w_{yy}\phi_{xx} - 2w_{xy}\phi_{xy}, \tag{2.1}$$

$$\nabla^4\phi + \rho w_{xx} = (w_{xy})^2 - w_{xx}w_{yy}, \tag{2.2}$$

where ∇^4 denotes the two dimensional biharmonic operator, $x \in \mathbb{R}$ is the axial and $y \in [0, 2\pi R]$ is the circumferential coordinate. The parameters appearing in (2.1) and (2.2) are the curvature $\rho := 1/R$,

$$\kappa^2 := t^2/12(1-\nu^2),$$

where ν is Poisson's ratio and the bifurcation parameter

$$\lambda := P/Et,$$

where P is the compressive axial load applied per unit length and E is Young's modulus. Equations (2.1) and (2.2) are supplemented with periodic boundary conditions in y and asymptotic boundary conditions in the axial direction x: that w, ϕ and their x derivatives tend to zero (the fundamental solution) as $x \to \pm\infty$.

A standard periodic analysis of the von Kármán–Donnell equations is to seek the minimum load and corresponding axial and circumferential wavelengths such that a bifurcation occurs. Waves for which this is satisfied appear on the Koiter circle (Koiter 1945). For a discussion of mode interaction on the Koiter circle for the von Kármán–Donnell equations see Hunt & Lucena Neto (1991).

As stated in the introduction we perform a Galerkin approximation in the circumferential direction. We seek even periodic solutions to the von Kármán–Donnell equations in the circumferential direction y and hence use the cosine functions $\cos(m\rho y)$, $m \in \mathbb{N} \cup \{0\}$ as the basis functions in the Galerkin approximation.

The system (2.1) and (2.2) has a rich structure of symmetries (see Golubitsky *et al.* 1984; Hunt 1986; Hunt *et al.* 1986; Wohlever & Healey 1995). In order to examine deformations of the cylinder that remain within the subspace corresponding to invariance under rotation through $2\pi/s$, we introduce a seed mode $\psi_1^s = \cos(s\rho y)$ and let

$$w(y) = \sum_{m=0}^{\infty} a_m \psi_m^s, \quad \phi(y) = \sum_{m=0}^{\infty} b_m \psi_m^s,$$

where

$$\psi_m^s = \cos(ms\rho y), \quad m \in \mathbb{N} \cup \{0\}, \quad s \in \mathbb{N}.$$

Substituting into the von Kármán–Donnell equations, taking the L^2 innerproduct and expanding the nonlinear terms we find the following system of ODEs:

$$\kappa^2\left\{\frac{\partial^4 a_m}{\partial x^4} - 2s^2\rho^2 m^2\frac{\partial^2 a_m}{\partial x^2} + s^4\rho^4 m^4 a_m\right\} + \lambda\frac{\partial^2 a_m}{\partial x^2} - \rho\frac{\partial^2 b_m}{\partial x^2}$$
$$= s^2\rho^2\chi\left\{-\frac{1}{2}\tilde{\sum_{k,\ell,m}}\left(\ell^2\frac{\partial^2 a_k}{\partial x^2}b_\ell + k^2 a_k\frac{\partial^2 b_\ell}{\partial x^2}\right) - \sum_{k,\ell,m}{}' k\ell\frac{\partial a_k}{\partial x}\frac{\partial b_\ell}{\partial x} + \sum_{k,\ell,m}{}'' k\ell\frac{\partial a_k}{\partial x}\frac{\partial b_\ell}{\partial x}\right\}, \tag{2.3}$$

$$\frac{\partial^4 b_m}{\partial x^4} - 2s^2\rho^2 m^2 \frac{\partial^2 b_m}{\partial x^2} + s^4\rho^4 m^4 b_m + \rho\frac{\partial^2 a_m}{\partial x^2}$$
$$= s^2\rho^2 \tfrac{1}{2}\chi\left\{\sum_{k,\ell,m}^{\sim} \ell^2 \frac{\partial^2 a_k}{\partial x^2} a_\ell + \sum_{k,\ell,m}^{\prime} k\ell \frac{\partial a_k}{\partial x}\frac{\partial a_\ell}{\partial x} - \sum_{k,\ell,m}^{\prime\prime} k\ell \frac{\partial a_k}{\partial x}\frac{\partial a_\ell}{\partial x}\right\}, \tag{2.4}$$

for $m = 0, 1, 2, \ldots, \infty$, and where

$$\chi = \begin{cases} \frac{1}{2}, & m = 0, \\ 1, & \text{otherwise.} \end{cases}$$

The summations terms in (2.3) and (2.4) are defined for a given value of m by:

$$\sum_{k,\ell,m}^{\sim} := \sum_{\substack{k+\ell-m=0,\ k-\ell+m=0 \\ k-\ell-m=0,\ k,\ell\in\mathbb{N}\cup 0}}, \quad \sum_{k,\ell,m}^{\prime} := \sum_{\substack{k-\ell+m=0,\ k-\ell-m=0 \\ k,\ell\in\mathbb{N}\cup 0}}, \quad \sum_{k,\ell,m}^{\prime\prime} := \sum_{\substack{k+\ell-m=0 \\ k,\ell\in\mathbb{N}\cup 0}}.$$

The Galerkin approximation is formed by taking equations (2.3) and (2.4) for $m = 0, \ldots, M-1$ only, for some finite M. Taking $s = 1$ corresponds to a standard Galerkin approximation discussed, for example, in Gottlieb & Orszag (1977), whereas $s > 1$ corresponds to seeking a solution in a specific subspace corresponding to the circumferential wave number s.

Note that there is a 'degeneracy' in the equations for the zero mode ($m = 0$) such that these could be solved with initial conditions for

$$\frac{\partial^2 a_0}{\partial x^2}, \frac{\partial^3 a_0}{\partial x^3}, \frac{\partial^2 b_0}{\partial x^2}, \frac{\partial^3 b_0}{\partial x^3},$$

independently of the initial conditions for a_0, b_0 and the first derivatives. This corresponds to a trivial translational symmetry in the problem.

The single mode approximation found by taking $M = 1$ ($m = 0$) in (2.3) and (2.4) yields a linear system for which there are no homoclinic solutions. Thus the simplest approximation for which we may expect to find a homoclinic orbit is the two mode approximation found by taking $M = 2$ ($m = 0, 1$) in (2.3) and (2.4). We truncate the system of ODEs (2.3) and (2.4) to a large finite interval, $x \in [0, T]$ and solve as a boundary value problem over half the length of the cylinder.

Boundary conditions are chosen at $x = 0$ to project out a linear approximation to the stable manifold of $a_k = b_k = 0$. This imposes the condition that the solution would leave the zero state along an unstable direction, as appropriate for the infinite length case (see, for example, Beyn 1990).

At $x = T$ we impose appropriate boundary conditions on the Fourier modes for solving over half the length of the cylinder. A number of choices, for example those exhibiting complete axial symmetry, are available. Here we attempt to match experiments by seeking solutions that we call (axially) *cross-symmetric*. These are solutions $w(x, y)$, $\phi(x, y)$ which are invariant under the simultaneous application of $x \to 2T - x$ and $y \to y + \pi R/s$, but are in fact produced by symmetric conditions for the even-numbered modes and anti-symmetric conditions for the odd-numbered modes. An example of a cross-symmetric solution may be seen in the contour plot of figure 5. In contrast we refer to (axially) *symmetric* solutions as solutions which are symmetric in y about $x = T$ (i.e. $w(x, y)$ and $\phi(x, y)$ are invariant under $x \to 2T - x$ and $y \to y$).

In the parameter regime of interest the linearization of (2.3), (2.4) about the trivial solution yields four zero eigenvalues, $(4M - 2)$ eigenvalues in the left half plane and

$(4M-2)$ eigenvalues in the right half plane. The non–zero eigenvalues occur in complex conjugate pairs. We see from the linearization that the equilibrium is not hyperbolic and instead has a centre-stable and centre-unstable manifolds. However, by inspection it is clear that these, the zero eigenvalues, and associated directions are a consequence of the 'degeneracy' in the zero mode. Thus, although we solve for the full system we consider the reduced linear problem for computing the projection boundary conditions at $x=0$. The degeneracy in the zero mode forces the boundary conditions for that mode to be modified. The boundary value problem we solve is as follows. Rescale (2.3), (2.4) to the interval $[0,1]$ and solve subject to the following.

(i) Boundary conditions at $x=0$ for the zero mode $(m=0)$

$$a_0(0)=0, \quad \frac{\partial a_0(0)}{\partial x}=0, \quad b_0(0)=0, \quad \frac{\partial b_0(0)}{\partial x}=0. \tag{2.5}$$

These four conditions impose that the solution is homoclinic to the flat equilibrium state, rather than some translate of it.

(ii) Standard projection boundary conditions for the non-zero modes $(m>0)$ at $x=0$:

$$L_\mathrm{s}(U(0)-0)=0, \tag{2.6}$$

where

$$U(0):=\left[a_1(0),\dots,\frac{\partial^3 a_1(0)}{\partial x^3},\dots,a_{M-1}(0),\dots,\frac{\partial^3 a_{M-1}(0)}{\partial x^3},\right.$$
$$\left.b_1(0),\dots,\frac{\partial^3 b_1(0)}{\partial x^3},\dots,b_{M-1}(0),\dots,\frac{\partial^3 b_{M-1}(0)}{\partial x^3}\right]^\mathrm{T}$$

and L_s is the matrix whose rows span the left eigenspace of the corresponding stable (s) eigenvalues. This gives $4M-2$ conditions.

(iii) Cross-symmetric section boundary conditions at $x=1$ given by

$$\frac{\partial^2 a_0(1)}{\partial x^2}=0, \quad \frac{\partial^2 b_0(1)}{\partial x^2}=0, \tag{2.7}$$

$$\frac{\partial^i a_m(1)}{\partial x^i}=0, \quad \frac{\partial^{i+2} a_m(1)}{\partial x^{i+2}}=0, \quad \frac{\partial^i a_m(1)}{\partial x^i}=0, \quad \frac{\partial^{i+2} b_m(1)}{\partial x^{i+2}}=0, \tag{2.8}$$

for $m=1,\dots,M-1$ and $i=0$ if m is even and $i=1$ if m is odd. These give a further $4M-2$ boundary conditions.

In total we have $8M$ boundary conditions for $8M$ equations, which is the appropriate number for a well-posed problem. This boundary value problem can be solved by a regular continuation code, such as AUTO (Doedel & Kernevez 1986), to compute load–deflection bifurcation diagrams as in figure 2.

Initial approximations for the boundary value problems are found using a mixture of shooting and homotopy techniques. The details of the boundary conditions and the numerical techniques used for initial approximations will be discussed in later work where symmetric and other forms of solution will be considered.

3. Numerical Results

To compare with the experiments of Yamaki (1984) and the asymptotic analysis of Hunt & Lucena Neto (1993) calculations were performed for a shell with

$$\rho=0.01\,\mathrm{mm}^{-1}, \quad t=0.247\,\mathrm{mm}, \quad \nu=0.3, \quad E=5.56\,\mathrm{GPa}.$$

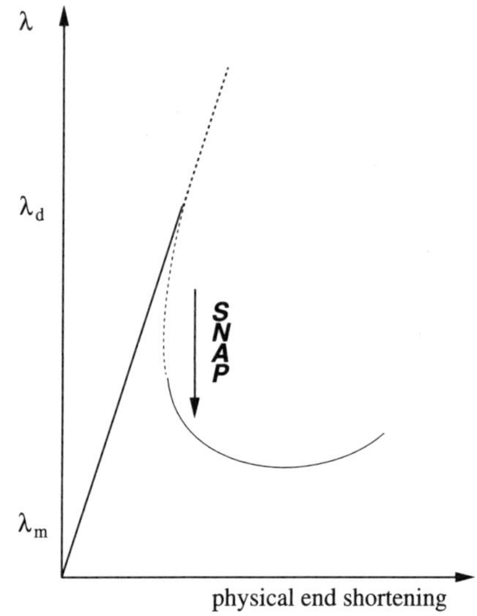

Figure 1. Schematic representation of expected response. The straight line represents the unbuckled state and the curved line the buckled state. Solid lines are stable, and dashed lines unstable, equilibrium states under rigid loading conditions.

Yamaki's experiments were carried out for clamped cylindrical shells —we consider here results from the longest of such shells, length $L = 160.9\,\mathrm{mm}$. Note that even this could hardly be described as long, since its aspect ratio (length to diameter) is only $L/2R \approx 0.8$. Yamaki found buckle patterns which were both symmetric and cross-symmetric; the solutions presented here will all be cross-symmetric.

We compare with experiments the minimum (W_{min}) and maximum (W_{max}) values of the displacement w, and the ratios

$$\lambda_{\mathrm{m}}/\lambda_{\mathrm{d}}, \quad \beta = \frac{\text{axial wavelength at } \lambda_{\mathrm{m}}}{\text{circumferential wavelength at } \lambda_{\mathrm{m}}},$$

where λ_{d} denotes the smallest value of λ at which the fundamental solution bifurcates, and λ_{m} corresponds to the minimum post buckling load. For the Yamaki cylinder described above it is easily shown that $\lambda_{\mathrm{d}} \approx 1.494912 \times 10^{-3}$. For the numerical simulations, λ_{m} was taken to be the first limit point on the branch of homoclinics as the load was decreased (corresponding to the minimum of the curves in figure 2). For the *aspect ratio*, β, only the axial wavelength was required from the simulations, since the circumferential wave number and hence corresponding wavelength were known. The axial wavelength was defined to be twice the axial distance between a maximum and a minimum, that is twice the distance between the point P and Q on the contour plot (figure 5). This quantity was estimated from the line $y = 0$.

In the bifurcation diagram (figure 2), we plot the loading parameter, λ, against a measure of the end shortening defined by

$$\int_{-\infty}^{\infty} |w_x(x,y)|\,\mathrm{d}x = \int_{-\infty}^{\infty} \left| \sum_{k=0}^{\infty} \frac{\partial}{\partial x}(a_k(x)\cos(ks\rho y)) \right| \mathrm{d}x,$$

which in general varies with y. For simplicity we take $y = 0$ and measure over the

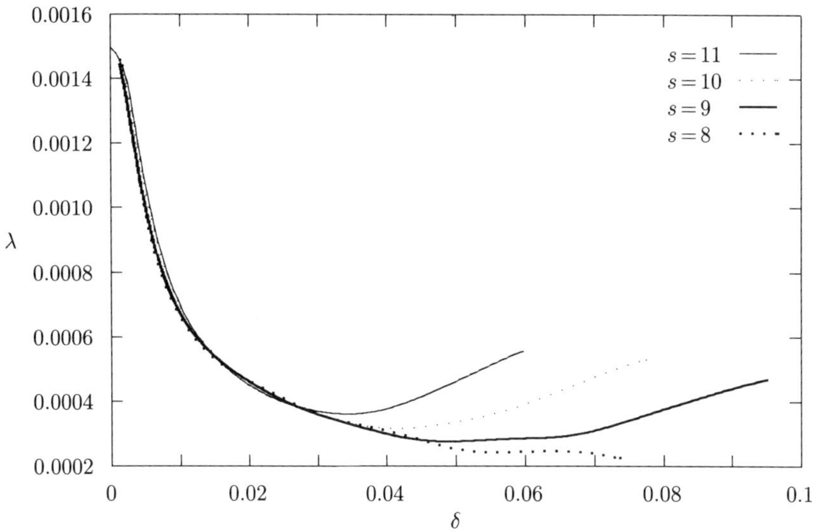

Figure 2. Bifurcation diagram showing the load parameter λ against a measure of end-shortening δ for wave numbers s = 8, 9, 10 and 11.

Table 1. *Convergence of the minimum load λ_{m}, ratio of wavelengths β and the minimum and maximum displacements W_{min} and W_{max} as the number of modes in the approximation is increased*

$s = 11$	$M = 2$	$M = 3$	$M = 4$	$M = 5$	$M = 6$
$\lambda_{\mathrm{m}}/\lambda_{\mathrm{d}}$	0.1515	0.1938	0.2321	0.2436	0.2418
β	2.123	1.819	1.571	1.789	1.741
W_{min}	−0.742	−0.946	−1.018	−0.839	−0.866
W_{max}	1.622	1.959	2.192	1.919	1.966

half length of the cylinder to get,

$$\delta = \int_0^{\mathrm{T}} \left| \sum_{k=0}^{M-1} \frac{\partial a_k}{\partial x} \right| \mathrm{d}x. \tag{3.1}$$

We use this as a *measure* of the end-shortening, but note that it bears no direct relation to *actual* end shortening of a specimen cylinder: more physically meaningful measures will be considered in future work. As the pure-squash component of δ is absent from the present formulation, the 'snapback' phenomenon indicated in the schematic diagram of figure 1 is only implied in the plotted bifurcation diagram of figure 2. The λ axis of figure 2 thus corresponds to the sloped line in the schematic figure 1. Note that in practice one would only expect to see the stable portions of the curves, as for example in Yamaki's experiments.

We note that in the computations below the half length of the cylinder T was either taken to be $T = 300$ or, for direct comparison with Yamaki, $T = 100$. The values for λ_{m}, W_{min}, W_{max} and β were found to be essentially independent of the length of the interval T solved over in accordance with results of Beyn (1990), Schecter (1993) and

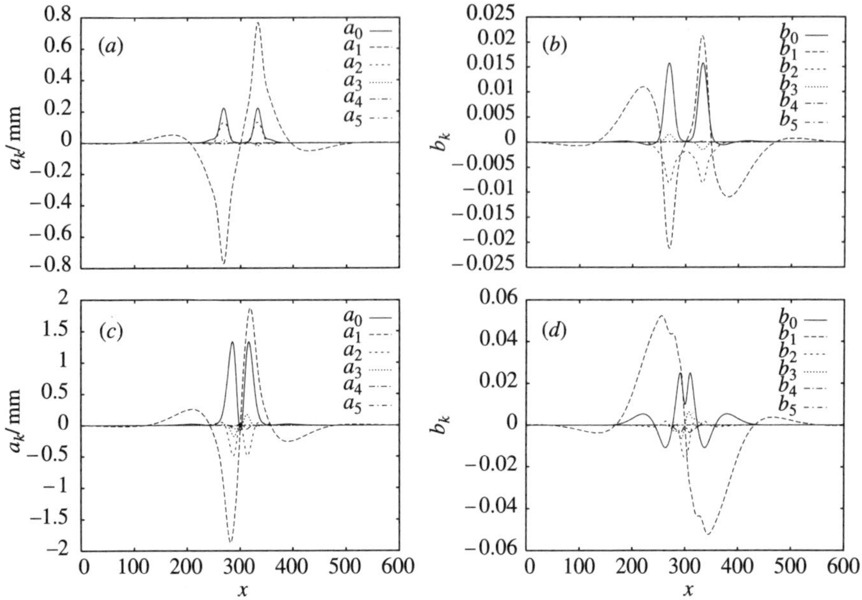

Figure 3. Fourier modes a_k and b_k for the displacement, w, and stress function, ϕ, respectively, at $\lambda = 5 \times 10^{-4}$. (*a*), (*b*) As the load is decreased from $\lambda_{\rm d}$ before the minimum load $\lambda_{\rm m}$ and (*c*), (*d*) after the minimum load $\lambda_{\rm m}$.

Sandstede (1996). In the figures that follow, x is plotted on $[0, 2T]$, y is plotted on $[0, 2\pi R]$ and x, y, $W_{\rm min}$, $W_{\rm max}$ and the Fourier coefficients a_k are measured in mm.

(*a*) *Results*

We compare our results to the cross-symmetric results of Yamaki (1984). For the shell under consideration Yamaki describes in some detail the case of 11 circumferential waves ($s = 11$) close to the minimum load $\lambda_{\rm m}$. The case $s = 11$ was also considered by Hunt & Lucena Neto (1993); although dependent on R/t, this is typical of the number of circumferential modes observed experimentally (recall the quote from Donnell & Wan (1950) in the introduction). We present results for $s = 8$, 9, 10 and 11 and consider in greater detail the case $s = 11$.

First we examine convergence of the numerics as the number of circumferential modes used in the approximation is increased. The computations were performed for wave number $s = 11$. In table 1 we present values of $\lambda_{\rm m}/\lambda_{\rm d}$, β, and minimum and maximum values of the displacement, $W_{\rm min}$ and $W_{\rm max}$, as M is increased to a maximum of six. We see that $M = 6$ appears to give convergence to one decimal place in all four quantities. All further results presented were computed with $M = 6$.

In figure 2 we have plotted the bifurcation diagram of load, λ, versus end shortening, δ, for $s = 8$, 9, 10 and 11. Note from the diagram that as the circumferential wave number decreases, the minimum load $\lambda_{\rm m}$ decreases and the end shortening at this minimum load increases. This corresponds with experimental evidence depicted in figure 3.52e (*a*) of Yamaki (1984). Unlike the experimental solutions, our numerical solutions do not 'feel' their stability, and hence we are able to compute the bifurcation curves back to the originating bifurcation point $\lambda_{\rm d}$.

We next present detailed results for wave number $s = 11$. Figure 3 shows the

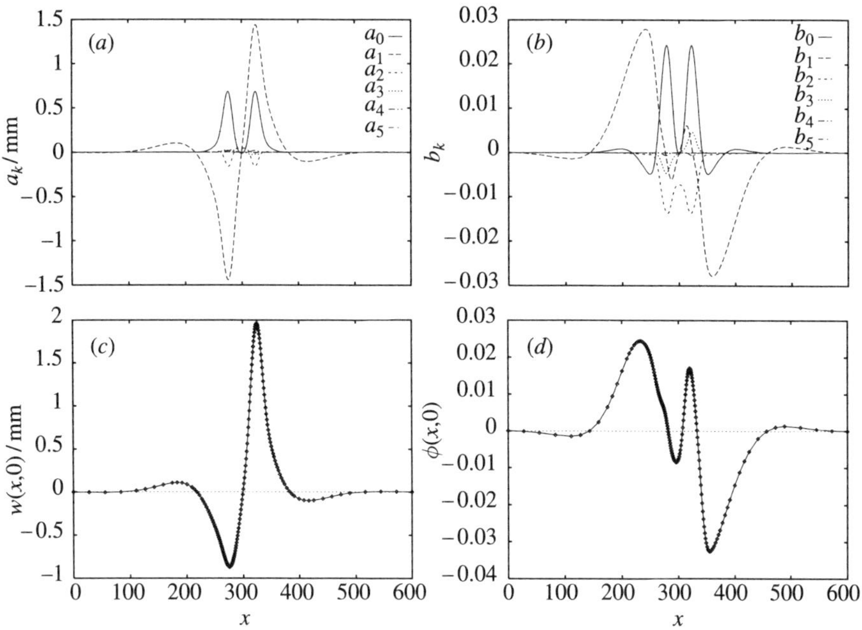

Figure 4. At the minimum load λ_m. (*a*) Fourier modes a_k for displacement w; (*b*) Fourier modes b_k for the stress function ϕ; (*c*) Reconstructed solution $w(x,0)$; and (*d*) $\phi(x,0)$.

Fourier modes a_k, b_k, $(k = 0,1,2,3,4,5)$ for the displacement w and stress function ϕ, respectively, before and after the minimum load λ_m. In (*a*) and (*b*) a_k and b_k are plotted at $\lambda = 5 \times 10^{-4}$ before the minimum load λ_m is reached (decreasing λ from λ_d) and in (*c*) and (*d*) a_k and b_k are re-plotted at the same value of λ but after passing through λ_m. Note that the solutions appear quite different. Comparing (*a*) to (*c*) and (*b*) to (*d*) shows that the magnitude of both the Fourier modes a_k and b_k is larger having passed through λ_m and that the Fourier modes appear more localized.

Figure 4 depicts the solution when the load λ is at the minimum load λ_m. In figure 4*a* we have plotted the Fourier modes a_k for the displacement w and in (*b*) the modes b_k for the stress function ϕ. In figure 4*c* we have reconstructed the solution $w(x,y)$ at $y = 0$ and in (*d*) we have plotted the $\phi(x,y)$ at $y = 0$.

Figure 5 shows the contour map of the displacement, w, plotted over the cylinder at the minimum load λ_m, with the values of certain contours marked. The cross-symmetric nature of the solution over the whole cylinder is clearly evident from this plot. The circumferential wave number, s, is easily verified as 11 and we have indicated on the diagram half the axial wavelength—the distance between P and Q. The reader may wish to compare this plot to the contour plots in Yamaki (1984, figure 3.52e (*b*)).

In figure 6 we have reconstructed the full solution $w(x,y)$ and in figure 7 we have reconstructed the full solution $\phi(w,y)$ at the minimum load λ_m for $s = 11$. In each case contours are plotted over the deformed cylinder, with the colour bar indicating the magnitude of w or ϕ. Note that the length of the cylinder is 200 mm which is comparable to the length of the Yamaki cylinder.

We now consider the *quantitative* agreement with the experiments. Table 2 presents a full comparison with the experimental and fully periodic numerical results of Yama-

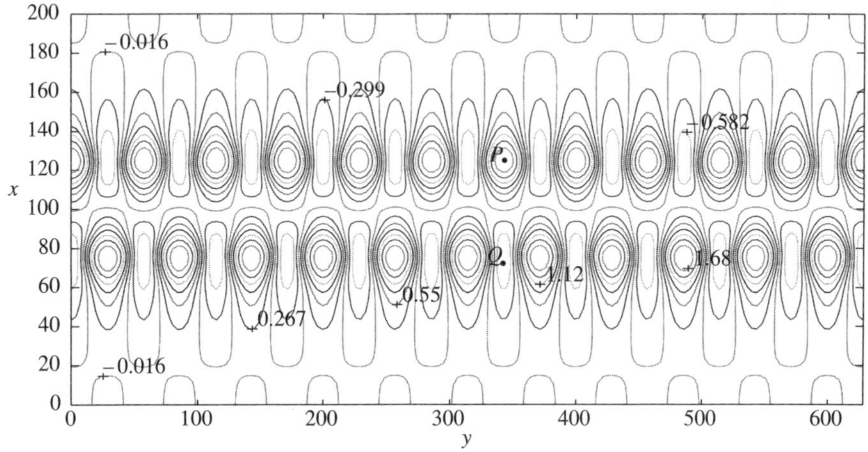

Figure 5. Contour plot for the cross-symmetric form of solution of the outward buckling displacement, w, at the minimum value of the buckling load parameter $\lambda = \lambda_m$. $x \in [0, 2T]$ is the axial coordinate and $y \in [0, 2\pi R]$ is the circumferential coordinate. Also indicated is half axial wavelength (PQ): details in text.

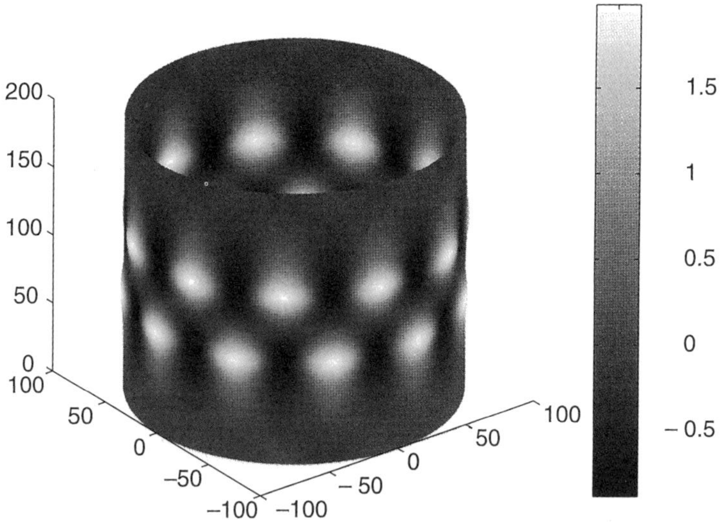

Figure 6. Reconstructed displacement $w(x, y)$ plotted over the deformed cylinder for $s = 11$.

ki (1984), and with the asymptotic results of Hunt & Lucena Neto (1993) based upon the Maxwell concept. For the seed modes $s = 10$ and $s = 11$ we see good agreement in the minimum load λ_m and also in the minimum and maximum displacements W_{min} and W_{max}. However, as the wave number is decreased our results and those of Hunt & Lucena Neto (1993) tend to drift from the experimental values; there are also growing differences in the aspect ratio, β. For $s = 9$ and $s = 8$, the results indicate that the six modes a_0 to a_5 may not be enough for convergence. Boundary conditions would also be expected to have a greater part to play for the longer wavelength modes.

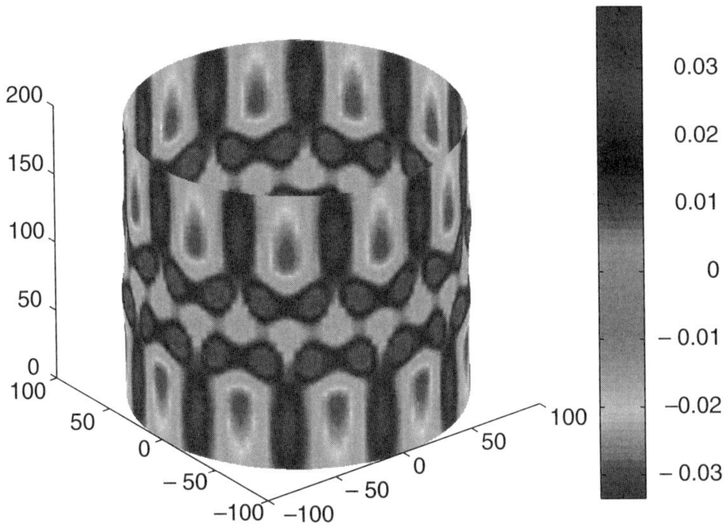

Figure 7. Reconstructed stress function $\phi(x, y)$ plotted over the deformed cylinder for $s = 11$.

Table 2. *Comparison of results*

	s	present $M = 6$	(Yamaki 1984) experimental	periodic	(Hunt & Lucena Neto 1993) asymptotics
λ_m/λ_d	11	0.242	0.24	0.26	0.24
	10	0.213	0.21	0.22	0.21
	9	0.185	0.17	0.17	0.18
	8	0.163	0.14	0.12	0.16
β	11	1.74	1.43[a]	1.54[a]	1.49
	10	1.84			1.51
	9	1.96			1.56
	8	2.13	1.37[a]	1.42[a]	1.60
W_{min}	11	−0.866	−0.9	−0.89	−1.09
	10	−1.075	−1.0	−1.24	−1.36
	9	−1.411	−1.56	−1.56	−1.68
	8	−1.894	−1.9	−1.90	−2.15
W_{max}	11	1.966	1.9	1.90	2.47
	10	2.384	2.6	2.52	3.01
	9	2.852	3.2	3.21	3.73
	8	3.291	4.0	4.03	4.77

[a] Indicates that this was not at a minimum load.

A plausible explanation for why the wavelengths we observe numerically do not match those of experiments is that there is a large sensitivity of β to the value of λ near λ_m and that the experimental wavelengths given in the table are taken for values of the load λ close to, but not exactly at, the minimum load λ_m (compare figure 3.52e (*b*) and table 3.7 pages 231 and 232 in Yamaki (1984)). This sensitivity can be verified from our numerics. Figure 8 shows the load, λ, plotted against the

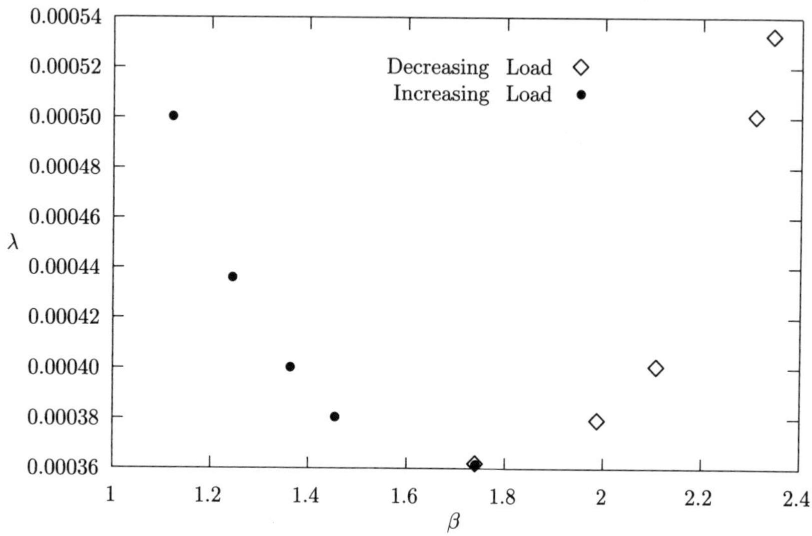

Figure 8. Plot of β as the load is decreased ($\diamond$) to the minimum value λ_m and is increased ($\bullet$) past λ_m.

buckle pattern aspect ratio, β, for $s = 11$ with $M = 6$. As λ is decreased from the bifurcation value λ_d we see that β also decreases. As λ continues through the limit point and starts to increase, β continues to decrease. What is clear is that close to the limit point a small variation in λ leads to a large variation in β. Indeed the curve appears to be well fitted (in the least squares sense) close to the minimum λ_m by a quartic

$$\lambda_p(\beta) = 0.0006\beta^4 - 0.004\beta^3 + 0.01\beta^2 - 0.0121\beta + 0.0058, \tag{3.2}$$

in which case the points λ (shown in figure 8) are given by $\lambda = \lambda_p + \epsilon$ where the error ϵ has a maximum value of $\epsilon = 3.053 \times 10^{-6}$. Furthermore we can test this against the experiments: Yamaki (1984) found that $\beta = 1.43$, which from (3.2) suggests that a load of $\lambda \approx 3.8501 \times 10^{-4}$; in fact the experimental value was $\lambda = 3.743 \times 10^{-4}$. The fast variation of the axial wavelength around the minimum value indicates that there is a large scope for error in estimating this particular parameter and that the data for the Yamaki cylinder is for a value of λ just past the minimum load λ_m.

4. Discussion and conclusion

This paper demonstrates excellent qualitative and quantitative agreement between numerics and experiments. The agreement is particularly good considering our analysis is for an infinitely long cylinder, whereas the comparison is with a moderately long cylinder of aspect ratio (length to diameter) of approximately 0.8. Results are essentially independent of chosen (structural) boundary conditions, the only criterion being that at the boundaries the solution lies either on the stable, or the unstable, manifold. The experimental displacements and the minimum load values are well matched by the numerics. We have shown that the axial wavelength measured by β varies rapidly with λ around the minimum load λ_m, this sensitivity explaining the discrepancy in the values of β.

Particularly good agreement is found for circumferential wave numbers $s = 10$ and

$s = 11$; however, for $s = 8$ and $s = 9$ a slight discrepancy between the numerical and experimental values is observed. Preliminary evidence suggests that, although convergence is achieved for $s = 11$, more modes are required to fully resolve the solution for $s \leqslant 9$. Indeed Yamaki (1984) states that (p. 254) that 'Some discrepancies will be seen for large deformations of long shells ... which may be attributed to the insufficient number of unknown parameters a_{mn} retained in the calculation'. The large deflection cases correspond exactly to the lower wave numbers $s = 8$ and $s = 9$. Thus given the excellent agreement with six modes for the higher wave-number cases we expect improved results with more modes for the lower wave numbers. It must also be remembered that these results are taken at the minimum load. From AUTO this minimum is found accurately (using the limit point detection facility) whereas it has to be estimated both experimentally and numerically in Yamaki (1984). This is certainly another factor in the observed discrepancies.

Our results compare favourably with the asymptotic results of Hunt & Lucena Neto (1993) who used a 16-mode approximation of fully periodic modes, together with an argument based on the Maxwell critical load, to distinguish a localized minimum. In particular we seem to capture the minimum and maximum displacement much better. This is not surprising since for the cross-symmetric solutions the maximum displacement occurs away from the mid-length of the cylinder. Note that Hunt & Lucena Neto also comment that more modes were required for lower wave numbers.

Our findings so far are preliminary and a more complete analysis including symmetric buckling patterns, and an investigation of the Maxwell load hypothesis, will form the subject of future work. We also remark, based on preliminary unreported computations and on previous experience with simpler Hamiltonian systems (Devaney 1976; Champneys & Toland 1993; Hunt *et al.* 1989) that for each mode number, s, the existence of one localized solution implies infinitely many other localized solutions. These are essentially multiple copies of the symmetric and anti-symmetric solutions separated by finite axial distances. The investigation of these multi-modal solutions is also left to future work.

One advantage of our approach is that fewer modes are required to capture the localized buckling than was used in Yamaki (1984). For $s = 10$ and $s = 11$ our six-mode approximation ($M = 6$) is as accurate as his 20-mode approximation. Another advantage of the numerical procedures used here as opposed to a finite element analysis (see, for example, Wohlever & Healey 1995), is that our methods are well suited to exploring multiplicities of solution and understanding the structure not just from the point of view of symmetries, but also from the inherent multiplicities associated with homoclinic bifurcations (see Hunt *et al.* (1997) for a comparison with finite elements in a much simpler problem).

The work in this paper lends weight to the description of the localized buckling of long thin axially compressed cylinders as a homoclinic phenomenon. We have obtained excellent agreement with experimental work on the longest elastic cylinder for which detailed experimental results could be found, with only a few circumferential modes in the numerical approximation. The localization occurs naturally as a solution to the von Kármán–Donnell equations, independently of any imperfection in the cylinder, and demonstrates that the corresponding asymptotic boundary conditions are natural boundary conditions for long cylinders.

The authors thank E. Doedel (Concordia University) for his help in customizing the continuation code AUTO. The research of G.J.L. was supported by a research grant from the EPSRC under the Applied Nonlinear Mathematics Initiative.

References

Beyn, W.-J. 1990 The numerical computation of connecting orbits in dynamical systems. *IMA J. Numer. Analys.* **9**, 379–405.

Champneys, A. R. & Toland, J. F 1993 Bifurcation of a plethora of multi-modal homoclinic orbits fo autonomous Hamiltonian sytems. *Nonlinearity* **6**, 665–772.

Devaney, R. L. 1976 Homoclinic orbits in Hamiltonian systems. *J. Differ. Equat.* **21**, 431–438.

Doedel, E. & Kernevez, J. P. 1986 AUTO: Software for continuation problems in ordinary differential equations with applications. California Institute of Technology Applied Mathematics Technical Report.

Donnell, L. H. 1934 A new theory for buckling of thin cylinders under axial compression and bending. *Trans. ASME Aero. Engng* **56**, 795–806.

Donnell, L. H. & Wan, C. C. 1950 Effect of imperfections on buckling of thin cylinders and columns under axial compression. *J. Appl. Mech.* **49**, 73–83.

Friedman, M. J. & Doedel, E. J. 1994 Numerical computational of invariant manifolds connecting fixed points. *SIAM J. Numer. Analys.* **28**, 789—808.

Golubitsky, M., Marsden, J. & Schaeffer, D. 1984 Bifurcation problems with hidden symmetries. In *Partial differential equations and dynamical systems* (ed. W. Fitzgibbon), pp. 181–210. Boston, MA: Pitman.

Gottlieb, D & Orszag, S. A. 1977 *Numerical analysis of spectral methods: theory and applications.* Philadelphia, PA: SIAM.

Hagstrom, T. M. & Keller, H. B. 1987 Asymptotic boundary conditions and numerical methods for nonlinear elliptic problems on unbounded domains. *Math. Comp.* **48**, 449–470.

Hoff, N., Madsen, W. A. & Mayers, J. 1966 Post-buckling equilibrium of axially compressed circular cylindrical shells. *AIAA Jl* **4**, 126–133.

Hunt, G. W. 1986 Hidden (a)symmetries of elastic and plastic bifurcation. *Appl. Mech. Rev.* **39**, 1165–1186.

Hunt, G. W. & Lucena Neto, E. 1991 Localized buckling in long axially loaded cylindrical shells. *J. Mech. Phys. Solids* **39**, 881–894.

Hunt, G. W. & Lucena Neto, E. 1993 Maxwell critical loads for axially loaded cylindrical shells. *Trans. ASME* **60**, 702–706.

Hunt, G. W., Bolt, H. M. & Thompson, J. M. T. 1989 Structural localisation phenomena and the dynamical phase-space analogy. *Proc. R. Soc. Lond.* A **425**, 245–267.

Hunt, G. W. Lawther, R. & Providência e Costa, P. 1997 Finite element modelling of spatially chaotic structures. *Int. J. Numer. Meth. Engng.* **40**, 2237–2256.

Hunt, G. W., Williams, K. A J. & Cowell, R. G. 1986 Hidden symmetry concepts in the elastic buckling of axially loaded cylinders. *Int. J. Solids Structures* **22**, 1501–1515.

Kirchgässner, K. 1982 Wave solutions of reversible systems and applications. *J. Differ. Equat.* **45**, 113–127.

Koiter, W. T. 1945 On the stability of elastic equilibrium. Ph.D. thesis, University of Delft. (Engl. Transl. Tech. Rep. AFFDL-TR-70-25 Air Force Flight Dyn. Lab. 1970.)

Mielke, A. 1991 *Hamiltonian and Lagrangian flows on center manifolds: with applications to elliptic variational problems.* Springer Lecture Notes in Mathematics, vol. 1489. Berlin: Springer.

Sandstede, B. 1997 Convergence estimates for the numerical approximation of homoclinic solutions. *IMA J. Numer. Analysis* **17**, 437–462.

Schecter, S. 1993 Numerical computation of saddle–node homoclinic bifurcation points. *SIAM J. Numer. Analysis* **30**, 1155–1178.

Wohlever, J. C. & Healey, T. J. 1995 A group theoretic approach to the global-bifurcation analysis of an axially compressed cylindrical-shell. *Comp. Meth. Appl. Mech. Engng* **122**, 315–349.

Yamaki, N. 1984 *Elastic stability of circular cylindrical shells.* Applied Mathematics and Mechanics Series, vol. 27. New York: Elsevier.

Yoshimura, Y. 1930 On the mechanism of buckling of a circular clindrical shell under axial compression. NACA Technical Memorandum.

ELSEVIER

Comput. Methods Appl. Mech. Engrg. 170 (1999) 239–251

Computer methods in applied mechanics and engineering

Homoclinic and heteroclinic orbits underlying the post-buckling of axially-compressed cylindrical shells

G.W. Hunt[a,*], G.J. Lord[b], A.R. Champneys[c]

[a]*Department of Mechanical Engineering, University of Bath, Bath, UK*
[b]*National Physical Laboratory, Teddington, UK*
[c]*Department of Engineering Mathematics, University of Bristol, Bristol, UK*

Abstract

A structural system with an unstable post-buckling response that subsequently restabilizes has the potential to exhibit homoclinic connections from the fundamental equilibrium state to itself over a range of loads, and heteroclinic connections between fundamental and periodic equilibrium states over a different (smaller) range of loads. It is argued that such equilibrium configurations are important in the interpretation of observed behaviour, and govern the minimum possible post-buckling loads.

To illustrate this, the classical problem of a long thin axially-compressed cylindrical shell is revisited from three different perspectives: asymptotic conjecture, analogy with nonlinear dynamics, and numerical continuation analysis of a partial spectral decomposition of the underlying equilibrium equations. The nonlinear dynamics analogy demonstrates that the structure of the heteroclinic connections is more complicated than that indicated by the asymptotics: this is confirmed by the numerics. However, when the asymptotic portrayal is compared to the numerics, it turns out to be surprisingly accurate in its Maxwell-load prediction of the practically-significant first minimum to appear in the post-buckling regime.

1. Introduction

Buckling of elastic structures at symmetric bifurcation points is broadly classifiable as either supercritical or subcritical, depending on whether the bifurcation point itself is stable or unstable [1,2]. For a structure that is long in comparison with the wavelength of the buckle pattern, quite different forms of behaviour would be expected to emanate from each. A supercritical bifurcation typically leads to a post-buckling response that is periodic along the length, but subcritical responses are likely to localize, such that the buckle pattern is spread effectively only over a small range of a spatial length variable x. For such responses, provided the structure is sufficiently long and buckling does not occur close to the boundaries, x may be regarded locally as ranging over an infinite 'time-like' domain. Hence, employing a dynamical systems analogy, as in [3], a localised response represents a so-called *homoclinic connection*, from the unbuckled (fundamental) equilibrium state to itself. Near a subcritical bifurcation spatially periodic responses may also exist, but they require more energy to trigger and hence will not appear in a physical test. Analytically and numerically however, periodic solutions abound, and as a result they often mask the appearance of the localization.

There is a developing body of literature on localization problems in elastic systems, with several recent extensions into related viscous and visco-elastic behaviour (see for example [4] and [5] for early work and [6] for a recent collection). In fact, the situation is often yet more complicated because, in dynamical systems terms, the structural bifurcation point represents a subcritical Hamiltonian–Hopf bifurcation and it is known that such a bifurcation gives rise to infinitely many additional complex paths of 'multi-humped' localised responses (see e.g.

* Corresponding author.

0045-7825/99/$ – see front matter
PII: S0045-7825(98)00197-2

240 G.W. Hunt et al. / Comput. Methods Appl. Mech. Engrg. 170 (1999) 239–251

[7,8] and Section 4 below for more details). Largely, these extra localised modes are not important, and it is the primary (or ground state) mode that is physically stable under appropriate loading conditions. However, as we see in Section 4, certain of these multi-humped states do play a role in the mechanism we wish to uncover.

Specifically, we are thinking of the following situation. If a system has the potential to first destabilize and then restabilize, such that the underlying potential energy function at the critical bifurcation point has the form of a local maximum within a global minimum, both localization and periodicity can appear in the same topology. Under parametric change in load the possibility then exists of a connection between the fundamental equilibrium state and a periodically buckled equilibrium state (see Fig. 2 below). In dynamical terms this is a *heteroclinic connection* between one equilibrium state and another. An earlier asymptotic approach [9] links the phenomenon to the classical *Maxwell load* of the periodic response, but to our knowledge, examination in the full spatial context has up to now been lacking. There is, however, a broad class of problems to which it should apply, including the example worked out in detail here—the axially-compressed cylindrical shell. For such systems, a mechanism for explaining the postbuckling behaviour is fundamentally important since the effective (observed) buckling load may be much smaller than the bifurcation (predicted) load. In the case of the cylinder the effective buckling load may be as little as a quarter of the bifurcation load.

After presenting in Section 2 a succinct derivation of energetic and equilibrium equations for a cylindrical shell, we provide an explanation of the homoclinic/heteroclinic mechanism on three different levels. First, in Section 3, we review a multiple scale asymptotic portrayal, the so-called 'slow-space' description, in which the heteroclinic connection appears at a single distinct load. Next, in Section 4, by using a generic dynamical systems unfolding of the assumed global bifurcation sequence, it is demonstrated that heteroclinic connections will, in fact, range over a sequence of loads. Moreover, the bifurcation sequence from the homoclinic to the heteroclinic response causes a succession of 'compartmentalized' buckles via an infinite sequence of snaking destabilizing and restabilizing folds. Finally, in Section 5, the preceding theory is confirmed by numerical continuation analysis of a discretization of the equilibrium equations of the cylinder. Section 6 draws conclusions. We find that the slow space portrayal allied with its dynamical systems interpretation is remarkably successful, predicting not only appearance of the heteroclinic connection itself, but also its close relationship with the practically-important first minimum load encountered on the post-buckling equilibrium path. The asymptotics show, in effect, that this minimum load can be interpreted as the classical critical Maxwell load of the periodic potential energy function.

2. Formulation: von Kármán–Donnell equations and total potential energy

The classical *equilibrium-equation* formulation for a thin (shallow) cylindrical shell of radius R and thickness t usually starts with governing nonlinear von Kármán–Donnell differential equations:

$$\kappa^2\nabla^4 w + \lambda w_{xx} - \rho\phi_{xx} = w_{xx}\phi_{yy} + w_{yy}\phi_{xx} - 2w_{xy}\phi_{xy} \tag{1}$$

$$\nabla^4\phi + \rho w_{xx} = (w_{xy})^2 - w_{xx}w_{yy}\,, \tag{2}$$

where ∇^4 is the two-dimensional bi-harmonic operator, $x \in \mathbb{R}$ is the axial and $y \in [0, 2\pi R)$ is the circumferential co-ordinate, w is the radial displacement measured from a non-trivial (fundamental) unbuckled state, and ϕ is a stress function [1]. Eqs. (1) and (2) are supplemented by suitable boundary conditions (e.g. for a solution that localizes in x; w, ϕ and their x derivatives may tend to zero, while natural periodic conditions can be assumed in y). Parameters appearing are the curvature $\rho = 1/R$, the geometric constant,

$$\kappa^2 = t^2/12(1-\nu^2)\,,$$

where ν is Poisson's ratio, and the bifurcation parameter,

$$\lambda = P/Et\,,$$

where P is the compressive axial load applied per unit length and E is Young's modulus. Here, (1) is an equilibrium and (2) a compatibility equation.

Alternatively, a *variational* formulation of the problem is obtained upon writing a total potential energy functional $\mathscr{V}$, comprising bending energy,

$$U_B = \frac{Et^3}{(1-\nu^2)} \int_0^L \int_0^{2\pi R} \{(\nabla^2 w)^2 - 2(1-\nu)[w_{xx}w_{yy} - (w_{xy})^2]\}\, dx\, dy\,, \tag{3}$$

the membrane (stretching) energy,

$$U_M = \frac{Et}{2} \int_0^L \int_0^{2\pi R} \{(\nabla^2 \phi)^2 - 2(1+\nu)[\phi_{xx}\phi_{yy} - (\phi_{xy})^2]\}\, dx\, dy\,, \tag{4}$$

and the first-order work done by the axial compressive load λ,

$$W = -\frac{Et\lambda}{2} \int_0^L \int_0^{2\pi R} (w_x)^2\, dx\, dy\,. \tag{5}$$

As they take no account of the constraint of compatibility, these functionals cannot be used directly to find minimum energy configurations. However, a Lagrange multiplier introduced on the compatibility equation (2) gives the usable form:

$$\begin{aligned}\mathscr{V} = {} & \frac{Et^3}{24(1-\nu^2)} \int_0^L \int_0^{2\pi R} \{(\nabla^2 w)^2 - 2(1-\nu)[w_{xx}w_{yy} - (w_{xy})^2]\}\, dx\, dy \\ & + \frac{Et}{2} \int_0^L \int_0^{2\pi R} \{(\nabla^2 \phi)^2 - 2(1+\nu)[\phi_{xx}\phi_{yy} - (\phi_{xy})^2]\}\, dx\, dy \\ & - \frac{Et\lambda}{2} \int_0^L \int_0^{2\pi R} (w_x)^2\, dx\, dy - Et \int_0^L \int_0^{2\pi R} \phi[\nabla^4 \phi + \rho w_{xx} - (w_{xy})^2 + w_{xx}w_{yy}]\, dx\, dy\,. \end{aligned} \tag{6}$$

Minimization with respect to w then gives the equilibrium equation (1) and with respect to ϕ the compatibility equation (2).

Since we are not considering finite-length or boundary effects, much of what follows leans heavily on regarding a spatial variable (for the cylinder, the axial co-ordinate x) as time-like and then using energy properties found in the formalizations of Lagrangian and Hamiltonian dynamical systems (see [10] for some general theorems in this direction).

3. Asymptotic representation

The equilibrium behaviour of a typical one degree-of-freedom (1 d.o.f.) conservative structural system with a fundamental path which loses stability at a subcritical bifurcation point is shown schematically in Fig. 1. If $V(A, P)$ represents the total potential energy, where A measures the degree of freedom and P the load,

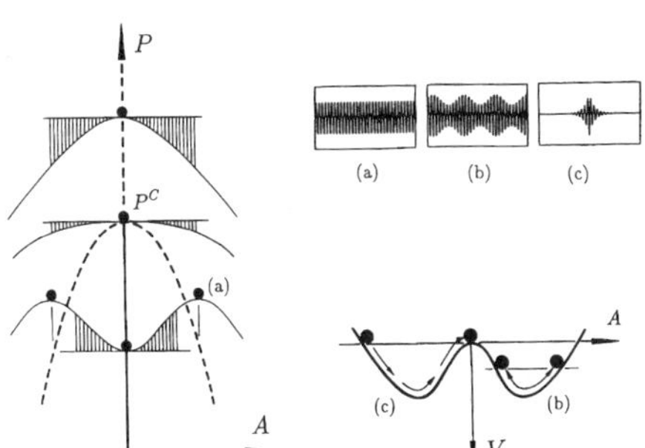

Fig. 1. Left: typical subcritical 1 d.o.f. response. Right: asymptotic portrayal in slow space: (a) periodic; (b) modulated; (c) localized.

equilibrium at different but constant values of P is given by $\partial V/\partial A = 0$. The shape of the V-surface then is as shown in Fig. 1, sketched for three typical P-values below, at, and above the critical bifurcation load P^C. For $P < P^C$ we find a surface with a minimum of V flanked by two maxima, evolving under increasing load to one with just a single maximum [2].

This single d.o.f. picture can be applied to a long structure that buckles periodically, using Rayleigh–Ritz or some other spectral decomposition combined with Liapunov–Schmidt reduction (see e.g. [11]). The assumption of periodic fluctuation in space effectively removes any spatial co-ordinates from consideration, giving again simply $V = V(A, P)$ for the energy, where A is now the amplitude of the major harmonic component. Localization or other modulation of the buckle pattern demands reintroduction of spatial variables, and this can be done either by reverting to the full von Kármán–Donnell equations, or in a useful asymptotic sense by assuming that A is slowly varying in space. This view is strictly only valid close to the critical point where $P = P^C$, and it remains one of the central aims of this paper to explore the extent of its validity by comparing predictions with numerical solutions to the full equations.

3.1. Localization—Bifurcation of a homoclinic connection

The spatial variable can be reintroduced in an asymptotic sense by writing $X = xs$, where s is a perturbation parameter measuring movement away from the critical point (e.g. $s = (P^C - P)^\alpha$ for some α). As a canonical example, consider the dimensionless model of a compressed strut on a destabilizing quadratic (Winkler) foundation, described by the fourth-order ordinary differential equation for lateral displacement w in terms of axial coordinate x [3,12]

$$\ddddot{w} + p\ddot{w} + w - w^2 = 0\,, \tag{7}$$

where p represents the loading parameter with a critical value of $p = 2$. Dots denote differentiation with respect to 'fast' space x. Taking the ansatz $w(x) = A(X)\cos x$ with $X = x\sqrt{(2-p)}$, then, to lowest order, (7) reduces to the second-order differential equation in A,

$$4A'' - A + \frac{19}{18}A^3 = 0\,, \tag{8}$$

the prime denoting differentiation with respect to 'slow' space X [5,3]. Note that (8) has the exact pair of homoclinic solutions

$$A(X) = \pm\sqrt{36/19}\,\mathrm{sech}(X/2)\,. \tag{9}$$

However, it is more instructive for the present arguments to view the equation energetically.

In a Lagrangian interpretation of Eq. (8), spatial potential energy V, expressed by

$$V = \frac{1}{2}A^2 - \frac{19}{72}A^4\,, \tag{10}$$

is independent of the 'time' measure X; V thus turns out to be the same as the 1 d.o.f. V for a periodic fluctuation of the same system. The 'velocity' dependent or kinetic component of the energy,

$$T = -2(A')^2\,, \tag{11}$$

involves the first derivative A'. As T is negative definite, the system responds like a ball rolling with conserved total energy $\mathcal{H} = T + V$ in the upside-down version of V, as seen on the right of Fig. 1. For a fully localized solution, the level of V where A is a maximum and hence $A' = T = 0$, must thus match that of the fundamental (unbuckled) path.

3.2. Cylindrical shell formulation

Both periodic and localized behaviour of axially-compressed cylindrical shells are known to involve more than just one d.o.f. The nearest equivalent set of equations to (8) for the cylinder problem appears to comprise the three linked nonlinear equations,

$$
\begin{aligned}
&8(1-\beta^2)^2\kappa^2 A'' - 2A + 3\rho\beta^2 BC = 0\\
&16(1+\beta^2)^2\kappa^2 B'' - 4(1+\beta^2)^2 B + 3\rho\beta^2 AC = 0\\
&8(1-\beta^2)^2\kappa^2 C'' - 2\beta^4 C + 3\rho\beta^2 AB = 0
\end{aligned}
\tag{12}
$$

which were obtained in [13] by taking leading-order terms based on the assumed form for w and ϕ,

$$
\begin{aligned}
&w = A(X)\cos(\gamma x)\cos(\beta\gamma y) + B(X)\cos((1+\beta^2)\gamma x) + C(X)\cos(\beta^2\gamma x)\cos(\beta\gamma y)\,;\\
&f = \kappa A(X)\cos(\gamma x)\cos(\beta\gamma y) + \kappa B(X)\cos((1+\beta^2)\gamma x) + \kappa C(X)\cos(\beta^2\gamma x)\cos(\beta\gamma y)\,.
\end{aligned}
\tag{13}
$$

The form of (13) is obtained by taking the three interacting modes on the classical *Koiter circle* [1] which describes the neutral stability curve in (γ, β)-space where γ is an axial wavenumber and β an aspect ratio of axial to circumferential wavelength. Corresponding V and T functions are

$$
\begin{aligned}
&V = A^2 + 2(1+\beta^2)^2 B^2 + \beta^4 C^2 - 3\rho\beta^2 ABC\,,\\
&T = -4\kappa^2[(1-\beta^2)^2 A'^2 + 2(1+\beta^2)^2 B'^2 + (1-\beta^2)^2 C'^2]\,.
\end{aligned}
\tag{14}
$$

Again, with $T = 0$, V is identical to the integrated total structural potential energy obtained from spectral decomposition combined with Liapunov–Schmidt reduction to the three active degrees of freedom.

In [13] it was shown by numerical shooting that the system (12) possesses a homoclinic solution that is similar in character (i.e. being symmetric, positive and monotonic up to its point of symmetry) to the positive solution (9) of (8).

3.3. Maxwell load hypothesis—heteroclinic connnection

If the slow space concept of Fig. 1 is extended to a system that first destabilizes and then restabilizes, as shown in Fig. 2, several interesting features emerge. Here, periodic states for which $\partial V/\partial A = 0$ are shown as black circles, while maximum amplitudes of the localized (homoclinic) solutions, where $A' = T = 0$, are shown as open circles. We see that the minimum post-buckling load for the localized solution must always appear above that for the periodic solution. Moreover, at the minimum localized post-buckling load, the two solutions must be one and the same; the periodic state is exactly that which matches the V level of the fundamental path, $A = 0$. Interestingly, this coincides with the classical Maxwell critical load, relevant for stochastic systems governed by basic underlying disturbances or excitations, where the state of stable equilibrium is associated only with the global minimum of V, rather than local minima (see for example [14, pp. 306–310]); according to the Maxwell convention, states on the fundamental path between P^M and P^D are unstable.

Although having three rather than one essential d.o.f. in its slow-space formulation, it is interesting to extend these ideas to the cylinder problem. Hunt and Lucena Neto [9] compare Maxwell critical loads obtained from multi-dimensional Rayleigh–Ritz models which are periodic both circumferentially and axially, with experimen-

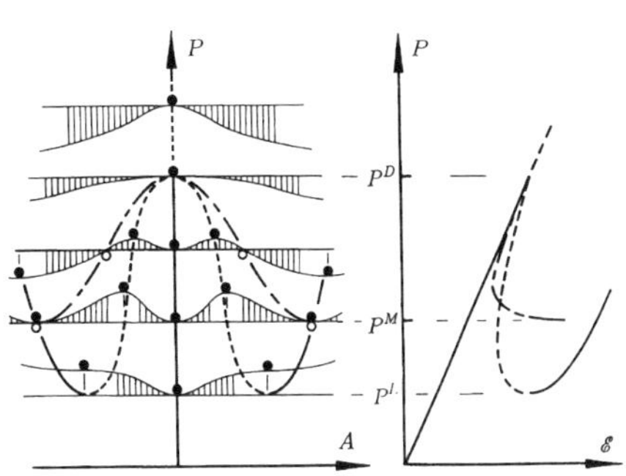

Fig. 2. Asymptotic view of Fig. 1, extended to restabilization. ℰ: corresponding deflection of load P. P^D: critical eigenvalue load. P^M: Maxwell load. P^L: minimum post-buckling load for periodic responses.

tal results due to Yamaki [15]. It was found that, not only do the Maxwell loads thus obtained match the observed minimum post-buckling loads with surprising accuracy, but so too do the axial wavelengths of the buckle pattern, which in the Rayleigh–Ritz formulation are allowed to choose themselves by direct energy minimization.

As, over the large deflections involved, the system might well have diverged significantly from the slow-space asymptotic portrayal, these results are sensibly viewed with suspicion. The dynamical view of the following section, for example, demonstrates that the Maxwell criterion must operate over a range of loads, rather than the single load of Fig. 2. It is also debatable whether the cylinder of Yamaki [15] is long enough for the structural boundary conditions to be clear of influence. Nevertheless, we now have good agreement between numerics and experiments [16,17], and it remains a useful exercise to extend the numerics to capture periodic solutions of any wavelength. This gives a second check on the Maxwell criterion and related predictions, which can be operated entirely independently of all experimental issues.

The numerics, presented in Section 5 below, thus provide stronger evidence for both the existence of heteroclinic connections and for their role in linking the Maxwell critical load for periodic solutions with the minimum post-buckling load for localized behaviour. First we apply dynamical systems arguments to put our tentative statements about heteroclinic connections, based on a 1 d.o.f. analogy, on a more sound footing.

4. Dynamical systems analogy

We consider an extension to arbitrary d.o.f. Hamiltonian reversible dynamical systems of the concept of the heteroclinic connection in the 1 d.o.f. analogy embodied in Figs. 1 and 2. Here, *reversibility* defines the existence of a linear transformation R that fixes half the phase space variables (the co-ordinate variables of a classical Hamiltonian system) and negates the others (the momentum variables), such that the system is invariant under R and a reversal of time [18]. Most structural systems have this notion of reversibility; see Section 5 for the reversibilities of the von Kármán–Donnell equations.

Recall, by analogy with the strut model (7) which is equivalent to a 2 d.o.f. Hamiltonian system, that the structural bifurcation at $P=P^C$ corresponds to a *Hamiltonian Hopf* bifurcation [19] of the trivial equilibrium $w=0$. At such a point, structural instability occurs when the linearization about the trivial equilibrium involves a complex quadruple of eigenvalues $\pm\mu\pm i\omega$ (a saddle focus) becoming pure imaginary (a centre) as $\mu\to 0$. Under the conditions of subcriticality, then it is known that a pair of reversible homoclinic orbits bifurcate into the region where the equilibrium is a saddle-focus. For the strut model, (7), these two solutions are those expressed asymptotically as (9). In general, they represent two sine waves 180° out of phase modulated by a sech-like profile. One of these solutions is identified as being the primary localised mode of buckling as the parameter P moves one away from the critical load. In fact, under generic *transversality* hypotheses, given a primary homoclinic orbit to a saddle-focus, there will be an infinite number of *N-pulse* homoclinic orbits which are like multiple copies of these primaries separated by a finite number of oscillations; see [20] for a review. Note that none of these more complex localised buckling modes will bifurcate from the trivial solution, but instead will remain of finite amplitude and form folds with respect to P for $P<P^C$. In general these solutions are not of interest because they are unlikely to be physically stable [21] and would require greater triggering energy than the primary localised mode.

We now suppose that at some other parameter value the primary homoclinic orbit 'collides' with a finite amplitude periodic orbit thus forming an equilibrium-to-periodic heteroclinic connection. We consider only Hamiltonian reversible system of the lowest possible dimension to describe an unfolding of this scenario in a structurally stable way, namely 2 d.o.f. (4 dimensions of phase space). Although, this may seem restrictive, note that there is a so-called Homoclinic Center Manifold Theorem [22], which loosely speaking states that four dimensions is enough to explain this phenomenon occurring in higher (even, in some circumstances, infinite) dimensional systems.

Therefore, consider a non-integrable, 2 d.o.f. Hamiltonian system with a saddle focus equilibrium O, which is in the fixed point set of a reversibility, whose unstable manifold contains a heteroclinic connection with a hyperbolic reversible periodic orbit L. Without loss of generality, suppose that both the equilibrium and the periodic orbit lie in $H(0)$, the zero level set of the Hamiltonian function. Furthermore, within $H(0)$, take a Poincaré section Π which contains O and is transverse to L. Note that this will not be a nondegenerate Poincaré section since it cannot be transverse to the flow near O; however, this is unimportant, provided that Π is chosen to be transverse to the flow in the stable and unstable manifolds of O. Then the intersections with Π of the stable and unstable manifolds of both O and L will all be one-dimensional. Strictly speaking, in order to choose such a Poincaré section it is necessary to assume that these stable and unstable manifolds do not intersect non-transversally, in which case we would have to argue more delicately. For the purposes of illustration though, we shall pretend that Π is globally defined even through homoclinic tangencies. Furthermore, choose Π such that it contains the portion within $H(0)$ of the fixed-point set of the reversibility, the so-called symmetric section S (see Fig. 3).

Our first observation from such a picture is that a heteroclinic connection between O and L persists under parametric perturbation. Hence, in order to consider a bifurcation phenomenon we have to consider a tangency between the unstable manifold of O, $W^u(O)$, and the stable manifold of L, $W^s(L)$.

Fig. 3 depicts the situation when $W^u(O)$ and $W^s(L)$ 'pass through' each other under parameter variation, including two successive heteroclinic tangencies (in panels (b) and (d)). Clearly, the complete unfolding of this bifurcation would lead to a complex story involving periodic, homoclinic and heteroclinic orbits of various kinds. Here, we focus only on the implications for the simplest homoclinic orbits to O, by which we mean orbits that spend a unique time interval outside a small neighbourhood of O. The existence of such homoclinic orbits can be inferred from points of intersection of $W^u(O) \cap \Pi$ and S, since, by reversibility, such points must be in the stable manifold of O also. This is illustrated in Fig. 3 by S acting like a mirror. Whenever the solid line $W^u(O)$ intersects S, its dashed mirror-image $W^s(O)$ intersects at the same point. The geometry of the heteroclinic tangle depicted in Fig. 3(c) (which can be rigorously established via the Lambda Lemma [23, Sect. 5.2] and the assumed sequence of heteroclinic tangencies) implies that for parameter values between those of the two tangencies, there will be an infinity of such intersection points. Moreover, information may be gleaned on the creation and destruction of the corresponding orbits, as is depicted in a bifurcation diagram in Fig. 4.

Fig. 4 shows that the infinity of simplest homoclinic orbits lie on a single wiggly curve. As the solution progresses up this curve it undergoes a succession of fold points which correspond to points of tangency between $W^u(O)$ and S. In so doing, the homoclinic orbit gains an extra 'hump' by passing one more time in the neighbourhood of L. The wiggly curve accumulates on the parameter values of the two heteroclinic tangencies from the left in both cases, given the parameter dependence depicted. Hence, viewing the bifurcation from left to right, the first tangency suddenly creates infinitely many homoclinic orbits, whereas the second one only starts

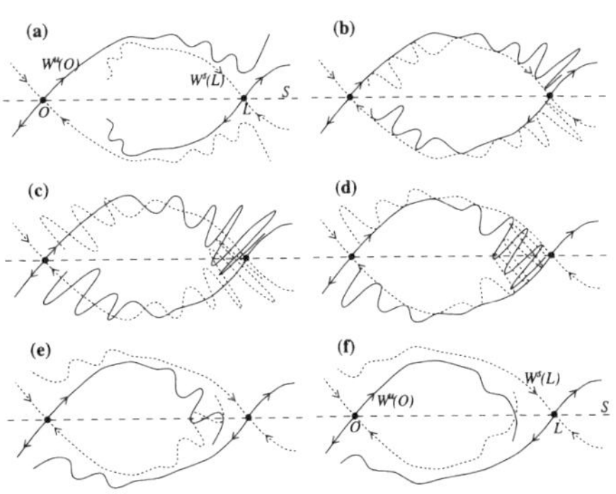

Fig. 3. Illustrating the parameter unfolding of two successive heteroclinic tangencies between a saddle-focus equilibrium O and a saddle-type periodic orbit L for a four-dimensional reversible Hamiltonian system. The picture is drawn schematically by taking a formal Poincaré section within the zero level set of the Hamiltonian function (see text for details). Here, S is the symmetric section, i.e. fixed point set of the reversibility, and unstable and stable manifolds are depicted respectively by solid and broken lines.

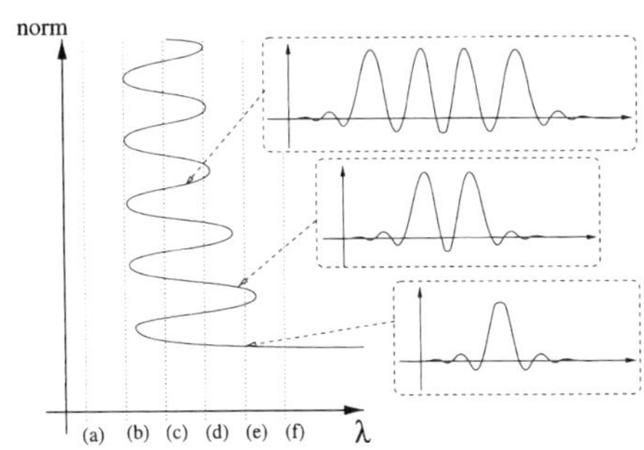

Fig. 4. Bifurcation diagram of L^2-norm against bifurcation parameter λ of homoclinic solutions to O implied by the unfolding of successive heteroclinic tangencies between O and the periodic orbit L. Parameter values (a)–(f) refer to the corresponding panels of Fig. 3.

the process of their destruction. Alternatively, viewing the bifurcation diagram from right to left, the overall effect of the bifurcation sequence is to remove a simple 'one-humped' homoclinic orbit, by repeatedly snaking back and forth in the bifurcation diagram, gaining more and more humps in the process, until it is indistinguishable from a O to L heteroclinic connection.

Finally in this section, note that the assumption of two successive heteroclinic tangencies is reasonable, since this would occur in small perturbations of completely integrable Hamiltonian systems containing a structurally unstable connection formed by the identification of $W^u(O)$ and $W^s(L)$. (See [24] for a similar assumption on successive *homoclinic* tangencies occurring in another context.) Indeed, in work to be presented elsewhere [25], it will be shown that such a perturbation from an integrable heteroclinic connection occurs in an unfolding of the boundary between supercritical and subcritical Hamiltonian–Hopf bifurcation points.

5. Numerical verification

5.1. Formulation

Numerical confirmation of the phenomena just outlined for the cylindrical shell requires careful formulation, taking care to compute the appropriate solutions among the infinite possibilities for a (theoretically infinitely) long cylinder. In our earlier work [16,17], an intensive examination of the homoclinic solutions was carried out by decomposing the circumferential component of the von Kármán–Donnell PDEs into M spectral modes via a Galerkin procedure, reducing them to $8M$ coupled nonlinear ODEs. There, we typically only computed localised buckling modes from their bifurcation point to the first minimum of load with respect to end shortening. Here, we present results beyond this first minimum in order to illustrate their connection with the analysis of Sections 3 and 4, and in particular demonstate how this first minimum is linked to the existence of a heteroclinic connection.

We discretize the von Kármán–Donnell equations (1) and (2) in such a way as to exploit the natural symmetries in the problem. Experimentally, a well-defined number, s, of waves is observed circumferentially [1,26,15] in the buckled deformation, corresponding to an invariance under rotation of $2\pi/s$. Hence, we write

$$w(x, y) = \sum_{m=0}^{\infty} a_m(x)\cos(ms\rho y)\,; \qquad \phi(x, y) = \sum_{m=0}^{\infty} b_m(x)\cos(ms\rho y)\,, \quad s \in \mathbb{N}\,,$$

and refer to $\cos(s\rho y)$ as the *seed mode*. Substituting into the von Kármán–Donnell equations and taking the L^2 inner product with $\cos(ms\rho y)$ we obtain a system of nonlinear ODEs for the Fourier modes a_m and b_m for $m = 0, \ldots, \infty$. The Galerkin approximation may then be formed by taking $m = 0, \ldots, M-1$ for some finite M giving a system of $8M$ first-order ordinary differential equations. Note that $s = 1$ corresponds to the standard Galerkin (or Rayleigh–Ritz) approximation.

There is a further symmetry which plays an important role in localized buckling. Experimentally, the observed patterns tend to be either *symmetric* or, more frequently, *cross-symmetric*. A solution to the von Kármán–Donnell equations that is symmetric about a section at $x = T$ satisfies

$$w(x, y) = w(2T - x, y) \quad \text{and} \quad \phi(x, y) = \phi(2T - x, y)\,, \tag{15}$$

whereas a solution which is cross-symmetric about a section at $x = T$ satisfies for some seed s

$$w(x, y) = w(2T - x, y + \pi R/s) \quad \text{and} \quad \phi(x, y) = \phi(2T - x, y + \pi R/s)\,. \tag{16}$$

It is not difficult to see that the symmetries defined by (15) and (16) imply a reversibility as described in Section 4. Thus, we may define a fixed point sets, $\mathcal{S}$, of the reversibility forming $4M$-dimensional sub-manifolds of phase space $\mathbb{R}^{8M}$. We can then impose boundary conditions such that the solution we seek has $\mathcal{S}$ at its midpoint.

In fact, symmetric solutions are observed far less frequently in experiments than cross-symmetric solutions. Furthermore, when symmetric solutions do occur, they are generally for short cylinders and are less localized (cf. [15,27,28]). Note also that there is a degeneracy in the equations for the zero mode ($m = 0$) such that these may be solved for a_0'', a_0''', b_0'', b_0''' independently of the initial conditions for a_0, a_0', b_0, b_0'. This corresponds to a trivial translational symmetry in the problem (sometimes called a rigid body mode).

The numerical method for computing the homoclinic orbits is described in detail in [17]. We solve for (cross)-symmetric homoclinic solutions as a boundary value problem on a truncated domain with boundary conditions which at one end place the solution in the linearized unstable manifold of the trivial solution (see [29]) and at the other end exploit the boundary conditions from the section $\mathscr{S}$. To compute periodic solutions we again solve a boundary value problem with section $\mathscr{S}$ boundary conditions imposed at both ends.

In the computations below we have used a shell with the following properties:

$$R = 100\,\text{mm}, \qquad t = 0.247\,\text{mm}, \qquad \nu = 0.3, \qquad E = 5.56\,\text{GPa}\,, \tag{17}$$

which correspond to the experimental values taken by Yamaki [15]. For the sake of clarity we only describe the cross-symmetric case and concentrate on one particular seed mode: $s = 11$. Similar results hold in the symmetric case and with other seed modes.

Numerically, we have found that it is convenient to take $M = 6$ as the best compromise between accuracy and efficiency. Qualitatively, similar results have been obtained for smaller and larger M. Convergence as both M and the cylinder length are increased was discussed in [17]; wider convergence questions are addressed in [30].

The equations are then solved in the continuation code AUTO [31], under parametric variation of the loading parameter $\lambda = P/Et$. The result from this is shown in Fig. 5 in the space of λ against a measure of end-shortening given by the arclength of the solution computed to the section $\mathscr{S}$ at the line given by $y = 0$ on the cylinder:

$$\text{arclength} = \int_0^T \left[1 + \left(\sum_{k=0}^{M-1} \frac{\partial a_k}{\partial x}\right)^2\right]^{1/2} \mathrm{d}x\,.$$

The fluctuating nature of the post-buckling curve signifies sequential localized buckling in qualitatively the same manner as in Fig. 4, as predicted by the dynamical system analysis. Fig. 6 presents several solutions to the

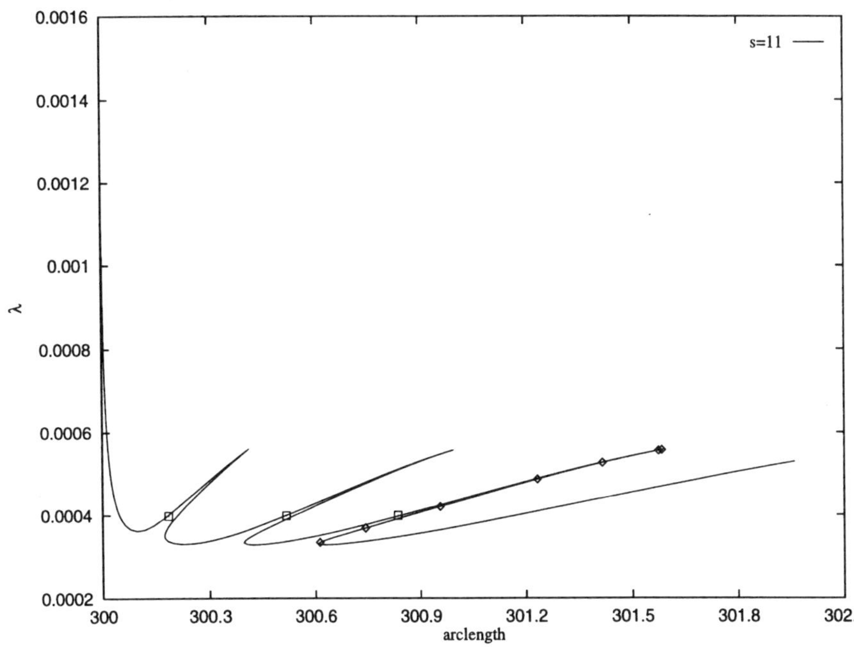

Fig. 5. The localised buckling path plotting the loading parameter λ against the 'arclength' measure of end shortening. Squares show the states plotted in Fig. 6. Diamonds indicate points at which the corresponding periodic orbit has been 'chopped out' (see Fig. 7).

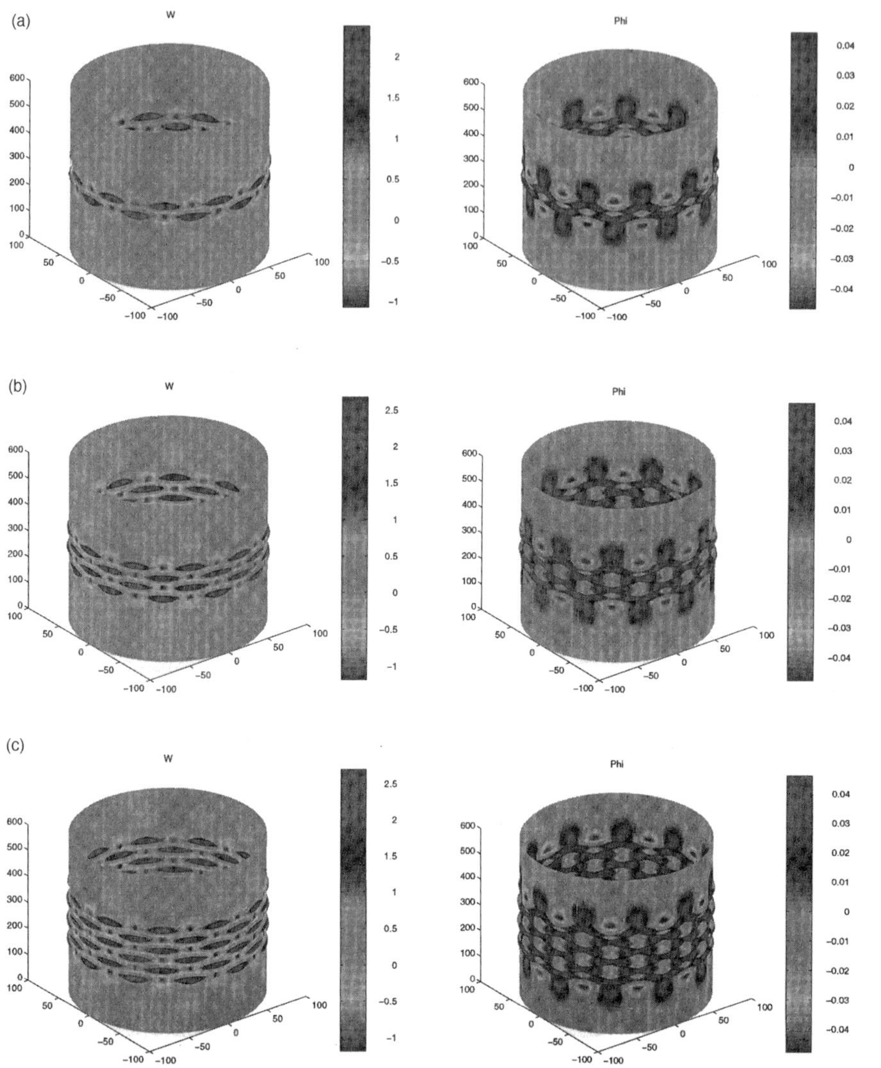

Fig. 6. Deflection $w(x, y)$ and stress function $\phi(x, y)$ reconstructed from the numerical solutions on the localised buckling path at $\lambda = 4.0 \times 10^{-4}$. (a) Near first minimum. (b) Near second minimum. (c) Near third minimum. After each pass through a maximum the solution picks up an additional cell.

equations along this path. For the initial post-buckling, a localization involves the single diamond pattern shown in Fig. 6(a). It restabilizes at the first minimum post-buckling load, whereupon under increasing load it then meets a second instability at a maximum limit point, signifying a second localized buckle concatenating with the first. This results in the addition of a further cell to the buckle pattern (Fig. 6(b)). The process continues to include any number of cells, three being shown in Fig. 6(c). One interpretation of this sequence of minima and maxima is thus that, starting from each minimum, we are essentially buckling an already deformed shell—each new buckle simply adds a new cell to the deformation. The limit of an infinite number of cells then represents heteroclinic connection from the flat state to periodic. It is, therefore, instructive to search for the periodic solution to which this connection occurs. Note that the homoclinic to periodic connection occurs over a range of loads (between the limiting values of the sequences of maxima and minima along the buckling curve). To test the Maxwell load hypothesis, it would be interesting to see if there is some energetic criterion that picks out the periodic solution to which this connection is made, and to see how this relates to the first minimum on the branch of the localised buckling mode.

Periodic equilibrium states can be found from these buckled forms by 'chopping out' the central section of the buckle pattern of Fig. 6(c) and forcing periodic boundary conditions. Although the chopped-out section is not precisely periodic, in our experience one step of continuation is usually enough to enforce it. Points at which the periodic solution is examined in this way are indicated by a diamond.

Different periodic states can arise depending on the boundary conditions; the full allowable range at any load level can be explored by continuation in the period while holding the load constant. Typical results at three different load levels are shown in Fig. 7 on a plot of period against 'scaled' total energy (energy of Eq. (6), divided by period, to give energy per unit length). Periodic solutions in Fig. 7 marked by a diamond are those chopped from the homoclinic orbits of Fig. 5; of these, the smallest period (negative energy) solution corresponds to a maximum of the wiggle of Fig. 5, while the largest period (positive energy) solution corresponds to a minimum. To avoid clutter we have plotted this curve at the three load values of $\lambda = 3.3331 \times 10^{-4}$, $\lambda = 5.5676 \times 10^{-4}$, and $\lambda = 3.6799 \times 10^{-4}$. We note that the last of these is close to, but just above, the first minimum load encountered on the post-buckling path, for which $\lambda = 3.6146 \times 10^{-4}$.

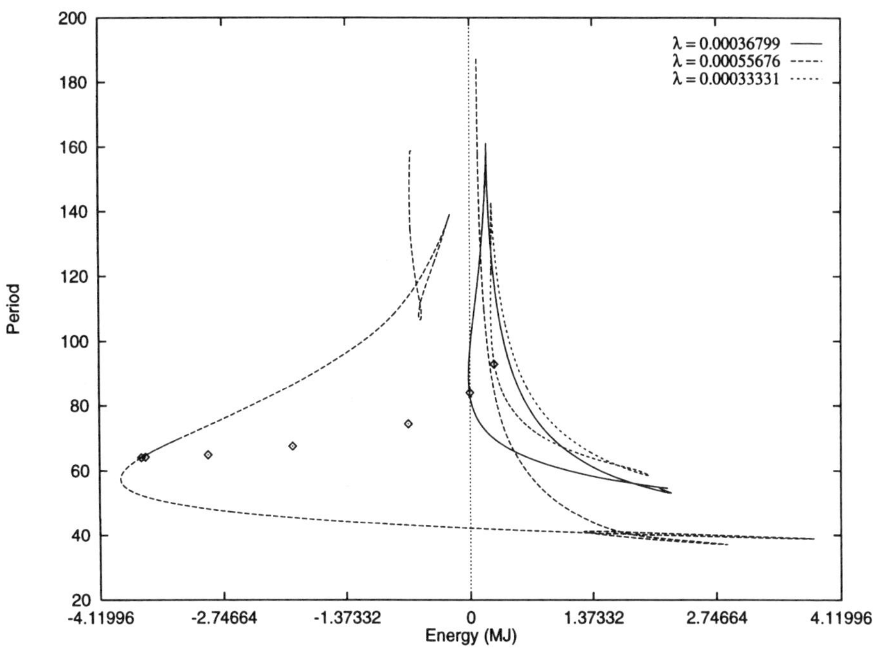

Fig. 7. Curves of periodic solutions for $s = 11$ versus 'scaled' potential energy ($\mathscr{V}$ of (6) divided by the period of the orbit). Also plotted as diamonds are the periodic orbits (obtained from the 'chopped out' middle sections of the homoclinic orbits of Fig. 5) to which a heteroclinic connection occurs.

5.2. Comparisons

Interest naturally falls on the solution with minimum energy, at the leftmost point of each enclosed curve, as being that most likely to be linked with the localized solution that would arise in practice. In each of the curves we have computed, the periodic orbit associated with heteroclinic connection is at, or close to, this minimum energy solution.

We are particularly interested in the point at which the minimum energy is zero, predicted by the slow-space formulation as the Maxwell load below which no localized solution can occur. We see that the connection to the periodic with zero energy would occur at almost exactly the same value of the load as the first post-buckled minimum; the curve for $\lambda = 3.6799 \times 10^{-4}$ just crosses into the negative energy regime, so zero minimum energy would be found at a load just less than this value. Comparing values, we see that the first minimum occurs at $\lambda_M/\lambda_D = 0.242$ and the near-zero energy periodic solution at $\lambda_M/\lambda_D = 0.246$. Hunt and Lucena Neto [9] give both the Maxwell load computed from periodic Rayleigh–Ritz analysis, and the minimum load reached experimentally, for $s = 11$ at $\lambda_M/\lambda_D = 0.24$. Thus, the numerics suggest that the load at the first post-buckled minimum of a *localized* solution may be predicted from the energy of a *periodic* solution (given by the classical Maxwell load).

It is interesting also to compare aspect ratios β (ratio of axial to circumferential wavelengths) for the periodic solutions. The 'chopped out' period of the near-zero energy curve of Fig. 7 gives $\beta = 1.47$. This compares favourably with the Maxwell load value of $\beta = 1.49$ for the Rayleigh–Ritz solution of Hunt and Lucena Neto [9] and the best estimate of an experimental value of $\beta = 1.43$, taken from surface maps of the buckled cylinder found in [15].

6. Concluding remarks

At this point it is a good idea to relate back to what might be observed in a realistic cylinder experiment. The form of load versus end-shortening diagram seen in Fig. 5 would not be expected in reality. In actual experiments under controlled end shortening (see for example [15, Fig. 3.52e, p. 231]) the initially unbuckled shell destabilizes and undergoes a snap response into the buckled pattern at a load less than the classical bifurcation point; the stabilized buckling pattern that appears would typically be held at a load above the first minimum post-buckling load. Under further application of end-shortening the now-buckled shell would pass through the first minimum and start up towards the first maximum of Fig. 5. At some point along that curve the shell would lose stability and bifurcate typically into a mode in which the number of circumferential waves is reduced by one, equivalent to a change in the seed mode from s to $s-1$. In the work of Riks et al. [32] and Riks and Rankin [33], a combination of statics and dynamics was used to attempt to form a numerical picture using finite elements of this phenomenon, allowing the shell to buckle dynamically to find the wave number s. In contrast, for simplicity we have fixed the subspace s in which we are working to examine the buckling phenomena, and used these numerics to confirm predictions from the asymptotic analysis and dynamical systems analogy. Although the results presented here are for a single seed mode and for the cross-symmetric form of solution only, preliminary results indicate similar pictures for other seed modes and the symmetric case.

In particular, we have shown that the concept of the Maxwell load obtained from the asymptotic approximation gives a good prediction of the first minimum post-buckling load. By considering the smallest allowable d.o.f. system that is capable of capturing the full dynamics, a more complete picture is obtained from the dynamic analogy. This demonstrates that an infinity of further minima lie between the first minimum and that of the periodic response, effectively bridging the gap between the localized and periodic responses of Fig. 2. The outcome is an oscillatory form of bifurcation diagram—between two well-defined values with an associated heteroclinic tangle within the range—that must be seen as typical for systems that first destabilize and then restabilize. Our numerics confirm this form of behaviour in the von Kármán–Donnell equations.

References

[1] W.T. Koiter, On the stability of elastic equilibrium, Ph.D. Thesis, University of Delft. English Translation: Tech. Rep. AFFDL-TR-70-25 Air Force Flight Dyn. Lab. 1970, 1945.

[2] J.M.T. Thompson and G.W. Hunt, A General Theory of Elastic Instability (Wiley, London, 1973).

[3] G.W. Hunt, H.M. Bolt and J.M.T. Thompson, Structural localisation phenomena and the dynamical phase-space analogy, Proc. R. Soc. Lond. A 425 (1989) 245–267.

[4] V. Tvergaard and A. Needleman, On the localisation of buckle patterns, ASME J. Appl. Mech. 47 (1980) 613–619.

[5] M. Potier-Ferry, Amplitude modulation, phase modulation and localisation of buckle patters, in: J.M.T. Thompson and G.W. Hunt, eds., COLLAPSE: The Buckling of Structures in Theory and Practice (Cambridge University Press, 1983).

[6] A.R. Champneys, G.W. Hunt and J.M.T. Thomspon, eds., Localisation and solitary waves in solid mechanics, Special Issue of Phil. Trans. Roy. Soc. Lond. A 355 (1997) 2073–2213.

[7] B. Buffoni, A.R. Champneys and J.F. Toland, Bifurcation and coalescence of a plethora of homoclinic orbits for a Hamiltonian system, J. Dyn. Diff. Eq. 8 (1996) 221–281.

[8] G. Iooss and M.C. Peroueme, Perturbed homoclinic solutions in reversible 1:1 resonance vector fields, J. Diff. Eq. 102 (1993) 62–88.

[9] G.W. Hunt and E. Lucena Neto, Maxwell critical loads for axially-loaded cylindrical shells, ASME J. Appl. Mech. 60(3) (1993) 702–706.

[10] A. Mielke, Hamiltonian and Lagrangian flows on center manifolds: with applications to elliptic variational problems, Lecture Notes in Mathematics; 1489 (Springer-Verlag, Berlin, 1991).

[11] M. Golubitsky and D. Schaeffer, Singularities and Groups in Bifurcation Theory, Vol. 1, Applied Mathematical Sciences; 51 (Springer-Verlag, Berlin, 1985).

[12] G.W. Hunt and M.K. Wadee, Comparative Lagrangian formulations for localised buckling, Proc. R. Soc. Lond. A 434 (1991) 485–502.

[13] G.W. Hunt and E. Lucena Neto, Localized buckling in long axially-loaded cylindrical shells, J. Mech. Phys. Solids 39(7) (1991) 881–894.

[14] E. Zeeman, Catastrophe Theory: Selected Papers 1972–1977 (Addison-Wesley, London, 1977).

[15] N. Yamaki, Elastic Stability of Circular Cylindrical Shells, Vol. 27 of Applied Mathematics and Mechanics (Elsevier, 1984).

[16] G.J. Lord, A.R. Champneys and G.W. Hunt, Computation of localized post buckling in long axially-compressed cylindrical shells, in: Theme Issue on Localization and Solitary Waves in Solid Mechanics, Phil. Trans. Roy. Soc. Lond. A 355 (1997) 2137–2150.

[17] G.J. Lord, A.R. Champneys and G.W. Hunt, Computation of homoclinic orbits in partial differential equations: an application to cylindrical shell buckling, SIAM J. Sci. Comp., to appear.

[18] R. Devaney, Reversible diffeomorphisms and flows, Trans. Amer. Math. Soc. 218 (1976) 89–113.

[19] J. van der Meer, The Hamiltonian Hopf bifurcation, Lecture Notes in Mathematics, 1160 (Springer-Verlag, Berlin, 1985).

[20] A.R. Champneys, Homoclinic orbits in reversible systems and their applications in mechanics, fluids and optics, Physica D 112 (1998) 158–186.

[21] B. Sandstede, Instability of localised buckling modes in a mode-dimensional strut model, Phil. Trans. Roy. Soc. Lond. A 355 (1997) 2083–2097.

[22] B. Sandstede, Center manifolds for homoclinic solutions, Technical report, Weierstraß Instut für Angewandte Analysis und Stochastik. Preprint No. 186, 1995.

[23] J. Guckenheimer and P. Holmes, Nonlinear Oscillations, Dynamical Systems and Bifurcations of Vector Fields, Applied Mathematical Sciences, 42 (Springer-Verlag, New York, USA, 1983).

[24] P. Hirschberg and C. Laing, Successive homoclinic tangencies to a limit cycle, Physica D 89 (1995) 1–14.

[25] P.D. Woods and A.R. Champneys, Heteroclinic tangles and home-clinic shaking in the unfolding of a degenerate reversible Hamiltonian Hopf bifurcation, Physica D, submitted.

[26] L.H. Donnell and C.C. Wan, Effect of imperfections on buckling of thin cylinders and columns under axial compression, J. Appl. Mech. APM-14(49) (1950) 73–83.

[27] M. Eßlinger, Hochgeschwindigkeitsaufnahmen vom beulvorgang dünnwandiger, axialbelasteter zylinder, Der Stahlbau (1970).

[28] M. Eßlinger and B. Geier, Gerechnete nachbeullasten als untere grenze der experimentallen axialen beullasten von kreiszylindern, Der Stahlbau (1972).

[29] W.-J. Beyn, The numerical computation of connecting orbits in dynamical systems, IMA J. Numer. Anal. 9 (1990) 379–405.

[30] G.J. Lord, D. Peterhof, B. Sandstede and A. Scheel, Numerical computation of solitary waves in semilinear elliptic problems on infinite cylinders, SIAM J. Numer. Anal., submitted.

[31] E. Doedel and J.P. Kernevez, (1986), AUTO: Software for continuation problems in ordinary differential equations with applications, Technical report, California Institute of Technology. Applied Mathematics Technical Report, 1986.

[32] E. Riks, C.C. Rankin and F.A. Brogan, On the solution of mode jumping phenomena in thin-walled shell structures, Comput. Methods Appl. Mech. Engrg. 136(1) (1996) 59–92.

[33] E. Riks and C.C. Rankin, Computer simulation of the buckling behavior of thin shells under quasi static loads, in: M. Kleiber, E. Onate and O. Zienkiewicz, eds., Computer Methods in Applied Mechanics, to appear.

Chapter 6

Other buckling problems

In this chapter we present three further articles by G.W. HUNT and co-workers on applied buckling problems in which the methodology of this book can be applied. The first of these is a model of a sandwich structure used in light-weight manufacture, the second is a simplified model of pipeline uplift and the final one is a more phenomenological study of novel models for geological folding.

Localization and mode interaction in sandwich structures. WADEE, M.A. & HUNT G.W. *Proc. Roy. Soc. Lond. A* **454** (1998) 1197-1216.

Homoclinic and heteroclinic solutions of upheaval buckling. HUNT, G.W. & BLACKMORE, A. *Phil. Trans. Roy. Soc. A* **355** (1997) 2185-2297.

Folding processes and solitary waves in structural geology. HUNT, G.W. MÜHLHAUS, H.-B. & WHITING, A.I.M. *Phil. Trans. Roy. Soc. A* **355** (1997) 2197-2213.

The Royal Society

Localization and mode interaction in sandwich structures

By Giles W. Hunt and M. Ahmer Wadee
Department of Mechanical Engineering, Faculty of Engineering and Design, University of Bath, Claverton Down, Bath BA2 7AY, UK

Received 12 March 1997; accepted 6 August 1997

Compressed sandwich structures, comprising two stiff face plates separated by a softer core material, while designed principally as efficient integral structures, can lose this quality when faces buckle locally. Interaction between overall (Euler) buckling and local buckling of one face suggests that failure will localize into the centre. A variational formulation, leading to a pair of nonlinear differential equations subject to integral constraints, describes the post-buckling response. These are solved by a combination of numerical shooting and continuation techniques, such that the response far into the unstable post-buckling regime can be portrayed. Solutions with both linear and nonlinear constitutive core relations are compared with the results of an engineering (body-force) approach, and with those of earlier (periodic) Rayleigh–Ritz analyses. The latter demonstrate the extra destabilization that comes with localization.

Keywords: aperiodic buckling; interactive buckling; laminated structures; composite structures; localization; numerical continuation

1. Introduction

Sandwich construction is increasingly popular as a provider of structural strength combined with weight efficiency. However, precisely because they are both specialized and efficient, sandwich structures are liable to exhibit complicated collapse mechanisms. One of the classic dangers of structural optimization is that failure may occur in more than one mode simultaneously (Thompson & Hunt 1973), and it seems likely that sandwich systems may fall into this category. It is well known from classical work (Allen 1969), that although designed as integral structures, compressed sandwich panels sometimes fail by local buckling or wrinkling of face plates or faces. This is conventionally seen as separate from overall buckling, being perhaps treated along the lines of a strut on an elastic foundation. Where *mode interaction* has been considered it is only in the sense of Rayleigh–Ritz analysis, with periodic buckling of the faces describing local buckling (Hunt *et al.* 1988).

Figure 1 shows a typical panel, comprising thin metal faces separated by a lighter (foam) core material, having failed in compression. It is clear from this photograph that local buckling, that has *localized* to the centre of one face, has contributed significantly to loss of integrity in the overall sense; local failure of the face (here as a plastic mechanism) interacts with overall failure as it loses stiffness and shortens. Such interactions have occupied some attention in the literature (Koiter & Pignataro 1976; Budiansky & Hutchinson 1979; Hunt 1981) but as far as we are aware have

Figure 1. A sandwich panel showing localized failure contributing to overall instability. For clarity, dotted lines have been superimposed on the edges of the faces.

never been fully formulated for sandwich systems. Only now, with modern algebraic manipulation and numerical techniques does this seem viable. The problem is inherently complex for the following reason.

Classical sandwich theory (Allen 1969), unlike standard engineers' bending theory, allows shearing displacement in the core; the interactive buckle of figure 1 clearly could not take place under the constraint of infinite core shear stiffness. The wrinkling or buckling of a single face is similar to that of a strut on elastic foundation, with the core material providing the foundation support. But the simplest such model, employing the so-called *Winkler* foundation and leading to a fourth-order governing differential equation, takes the foundation as comprising translational springs alone; shearing resistance is ignored entirely. In the context of a sandwich structure this would deny the system overall integrity, leaving it free to respond as two separate faces buckling in sympathy. Inclusion of shear in the foundation is known to complicate matters considerably; integro-differential equations would be expected (Hunt *et al.* 1997), and at the very least there would be an increase in the order of the underlying governing equation. Such complications are apparently unavoidable and appear naturally in the formulation below.

The formulation starts with a brief description of the classical linear eigenvalue result for overall buckling (Allen 1969), involving two degrees–of–freedom giving, independently, sway and tilt components of a half-sine wave buckling mode. To model the local buckling of a single face, the same mode-shape is then introduced as a parametric element into the formulation for a strut on elastic foundation. A linear governing differential equation with non-constant coefficients is the result (see also Luongo 1991). This heuristic model is followed by a fully interactive formulation, in which overall buckling is represented by the same two sway and tilt components of a long wave mode, and the form of local buckling is found by application of the calculus of variations on the two functions $w(x)$ and $u(x)$ representing, respectively, lateral and in-plane displacements of a single face. Both formulations are enhanced

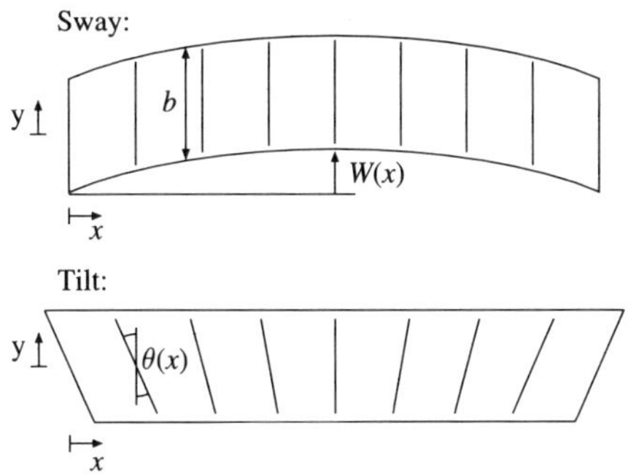

Figure 2. Sway and tilt components of overall mode.

with appropriate nonlinearities for the core material, and results are presented which apparently compare well with the form of figure 1.

2. Modal descriptions

(*a*) *Overall buckling mode*

Previous work on sandwich structures has shown first that overall buckling precedes localized failure, and second that, contrasting with the normal assumption of engineers' bending theory, shearing displacement in the core carries a significant effect. We allow for this in the classical manner of Allen (1969) by introducing separate 'sway' and 'tilt' components of overall buckling in the two degree–of–freedom formulation shown in figure 2, where, from simple linear theory, lateral displacement $W(x)$ and rotation $\theta(x)$ are represented by

$$\begin{aligned} W(x) &= q_s L \sin(\pi x/L), \\ \theta(x) &= q_t \pi \cos(\pi x/L). \end{aligned} \tag{2.1}$$

A linear eigenvalue analysis can determine the ratio of amplitudes q_s/q_t. Shear angles from the two components combine to give the shear strain,

$$\gamma_{xy} = \frac{\partial W}{\partial x} - \theta = (q_s - q_t)\pi \cos \frac{\pi x}{L}. \tag{2.2}$$

Standard engineers' bending theory in which shear strain is taken to be absent—the Euler–Bernoulli beam approximation—is therefore represented by the constraint $q_s = q_t$. The effect of allowing shear in the core, when compared with the fully integral structure, is typically to reduce the Euler (critical buckling) load by the order of 10%. While both components contribute to shear in the core, only the tilt, represented by q_t, contributes to compressive strain energy in the faces; it is thus only q_t that appears in the simplified description of face buckling that follows.

(*b*) *Body force approach to localized buckling*

Figure 1 shows that the local mode is aperiodic and apparently confined to the middle of the strut. The buckling of a single face supported by the core might be modelled in the first instance as a strut of bending stiffness, EI, resting on a nonlinear elastic foundation, as seen in figure 3, with core shear associated purely with the local displacement $w(x)$ being neglected. If the resisting springs provide a force of

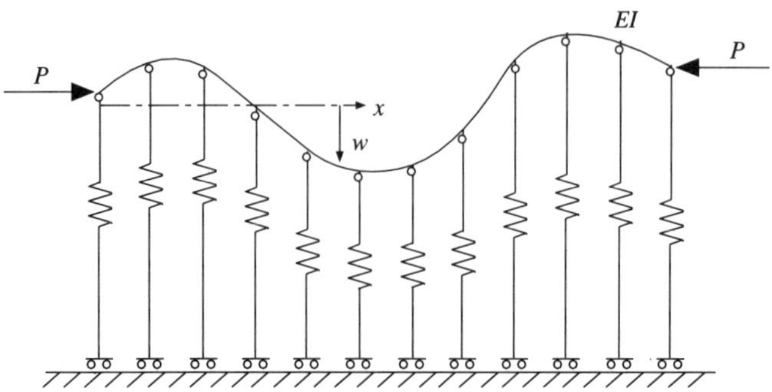

Figure 3. The strut on an elastic foundation.

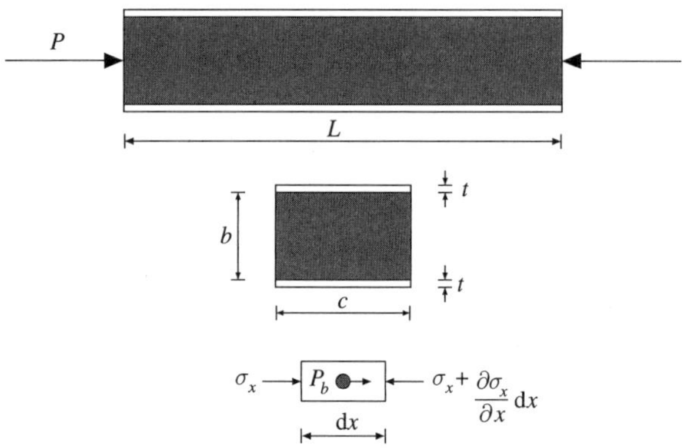

Figure 4. The sandwich strut and its cross-section.

$F = kw - k_1w^2 + k_2w^3$ per unit length of the strut, the response of figure 3 could be described by the equation,

$$EI\ddddot{w} + P\ddot{w} + kw - k_1w^2 + k_2w^3 = 0 \tag{2.3}$$

(Potier-Ferry 1983; Hunt & Wadee 1991), dots denoting differentiation with respect to x. For the sandwich face, however, this classical constant-load formulation no longer suffices. The tilt component of the overall mode adds a long wave variation to the horizontal stress, the effect of which can be modelled as a variable body force (Luongo 1991).

The tilt component alone generates a variable axial strain ε_x in the compression face of the system of figure 4 of

$$\varepsilon_x = \dot{u}_{\mathrm{t}} = \tfrac{1}{2}b\dot{\theta} = -q_{\mathrm{t}}\frac{b\pi^2}{2L}\sin\frac{\pi x}{L}, \tag{2.4}$$

u_{t} being the associated horizontal displacement. Hooke's law then gives the stress σ_x in the face as

$$\sigma_x = E\varepsilon_x = \frac{Eb\pi^2}{2L}q_{\mathrm{t}}\sin\frac{\pi x}{L}, \tag{2.5}$$

where E is the Young's modulus of the face material.

The same stress state can be produced with a variable body force. Equilibrium of the differential face element of figure 4 gives

$$P_{\mathrm{b}}\,\mathrm{d}x = ct\frac{\partial\sigma_x}{\partial x}\,\mathrm{d}x. \tag{2.6}$$

A replacement total load at a position x on the compression face is then given by,

$$P_{\mathrm{total}} = \tfrac{1}{2}P + \int_0^x P_{\mathrm{b}}(x)\,\mathrm{d}x = \tfrac{1}{2}P + P_0 \sin\frac{\pi x}{L}, \tag{2.7}$$

where

$$P_0 = (Etcb\pi^2/2L)q_{\mathrm{t}}. \tag{2.8}$$

If this variable load is added to the classical formulation the following differential-equation results:

$$EI\ddddot{w} + \tfrac{1}{2}P\ddot{w} + P_0\left(\sin\frac{\pi x}{L}\ddot{w} + \frac{\pi}{L}\cos\frac{\pi x}{L}\dot{w}\right) + kw - k_1w^2 + k_2w^3 = 0, \tag{2.9}$$

EI being the flexural rigidity of one face and k being the transverse stiffness of the core. Nonlinearities ($k_1, k_2 \neq 0$) are included in the core to model real situations (Gibson & Ashby 1982; Gibson *et al.* 1982), and such that the model is consistent with the fully interactive model of the next section.

This is not a full interactive buckling model in the sense that the relative magnitudes of overall and local buckling cannot be determined directly; both P and P_0 appear as undefined parameters. It does, however, provide a useful check to the fully interactive buckling model that follows.

(*c*) *Non-dimensionalization*

It is useful at this point to rescale the equations in terms of non-dimensional variables. This process is vital if the equations are to be successfully analysed by the numerical continuation package *AUTO94* (Doedel *et al.* 1995), which is only programmed in non-dimensional terms over length $\tilde{x} = [0, 1]$. The scalings used for the principal variables are as follows:

$$\tilde{x} \rightarrow \frac{2}{L}x, \quad \tilde{w} \rightarrow \frac{2}{L}w. \tag{2.10}$$

For the load and the material parameters:

$$\tilde{P} \rightarrow \frac{PL^2}{8EI}, \quad \tilde{P}_0 \rightarrow \frac{P_0L^2}{8EI}, \quad \tilde{k} \rightarrow \frac{kL^4}{16EI}, \quad \tilde{k}_1 \rightarrow \frac{k_1L^5}{32EI}, \quad \tilde{k}_2 \rightarrow \frac{k_2L^6}{64EI}. \tag{2.11}$$

The scaling of the spatial variables is equivalent to setting $L = 2$:

$$\ddddot{\tilde{w}} + \tilde{P}\ddot{\tilde{w}} + 2\tilde{P}_0(\sin\tfrac{1}{2}\pi\tilde{x}\ddot{\tilde{w}} + \tfrac{1}{2}\pi\cos\tfrac{1}{2}\pi\tilde{x}\dot{\tilde{w}}) + \tilde{k}\tilde{w} - \tilde{k}_1\tilde{w}^2 + \tilde{k}_2\tilde{w}^3 = 0. \tag{2.12}$$

3. Interactive buckling model

The sandwich panel is modelled as two thin plates or faces separated by a softer isotropic core material, with the dimensions of figure 4. Young's moduli and Poisson's ratio of the face and core materials are E and ν, and E_{c} and ν_{c}, respectively. Loading is provided by an axial compression P acting at the midsection as shown, with rigid end plates to ensure that the load transfers directly and equally to the faces. Shear in the core is permitted to enable local buckling (figure 1) and has an associated shear modulus G_{c}.

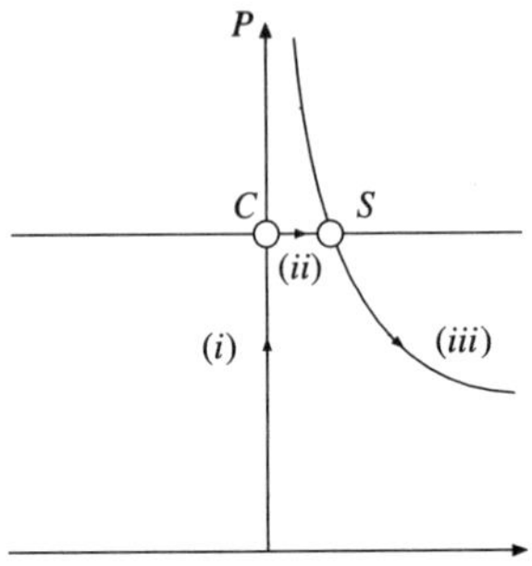

Figure 5. Typical equilibrium diagram for sandwich struts: (*i*) fundamental path; (*ii*) critical path of overall buckling; (*iii*) secondary path of localized buckling. C, critical bifurcation; S, secondary bifurcation.

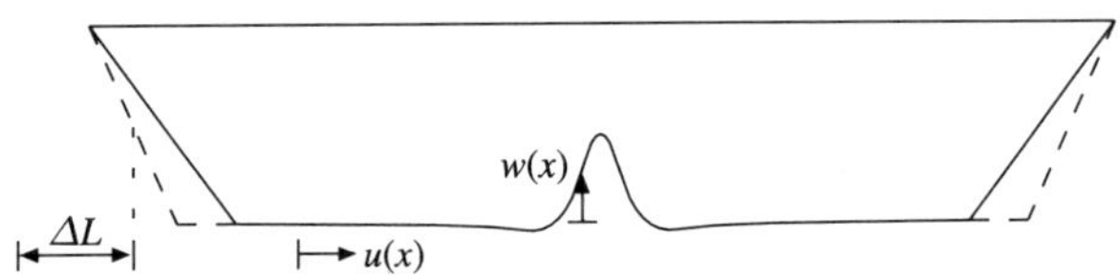

Figure 6. Displacement functions used to model localized buckling.

Overall buckling is represented by the two degrees–of–freedom of sway and tilt of figure 2 whereas no preconceived shape is assumed for local buckling, which remains free to choose its minimum energy configuration via the calculus of variations. The model thus involves the shear angle of the core, but unlike earlier formulations has the freedom for (one of the) faces to compress, buckle locally and localize. We proceed to find the total potential energy of the deflected system.

The expected behaviour of the panel in compression is for it to follow the pure squash fundamental path (figure 5(*i*)) and then buckle in the overall mode (figure 5(*ii*)), causing differential compression in the faces. This is liable to be followed closely by a secondary bifurcation, in which the face under the greater compression buckles in a second *localized* mode (figure 5(*iii*)). The combined outcome is then a highly unstable post-buckling response.

(*a*) *Face-plate displacements*

Along with the sway and tilt contributions from the overall mode, local buckling involves displacement of one face relative to the other. Shown along with the tilt component of overall buckling in figure 6, $w(x)$ is hence defined as the displacement of the buckled face perpendicular to, and $u(x)$ the corresponding displacement parallel to, the unbuckled face. There is also a pure squash strain component of both faces, Δ, which is introduced as a degree of freedom.

(*b*) *Strain energy*

In a general state of deflection there are three components of strain energy, from pure bending in the faces alone (U_{b}), membrane stretching in the faces from overall (integral) structural action (U_{m}); and core energy (U_{c}).

(i) *Bending energy*

The bending energy components arise from the overall bending of both faces, coupled with local bending of the bottom face of figure 6. Linear curvature expressions suffice, giving

$$U_{\mathrm{b}} = \tfrac{1}{2}EI\int_0^L \left(2q_{\mathrm{s}}^2\frac{\pi^4}{L^2}\sin^2\frac{\pi x}{L} + \ddot{w}^2\right)\mathrm{d}x, \tag{3.1}$$

where EI is equivalent to the flexural rigidity of a single face plate, given by,

$$EI = (Ect^3/12(1-\nu^2)). \tag{3.2}$$

(ii) *Membrane energy*

For U_{m}, the top face simply contributes the tensile equivalent of the compressive strain resulting from overall buckling, equation (2.4), along with the component of pure squash

$$\varepsilon_{x\mathrm{t}} = q_{\mathrm{t}}\frac{b\pi^2}{2L}\sin\frac{\pi x}{L} - \Delta, \tag{3.3}$$

while the bottom face adds as extra contributions the corresponding strains $\mathrm{d}u/\mathrm{d}x$ and $\frac{1}{2}(\mathrm{d}w/\mathrm{d}x)^2$, terms familiar from standard von Kármán large-deflection plate theory:

$$\varepsilon_{x\mathrm{b}} = -q_{\mathrm{t}}\frac{b\pi^2}{2L}\sin\frac{\pi x}{L} - \Delta + \dot{u} + \tfrac{1}{2}\dot{w}^2. \tag{3.4}$$

The membrane strain energy is therefore:

$$\begin{aligned} U_{\mathrm{m}} &= \tfrac{1}{2}Etc\int_0^L (\varepsilon_{x\mathrm{t}}^2 + \varepsilon_{x\mathrm{b}}^2)\,\mathrm{d}x \\ &= \tfrac{1}{2}Etc\int_0^L \Bigg[2\Delta^2 + q_{\mathrm{t}}^2\frac{b^2\pi^4}{2L^2}\sin^2\frac{\pi x}{L} + \dot{u}^2 + \tfrac{1}{4}\dot{w}^4 \\ &\qquad - \left(2\Delta + q_{\mathrm{t}}\frac{b\pi^2}{L}\sin\frac{\pi x}{L}\right)\left(\dot{u} + \tfrac{1}{2}\dot{w}^2\right) + \dot{u}\dot{w}^2\Bigg]\mathrm{d}x. \end{aligned} \tag{3.5}$$

(iii) *Core energy*

In general the core provides all the transverse and shear resistance but only some of the longitudinal resistance, most of which comes from the faces. Analysis is somewhat simplified if it is assumed that the faces provide all of the latter. Plane stress expressions are readily derivable which include a longitudinal strain component in the core, but not much is lost conceptually if this is ignored. To match the assumed displacements $w(x)$ and $u(x)$ on the bottom face and zero on the top face of figure 6, displacements $w_{\mathrm{c}}(x,y)$ and $u_{\mathrm{c}}(x,y)$ must vary through the core. If x and y are defined as in figure 2, linear variations with y are assumed as follows:

$$u_{\mathrm{c}}(x,y) = \left(\frac{b-2y}{2b}\right)u(x), \tag{3.6}$$

$$w_{\mathrm{c}}(x,y) = \left(\frac{b-2y}{2b}\right)w(x). \tag{3.7}$$

Under the assumption that ε_x is zero in the core, and with the addition of the component (2.2) from overall buckling, the remaining core strains can be written

$$\varepsilon_y = \frac{\partial w_{\mathrm{c}}}{\partial y} = -\frac{w}{b}, \tag{3.8}$$

$$\gamma_{xy} = (q_{\mathrm{s}} - q_{\mathrm{t}})\pi\cos\frac{\pi x}{L} + \frac{\partial w_{\mathrm{c}}}{\partial x} + \frac{\partial u_{\mathrm{c}}}{\partial y} = (q_{\mathrm{s}} - q_{\mathrm{t}})\pi\cos\frac{\pi x}{L} + \left(\frac{b-2y}{2b}\right)\dot{w} - \frac{u}{b}. \tag{3.9}$$

The strain energy of the core is thus

$$U_{\mathrm{c}} = \frac{E_{\mathrm{c}}c}{2(1-\nu_{\mathrm{c}}^2)}\int_0^L \int_{-b/2}^{b/2} \varepsilon_y^2 \,\mathrm{d}y\mathrm{d}x + \tfrac{1}{2}G_{\mathrm{c}}c\int_0^L \int_{-b/2}^{b/2} \gamma_{xy}^2 \,\mathrm{d}y\mathrm{d}x. \tag{3.10}$$

Substituting the appropriate expressions and integrating with respect to y, we obtain

$$U_{\mathrm{c}} = \int_0^L \Bigg\{ \tfrac{1}{2}kw^2 + \tfrac{1}{2}G_{\mathrm{c}}cb\bigg[(q_{\mathrm{s}} - q_{\mathrm{t}})^2\pi^2\cos^2\frac{\pi x}{L} + (q_{\mathrm{s}} - q_{\mathrm{t}})\pi\cos\frac{\pi x}{L}\dot{w} + \tfrac{1}{3}\dot{w}^2 + \frac{u^2}{b^2} - \frac{u\dot{w}}{b} - 2u\frac{(q_{\mathrm{s}} - q_{\mathrm{t}})}{b}\pi\cos\frac{\pi x}{L}\bigg]\Bigg\}\,\mathrm{d}x, \tag{3.11}$$

where k is introduced as the linear transverse core stiffness:

$$k = \frac{E_{\mathrm{c}}c}{(1-\nu_{\mathrm{c}}^2)b}. \tag{3.12}$$

We note that the introduction of shear interactions makes the model similar to a strut on a Pasternak foundation (Kerr 1964), which accounts for both transverse and shear resistance.

(iv) *Core material nonlinearities*

It has been found experimentally that cellular core materials such as foams and balsa wood do not exhibit linear elastic behaviour when in compression. Their actual behaviour is elastoplastic, and is primarily concerned with the response of the cells in compression at the microscopic level (Gibson *et al.* 1982; Gibson & Ashby 1982). At first the cells behave in linearly elastic manner (figure 7*a*). After a certain strain, the cell walls buckle elastically causing a loss of compressive stiffness, sometimes to an extent where it becomes negative (figure 7*b*). However, as the buckles become larger and the opposite walls of the cell are so close that they begin to bear up against each other (figure 7*c*); the stiffness then rises again to values close to the linear elastic value. Although no plasticity effects are included, for modelling purposes, the transverse resistance of the core, F, involves both a softening quadratic nonlinearity (k_1) to model the initial loss of stiffness, and a hardening cubic nonlinearity (k_2) to account for the subsequent increase. The same terms are included in the body force representation of equation (2.9).

The nonlinear transverse resistance (F) and the additional nonlinear core strain energy (U_F) are thus

$$F = kw - k_1w^2 + k_2w^3, \tag{3.13}$$

$$U_F = \int_0^L (-\tfrac{1}{3}k_1w^3 + \tfrac{1}{4}k_2w^4)\,\mathrm{d}x. \tag{3.14}$$

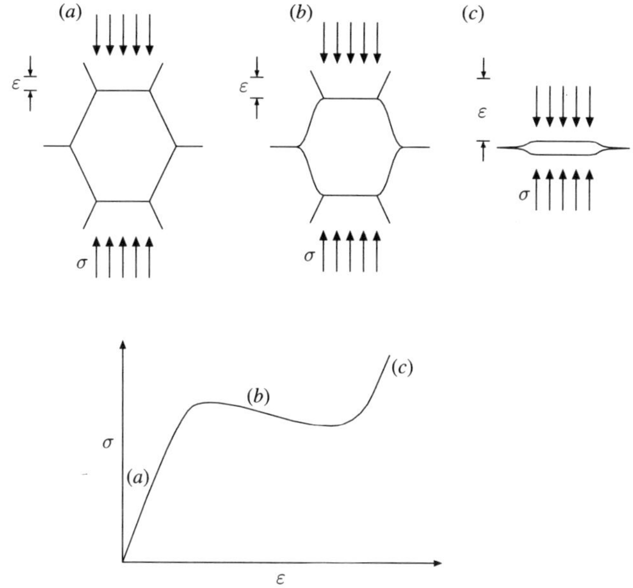

Figure 7. Core material behaviour in compression at the microscopic cell level.

(c) *Work done by load*

The final component of energy to be identified is the work done by the load. The overall tilt mode contributes nothing to the corresponding deflection $\mathcal{E}$ of a load P applied half way between the faces (figure 4), while the contribution from the local buckling of a single face is clearly just one half of the end shortening of the face itself. Together with the contributions from pure squash and sway from overall buckling (figure 2) this gives the work done as

$$P\mathcal{E} = P\int_0^L \left(\tfrac{1}{2}q_s^2\pi^2\cos^2\frac{\pi x}{L} - \tfrac{1}{2}\dot{u} + \Delta\right)\mathrm{d}x. \tag{3.15}$$

(d) *Potential energy functional*

The total potential energy of the complete structure,

$$\begin{aligned} V = \int_0^L \Bigg\{ & \tfrac{1}{2}G_c cb\Bigg[(q_s - q_t)^2\pi^2\cos^2\frac{\pi x}{L} + (q_s - q_t)\pi\cos\frac{\pi x}{L}\dot{w} \\ & -2\frac{(q_s - q_t)}{b}\pi\cos\frac{\pi x}{L}u + \frac{u^2}{b^2} + \tfrac{1}{3}\dot{w}^2 - \frac{u\dot{w}}{b}\Bigg] + \tfrac{1}{2}kw^2 - \tfrac{1}{3}k_1w^3 + \tfrac{1}{4}k_2w^4 \\ & +\tfrac{1}{2}Etc\Bigg[2\Delta^2 + \tfrac{1}{4}\dot{w}^4 + q_t^2\frac{b^2\pi^4}{2L^2}\sin^2\frac{\pi x}{L} + \dot{u}^2 + \dot{u}\dot{w}^2 \\ & -\left(2\Delta + q_t\frac{b\pi^2}{L}\sin\frac{\pi x}{L}\right)\left(\dot{u} + \tfrac{1}{2}\dot{w}^2\right)\Bigg] \\ & +\tfrac{1}{2}EI\left(\ddot{w}^2 + 2q_s^2\frac{\pi^4}{L^2}\sin^2\frac{\pi x}{L}\right) - P\left(\tfrac{1}{2}q_s^2\pi^2\cos^2\frac{\pi x}{L} - \tfrac{1}{2}\dot{u} + \Delta\right)\Bigg\}\,\mathrm{d}x, \end{aligned} \tag{3.16}$$

is given by the summation of all the strain energy terms: U_c, U_F, U_m and U_b, minus the work done by the load, $P\mathcal{E}$ (Thompson & Hunt 1973). This is in a sense a hybrid form, the integrand being first a function of three generalized coordinates q_s, q_t, and Δ and secondly a functional involving the two functions $w(x)$ and $u(x)$.

(e) *Linear eigenvalue analysis for overall buckling*

A simple linear eigenvalue analysis yields the critical load for overall buckling P^C, arising on the pure squash fundamental path F at which $q_s = q_t = w = u = 0$. The potential energy is non-diagonalized with respect to q_s and q_t, and so this critical load occurs when the matrix

$$\boldsymbol{V}_{ij}^{F} = \begin{pmatrix} V_{ss}^{F} & V_{st}^{F} \\ V_{ts}^{F} & V_{tt}^{F} \end{pmatrix}, \tag{3.17}$$

where

$$V_{ss}^{F} = \left.\frac{\partial^2 V}{\partial q_s^2}\right|^{F} = \frac{\pi^4 EI}{L} + \tfrac{1}{2}G_c cb\pi^2 L - P^F\frac{\pi^2 L}{2}, \tag{3.18}$$

$$V_{tt}^{F} = \left.\frac{\partial^2 V}{\partial q_t^2}\right|^{F} = \frac{Etcb^2\pi^4}{4L} + \tfrac{1}{2}G_c cb\pi^2 L, \tag{3.19}$$

$$V_{st}^{F} = V_{ts}^{F} = \left.\frac{\partial^2 V}{\partial q_s \partial q_t}\right|^{F} = -\tfrac{1}{2}G_c cb\pi^2 L. \tag{3.20}$$

is singular, subscripts denoting partial differentiation with respect to either q_s or q_t. Therefore at the critical point, C,

$$\boldsymbol{V}_{ij}^{C} = \begin{vmatrix} V_{ss}^{C} & V_{st}^{C} \\ V_{ts}^{C} & V_{tt}^{C} \end{vmatrix} = 0, \tag{3.21}$$

and the critical load for overall buckling is obtained:

$$P^C = G_c cb + \frac{2\pi^2 EI}{L^2} - \left[\frac{G_c^2 cb}{(Etb\pi^2/2L^2) + G_c}\right], \tag{3.22}$$

which agrees with the classical eigenvalue analysis of Allen (1969).

(f) *Equilibrium equations*

This section deals with the analysis of the potential energy functional (equation (3.16)) in terms of the variational formulation. Firstly, the variations of w and u are considered, leading to a pair of coupled ordinary differential equations. Then the equilibrium equations arising from minimizing the potential energy with respect to the sway and tilt degrees of freedom are formulated. These equations will form the basis for numerical runs to determine the post-buckling behaviour.

(i) *Lagrangian*

Equation (3.16) is an action integral representing the total potential energy of the physical system. For equilibrium V must be stationary, and so the calculus of variations is used to find the stationary point of the integral. The analysis that follows is a summary of this application of the calculus of variations. Consider the Lagrangian ($\mathcal{L}$) of the form

$$V = \int_0^L \mathcal{L}(\ddot{w}, \dot{w}, w, \dot{u}, u, x)\,\mathrm{d}x. \tag{3.23}$$

As spatial systems are by nature reversible, they can be considered, in a dynamical systems sense as Hamiltonian, and Hamilton's principle states that the integral of the Lagrangian, (V), is stationary with respect to the spatial variable, x. Conditions for V to be stationary requires the first variation of V, and this is given by the following expression (Fox 1987):

$$\delta V = \int_0^L \left[\frac{\partial \mathcal{L}}{\partial \ddot{w}} \delta \ddot{w} + \frac{\partial \mathcal{L}}{\partial \dot{w}} \delta \dot{w} + \frac{\partial \mathcal{L}}{\partial w} \delta w + \frac{\partial \mathcal{L}}{\partial \dot{u}} \delta \dot{u} + \frac{\partial \mathcal{L}}{\partial u} \delta u \right] \mathrm{d}x. \tag{3.24}$$

By substituting the appropriate terms, the following expression is obtained:

$$\begin{aligned} \delta V = \int_0^L \Bigg\{ & EI\ddot{w}\delta\ddot{w} + \Bigg[\tfrac{1}{2}Etc \left(\dot{w}^3 + 2\dot{u}\dot{w} - 2\Delta\dot{w} - q_{\mathrm{t}} \frac{b\pi^2}{L} \sin\frac{\pi x}{L} \dot{w} \right) \\ & + \tfrac{1}{2}G_{\mathrm{c}}cb \left(\tfrac{2}{3}\dot{w} - \frac{u}{b} + (q_{\mathrm{s}} - q_{\mathrm{t}})\pi \cos\frac{\pi x}{L} \right) \Bigg] \delta\dot{w} + [kw - k_1 w^2 + k_2 w^3]\delta w \\ & + \left[\tfrac{1}{2}Etc \left(2\dot{u} + \dot{w}^2 - 2\Delta - q_{\mathrm{t}} \frac{b\pi^2}{L} \sin\frac{\pi x}{L} \right) + \tfrac{1}{2}P \right] \delta\dot{u} \\ & + \left[\tfrac{1}{2}G_{\mathrm{c}}c \left(\frac{2u}{b} - \dot{w} - 2(q_{\mathrm{s}} - q_{\mathrm{t}})\pi \cos\frac{\pi x}{L} \right) \right] \delta u \Bigg\} \mathrm{d}x. \end{aligned} \tag{3.25}$$

For V to be stationary δV must disappear for all δw and δu, which means that two separate equations arise from V. First, however, two new parameters are introduced to encapsulate the material properties of the strut:

$$D = \tfrac{1}{2}Etc, \quad G = \tfrac{1}{2}G_{\mathrm{c}}cb. \tag{3.26}$$

Integrating by parts by changing the order of the differentiation, i.e. $\delta\dot{w} = (\mathrm{d}/\mathrm{d}x)\delta w$ and $\delta\ddot{w} = (\mathrm{d}/\mathrm{d}x)\delta\dot{w}$, we obtain

$$\begin{aligned} \delta V = & \Big[EI\ddot{w}\delta\dot{w} \Big]_0^L + \Bigg[\Bigg(EI\dddot{w} + D\left(\dot{w}^3 + 2\dot{u}\dot{w} - 2\Delta\dot{w} - q_{\mathrm{t}}\frac{b\pi^2}{L}\sin\frac{\pi x}{L}\dot{w} \right) \\ & + G\left(\tfrac{2}{3}\dot{w} - \frac{u}{b} + (q_{\mathrm{s}} - q_{\mathrm{t}})\pi\cos\frac{\pi x}{L} \right) \Bigg) \delta w \Bigg]_0^L \\ & + \left[\left(D\left(2\dot{u} + \dot{w}^2 - 2\Delta - q_{\mathrm{t}}\frac{b\pi^2}{L}\sin\frac{\pi x}{L} \right) + \tfrac{1}{2}P \right) \delta u \right]_0^L \\ & + \int_0^L \Bigg\{ \Bigg[EI\ddddot{w} + D\Bigg(2\Delta\ddot{w} + q_{\mathrm{t}}\frac{b\pi^2}{L}\left(\sin\frac{\pi x}{L}\ddot{w} + \frac{\pi}{L}\cos\frac{\pi x}{L}\dot{w} \right) \\ & - (2\ddot{u}\dot{w} + 2\dot{u}\ddot{w} + 3\dot{w}^2\ddot{w}) \Bigg) + G\left(\frac{\dot{u}}{b} - \tfrac{2}{3}\ddot{w} + (q_{\mathrm{s}} - q_{\mathrm{t}})\frac{\pi^2}{L}\sin\frac{\pi x}{L} \right) \\ & + kw - k_1 w^2 + k_2 w^3 \Bigg] \delta w + \Bigg[-2D\ddot{u} + \frac{2G}{b^2}u - \frac{G}{b}\dot{w} - 2D\dot{w}\ddot{w} \\ & - \pi\cos\frac{\pi x}{L}\left(2G\frac{(q_{\mathrm{s}} - q_{\mathrm{t}})}{b} - Dq_{\mathrm{t}}\frac{b\pi^2}{L^2} \right) \Bigg] \delta u \Bigg\} \mathrm{d}x. \end{aligned} \tag{3.27}$$

With the assumption of simple supports at the ends of the buckled faces, the physical boundary conditions

$$w(0) = w(L) = 0, \quad \ddot{w}(0) = \ddot{w}(L) = 0, \tag{3.28}$$

eliminate the first two terms in square brackets. Similarly, matching the applied stress at the ends,

$$-P/(4D) = \dot{u}(0) + \tfrac{1}{2}\dot{w}^2(0) - \Delta = \dot{u}(L) + \tfrac{1}{2}\dot{w}^2(L) - \Delta, \tag{3.29}$$

eliminates the third term. For $\delta V = 0$ this leaves the integrand, which also must vanish to give the *Euler–Lagrange equations*:

$$EI\ddddot{w} + D\left[2\Delta\ddot{w} + q_t\frac{b\pi^2}{L}\left(\sin\frac{\pi x}{L}\ddot{w} + \frac{\pi}{L}\cos\frac{\pi x}{L}\dot{w}\right) - (2\ddot{u}\ddot{w} + 2\dot{u}\dddot{w} + 3\dot{w}^2\ddot{w})\right]$$

$$+G\left[\frac{\dot{u}}{b} - \tfrac{2}{3}\ddot{w} + (q_s - q_t)\frac{\pi^2}{L}\sin\frac{\pi x}{L}\right] + kw - k_1w^2 + k_2w^3 = 0, \tag{3.30}$$

$$D\ddot{u} - \frac{G}{b^2}u + \frac{G}{2b}\dot{w} + D\dot{w}\ddot{w} + \pi\cos\frac{\pi x}{L}\left(G\frac{(q_s - q_t)}{b} - Dq_t\frac{b\pi^2}{2L^2}\right) = 0. \tag{3.31}$$

This is a pair of coupled differential equations, equivalent to a single equation of sixth order in x.

As well as δV being zero for any δw and δu, for equilibrium, V must also be stationary with respect to the degrees of freedom, q_s, q_t and Δ. Thus,

$$\frac{\partial V}{\partial q_s} = \frac{\pi^4 EI}{L}q_s - \tfrac{1}{2}P\pi^2 Lq_s + G\pi^2 L(q_s - q_t) + \int_0^L G\pi\cos\frac{\pi x}{L}\left(\dot{w} - \frac{2}{b}u\right)\mathrm{d}x = 0, \tag{3.32}$$

$$\frac{\partial V}{\partial q_t} = \frac{D\pi^4 b^2}{2L}q_t - G\pi^2 L(q_s - q_t) + \int_0^L \left[G\pi\cos\frac{\pi x}{L}\left(\frac{2}{b}u - \dot{w}\right) - D\frac{b\pi^2}{L}\sin\frac{\pi x}{L}(\dot{u} + \tfrac{1}{2}\dot{w}^2)\right]\mathrm{d}x = 0, \tag{3.33}$$

$$\frac{\partial V}{\partial \Delta} = 4D\Delta L - PL - \int_0^L (2\dot{u} + \dot{w}^2)\,\mathrm{d}x = 0, \tag{3.34}$$

must also be satisfied. Rearranging these to make P, q_s and Δ the subject of the equations, the following relationships are obtained:

$$P = \frac{2\pi^2 EI}{L^2} + 2G\left(\frac{q_s - q_t}{q_s}\right) + \frac{2G}{q_s\pi L}\int_0^L \cos\frac{\pi x}{L}\left(\dot{w} - \frac{2}{b}u\right)\mathrm{d}x, \tag{3.35}$$

$$q_s = q_t\left(\frac{D\pi^2 b^2}{2GL^2} + 1\right) + \int_0^L \left[\frac{1}{\pi L}\cos\frac{\pi x}{L}\left(\frac{2}{b}u - \dot{w}\right) - \frac{Db}{GL^2}\sin\frac{\pi x}{L}(\dot{u} + \tfrac{1}{2}\dot{w}^2)\right]\mathrm{d}x, \tag{3.36}$$

$$\Delta = \frac{P}{4D} + \frac{1}{2L}\int_0^L (\dot{u} + \tfrac{1}{2}\dot{w}^2)\,\mathrm{d}x. \tag{3.37}$$

(*g*) *Non-dimensionalization*

As for the earlier body force model, these equations can be non-dimensionalized. Scalings for x, w, k, k_1 and k_2 are the same as in equations (2.10) and (2.11). For the remainder, the following new definitions apply:

$$\tilde{u} \to \frac{2}{L}u, \quad \tilde{D} \to \frac{DL^2}{8EI}, \quad \tilde{G} \to \frac{GL^3}{8EIb}, \quad \tilde{P} \to \frac{P}{2G}. \tag{3.38}$$

For the degrees-of-freedom the following scalings are used:

$$\tilde{q}_{\mathrm{s}} \to \frac{b\pi^2}{L} q_{\mathrm{s}}, \quad \tilde{q}_{\mathrm{t}} \to \frac{b\pi^2}{L} q_{\mathrm{t}}. \tag{3.39}$$

Also, two new parameters $\tilde{\phi}$, and $\tilde{s}$ are defined thus

$$\tilde{\phi} = \frac{b}{L}, \quad \tilde{s} = \frac{2\tilde{G}}{\tilde{D}\tilde{\phi}\pi^2}. \tag{3.40}$$

The non-dimensional versions of the governing equations are

$$\begin{aligned} \ddddot{\tilde{w}} + 2\tilde{D}[2\Delta\ddot{\tilde{w}} + \tilde{q}_{\mathrm{t}}(\sin\tfrac{1}{2}\pi\tilde{x}\ddot{\tilde{w}} + \tfrac{1}{2}\pi\cos\tfrac{1}{2}\pi\tilde{x}\dot{\tilde{w}}) - (2\ddot{\tilde{u}}\dot{\tilde{w}} + 2\dot{\tilde{u}}\ddot{\tilde{w}} + 3\dot{\tilde{w}}^2\ddot{\tilde{w}})] \\ +\tilde{G}[\dot{\tilde{u}} - \tfrac{4}{3}\tilde{\phi}\ddot{\tilde{w}} + (\tilde{q}_{\mathrm{s}} - \tilde{q}_{\mathrm{t}})\sin\tfrac{1}{2}\pi\tilde{x}] + \tilde{k}\tilde{w} - \tilde{k}_1\tilde{w}^2 + \tilde{k}_2\tilde{w}^3 = 0, \end{aligned} \tag{3.41}$$

$$\ddot{\tilde{u}} - \tfrac{1}{8}\tilde{s}\pi^2(\tilde{u} - \tilde{\phi}\dot{\tilde{w}}) + \dot{\tilde{w}}\ddot{\tilde{w}} + [\tilde{s}(\tilde{q}_{\mathrm{s}} - \tilde{q}_{\mathrm{t}}) - \tilde{q}_{\mathrm{t}}]\tfrac{1}{4}\pi\cos\tfrac{1}{2}\pi\tilde{x} = 0, \tag{3.42}$$

$$\tilde{s}(\tilde{q}_{\mathrm{s}} - \tilde{q}_{\mathrm{t}}) = \tilde{q}_{\mathrm{t}} + \int_0^2 [\tfrac{1}{2}\tilde{s}\pi\cos\tfrac{1}{2}\pi\tilde{x}(\tilde{u} - \tilde{\phi}\dot{\tilde{w}}) - \sin\tfrac{1}{2}\pi\tilde{x}(\dot{\tilde{u}} + \tfrac{1}{2}\dot{\tilde{w}}^2)]\,\mathrm{d}\tilde{x}, \tag{3.43}$$

$$\tilde{P} = \frac{\pi^2}{8\tilde{G}\tilde{\phi}} + \left(\frac{\tilde{q}_{\mathrm{s}} - \tilde{q}_{\mathrm{t}}}{\tilde{q}_{\mathrm{s}}}\right) + \frac{1}{\tilde{q}_{\mathrm{s}}}\int_0^2 \tfrac{1}{2}\pi\cos\tfrac{1}{2}\pi\tilde{x}(\tilde{\phi}\dot{\tilde{w}} - \tilde{u})\,\mathrm{d}\tilde{x}, \tag{3.44}$$

$$\Delta = \left(\frac{\tilde{G}\tilde{\phi}}{2\tilde{D}}\right)\tilde{P} + \frac{1}{4}\int_0^2 (\dot{\tilde{u}} + \tfrac{1}{2}\dot{\tilde{w}}^2)\,\mathrm{d}\tilde{x}. \tag{3.45}$$

(*h*) *Secondary bifurcation for local buckling*

At the secondary bifurcation point, the relationships between $\tilde{q}_{\mathrm{s}}$ and $\tilde{q}_{\mathrm{t}}$, $\tilde{P}$ and Δ simplify to

$$\tilde{q}_{\mathrm{s}} - \tilde{q}_{\mathrm{t}} = \frac{\tilde{q}_{\mathrm{t}}}{\tilde{s}}, \quad \tilde{P} = \tilde{P}^{\mathrm{C}} = \frac{\pi^2}{8\tilde{G}\tilde{\phi}} + \frac{1}{1+\tilde{s}}, \quad \Delta = \left(\frac{\tilde{G}\tilde{\phi}}{2\tilde{D}}\right)\tilde{P}^{\mathrm{C}}. \tag{3.46}$$

Differential equations, (3.41) and (3.42), can also be linearized to give

$$\ddddot{\tilde{w}} + 2\tilde{G}\tilde{\phi}\ddot{\tilde{w}}[\tilde{P}^{\mathrm{C}} - \tfrac{2}{3}] + 2\tilde{D}\tilde{q}_{\mathrm{t}}[\sin\tfrac{1}{2}\pi\tilde{x}\ddot{\tilde{w}} + \tfrac{1}{2}\pi\cos\tfrac{1}{2}\pi\tilde{x}\dot{\tilde{w}}] + \tilde{G}[\dot{\tilde{u}} + (\tilde{q}_{\mathrm{t}}/\tilde{s})\sin\tfrac{1}{2}\pi\tilde{x}] + \tilde{k}\tilde{w} = 0, \tag{3.47}$$

$$\ddot{\tilde{u}} - \tfrac{1}{8}\tilde{s}\pi^2(\tilde{u} - \tilde{\phi}\dot{\tilde{w}}) = 0, \tag{3.48}$$

describing the face under the greater compression. As these equations are non-autonomous the position of the secondary bifurcation is best determined numerically. We thus vary q_{t} until a non-trivial solution with a slight hump at the centre is found. Physically, this occurs when the effective load at midspan is greater than the local buckling load.

4. Numerical experimentation

Having formulated the necessary equations, we now present the fully interactive solutions found from the numerical continuation package *AUTO94* (Doedel *et al.* 1995). Simply-supported boundary conditions (3.28) are chosen at the ends of the face in question, and the symmetric-section condition at the centre $\tilde{x} = 1$:

$$\dot{w} = \dddot{w} = u = 0, \tag{4.1}$$

Table 1. *Comparison of the position of secondary bifurcation for the interactive model and earlier Rayleigh–Ritz results*

(For $L = 508$ mm the periodic results are interpolated from graphical data.)

L/mm	measure at secondary bifurcation	periodic Rayleigh–Ritz model (Da Silva 1989)	interactive localized buckling model
50.8	q_s	0.140	0.116
	$\mathcal{E}/L$	0.053	0.038
508	q_s	0.46	0.304
	$\mathcal{E}/L$	0.53	0.228

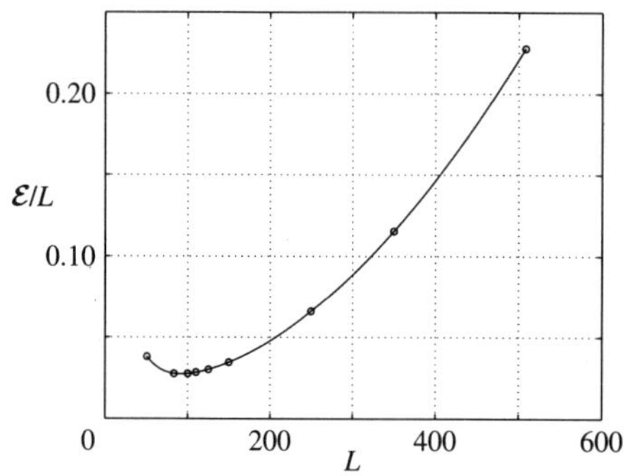

Figure 8. Relative end shortening at secondary bifurcation.

ensures that overall symmetry is not lost. The solutions are presented as follows: first, the position of the secondary bifurcation on the critical path for overall buckling (see figure 5) is found for a range of different lengths; secondly, for the 'worst case' in which this occurs closest to the initial bifurcation, the full post-buckling response is determined for the separate cases of linear, quadratic and cubic core transverse resistances; finally, results of the interactive model are compared with those found from the body force approach at the same q_t value.

(*a*) *Numerical results*

We choose a set of practical sandwich component properties from the literature (Hunt *et al.* 1988):

face Young's modulus :	$E = 68947.57\ \mathrm{N\,mm^{-2}}$,
core Young's modulus :	$E_c = 198.57\ \mathrm{N\,mm^{-2}}$,
core shear modulus :	$G_c = 82.74\ \mathrm{N\,mm^{-2}}$,
face thickness :	$t = 0.508$ mm,
core thickness :	$b = 5.08$ mm,
strut width :	$c = 254$ mm,
face Poisson's ratio :	$\nu = 0.3$,
core Poisson's ratio :	$\nu_c = 0.2$,

with the length of the structure (L) left as a possible parameter to vary.

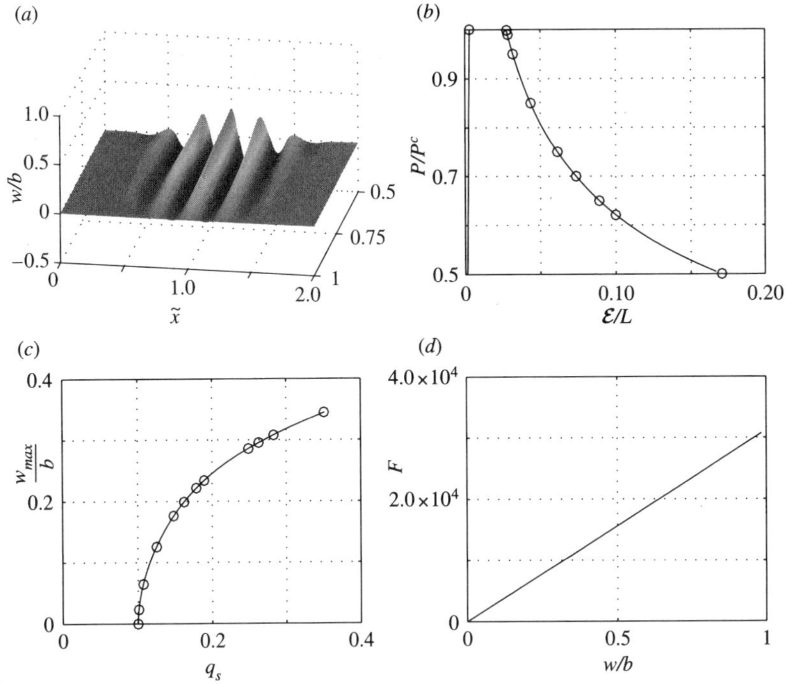

Figure 9. Post-buckling response for strut of $L = 100$ mm with linear core: (*a*) localized buckle profiles; (*b*) load–end shortening curve; (*c*) relative magnitudes of overall and localized modal amplitudes; (d) transverse resistance of core.

(i) *Location of secondary bifurcation*

Although linear, the secondary bifurcation equations (3.47) and (3.48) are non-autonomous and most easily handled numerically. Solutions are first sought at constant q_t by shooting for the symmetric section (4.1) using a symplectic Runge–Kutta scheme. These are then ported into *AUTO94* so that numerical continuation can be used to reduce parametrically the value of q_t until no solution can be found. This defines the position of the secondary bifurcation point. Figure 8 shows the relation between total relative end shortening at the position of secondary bifurcation,

$$(\mathcal{E}/L) = \tfrac{1}{4}q_s^2\pi^2 + \Delta, \tag{4.2}$$

with length. Table 1 compares this quantitatively with similar results from periodic (Rayleigh–Ritz) formulations, for two struts of the same dimensions as above with lengths of 50.8 mm and 508 mm (Da Silva 1989). In each case the secondary bifurcation that leads to localization occurs well inside that of the periodic model.

Over the range $50 \leqslant L \leqslant 150$, the end shortening at secondary bifurcation is relatively small. The curve reaches a minimum at approximately $L = 100$ mm, and we therefore choose this length for detailed post-buckling study below. This has quite a realistic L/b ratio of around twenty, which for a strut is not overly slender nor stocky.

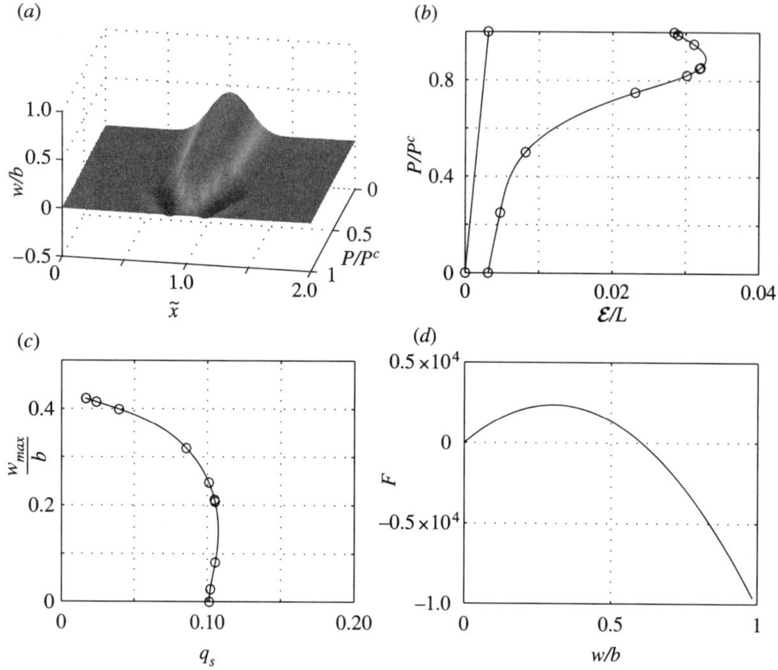

Figure 10. Post-buckling response for strut of $L = 100$ mm with quadratic core: (*a*) localized buckle profiles; (*b*) load–end shortening curve; (*c*) relative magnitudes of overall and localized modal amplitudes; (*d*) transverse resistance of core.

(ii) *Full post-buckling study*

The near-optimum length of $L = 100$ mm is next reviewed in post-buckling regime by solving equations (3.41) and (3.42) subject to the integral constraints (3.43), (3.44) and (3.45). Different results are presented for differing core resistance properties F as expressed in the non-dimensional form of equation (3.13). In the first case, the core is assumed to remain linearly elastic, in the second this stiffness is augmented by a destabilizing quadratic term, and in the third a restabilizing cubic term is introduced. Results are shown for through-core displacements of the order of $\frac{1}{2}b$.

Figure 9 details the response for a linearly elastic core: $F = kw$, where $k = 10341\ \mathrm{N\,mm^{-2}}$. In figure 9*a* the variation of the buckle profile of the face under greater compression is shown as load reduces after the secondary bifurcation at $P = P^{\mathrm{C}}$. Figure 9*b* shows the variation of the total end-shortening of the strut with load, which indicates a rapid deterioration in load-carrying capacity in the post-buckling range (with the straight line representing the pre-buckling path). Figure 9*b* levels out at $P \approx 0.3P^{\mathrm{C}}$, but deformations are by this stage so large as to be unrealistic. Figure 9*d* shows the core constitutive relationship.

The introduction of a destabilizing quadratic nonlinearity models core materials that initially soften under compression: $F = kw - k_1w^2$. Figure 10 shows results for a typical value of $k_1 = 6728\ \mathrm{N\,mm^{-3}}$. Figure 10*a* is comparable to figure 9*a* but with an exaggerated central spike, as would be expected. At about $P \leqslant 0.9P^{\mathrm{C}}$) a region of *snap-back* behaviour is initiated, where the load–end shortening relation

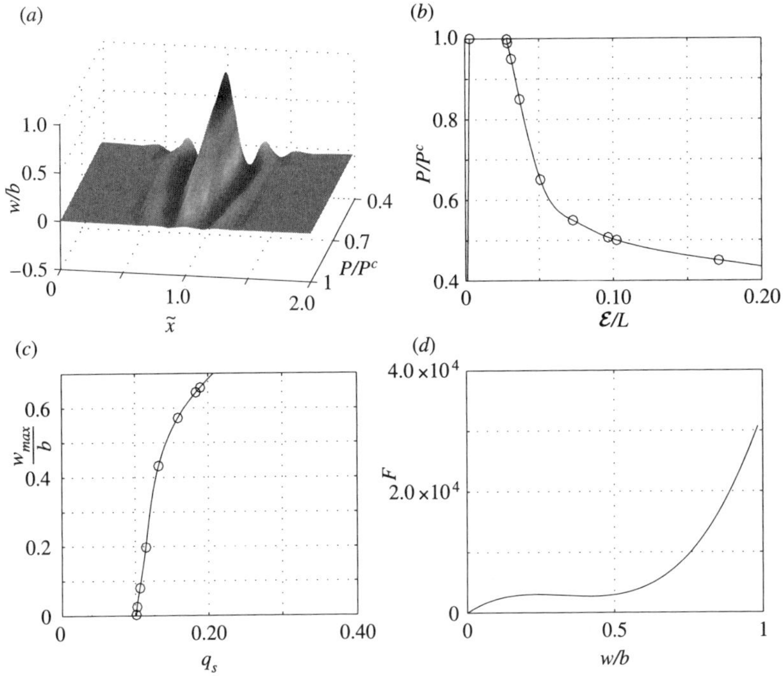

Figure 11. Post-buckling response for strut of $L = 100$ mm with cubic core: (*a*) localized buckle profiles; (*b*) load–end shortening curve; (*c*) relative magnitudes of overall and localized modal amplitudes; (*d*) transverse resistance of core.

of figure 10*b* has a positive slope, indicating instability under rigid as well as dead loading. Figure 10*d* shows, however, that large values of the maximum amplitude of w lead to a regime of negative core stiffness, which might be considered impractical.

To offset this and reflect the experimental evidence, a restabilizing cubic term is next introduced: $F = kw - k_1 w^2 + k_2 w^3$.

Figure 11 shows the results for the same k and k_1 as before, and $k_2 = 1346$ N mm^{-4}. Figure 11*a* has qualitatively the same features as figures 9*a* and 10*a*, in that the central amplitude begins to dominate as the load reduces. The load–end shortening relationship of figure 11*b* shares features of the linear and the quadratic cases. The strut restabilizes like the linear case, except that the cubic core has a sharper initial destabilization; for instance for the former when $\mathcal{E}/L = 0.05$, $P = 0.81P^{\mathrm{C}}$, but for the cubic case at the same $\mathcal{E}/L$ value, $P = 0.65P^{\mathrm{C}}$. This is of course an effect of the quadratic softening. The system restabilizes at $P \approx 0.40P^{\mathrm{C}}$, when the influence of the hardening cubic nonlinearity begins to dominate over the quadratic term. The destabilization–restabilization of the core is shown in figure 11*d*, which reflects qualitatively the form of figure 7.

(iii) *Comparison of body force approach with interactive buckling model*

By relating analogous forms for both P and q_{t}, comparisons between the body force approach and the results of the fully interactive model can be drawn up. Core shear appears in the latter but not the former, so $\tilde{P}$ of equation (2.12) in the body force

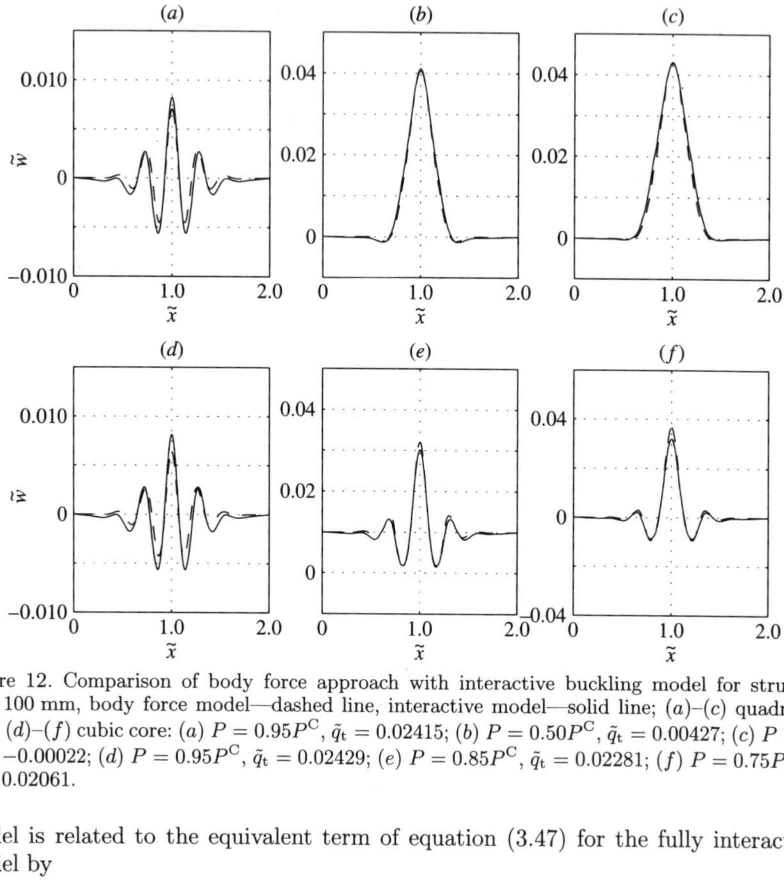

Figure 12. Comparison of body force approach with interactive buckling model for strut of $L = 100$ mm, body force model—dashed line, interactive model—solid line; (*a*)–(*c*) quadratic core, (*d*)–(*f*) cubic core: (*a*) $P = 0.95P^{\mathrm{C}}$, $\tilde{q}_{\mathrm{t}} = 0.02415$; (*b*) $P = 0.50P^{\mathrm{C}}$, $\tilde{q}_{\mathrm{t}} = 0.00427$; (*c*) $P = 0$, $\tilde{q}_{\mathrm{t}} = -0.00022$; (*d*) $P = 0.95P^{\mathrm{C}}$, $\tilde{q}_{\mathrm{t}} = 0.02429$; (*e*) $P = 0.85P^{\mathrm{C}}$, $\tilde{q}_{\mathrm{t}} = 0.02281$; (*f*) $P = 0.75P^{\mathrm{C}}G$, $\tilde{q}_{\mathrm{t}} = 0.02061$.

model is related to the equivalent term of equation (3.47) for the fully interactive model by

$$\tilde{P}^{\mathrm{C}}|_{\mathrm{BF}} = 2\tilde{G}\tilde{\phi}(\tilde{P}^{\mathrm{C}} - \tfrac{2}{3})|_{\mathrm{Int}}, \tag{4.3}$$

such that loads match at the secondary bifurcation point. Varying the values of P and q_{t} in equation (2.9) accordingly, the following observations are made from figure 12.

In view of the differences in the formulations, comparisons, for this geometry anyway, are satisfactory. For the maximum amplitude, wavelength and degree of localization, correlation is excellent in the case of the quadratic core, as seen in figures 12*a*–*c*. Agreement is good for the cubic case (figures 12*d*–*f*) although it tends to overestimate the maximum amplitude at lower loads; here the greater complexity of nonlinearity presumably makes it more difficult for the simplified body force approach to match the more complete interactive model.

5. Concluding remarks

This paper identifies an interactive form of buckling for a compressed sandwich panel which couples an overall instability with local buckling on one face. In contrast to earlier work (Hunt *et al.* 1988), localization of the latter mode is also allowed. This reflects the more general development described in a recent special issue of *Philo-*

sophical Transactions of the Royal Society (Champneys *et al.* 1997), in which spatial localization issues are seen in the context of homoclinic solutions to nonlinear differential equations. Qualitatively anyway, the model compares well with observed collapse of a typical sandwich panel (see figure 1) that involves overall, local and localized buckling. The localized buckle pattern first appears at a secondary bifurcation point which immediately destabilizes one face; as the deformation grows the buckle tends to concentrate at the strut midspan. Comparisons between the fully interactive model and a heuristic 'body-force' approach are excellent for a quadratic nonlinearity but slightly less good when a restabilizing cubic term is included; we put the discrepancy down to inadequacies in the body force modelling, in particular the absence of shear in the core from the interaction of the sway and tilt components of overall buckling, but more work is clearly required to clarify the issue. In comparison with the earlier periodic Rayleigh–Ritz analyses the secondary bifurcation occurs significantly earlier, and stability of the post-buckling response is correspondingly eroded.

Although linear eigenvalue results are shown to be unchanged from classical theory, the thorough instability of the process should sound a warning to designers. Imperfections in both forms of buckling mode would be expected to reduce significantly the load-carrying capacity, such that eigenvalue critical loads can never be obtained.

Financial support for M.A.W. was provided by a University of Bath studentship.

References

Allen, H. G. 1969 *Analysis and design of structural sandwich panels.* Oxford: Pergamon.

Budiansky, B. & Hutchinson, J. W. 1979 Buckling: progress and challenge. In *Trends in solid mechanics. Proc. Symp. dedicated to the 65th birthday of W. T. Koiter* (ed. J. F. Besseling & A. M. J. van der Heijden). Delft University Press.

Champneys, A. R., Hunt, G. W. & Thompson, J. M. T. 1997 Localization and solitary waves in solid mechanics. *Phil. Trans. R. Soc. Lond.* A **355**, 2073–2213.

Da Silva, L. S. 1989 Modal interactions in bending and buckling of sandwich structures. Ph.D. thesis, Imperial College of Science, Technology and Medicine, London.

Doedel, E. J., Wang, X. J. & Fairgrieve, T. F. 1995 *AUTO94: software for continuation and bifurcation problems in ordinary differential equations.* California Institute of Technology, CRPC-95-2.

Fox, C. 1987 *An introduction to the calculus of variations.* New York: Dover.

Gibson, L. J. & Ashby, M. F. 1982 The mechanics of three-dimensional cellular materials. *Proc. R. Soc. Lond.* A **382**, 43–59.

Gibson, L. J., Ashby, M. F., Schjaer, G. S. & Robertson, C. I. 1982 The mechanics of two-dimensional cellular materials. *Proc. R. Soc. Lond.* A **382**, 25–42.

Hunt, G. W. 1981 An algorithm for the nonlinear analysis of compound bifurcation. *Phil. Trans. R. Soc. Lond.* A **300**, 443–471.

Hunt, G. W. & Wadee, M. K. 1991 Comparative lagrangian formulations for localized buckling. *Proc. R. Soc. Lond.* A **434**, 485–502.

Hunt, G. W., Da Silva, L. S. & Manzocchi, G. M. E. 1988 Interactive buckling in sandwich structures. *Proc. R. Soc. Lond.* A **417**, 155–177.

Hunt, G. W., Mühlhaus, H.-B. & Whiting, A. I. M. 1997 Folding processes and solitary waves in structural geology: an overview. In *Localization and solitary waves in solid mechanics* (ed. A. R. Champneys, G. W. Hunt & J. M. T. Thompson). Special issue of *Phil. Trans. R. Soc. Lond.* A **355**, 2197–2213.

Kerr, A. D. 1964 Elastic and viscoelastic foundation models. *ASME J. Appl. Mech.* **31**, 491–498.

Koiter, W. T. & Pignataro, M. 1976 *A general theory for the interaction between local and overall buckling of stiffened panels.* WTHD 83. Delft: Delft University of Technology.

Luongo, A. 1991 On the amplitude modulation and localization phenomena in interactive buckling problems. *Int. J. Solids Struct.* **27**, 1943–1954.

Potier-Ferry, M. 1983 Amplitude modulation, phase modulation and localization of buckling patterns. In *Collapse: the buckling of structures in theory and practice* (ed. J. M. T. Thompson & G. W. Hunt). Cambridge University Press.

Thompson, J. M. T. & Hunt, G. W. 1973 *A general theory of elastic stability.* London: Wiley.

Homoclinic and heteroclinic solutions of upheaval buckling

BY G. W. HUNT[1] AND A. BLACKMORE[2]

[1]*Department of Mechanical Engineering, University of Bath, Bath BA2 7AY, UK*
[2]*Department of Engineering, University of Cambridge, Trumpington Street, Cambridge CB1 1PZ, UK*

Upheaval of a heavy flexible strut from a rigid bed is viewed as an initial-value problem and spatial kinetic and potential energy functions are consequently defined. Upheaval from a perfectly flat state is characterized by the simultaneous vanishing of both functions at the boundaries. Smooth (flat to flat) connection without contact over a step in the bed is thus deemed impossible. A linear non-homogeneous fourth-order ordinary differential equation governs in regions of separation, but not when contact with the bed is maintained. This piecewise property is enough to ensure that several kinds of homoclinic and heteroclinic solutions exist for prop and step imperfections. Application is to subsea pipelines and examples of competing solutions for a realistically proportioned finite-length experimental pipe are included.

1. Introduction

The problem of uplift or upheaval of a heavy compressed strut from a rigid bed, as well as being of practical importance to the off-shore engineering industry (Hobbs 1984; Taylor & Gan 1987; Ju & Kyriakides 1988), has some intriguing mathematical properties. It responds well to analytical shooting methods (Blackmore 1995) and in the spirit of this special issue exhibits both single and multiple-hump localized solutions (Hunt & Blackmore 1996). In the sense sometimes referred to as the dynamical systems analogy (Kirchhoff 1859; Hunt *et al.* 1989), this structural boundary-value problem (BVP) is usefully interpreted as a dynamical system, with initial-value (IVP) information and spatial energy characteristics.

The conventional analytical approach is to model the system as a BVP of unknown length. For a perfectly flat bed of finite stiffness, in the absence of boundary effects, there would exist inside the bed a flat (fundamental) equilibrium state, where the force in the bed just compensates for the weight of the pipe. Linear eigenvalue analysis shows that deflection (y) away from this state is by oscillation with exponentially growing amplitude. As the bed stiffness approaches infinity, both the amplitude and the wavelength of the oscillation tend to zero, so upheaval from a rigid bed is characterized by a non-zero deflection growing from a flat state at each end. Although this seems to imply that y and all its derivatives are zero at the ends, deflection is brought into the pipe and lift-off achieved by reactive shear loads, giving non-zero third derivatives ($\dddot{y}$), at the boundaries; y and the remaining derivatives ($\dot{y}$ and $\ddot{y}$) remain zero however. Techniques such as the finite element method experience considerable difficulties associated with unknown buckle lengths.

Phil. Trans. R. Soc. Lond. A (1997) **355**, 2185–2195
Printed in Great Britain

TEX Paper

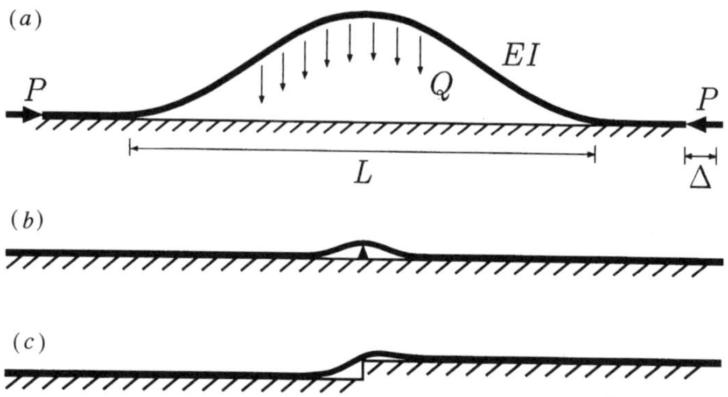

Figure 1. (*a*) Heavy pipe lifting from a flat bed. (*b*) Prop imperfection. (*c*) Step imperfection.

As a BVP the system is thus at first glance overdetermined. The uplifted shape can be characterized by a standard fourth-order linear ordinary differential equation (ODE), but is subject to three boundary conditions, $y = \dot{y} = \ddot{y} = 0$, at each end. This is traditionally resolved (Hobbs 1981; Taylor & Gan 1984) by including buckle length as an unknown and replacing three conditions at one end by two symmetric-section conditions, $\dot{y} = \dddot{y} = 0$, at the centre. However, the approach appears rather inflexible; it has difficulty in dealing with deflection over a step for example, where use of a symmetric-section condition is denied (Blackmore 1995).

For this and associated reasons we adopt shooting methods, whereby numerical integration of the governing ODE is undertaken with a set of four initial values (Champneys & Spence 1993). One or more of these can be varied until a target condition (e.g. the symmetric section as defined in § 3 *c* (iii)) is met at some previously unspecified point along the length; buckle lengths thus decide themselves. In this context, spatial analogies of kinetic and potential energy from dynamics prove most useful. Although the problem is driven by practical considerations, we take the opportunity to expand on some implications of modelling with a rigid bed. Solutions of regular periodic contact, or skipping, are identified, with homoclinic and heteroclinic connections emerging, depending on the type of imperfections, between both flat and periodic states. Alternative equilibrium states are interpreted for an experimental rig at present under development at Sheffield Hallam University (Tran 1994; Taylor & Tran 1996), aimed at exploring the relative effects of prop and step imperfections on subsea pipelines under compression due to hot oil or gas.

2. Practical considerations

The prototype system under consideration is shown in figure 1. Figure 1*a* has a finite-length inextensional elastic pipe of weight per unit length Q and bending stiffness EI, carrying a compressive load P, in a state of upheaval from a perfectly flat rigid bed. The separated length is denoted by L and the end-shortening (corresponding deflection of the load) by Δ. Upwards point load reactions (not shown) of magnitude $\frac{1}{2}QL$ occur at each lift-off point. Because Q acts to hold the pipe to the bed, an infinite (critical) load is required to buckle the perfect system of figure 1*a*. To describe the initiation of buckling more realistically, imperfections in the bed are therefore often introduced (Taylor & Gan 1986; Taylor & Tran 1993). Figures 1*b*,*c* show two typical types, the prop and the step. A shooting method applied to equa-

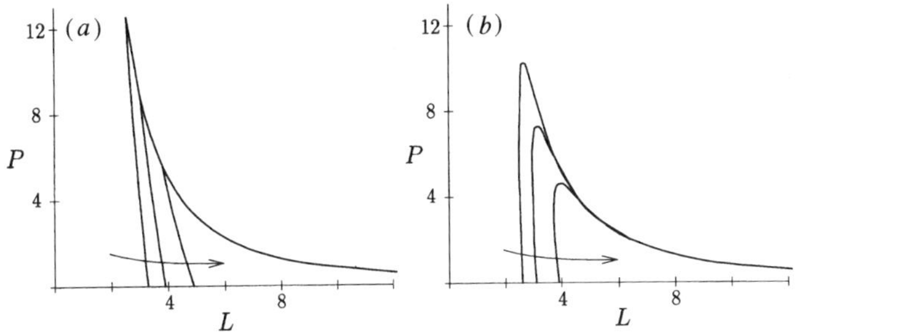

Figure 2. Comparative effects of (*a*) prop and (*b*) step imperfections. Arrows indicate increasing imperfection heights $h = 0.1$, 0.2 and 0.5.

tion (3.1) of the next section, with EI and Q each set to unity, leads to the response diagrams of figure 2, in which plots of P against L are shown for three different values for each imperfection.

Although it may not be strictly meaningful to compare quantitatively these two different types of imperfection, qualitative differences are plain to see. Each response goes through a maximum limiting load, with unbuckled states to the left (low L) and uplifted states to the right. However, the prop imperfection of figure 2*a* presents this as a sharp corner on each response curve, reflecting the fact that after upheaval contact with the prop is lost, whereas the step imperfection of figure 2*b* rounds the corners to genuine limit points as shown. It thus appears that a more profound destabilizing role can be attributed to the step than the prop.

Inherent in the concept of homoclinic connection is that of an infinite domain, with x extending to infinity in at least one direction as y approaches zero. Practical pipelines, on the other hand, are governed by 'active length', which friction dictates as being that which is capable of unloading into the buckled region; outside this length the pipe is essentially unaware that buckling has occurred. Active length is of considerable practical significance—in the presence of axial compressibility it governs how much effective shortening the uplifted region receives—but it changes as buckling progresses and is difficult to measure or estimate. Even conservative estimates put active lengths as greater than those of reasonably designed experiments, however, and we thus conclude that experiments of upheaval are unlikely to be free from boundary effects; these include skipping, in which the pipe periodically loses contact with the bed, as seen later. We note that shooting methods can be adapted for the practical case including friction, the associated forces being readily calculated as displacement progresses along the length (Blackmore & Hunt 1996).

3. Formulation

(*a*) *Differential equation*

The linearized governing equation for a strut of weight/unit length Q, bending stiffness EI and carrying an axial compression P (Timoshenko & Gere 1963) is,

$$EI\ddddot{y} + P\ddot{y} + Q = 0, \tag{3.1}$$

where dots denote differentiation with respect to the spatial axial coordinate x. To generalize this equation and find its eigenvalues we write $p = P/\sqrt{EIQ/\ell}$, $\tilde{y} = y/\ell$

and redefine dots to denote differentiation with respect to $\tilde{x} = x\sqrt[4]{Q/EI\ell}$, to give

$$\ddddot{\tilde{y}} + p\ddot{\tilde{y}} + 1 = 0, \tag{3.2}$$

where ℓ is a typical length measure, introduced merely to complete the non-dimensionalization; for all practical purposes, ℓ can be taken as unity. After writing $\tilde{y} = Ae^{\lambda\tilde{x}}$, the characteristic equation becomes

$$\lambda^4 + p\lambda^2 = 0, \tag{3.3}$$

which is of a degenerate form with two imaginary and two coincident zero eigenvalues. We note the similarity with the rotationally symmetric mode of the axially compressed cylinder (Lord *et al.*, this volume). The general solution is

$$\tilde{y} = a + b\tilde{x} + c\cos\sqrt{p}\tilde{x} + d\sin\sqrt{p}\tilde{x} - \frac{1}{2p}\tilde{x}^2, \tag{3.4}$$

where a, b, c and d are real constants which depend on boundary conditions (Hunt & Blackmore 1996). From this point, unless identified otherwise, all representations are dimensionless and we drop the tilde.

(b) *Spatial energies*

If equation (3.2) is viewed as a Lagrange equation, analogues to kinetic and potential energy can be written, respectively, as

$$\begin{aligned} T &= \dot{y}\dddot{y} + \tfrac{1}{2}p\dot{y}^2, \\ V &= y - \tfrac{1}{2}\ddot{y}^2. \end{aligned} \tag{3.5}$$

Whilst the Lagrangian $\mathcal{L} = T - V$ for such a system is not unique, this choice proves particularly convenient; these spatial energy functions are used extensively in the analysis that follows. The Hamiltonian nature of the system implies that $\mathcal{H} = T + V = \text{const.}$ over the spatial domain described by x.

(c) *Initial/boundary conditions*

No immediate distinction is made in the general description below of initial and boundary conditions applied to the above differential equation. However, a fourth-order system, when treated as a BVP, naturally suits two prescribed boundary conditions at each end. We identify in what follows a number of archetypal end conditions of practical importance which, depending on the type of solution under consideration, appear in different combinations as initial and target conditions for the numerical shooting method. Among these is the asymptotic boundary, which necessarily involves three separate conditions. Such a boundary is clearly more suited to IVP than BVP analysis, and this reinforces the selection of a shooting method based on a Runge–Kutta formulation, rather than a standard structural approach such as the finite element method.

(i) *Asymptotic condition*

If upheaval is from a flat rigid bed, the point of separation, which we choose to designate $y = 0$, is marked by the three conditions $y = \dot{y} = \ddot{y} = 0$. If, in addition, the vertical reactive force at lift-off is obliged to be upwards, this is accompanied by the further condition, $\dddot{y} > 0$.

By analogy with the system on the finite stiffness bed (Hunt & Blackmore 1996),

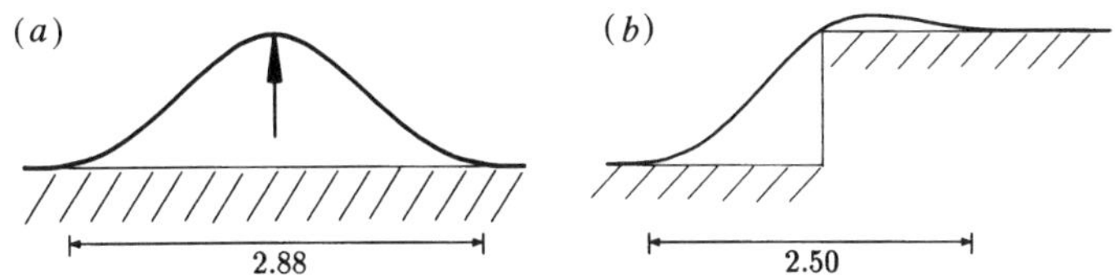

Figure 3. Draped solutions to non-dimensional equation (3.2) with $\ell = 1$, plotted at $p = 6$ over (*a*) prop of height $h = 0.1$; (*b*) step of height $h = 0.1$ (upheaval lengths are as indicated).

although it is met in finite rather than infinite x, we continue to refer to this as an *asymptotic condition*. It applies at the ends of a homoclinic connection. With y measured from the bed, the spatial energy functions (3.5) give $T = V = \mathcal{H} = 0$ at an asymptotic boundary point. Similar conditions also apply at the boundaries of a heteroclinic connection between different energy levels over a step, with $y \neq 0$ (and hence $\mathcal{H} \neq 0$) at one of the levels.

(ii) *Fixed-anchor point (FAP) condition*

For a fixed-anchor point or clamped condition, $y = \dot{y} = 0$. Also, for upwards lift-off from a flat bed, $\ddot{y} > 0$ at this point. When the bed is flat and level these conditions also apply at points of contact in the skipping mode (see § 3 *d* (iv)). Unlike point support at a corner (see figure 4 for instance), $\dot{y}$ is necessarily zero and a jump in $\dddot{y}$ due to the reactive force cannot then affect the total spatial energy $\mathcal{H}$, which is thus conserved through a skipping point of contact. Under experimental and certain practical conditions, FAPs exist where a pipe is clamped or restrained. At such a point, $T = 0$ and $\mathcal{H} = V = \frac{1}{2}\ddot{y}^2$.

(iii) *Symmetric-section condition*

Points on the so-called *symmetric-section* (Hunt & Wadee 1991) arise at the centre of an uplifted region over a flat bed. At such a point where $y = y_{\max}$, $\dot{y} = \dddot{y} = 0$ and hence $T = 0$ and $\mathcal{H} = V = y_{\max} - \frac{1}{2}\ddot{y}^2$.

(*d*) *Primary solutions*

More complicated shapes or behaviours can arise from combinations or modifications of the following forms, which again seem to carry some primary significance.

(i) *Draped shapes*

At zero or low axial loads, a pipe will drape over a prop or step as shown in figures 3*a*,*b*, respectively. In figure 3*a*, asymptotic conditions can only be met at both ends if the pipe rests symmetrically about the prop, such that the jump in $\dddot{y}$ at the prop does not result in a jump in the total spatial energy $\mathcal{H}$. To satisfy asymptotic conditions at each level of figure 3*b*, however, the pipe must lean against the step, as described later in § 3 *d* (iii).

(ii) *Homoclinic connection*

For a pipe resting initially on a prop, as load is increased equilibrium states can be found in which the pipe has lost contact with the prop. The deflected shape after upheaval then connects an asymptotic point to itself; with no possibility of input of energy, the hamiltonian $\mathcal{H}$ is conserved over the full range of x. While noting again that, in contrast to the classical form for a bed of finite stiffness, this occurs over a finite distance, we shall continue to refer to it as a *homoclinic connection*.

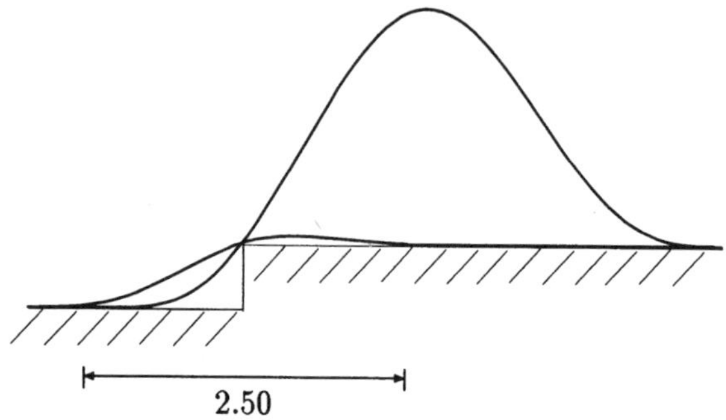

Figure 4. Pre- and post-upheaval leaning solutions to non-dimensional equation (3.2), with $\ell = 1$ at $p = 6$ and step height $h = 0.1$.

The concept of active length means that stable irregularly spaced multi-humped solutions can exist, with each hump outside the range of influence of the others. In the absence of friction stability is lost (see Sandstede, this volume), so if two humps are within range of one another, stability becomes an open question.

(iii) *Leaning solutions*

Analogously to the draped and homoclinic solutions over the prop, as the load is raised two equilibrium possibilities appear supported by a step, as shown in figure 4. In each case, the non-zero reaction at the corner of the step implies a jump in $\dddot{y}$, which couples with the non-zero $\dot{y}$ to give a jump in T. Enough spatial energy $\mathcal{H}$ is thus added or removed to enable asymptotic conditions to be satisfied at each level. These are *heteroclinic connections*, between different flat equilibrium states.

(iv) *Skipping solutions*

For a perfectly flat bed, skipping solutions exist that pass though FAP boundary conditions each time the pipe touches the bed. From the general solution (3.4), setting $y = \dot{y} = 0$ at $x = 0$ gives, directly,

$$a = -c = \frac{p\ddot{y}_0 + 1}{p^2} \quad \text{and} \quad b = -d\sqrt{p} = \frac{\dddot{y}_0}{p}. \tag{3.6}$$

However, $\ddot{y}_0$ and $\dddot{y}_0$ are not independent, but are linked by the symmetry of the flat bed. Symmetric-section conditions $\dot{y} = \dddot{y} = 0$ applied at $x = \frac{1}{2}L$ (half the skiplength) give

$$\tan\sqrt{p}\tfrac{1}{2}L = d/c \quad \text{and} \quad b = L/2p. \tag{3.7}$$

From (3.6), skiplength L is then

$$L = 2\dddot{y}_0, \tag{3.8}$$

while second derivatives $\ddot{y}_0$ and $\ddot{y}_{L/2}$ can be written in terms of $\dddot{y}_0$ as

$$p\ddot{y}_0 = \frac{\sqrt{p}\dddot{y}_0}{\tan\sqrt{p}\dddot{y}_0} - 1, \tag{3.9}$$

$$p\ddot{y}_{L/2} = \frac{\sqrt{p}\dddot{y}_0}{\sin\sqrt{p}\dddot{y}_0} - 1. \tag{3.10}$$

Small $\dddot{y}_0$ leads to negative $\ddot{y}_0$ and contravenes the FAP conditions, so vanishingly small amplitude skipping solutions are disallowed. Since $T_0 = T_{L/2} = 0$, $V_{L/2} = V_0$

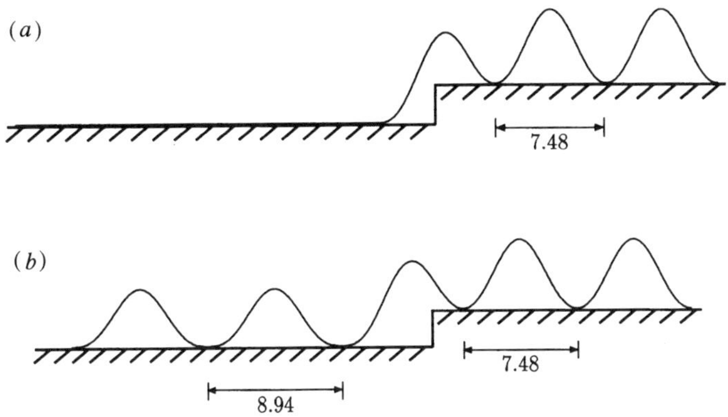

Figure 5. Heteroclinic connections involving skipping: (*a*) from a flat equilibrium state; (*b*) from one skipping state to another. Plotted for non-dimensional equation (3.2), with $\ell = 1$ at $p = 1$ and step height $h = 10$. Note that waves that differ only slightly on the upper level here connect with significantly different responses on the lower level.

and skipheight $y_{L/2}$ follows directly from the spatial potential energy V (3.5),

$$y_{L/2} = \tfrac{1}{2}(\ddot{y}_{L/2}^2 - \ddot{y}_0^2). \tag{3.11}$$

A skipping solution is typified by $\ddot{y}$ remaining continuous and positive at a point of contact, accompanied by a positive jump in $\dddot{y}$ associated with an upwards vertical reaction. For a perfectly flat bed, each skip is symmetrical about its centre and third derivatives $\dddot{y}$ at take-off and landing are equal and opposite; as only upwards reactions are allowed the first is necessarily positive and the second negative. Reactive forces at all points of contact are thus the same and complete regularity of behaviour (periodicity) is achieved.

Although the hamiltonian $\mathcal{H}$ is conserved over skipping points of contact, the very possibility of their presence provides other ways, apart from leaning, for a system to accommodate the jump in $\mathcal{H}$ associated with passage over a step. This is shown in figure 5, which demonstrates two valid forms of heteroclinic connection, from the flat fundamental state to a periodic skipping state, and from one skipping solution to another of a different frequency.

4. Characteristics of the step imperfection

Much work, experimental and theoretical, has been done on symmetric (prop) imperfections of various kinds (see, for example, Boer *et al.* 1986; Ju & Kyriakides 1988; Taylor & Tran 1993). Asymmetric imperfections like the step have received significantly less attention, however, despite the qualitatively more significant destabilizing role brought out in figure 2. Greater variety of solution is also possible. To demonstrate this we take the dimensions of a typical finite-length experimental rig with a step imperfection and impose upon it several of the possible solutions alluded to above.

(*a*) *Competing solutions*

For systems of finite length, or those governed by short active lengths, at least three types of solution compete for the most effective means of relieving a compressed state over a step. The leaning solution of figure 4 is one way of overcoming the difference in total spatial energy $\mathcal{H}$ between the levels. Asymptotic conditions can be satisfied at each point of contact and a heteroclinic connection established between the two flat equilibrium states. This is a truly localized solution, with the spatial position in x being governed entirely by the position of the step.

On the other hand, by skipping a pipe can carry a non-zero bending moment and hence a non-zero value of $\mathcal{H}$ back to a fixed anchor point. It is thus possible to connect over a step between an asymptotic boundary at a lower level and a skipping solution at an upper level, as seen in figure 5*a*, but not vice versa. (With y measured from the upper level, $\mathcal{H} = T + V = 0$, so at the lower level where y is negative and $T = 0$, $\ddot{y}^2 < 0$ from (3.5).) We refer to this as a semilocalized solution, heteroclinic between a flat and a periodic equilibrium state. Its spatial position in x depends on the position of the step only in the conditional sense that contact must be avoided, but is fixed by the distance from the FAP boundary.

Or finally, it may suit the boundaries and wavelengths involved for skipping solutions to apply on both sides of a step. As a skipping solution meets a step its periodicity is interrupted; periodicity at different wavelengths exists on each side of a step, as seen in figure 5*b*. Reactions closest to the step are both different from one another and from those of each periodic sequence. If FAP conditions are used at both ends of a relatively short pipe (as below) the problem is readily formulated in boundary-value terms. If, however, skipping is involved on each side of the step, different solutions with differing skiplengths may present themselves. Interesting mathematical questions, left for future study, are raised by long-but-finite lengths.

(*b*) *Experimental rig*

The following results are interpreted specifically for an experimental apparatus under development at Sheffield Hallam University under the direction of Dr N. W. Taylor. A steel pipe of wall thickness 1.6 mm and diameter 9.53 mm, 6 m in length, carries water at a slow flow rate. The ends are clamped so that all movement is restricted and FAP boundary conditions apply, and a 17 mm step is introduced at the mid-length. The water can be heated externally in a separate bath to induce buckling. The rig is an adaptation of one previously used for the study of prop imperfections (Taylor & Tran 1993; Tran 1994).

(*c*) *Friction and axial compressibility*

In genuine applications, friction between a pipe and the seabed, and axial compressibility, are clearly of prime significance. Via the concept of active length they together dictate how much elastic energy is available to unload into the uplifted region and hence determine the extent of the instability before restabilization.

For the effects of friction, consider the response curves of figure 6*a*. Here, for the dimensions of the above pipe and step height but over an active length governed by an assumed coefficient of friction $\phi = 0.2$, the variation of the load in the uplifted region, P, for the fully localized (leaning) solution is compared with that for the *control load*, P_0, found outside the active length. The latter varies linearly with temperature, while the difference in the two loads is taken up by friction. The curve for P has the form of figure 2*b* and continues to fall with L, but that for P_0 restabilizes, such that the indicated snap buckle at constant P_0 leads to a stable uplifted shape of just under 6 m in length—of the order of, but less than, the experimental length of 6 m; typically,

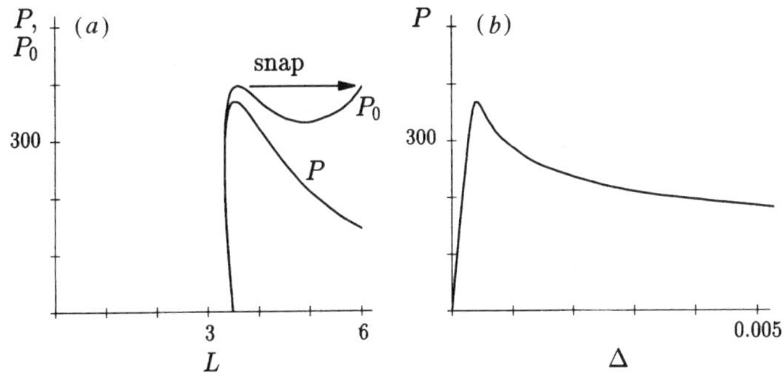

Figure 6. Response curves for the leaning solution. (*a*) Effects of friction under temperature control (active length *ca.* 100 m). (*b*) Load P plotted against end-shortening Δ under frictionless conditions (length = 6 m, constrained ends).

however, active lengths are found to be much greater than that of the experiment, of the order of 100 m at the destabilizing limit point for instance.

This suggests that practice and experiment may lead to quite different responses. The snap buckle of figure 6*a* depends critically on the axial compression over a long length unloading into the buckled region, and yet a typical experimental length is only of the order of the buckled length. Figure 6*b* shows load P plotted against the *total* end-shortening Δ for a frictionless bed and the length and step height in question. The latter comprises two components, EA-dependent shortening due to direct axial compression and EI-dependent shortening due to buckling. The axial (EA) component tends to tilt the curve to the right, with the degree of tilt increasing with total (active) length; the longer the pipe, the less stiff the initial response and the smaller the slope of the pre-buckling (fundamental) equilibrium path. The effect of this initial squash is to 'spring load' the system, such that *snap-back*, as exhibited, for example, by the axially compressed cylindrical shell (see figure 1 of Lord *et al.*, this volume), becomes a possibility. The horizontal snap buckle of figure 6*a* and consequent drop in P is then most closely represented, on the curve of figure 6*b*, by a downwards snap at constant Δ. Again, the longer the pipe, the greater the tilt on the curve and the more severe the snap-back.

For the length and step height considered here, however, the phenomenon of snap-back is clearly absent; the experimental length is simply too short. We hence conclude that the experimental leaning response, in contrast to that expected from a pipeline of practical dimensions laid over a step, is likely to be stable and undramatic.

The effect of end-constraint, either by friction or by clamping, is thus to inhibit the extreme instability nominally met by the compressed strut of infinite length; the infinite compressive strain energy available for transmission into the buckled region is rendered finite. After the maximum on the response curve, a jump may or may not take place to a reduced load state, the extent or otherwise of the jump depending crucially on either the active, or the actual experimental, length.

(*d*) *Comparative results*

Figure 7*a* compares fully and semilocalized solutions for the experimental apparatus of interest. The response curves are plotted up to the maximum possible length of 6 m, but, of course, for these localizing solutions they apply also for shorter lengths. In the loading region only the leaning solution exists, but in the post-maximum unloading region both solutions apparently are valid. However, figure 7*b* demon-

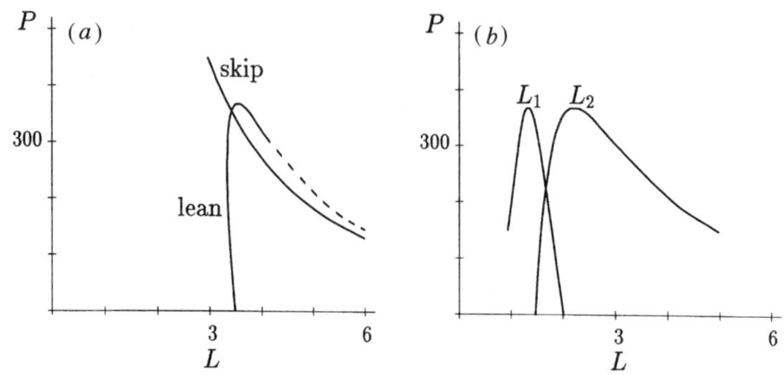

Figure 7. Response curves for the experimental data of § 4 *b*. (*a*) Comparison of fully localized (leaning) and semilocalized solutions. (*b*) Lower (L_1) and upper (L_2) separated lengths for the leaning solution.

strates a further constraint on the leaning solution, in that to maintain contact with the corner, the upper separated length L_2 must be less than the half length of 3 m; this invalidates the dashed portion of figure 7*a*. Although sought no further here, transitional solutions must exist in this region to transfer the path from the fully to the semilocalized state.

The localized solutions of figure 7 are valid for all lengths up to 6 m. Standard boundary-value solutions for fixed 6 m lengths are also available. Generally, skipping either on the upper level alone or on both the upper and lower levels provides further possibilities. For the length and load levels of interest here such responses complicate matters only at the higher loads of the 'skip' curve of figure 7*a*, where skiplengths are less than the upper length of 3 m. For moderately longer pipes, skipping becomes more of an option, with the number of possibilities for solution increasing with the experimental, or active, length. We note also that active lengths themselves could be considerably modified by skipping effects.

Actual detailed experiments are yet to be performed on the pipe in question and other factors would of course be involved in the choice of solution, not least imperfections in the pipe itself. However, on increasing temperature we hope to be able to identify more than one post-upheaval equilibrium state and, in particular, the transition from fully to semi- or unlocalized shape, as the limit of the (leaning) response curve of figure 7*a* is met. For longer experimental lengths, or the active lengths found in practice, full localization rather than skipping would be expected to provide the least stiff, and hence most likely, of the available options.

5. Concluding remarks

The paper takes a standard formulation for upheaval buckling and applies it in two new directions. The first is the study of the effects of asymmetric bed imperfections, typified by a step, rather than symmetric imperfections such as a prop of infiltrated material between the pipe and the bed (Taylor & Tran 1993). The extra significance of asymmetry is successfully demonstrated in figure 2. From a practical viewpoint it might be supposed that changes in seabed level are as fundamental to pipe modelling as the infiltration of foreign material, yet previous work relates almost exclusively to symmetric imperfections. The point is also made that as a source of bifurcation phenomena, asymmetry is again a richer pasture than symmetry.

The second, not entirely unconnected, new development is to explore the effects of finite lengths. Skipping solutions are computed for the first time and alternative solutions involving either asymptotic or fixed anchor point (FAP) conditions are compared for a typical experimental rig. A range of lengths and loads is identified over which a transition from fully localized to semilocalized behaviour must take place.

Finally, the dynamical systems analogy again proves most useful in identifying possible forms of localized, semilocalized and unlocalized periodic (skipping) solutions. We note that all the results presented here depend on the bed being flat and, apart from the step itself, level. For systems supported at finite positions rather than by a level bed—the so-called 'bed of nails' (Blackmore 1995)—many of the conclusions developed herein will not apply. More like the corner of a step than a flat bed, the latter allows non-zero first derivatives $\dot{y}$ and hence changes of spatial kinetic energy at discrete points of contact.

References

Blackmore, A. 1995 Dynamical systems analogy in upheaval buckling. Ph.D. thesis, Imperial College, University of London.

Blackmore, A. & Hunt, G. W. 1996 The dynamical phase space analogy as a tool for the upheaval buckling of pipelines. Preprint.

Boer, S., Hulsbergen, C. H., Richards, D. M., Klok, A. & Biaggi, J. P. 1986 Buckling considerations in the design of the gravel cover for a high-temperature oil line. In *Proc. 18th Annual OTC, Houston, Texas.* OTC 5294.

Champneys, A. R. & Spence, A. 1993 Hunting for homoclinic orbits for autonomous Hamiltonian sytems. *Adv. Comp. Maths* **1**, 81–108.

Hobbs, R. E. 1981 Pipeline buckling caused by axial loads. *J. Construct. Steel Res.* **1**, 2–10.

Hobbs, R. E. 1984 In-service buckling of heated pipelines. *ASCE J. Transp. Eng.* **110**, 175–189.

Hunt, G. W. & Blackmore, A. 1996 Principles of localized buckling for a strut on an elasto-plastic foundation. *ASME J. Appl. Mech.* **63**, 234–239.

Hunt, G. W. & Wadee, M. K. 1991 Comparative lagrangian formulations for localized buckling. *Proc. R. Soc. Lond.* A **434**, 485–502.

Hunt, G. W., Bolt, H. M. & Thompson, J. M. T. 1989 Structural localization phenomena and the dynamical phase-space analogy. *Proc. R. Soc. Lond.* A **425**, 245–267.

Ju, G. T. & Kyriakides, S. 1988 Thermal buckling of offshore pipelines. *ASME J. Offshore Mech. Arctic Eng.* **110**, 355–364.

Kirchhoff, G. 1859 Über das gleichgewicht und die bewegung eines unendlich dünnen elastischen stabes. *J. Mathematik (Crelle)* **56**, 285–313.

Taylor, N. & Gan, A. B. 1984 Regarding the buckling of pipelines subject to axial loading. *J. Construct. Steel Res.* **4**, 45–50.

Taylor, N. & Gan, A. B. 1986 Submarine pipeline buckling—imperfection studies. *Thin-Walled Structures* **4**, 295–323.

Taylor, N. & Gan, A. B. 1987 Refined modelling for the vertical buckling of submarine pipelines. *J. Construct. Steel Res.* **7**, 55–74.

Taylor, N. & Tran, V. 1993 Prop-imperfection subsea pipeline buckling. *Marine Struct.* **6**, 325–358.

Taylor, N. & Tran, V. 1996 Experimental and theoretical studies in subsea pipeline buckling. *Marine Struct.* **9**, 211–257.

Timoshenko, S. P. & Gere, J. M. 1963 *Theory of elastic stability*, 2nd edn. New York: McGraw-Hill.

Tran, V. C. 1994 Imperfect upheaval subsea pipeline buckling. Ph.D. thesis, Sheffield Hallam University, Sheffield.

Folding processes and solitary waves in structural geology

By G. W. Hunt[1], H.-B. Mühlhaus[2] and A. I. M. Whiting[3]
[1]*Department of Mechanical Engineering, University of Bath, Bath BA2 7AY, UK*
[2]*Division of Exploration and Mining, CSIRO, P.O. Box 437, Perth WA 6009, Australia*
[3]*Connell Wagner, Consulting Engineers, 385 St Paul's Terrace, Fortitude Valley, Brisbane Qld 4006, Australia*

Evolution of localized folding patterns in layered elastic and visco-elastic materials is reviewed in the context of compressed geological systems. The thin strut or plate embedded in a visco-elastic medium is used as an archetypal example to describe localized buckles which, in contrast to those from earlier formulations, appear in the absence of triggering imperfections. Structural and material effects are surveyed and important nonlinear characteristics are identified. A brief review of possible methods of analysis is conducted.

1. Introduction

The modelling of geological processes spanning perhaps hundreds of millions of years with well-posed differential equations is fraught with difficulty. There is a considerable gulf between, on the one hand, natural processes undergoing enumerable, large and small, continuous and discontinuous, influences as time progresses and, on the other, easily definable regulatory laws. Nonetheless, when an innovative shift in modelling brings about radical qualitative changes in response, interest is justifiably raised. The recent burst of interest in concepts of regularity and irregularity (fractals and chaos) as applied in the earth sciences (Kruhl 1994) is a good example of this progressive trend.

The process of folding under tectonic compression is an area of structural geology that is perhaps ready for such development. The pioneering work of Biot (1965) models such processes with elastic and viscous struts and plates resting on or within supporting (half-space) media (see figure 5), that themselves are subject to elastic, viscous, or visco-elastic laws. According to the demands and computing power of the time, Biot's modelling was largely restricted to linear differential equations and periodic harmonic deflected forms thus dominate the response. In a move towards physical reality, Mühlhaus (Mühlhaus 1993; Mühlaus *et al.* 1994) later introduced nonlinear terms associated with controlled end displacements rather than loads, but the response apparently remained periodic. This contrasts with observations in the field of more complex phenomena such as spatial localization and quasiperiodicity (Price & Cosgrove 1990), commonly associated with spatial chaos (Champneys & Toland 1993; Hunt *et al.* 1997).

Phil. Trans. R. Soc. Lond. A (1997) **355**, 2197–2213
Printed in Great Britain

TEX Paper

Such considerations are of real interest to organizations like the CSIRO of Australia, as there is known to be a tendency for mineral deposits to leach into areas of intensified localized folding. This suggests considerable scope for interaction between research workers of wide-ranging experience and expertise, from field geologists with knowledge of the constitutive makeup and nature of folding in real rock systems to numerical analysts at the cutting edge of modern applied mathematics. In such a broad college there is clearly room for contributions of different orders of complexity and rigour.

We attempt here an overview of the process of geological folding, with particular emphasis placed on the emergence of spatial forms that appear as solitary waves when seen at an instant in a slow evolutionary timeframe. We start from a modelling perspective and in the search for localization place the strut formulation of Biot (1965) on or within a *visco-elastic* medium. The elastic component thus allows an instantaneous or near-instantaneous initial response, followed by a slow evolution marking dissipation of stored energy into the viscous part of the support. The simplest mathematical setting for this is a nonlinear partial differential equation (PDE) that is fourth order in space and first order in time. Moreover, if the supporting medium is a half-space it has a non-local constitutive law and the equation is integro-differential.

The paper reviews several possible models for the localized folding of geological structures, from the immediately tractable to the presently intractable, and suggests a number of possible analytical and numerical approaches, from the heuristic to the more strictly rigorous. The emphasis is on localized solutions that occur naturally in 'perfect' systems, rather than those triggered by initial imperfections or perturbations that have received considerable attention, both theoretical and experimental, in the past (see, for example, Cobbold 1975, 1976; Abbasi & Mancktelow 1992; Mancktelow & Abbasi 1992). The scene is set by results from a crude truncation approach, with some asymptotic credentials but extended beyond its range of applicabililty. This foregoes strict rigour for the sake of otherwise unobtainable results and provides convincing qualitative time portraits. In the absence of rigorous single-dimensional methods for half-space formulations, it is introduced as a pointer for more respectable approaches that we hope will follow.

2. Geological evidence

Evidence of folding under in-plane compression appears in a geological context on many scales. Figure 1 shows a specimen of evaporite some 12 cm in length in which periodic folding at different wavelengths is clearly visible. Less obvious but still apparent are decaying fluctuations to the right end of the specimen, which bear a strong resemblance to the fourth-order homoclinic boundary condition described by four complex-conjugate eigenvalues seen elsewhere in this special issue (see, for example, Sandstede, this volume). Localized distortion, either with slowly decaying tails or severe, as in the box folding of figure 2, is thus a known geological feature. Evidence of quasi-periodicity also exists, as in figure 3, which demonstrates a number of interesting structural and material features including further examples of slow amplitude decay and evidence of folding on different wavelengths. Irregularity and self-similarity have recently been recognized as integral to the geological setting (Ord 1994). We note that in all of these cases the behaviour is liable to be significantly influenced by multi-layer (interactive) effects, which we ignore.

A Typical Fold

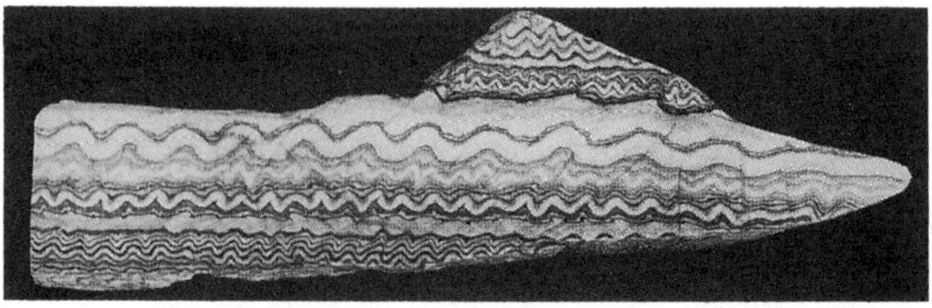

Figure 1. Periodic and decaying folds in a specimen of evaporite (courtesy of J. W. Cosgrove, Imperial College London).

Figure 2. Medium scale box folding seen from the beach at Bude, Cornwall.

Figure 3. Localization and quasi-periodicity in layered Mylor beds at Porthleven, Cornwall.

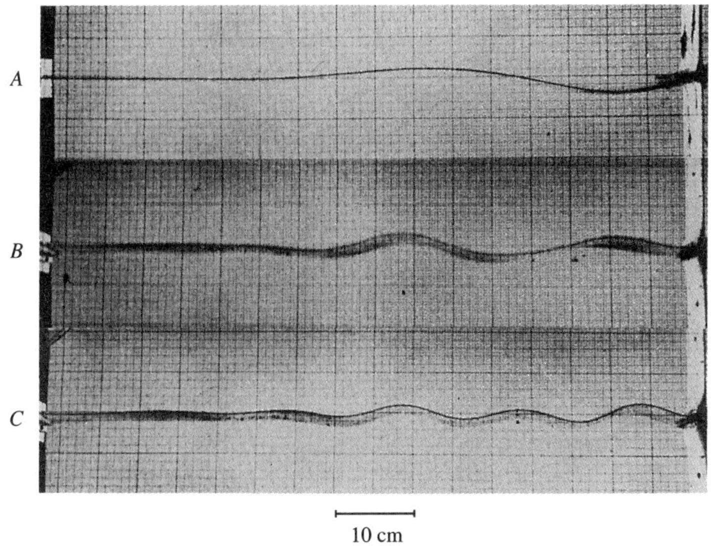

Figure 4. Experiments on an axially compressed thin elastic strip within a viscous medium (after Biot 1965).

3. Modelling considerations

Early attempts to put the folding of geological layers on a firm mathematical footing came from Biot (1965), who treated in some depth the buckling of elastic and viscous layers supported on or within elastic and viscous media. Linear differential equations were predominantly used and, as a result, periodicity tended to govern the responses. Spectral decomposition was central to the approach, leading via the concept of dispersion relations to a focus on the dominant or fastest growing wavelength of buckle. Experiments performed on elastic strips in a viscous oil, shown here in figure 4, are less clearly periodic, however, decaying amplitudes also being apparent. With the evidence also from figures 2 and 3, we can take it that aperiodicity is an accepted feature in the geological setting. While recent work on nonlinear elastic foundations has demonstrated the importance of localized solitary waves (Hunt & Wadee 1991; Champneys & Toland 1993), perfect viscous systems, even under the nonlinear constraint of fixed end displacement, remain periodic for all time (Mühlhaus 1993; Mühlhaus *et al.* 1994); as the compressive load drops to zero the deflected shape apparently evolves through a sequence of longer and longer periodic waves to the final result of a single long wave as $t \to \infty$.

It has, however, been pointed out by Price & Cosgrove (1990) that neither quantitatively nor qualitatively do the calculations from viscous bedding relations match observed phenomena. Indeed, it is argued that elasticity must be introduced. With this as our cue, we concentrate in what follows on elastic struts on linear and nonlinear visco-elastic foundations.

(*a*) *Structural effects*

Many types of structural form—struts, plates, shells and domes for instance—supported by various media, together with single and multi-layered sandwich structures, blocky structures and other similar archetypes, bear some geological relevance.

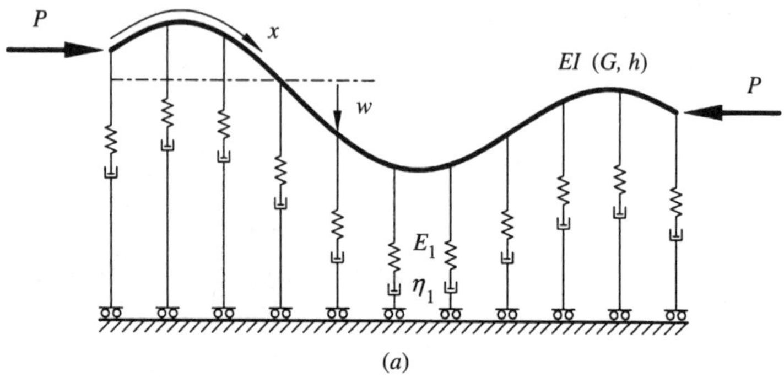

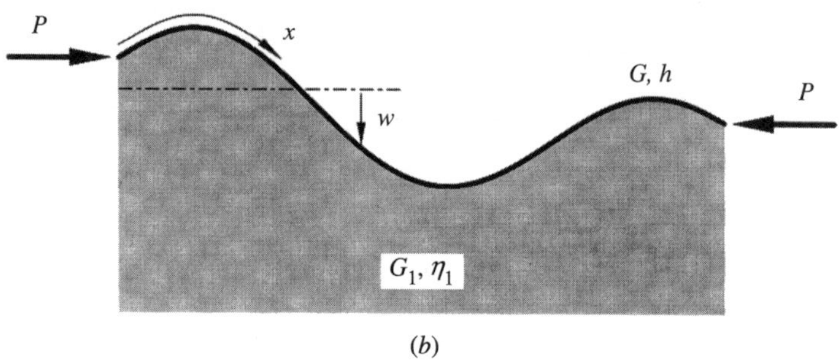

Figure 5. Embedded strut models. (*a*) Winkler foundation; (*b*) half-space bed.

Normal structural assumptions, those associated with thin plate or shell theory, for example, can be called into question. Contact, friction and separation or upheaval (Hunt & Blackmore, this volume) may play major roles. Here we concentrate on a thin strut or plate layer supported on or within elastic, viscous or visco-elastic media, as seen in figure 5. We take a two-dimensional view, with the same form assumed to extend to infinity in each direction in the third.

A strut made from an elastic material of Young's modulus E, of unit width, thickness h and cross-sectional second moment of area I, under a compressive load P and resting on a unspecified bed of vertical resisting force per unit length F, has the governing equation

$$EI[w''''(1-w'^2)^{-1} + 4w'''w''w'(1-w'^2)^{-2} + w''^3(1+3w'^2)(1-w'^2)^{-3}] + Pw''(1-w'^2)^{-3/2} + F = 0, \quad (3.1)$$

where primes denote differentiation with respect to the spatial coordinate x, measured along the length of the strut as seen in figure 5. If deflections are small, and hence nonlinear terms can be ignored, this becomes

$$EIw'''' + Pw'' + F = 0. \quad (3.2)$$

The single differential equation holds only for the disconnected Winkler foundation of figure 5*a*, where resistance to deflection into the bed is strictly local and vertical. For the infinite half-space of figure 5*b*, the situation is significantly different. Resistance now comes from shearing action. A wavy form in the strut causes the bed to shear and is resisted, but there is no resistance to a constant displacement w over infinite length

x. This introduces a wavelength dependence to the bed (Biot 1965) and historically has led to treatment by Fourier analysis.

The modelling implication of such effects is that a fourth-order differential equation in a single spatial variable x is no longer appropriate. The general half-space implies that non-local bedding relations are now expected; the force at a point x along the length depends not just on w at x, but on displacements at points other than x, and integro-differential equations govern. We note that Winkler foundations find limited application for granular, and possibly finite-depth, supporting media, but the half-space is generally considered the more realistic for geological applications. Other attempts to resolve this difficulty include hybrid supporting media like the Pasternak foundation (Kerr 1964).

(i) *Overburden pressure*

If, as in many geological situations, the force F of equations (3.1) and (3.2) contains a significant component of overburden pressure, at first glance it would seem that a layer could buckle under such pressure alone; pressure components top and bottom of the layer would tend to cancel each other, while that at the ends of the layer apparently provides the compressive force necessary for buckling. This, however, conflicts with common sense and is overcome by the following reasoning. Consider a small element $\mathrm{d}x$ of a layer of thickness h and unit width, bending through a small angle $\mathrm{d}\varphi$, as shown in figure 6. Under constant pressure p, tensile stretching of one face combined with compressive shortening of the other leads to an out-of-balance normal resisting force per unit length,

$$q = \frac{p(r + \frac{1}{2}h)\mathrm{d}\varphi - p(r - \frac{1}{2}h)\mathrm{d}\varphi}{r\mathrm{d}\varphi} = \frac{ph}{r}, \tag{3.3}$$

where r is the radius of bending curvature and h is the thickness of the layer. For the small deflection form of equation (3.2), $w'' = -1/r$, and the net effect is to introduce a term $-phw''$ into the governing differential equation that exactly cancels the perceived compressive force at the ends coming from the overburden pressure. The same conclusion is reached in a continuum context by Mühlhaus *et al.* (1998). The axial load P thus arises entirely from external events like tectonic collision.

(*b*) *Material effects*

The distinction between material and structural behaviour may at times be somewhat blurred, yet it remains a useful exercise to attempt to distinguish between local constitutive laws and structural response.

(i) *Constitutive relations*

The most abundant material phases in the Earth's crust and the upper mantle are quartz/feldspar and olivine, respectively. They deform permanently by a number of alternative, often competing, mechanisms, the most important in a geological/geophysical context being: (1) diffusional flow (Nabarro-Herring creep, Coble creep); (2) power-law creep by dislocation glide; (3) collapse at yield strength. Each can be described by a rate equation relating the strain rate $\dot{\gamma}$ to the stress τ, the temperature T and the structure of the material at that instant (see, for example, Frost & Ashby 1982; Estrin 1996), where dots denote differentiation with respect to time. In three-dimensional applications, the rate equations have to be complemented by a flow rule (in the simplest cases, the Prandtl–Reuss flow rule, see Estrin 1996)

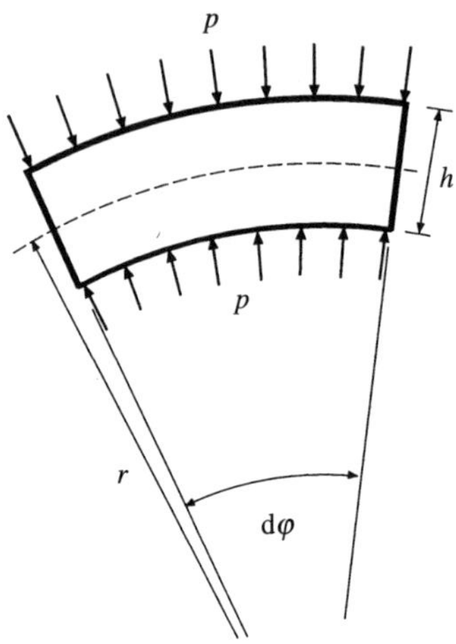

Figure 6. Elemental system.

and $\dot{\gamma}$ and τ are usually defined as

$$\dot{\gamma} = (2D_{ij}D_{ij})^{1/2} \quad \text{and} \quad \tau = (\tfrac{1}{2}\sigma'_{ij}\sigma'_{ij})^{1/2}, \tag{3.4}$$

where D_{ij} is the stretching and σ'_{ij} is the deviatoric part of Cauchy's stress tensor σ_{ij}. A rate law which captures all of the above mechanisms reads

$$\dot{\gamma} = \frac{a}{kT}\left(\frac{\tau}{\tau_0}\right)^n \exp\left(-\frac{Q - pV}{kT}\right). \tag{3.5}$$

Here τ_0 is the yield strength at zero Kelvin, $p = -\frac{1}{3}\sigma_{kk}$ is the pressure, Q is the activation energy at $p = 0$, V is the activation volume for diffusion, k is the Boltzmann constant and a is a parameter with the dimension of power which, in general, is pressure dependent, although this dependency only becomes significant for pressures larger than one tenth of the elastic bulk modulus. Pre-exponential temperature dependency is also mostly negligible and hence a/kT can be replaced in (3.5) by a constant reference strain rate $\dot{\gamma}_0$. The reference stress τ_0 may also depend on the activities of water and of oxygen, a_{H_2O} and a_{O_2}, respectively (Hobbs 1981; Ord & Hobbs 1989), but this dependency is very weak and is probably only important in zones of high gradients of the activities. If molar values for Q and V are to be used then k must be replaced by the gas constant $R = 8.314\ \mathrm{J\,mol^{-1}\,K^{-1}}$. Typical orders of magnitude for Q and V are 150–550 kJ mol^{-1} and $7\text{–}30 \times 10^{-6}\ \mathrm{m^3\,mol^{-1}}$, respectively. Lithostatic pressures approach Q/V in magnitude only for depths greater than 700–800 km. Hence, under crustal and upper mantle conditions (roughly taken as depths of up to 300 km), the pressure dependency of the exponent in (3.5) can safely be neglected. The power law exponent n is equal to unity for diffusional flow (Nabarro–Herring creep, Coble creep) and larger than unity for power-law creep, typical values for crustal and mantle material ranging between $2 < n < 3$.

If more detailed analysis of the deformation of geological structures is required, the influence of elastic deformations in the rate law must be considered. Here we are concerned with the evolution of buckling folds, which usually take place on spatio-temporal scales within which the temperature can be regarded as constant. Neglecting the pressure sensitivity as discussed above, the relevant rate law for constant temperature reads

$$\dot{\gamma} = \frac{\dot{\tau}}{\mu} + \frac{\tau}{\eta}, \tag{3.6}$$

where μ is the elastic shear modulus, the viscosity η is given by

$$\eta = \frac{\tau_0}{\dot{\gamma}_0}\left(\frac{\tau}{\tau_0}\right)^{1-n}. \tag{3.7}$$

For simplicity we assume in the following that $n = 1$ and

$$\tau = \hat{\tau}e^{\omega t} \quad \text{and} \quad \gamma = \hat{\gamma}e^{\omega t}. \tag{3.8}$$

Inserting into (3.6) and re-arranging yields the relation $\hat{\tau} = \hat{\mu}\hat{\gamma}$, where

$$\hat{\mu} = \frac{\omega\eta}{1 + \omega\eta/\mu}. \tag{3.9}$$

The material is predominantly elastic if $\omega\eta >> \mu$ and predominantly viscous if $\omega\eta << \mu$.

Biot (1965, p. 419) investigates the folding of an elastic plate in a viscous medium. The growth coefficient ω_d of the fastest growing periodic wave is obtained as

$$\omega_d = \frac{\sigma}{6\eta_1}\sqrt{\frac{\sigma}{\mu}}, \tag{3.10}$$

where η_1 is the viscosity of the embedding medium, μ is the shear modulus of the (incompressible) plate material and the axial stress $\sigma = P/h$ is the driving force of the instability. Inserting (3.10) into $\omega\eta/\mu$ gives the criterion

$$\frac{\eta}{6\eta_1}\left(\frac{\sigma}{\mu}\right)^{3/2} >> 1, \tag{3.11}$$

for predominantly elastic behaviour of the plate during folding.

The single layer folding model considered here comprises a thin, axially pre-stressed layer embedded in a visco-elastic medium. The early stages of the fold amplitude evolution can be understood as a small perturbation of the homogeneously stressed ground state. The tangent shear modulus is obtained in symbolic form as

$$\hat{\mu} = \frac{\mu\eta\dfrac{\partial}{\partial t}}{\mu + \eta\dfrac{\partial}{\partial t}}, \tag{3.12}$$

where viscosity η is given by (3.7) and $\tau = \sigma/(2\sqrt{2})$ for uniaxial plane strain, σ again being the magnitude of the axial pre-stress. The instantaneous bending stiffness of the strut in equation (3.1) is then

$$EI = \tfrac{1}{3}\hat{\mu}h^3. \tag{3.13}$$

The following section gives an interpretation of such basic rheologies in terms of structural elements.

(ii) *Simplified rheologies*

Standard rheological models are available for application in the geological context, three of the most common being shown in figure 7. The Maxwell fluid of figure 7*a*,

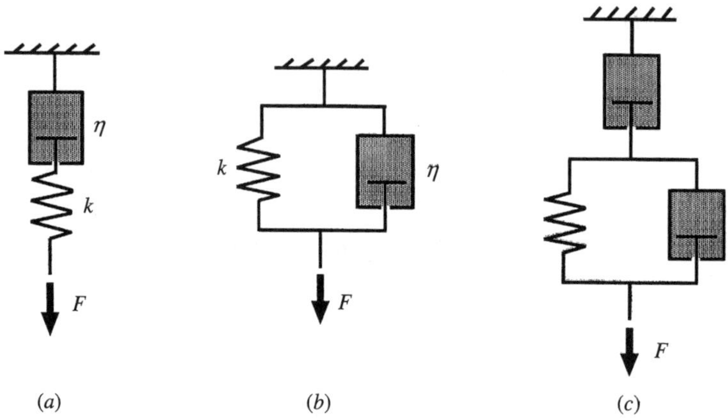

Figure 7. Rheological models and associated responses. (*a*) Maxwell fluid; (*b*) Kelvin–Voigt solid; (*c*) combined fluid.

comprising a spring and dashpot in series, is the most immediately relevant for our purposes. On application of dead load F it shows an instantaneous elastic displacement in the spring, followed by a constant rate of displacement in the dashpot. On application of a constant displacement the spring again responds instantly, but over time gradually unloads into the dashpot such that F falls off exponentially. The governing differential equation is

$$\dot{w} = \frac{1}{k}\dot{F} + \frac{1}{\eta}F, \tag{3.14}$$

where k is the spring stiffness and η is the dashpot viscosity (cf. equation (3.6)).

In contrast, the Kelvin–Voigt solid of figure 7*b*, comprising spring and dashpot in parallel, has the ability to store strain energy. It is represented by the differential equation

$$F = kw + \eta\dot{w}. \tag{3.15}$$

The combined fluid of figure 7*c* slows down the instantaneous elastic response of the Maxwell unit, at the expense of a governing differential equation of second order with the general form

$$F + p_1\dot{F} = q_1\dot{w} + q_2\ddot{w}, \tag{3.16}$$

where p_i and q_i are material constants relating to spring stiffnesses and dashpot viscosities. These are defined more completely in Roscoe (1950), who shows that a general arrangement of such linear springs and dashpots can be reduced to one of two possible canonical forms.

(iii) *Nonlinear models*

Nonlinearities of many kinds can be introduced to the single dimensional rheological models of figure 7. Figure 8 shows three typical types produced by simple geometric effects, all of which can occur naturally. The hardening response of figure 8*a* might be associated, for example, with closing of voids in porous medium, while the softening behaviour of figure 8*b* is similar to that of the shallow arch (Thompson & Hunt 1984). The bilinearity of figure 8*c* is produced by an elastic bifurcation, but

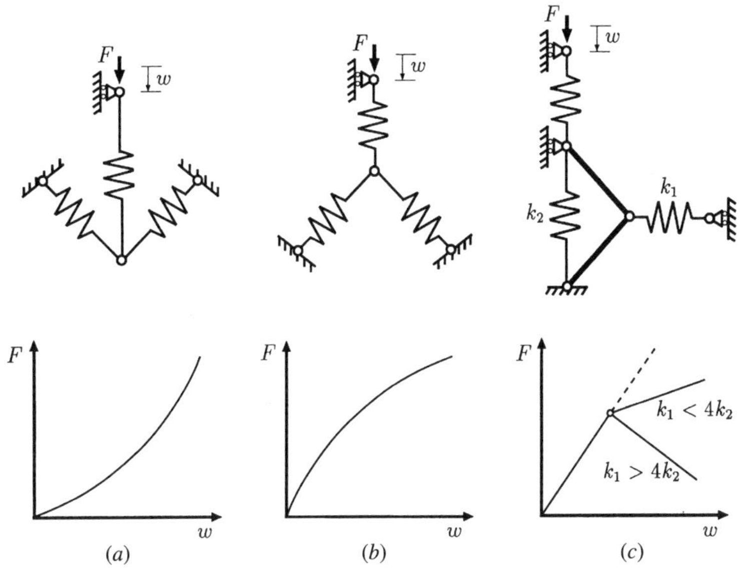

Figure 8. Nonlinear elastic models and associated responses: (*a*) hardening; (*b*) softening; (*c*) bilinear.

in the absence of unloading can be used to mimic elastoplasticity or quasi-brittle fracture (Hunt & Baker 1995).

For purely elastic systems, nonlinearities can be relatively straightforward to introduce and interpret, but when elastic and viscous parts are present it is necessary to identify just how the nonlinearity is to enter the modelling process. In an attempt to reproduce the softening of figure 8*b* in a Winkler foundation, we might, for example, introduce $F = w_e - w_e^3$ into the elastic part of the Maxwell element of figure 7*a*, w_e being the displacement in the spring alone; this can be taken either as a simple cubic law in its own right, or as the leading nonlinear term of a more general power expansion. The total deflection w is difficult to determine from this perspective, but is readily obtained if the relation is reversed to give $w_e = F + F^3 + \cdots$. This more convenient expansion may, however, have a reduced range of validity; unlike its counterpart it cannot, for instance, represent a response that continues over a maximum into a regime of negative stiffness (Whiting 1996).

We note in passing that a nonlinearity on the viscous part, or the whole of w as in Hunt *et al.* (1996*a*), would have a growing effect over time. The viscous part would continue to deflect and involve the nonlinearity to an ever-increasing extent.

(iv) *Half-space formulations*

Linear and nonlinear rheological models are readily adapted to Winkler foundations but not so easily to a half-space. Biot (1937) took a Fourier expansion for a general load on the boundary of an elastic half-space,

$$F(x) = \sum_{0}^{\infty} q_m \cos k_m x, \tag{3.17}$$

and passed it through the linear elastic plane strain equation ($\nabla^4 \Phi = 0$, where Φ is a stress function) to obtain a relation between q_m and corresponding amplitude of displacement a_{m} given by

$$w(x) = \sum_{0}^{\infty} a_{\mathrm{m}} \cos k_{\mathrm{m}} x, \tag{3.18}$$

as follows

$$q_{\mathrm{m}} = -4\mu_1 k_{\mathrm{m}} a_{\mathrm{m}}, \tag{3.19}$$

where μ_1 is the shear modulus of the foundation material. The appearance of wavenumber k_{m} marks the fact that the resisting force is wavelength dependent, a feature that is linked to the non-local nature of the bedding support; this contrasts directly with the Winkler formulation in which k_{m} would be absent. A similar exercise for a purely viscous foundation of viscosity η_1 leads to

$$q_{\mathrm{m}} = -4\eta_1 k_{\mathrm{m}} \dot{a}_{\mathrm{m}} \tag{3.20}$$

and, extending the characteristic of equation (3.6) to the linear Maxwell visco-elastic half-space (Hunt *et al.* 1996*a*),

$$\dot{q}_{\mathrm{m}} + r q_{\mathrm{m}} = -4\mu_1 k_{\mathrm{m}} \dot{a}_{\mathrm{m}}, \tag{3.21}$$

where $r = \mu_1/\eta_1$.

(v) *Elastica nonlinearities*

Mühlhaus (1993) suggests that the natural nonlinearities of large-deflection bending seen in equation (3.1) generate the following equation in tangent angle φ:

$$\varphi'''' + 2\varphi'' + \tfrac{3}{2}\varphi'^2\varphi'' + F = 0, \tag{3.22}$$

where Fourier decomposition would, in principle, allow the displacement w in the Maxwell relation (3.14) to be replaced by φ, giving the general non-dimensional form

$$\dot{F} + F = \dot{\varphi}. \tag{3.23}$$

Numerical studies of the elastica on Winkler foundation with a Maxwell law, useful for comparison purposes, are given by Whiting & Hunt (1997).

(c) *Single governing equation*

For Winkler media, equation (3.1), or its linearized equivalent (3.2) about $w = F = 0$, can be combined simply with constitutive relations such as (3.14) to provide a single governing PDE in one spatial and one time dimension. For example, if the compressive load P remains constant, after a little symbolic manipulation of (3.2) and (3.14), the single equation

$$EI\dot{w}'''' + P\dot{w}'' + k\dot{w} + \frac{k}{\eta}(EIw'''' + Pw'') = 0 \tag{3.24}$$

appears (Hunt *et al.* 1996*a*). Half-space formulations are not so straightforward, however, and lead, in general, to partial integro-differential equations; these are then open to analysis either by Fourier expansion as described above or by general application of Green's theorem. Such approaches work well for linearized systems (Biot 1965) and possibly also those with slowly varying amplitudes and phase. However, they are liable to be called into question when thoroughly localized responses are involved.

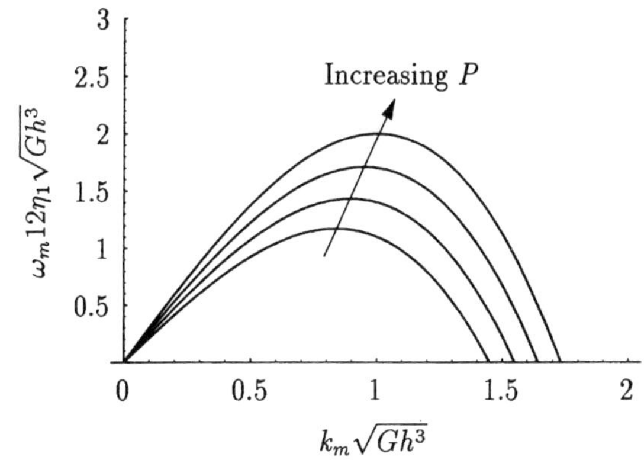

Figure 9. Dispersion relation for a strut on a viscous half-space.

4. Methods of analysis

Analytical and numerical methods for solving such differential and integro-differential equations are now briefly reviewed. Linear and nonlinear systems often demand quite different analytical approaches.

(*a*) *Linearized systems*

Biot's (1937) approach, expressed in Fourier transforms using equation (3.20) for an elastic layer of shear modulus μ and thickness h compressed by a load P and supported by a viscous half-space of viscosity η_1, leads to the dispersion relation of figure 9, where growth rate ω_{m} is given by the eigenmode, $a_{\mathrm{m}} = A_{\mathrm{m}}\mathrm{e}^{\omega_{\mathrm{m}}t}$. A finite range of wavelengths is involved, from the infinitely long ($k_{\mathrm{m}} = 0$), through the *dominant wavelength* of equation (3.10), for which ω_{m} is a maximum,

$$\omega_{\mathrm{d}} = \frac{P}{6\eta_1}k_{\mathrm{d}}, \quad \text{where} \quad k_{\mathrm{d}} = \frac{1}{h}\sqrt{\frac{P}{\mu h}} \tag{4.1}$$

(Biot 1965), and ending at a P-dependent short wavelength cut-off value. Plotted in non-dimensional form these curves superimpose on one another (Hunt *et al.* 1996*a*) and so it is possible to remove P from all consideration.

This contrasts with the (non-dimensionalized) dispersion relations for a visco-elastic medium of viscosity η_1 and shear modulus μ_1 shown in figure 10. Here the minimum critical buckling load $P^{\mathrm{C}}_{\min}$ and its corresponding mode $k_{\min}$ are given by

$$P^{\mathrm{C}}_{\min} = \frac{6}{k_{\min}}\mu_1, \quad \text{where} \quad k_{\min} = \frac{1}{h}\sqrt[3]{\frac{6\mu_1}{\mu}}, \tag{4.2}$$

$p = P/P^{\mathrm{C}}_{\min}$, $\tilde{k}_{\mathrm{m}} = k_{\mathrm{m}}/k_{\min}$, $\tilde{t} = rt$, $\tilde{r} = r/\omega_{\mathrm{d}}$ and $\tilde{\omega}_{\mathrm{m}}$ comes from the eigensolution $a_{\mathrm{m}} = A_{\mathrm{m}}\mathrm{e}^{\tilde{\omega}_{\mathrm{m}}\tilde{t}}$, as before. As p approaches the critical elastic buckling value $p^{\mathrm{C}} = 1$, the rate of growth of the dominant mode amplifies in a kind of resonance phenomenon, until at the critical load it grows infinitely fast and expresses the instantaneous elastic response in the critical buckling mode.

(*b*) *Nonlinear systems*

Dispersion relations are by their nature linear concepts, being based on comparisons between simple spectral components of the displacement pattern; interactions

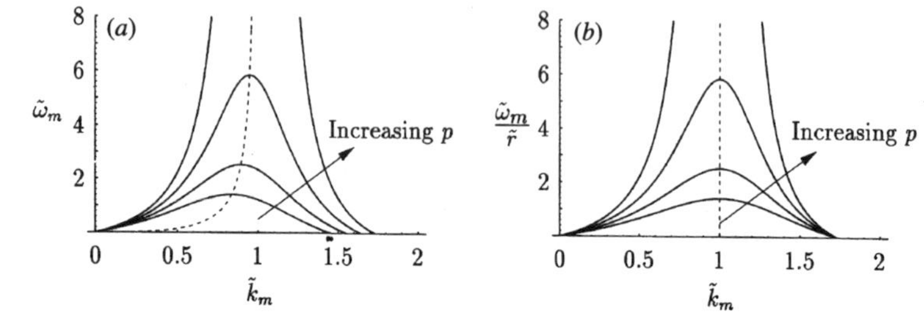

Figure 10. Dispersion relation for a strut on a visco-elastic half-space.

between components are never considered. Linear equations are never able to describe the evolution of localized forms that is of primary concern here: when such nonlinear features are involved, quite different analytical and numerical tools are required (see, for example, Beyn 1990; Champneys & Toland 1993). The added complexity of wavelength dependence, associated with the concept of a half-space, is again based on spectral decomposition and linear in concept. Nonlinear analysis can, however, sometimes draw on such a view by extending it in the sequential asymptotic sense, as, for example, in the use of harmonic fluctuations with slowly varying amplitude, phase and indeed wavelength. Provided the nonlinearities are polynomial and of moderate order, Fourier transformation methods are known to work well for viscous media (Mühlhaus *et al.* 1998), but are less useful for the severe localization associated with brittle (fracturing) structures.

(i) *Trial functions*

Trial functions, as seen, for example, in Galerkin or Rayleigh–Ritz formulations, might prove useful in overcoming problems of wavelength dependence; wavelengths are built in to the modelling process and have the potential to vary (see, for example, Wadee *et al.* 1997). However, Galerkin procedures involving wavelength variation are yet to be successfully devised for such problems, whereas Rayleigh–Ritz procedures depend on conservation of energy which is contravened for visco-elastic foundations. Although it is felt that useful analytical approaches could develop from such procedures, at present, under the banner of trial function methods, there only exists, as far as we are aware, a crude truncation approach (Hunt *et al.* 1996*a*). Results from this process should be treated with caution, taken over finite time perhaps merely as qualitative pointers towards possible behaviour; it does, however, provide reasonably convincing time portraits.

The process is described fully in Hunt *et al.* (1996*a*) and summarized in Hunt *et al.* (1996*b*). Under the constraint of constant end-shortening, trial functions are fed into a governing nonlinear PDE with a zero right-hand side, and enough coefficients of the resulting set of functions put to zero to match a number, typically seven, unknown variables. A simple marching algorithm then gives the evolution over time. Figure 11 shows a typical output based on the PDE

$$\dot{w}'''' + 3p\dot{w}'' + \dot{\overline{[2\beta(w - w^3)]}} + (w'''' + 3pw'') = 0. \tag{4.3}$$

Wavelength dependence is represented by the explicit appearance of β, as defined by the trial function

$$w = A \operatorname{sech} \alpha x \cos \beta x + B \operatorname{sech} \alpha x \tanh \alpha x \sin \beta x + Cx \operatorname{sech} \alpha x \sin \beta x + Dx \operatorname{sech} \alpha x \tanh \alpha x \cos \beta x. \tag{4.4}$$

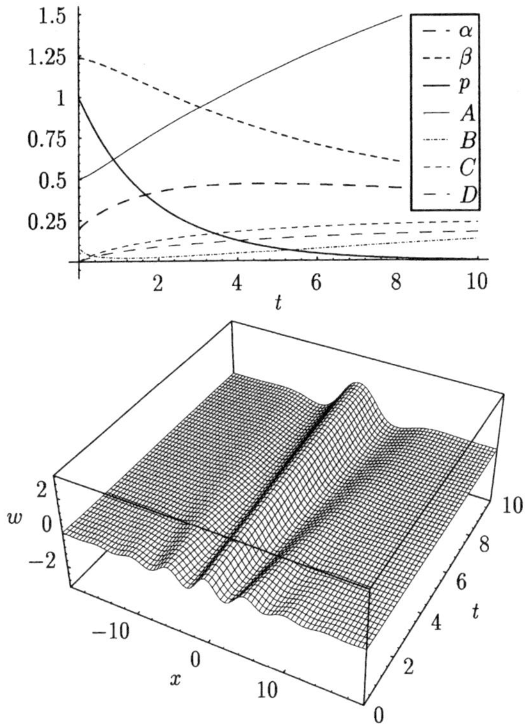

Figure 11. Seven-variable evolution of a localized shape under constant end displacement. Based on the elastic strut of equation (3.2) on a visco-elastic half-space with $EI = 1$, $P = 3p$ and $F = 2\beta(w - w^3)$. (See Hunt *et al.* 1996*a*.)

The free variables are A, B, C, D, α, β and p. Six first-order ODEs in time arise from setting the coefficients of $\operatorname{sech} \alpha x \cos \beta x$, $\operatorname{sech} \alpha x \tanh \alpha x \sin \beta x$, $\operatorname{sech}^3 \alpha x \cos \beta x$, $\operatorname{sech}^3 \alpha x \tanh \alpha x \sin \beta x$, $x \operatorname{sech} \alpha x \sin \beta x$ and $x \operatorname{sech} \alpha x \tanh \alpha x \cos \beta x$ to zero, with the seventh coming from the constraint of constant end-shortening. As the load p drops to zero, the localized wave pattern opens out to release the bending strain energy of the elastic strut into the foundation. The eventual outcome, not portrayed here, is a single long wave which, if length is taken to infinity, has zero amplitude.

(ii) *Numerical methods*

In the modern context, numerical methods provide the most obvious way of tackling the nonlinear governing equations that arise in modelling of geological systems. Yet there is little published work of which we are aware that relates directly to the solitary waveforms of interest here. Typically, although nonlinearities and localization are now routinely considered, the latter is usually taken in the sense of a thorough localization, as in brittle or quasi-brittle fracture (Bažant & Cedolin 1991; Leroy & Triantafillydis 1996). There remain a number of fundamental problems to be addressed. It is generally accepted in geological circles that half-space formulations are the most credible, but these carry the penalty of non-local regulatory laws leading in the first instance to integro-differential equations. In the linear context such equations can be reduced to wavelength dependent PDEs as discussed above (Biot 1965), but for station-based approaches such as the collocation method this is numerically awkward; the equation to be solved depends on the solution. Green's function methods suggest one possible way forward, yet these too are limited by being linear in concept.

The expansive alternative is to introduce a second spatial dimension as in a finite element (FE) formulation, but with time providing a third dimension this is likely to be expensive in computing power. Such formulations are popular (Lewis & Williams 1978; Williams *et al.* 1978) but may be problematical (Cobbold 1977). They inevitably suffer from the fact that a plate or a strut is long and thin, and must be subdivided into many elements to achieve the necessary accuracy for short wavelength deformation. The supporting medium, where deformation is minimal, would then require subdivision into something like n^2 (or n^3) elements, n being the number of beam elements. A combination of FE and boundary elements might prove more efficient, but boundary elements only work well if the nonlinearity in the supporting medium is unimportant. FE formulation might also require remeshing at larger amplitudes, because of mesh distortion. One recent relevant thesis is due to Breekman (1994), who models lithospheric and crustal elasto-viscoplastic deformation using finite elements.

We note finally that it is certainly possible that the problem is suited to analysis by wavelets, but this approach is yet to be explored. A second promising alternative is a particle-based, as opposed to mesh-based, approach to discretization (Mühlhaus & Hornby 1996).

5. Concluding remarks

The paper emphasizes that nonlinear PDEs and integro-differential equations lie at the heart of the problem of geological folding. The field is in its infancy: Winkler foundations can be formulated, but the more relevant half-space has not apparently been tackled rigorously in the nonlinear range. Where material softening is involved, it is demonstrated that spatial localization is likely to be central to the response over time. Trial function methods supply perhaps the best analytical way forward; the relatively short length scales and long time scales involved suggest that space and time may be able to be handled separately. Likely also to be significant to real geological systems are constitutive laws with hardening responses, or those that initially soften followed by hardening. Quasi-periodicity, seen in the field but not part of the present description, would then naturally enter the frame; such matters are, however, out of the remit of this special issue.

The review generally raises more questions than it answers. It is worthy of note that a recent UK EPSRC grant under the Applied Nonlinear Mathematics Initiative, to address some of the matters raised, commenced in January 1997.

Financial support for A.I.M.W. was generously provided by the University of Sydney, Australia, and the Overseas Research Students (ORS) Award Scheme, UK.

References

Abbasi, M. R. & Mancktelow, N. S. 1992 Single layer buckle folding in non-linear materials. I. Experimental study of fold development from an isolated initial perturbation. *J. Struct. Geol.* **14**, 85–104.

Bažant, Z. P. & Cedolin, L. 1991 *Stability of structures: elastic, inelastic, fracture, and damage theories.* Oxford University Press.

Beyn, W.-J. 1990 The numerical computation of connecting orbits in dynamical systems. *IMA J. Numer. Analys.* **9**, 379–405.

Biot, M. A. 1937 Bending of an infinite beam on an elastic foundation. *ASME J. Appl. Mech.* A **59**, 1–7.

Biot, M. A. 1965 *Mechanics of incremental deformation.* New York: Wiley.

Breekman, F. 1994 Tectonic modelling of thick-skinned compressional intraplate deformation. Ph.D. thesis, Vrije Universiteit, Amsterdam.

Champneys, A. R. & Toland, J. F. 1993 Bifurcation of a plethora of large amplitude homoclinic orbits for Hamiltonian systems. *Nonlinearity* **6**, 665–721.

Cobbold, P. R. 1975 Fold propatation in single embedded layers. *Tectonophys.* **27**, 333–351.

Cobbold, P. R. 1976 Fold shapes as functions of progressive strain. *Phil. Trans. R. Soc. Lond.* A **283**, 129–138.

Cobbold, P. R. 1977 Finite-element analysis of fold propatation—a problematic application? *Tectonophys.* **38**, 339–353.

Estrin, Y. 1996 Dislocation—density-related constitutive modelling. In *Unified constitutive laws of plastic deformation* (ed. A. S. Krausz & K. Krausz), pp. 69–106. New York: Academic.

Frost, H. J. & Ashby, M. F. 1982 *Deformation-mechanism maps.* New York: Pergamon.

Hobbs, B. E. 1981 The influence of metamorphic environment upon the deformation of minerals. *Tectonophys.* **78**, 335–383.

Hunt, G. W. & Baker, G. 1995 Principles of localization in the fracture of quasi-brittle structures. *J. Mech. Phys. Solids* **43**, 1127–1150.

Hunt, G. W. & Wadee, M. K. 1991 Comparative lagrangian formulations for localized buckling. *Proc. R. Soc. Lond.* A **434**, 485–502.

Hunt, G. W., Mühlhaus, H.-B. & Whiting, A. I. M. 1996*a* Evolution of localized folding for a thin elastic layer in a softening visco-elastic medium. *Pure Appl. Geophys.* **146**, 229–252.

Hunt, G. W., Mühlhaus, H.-B., Hobbs, B. & Ord, A. 1996*b* Localized folding of viscous layers. *Geologische Rundschau (Int. J. Earth Sciences)* **85**, 58–64.

Hunt, G. W., Lawther, R. & Providência e Costa, P. 1997 Finite element modelling of spatially-chaotic structures. *Int. J. Num. Meth. Engng* **40**, 2237–2256.

Kerr, A. D. 1964 Elastic and visco-elastic foundation models. *ASME J. Appl. Mech.* **31**, 491–498.

Kruhl, J. H. (ed.) 1994 *Fractals and dynamic systems in geoscience.* Berlin: Springer.

Leroy, Y. M. & Triantafillydis, N. 1996 Stability of a frictional, cohesive layer on a viscous substratum: variational formulation and asymptotic solution. *J. Geophys. Res.* **101**, 17 795–17 811.

Lewis, R. W. & Williams, J. R. 1978 A finite-element study of fold propagation in a viscous layer. *Tectonophys.* **48**, 263–283.

Mancktelow, N. S. & Abbasi, M. R. 1992 Single layer buckle folding in non-linear materials. II. Comparison between theory and experiment. *J. Struct. Geol.* **14**, 85–104.

Mühlhaus, H.-B. 1993 Evolution of elastic folds in plane strain. In *Modern approaches to plasticity* (ed. D. Kolymbas), pp. 734–765. New York: Elsevier.

Mühlhaus, H.-B. & Hornby, P. 1996 On the reality of antisymmetric stresses in fast granular flows. In *Proc. IUTAM Conf. on Granular and Porous Media, Cambridge, July 1996* (ed. Fleck & Cocks). Dordrecht: Kluwer.

Mühlhaus, H.-B., Hobbs, B. E. & Ord, A. 1994 The role of axial constraints on the evolution of folds in single layers. In *Computer methods and advances in geomechanics* (ed. H. J. Siriwardane & M. M. Zaman), vol. 1, pp. 223–231. Rotterdam: Balkema.

Mühlhaus, H.-B., Sakaguchi, H. & Hobbs, B. E. 1998 Evolution of 3D folds for a non-Newtonian layer in a viscous medium. *Proc. R. Soc. Lond.* A. (In the press.)

Ord, A. 1994 The fractal geometry of patterned structures in numerical models of rock formation. In *Fractals and dynamic systems in geoscience* (ed. J. H. Kruhl). Berlin: Springer.

Ord, A. & Hobbs, B. E. 1989 The strength of the continental crust, detachment zones and the development of plastic instabilities. *Tectonophys.* **158**, 269–289.

Price, N. J. & Cosgrove, J. W. 1990 *Analysis of geological structures.* Cambridge University Press.

Roscoe, R. 1950 Mechanical models for the representation of visco-elastic properties. *Br. J. Appl. Phys.* **1**, 171–173.

Thompson, J. M. T. & Hunt, G. W. 1984 *Elastic instability phenomena.* Chichester: Wiley.

Wadee, M. K., Hunt, G. W. & Whiting, A. I. M. 1997 Asymptotic and Rayleigh–Ritz routes to localized buckling solutions in an elastic instability problem. *Proc. R. Soc. Lond.* A **453**, 2085–2107.

Whiting, A. I. M. 1996 Localized buckling of an elastic strut in a visco-elastic medium. Ph.D. thesis, Imperial College, University of London.

Whiting, A. I. M. & Hunt, G. W. 1997 Evolution of nonperiodic forms in geological folds. *Math. Geol.* **29**, 705–723.

Williams, J. R., Lewis, R. W. & Zienkiewicz, O. C. 1978 A finite-element analysis of the role of initial perturbations in the folding of a single viscous layer. *Tectonophys.* **45**, 187–200.

Chapter 7

Solitary waves

Finally we look at a couple of papers that consider solitary wave solutions to evolution equations arising in solid mechanics. In many cases the search for solitary waves, in the appropriate *moving* co-ordinate frame, leads to the same analytical techniques and even the same equations as those governing localized static modes. The first article, by COLEMAN ET AL illustrates this point for a model of a rod in the presence of flexure and stretch. Note that the ordinary differential equations that arise and the phase plane analysis of them are similar in spirit to much of the statics of the preceding articles. The second article by CHEN & MCKENNA reviews some analytical results and presents new numerical work on a model of a suspended beam supported from one side only. The results touch upon some of the more complex issues associated with stability of solitary waves in these kinds of structural models. In contrast to the results of SANDSTEDE in Chapter 2 on the statics of a strut model, it appears that the multi-modal solitary waves may be stable in this case. Indeed the numerical results suggest that stable solitary waves may interact almost like solitons, although this model does not appear to be integrable.

Arch. Rational Mech. Anal. 129 (1995) 147–174. © Springer-Verlag 1995

On the Dynamics of Flexure and Stretch in the Theory of Elastic Rods

BERNARD D. COLEMAN, ELLIS H. DILL & DAVID SWIGON

Contents

1. Introduction

We here examine a theory of planar motions of naturally straight elastic rods intended for cases in which the strains relative to an undistorted configuration, $\mathfrak{C}^{\mathrm{u}}$, are small, although displacements and rotations may be large. The dynamical equations of the theory, which are complete to within an error of order two in a dimensionless measure of thickness, curvature, and extension, can be derived from Hamilton's principle by assuming that in appropriate units ψ and k, the densities of free energy and kinetic per unit of length in $\mathfrak{C}^{\mathrm{u}}$, have the form:

$$2\psi = (\theta_S)^2 + \varepsilon^2, \tag{1.1}$$

$$2k = (\theta_t)^2 + (x_t)^2 + (y_t)^2. \tag{1.2}$$

Subscripts denote derivatives; $x = x(S, t)$ and $y = y(S, t)$ are Cartesian coordinates at time t of the material point on the rod axis that has arc-length coordinate S when the rod is in the configuration $\mathfrak{C}^{\mathrm{u}}$. We write $s = s(S, t)$ for the arc-length coordinate at t; $\varepsilon = \varepsilon(S, t) = \partial s/\partial S - 1$ is the elongational strain; and $\theta = \theta(S, t)$ is the angle that the tangent to the axial curve makes with the x-axis.

The dynamical equations corresponding to (1.1) and (1.2) can be written

$$((1+\varepsilon)\cos\theta)_{tt} = (T\cos\theta - G\sin\theta)_{SS}, \tag{1.3a}$$

$$((1+\varepsilon)\sin\theta)_{tt} = (T\sin\theta + G\cos\theta)_{SS}, \tag{1.3b}$$

$$\theta_{tt} - \theta_{SS} = (1+\varepsilon)G, \tag{1.3c}$$

where $G = G(S,t)$ is the resultant of the shearing forces and $T = T(S,t)$ the resultant of the normal forces, on the surface $\mathscr{S}(S,t)$ formed at time t by the material points that are on the cross section that has arc-length coordinate S in $\mathfrak{C}^{\mathrm{u}}$. The shearing force G is a reactive parameter not given by constitutive relations; the tension T obeys a familiar constitutive equation that in the units we are using takes the form

$$T = \varepsilon. \tag{1.3d}$$

Thus, the governing equations of the present theory can be written as a system of three equations in three unknowns: θ, ε, G. The theory is the specialization to planar motions of the Kirchhoff-Clebsch theory of rods [1859, 1], [1862, 1], [1876, 1], [1883, 1] without the usual assumption of inextensibility, i.e., of temporal invariance of arc length.

In Sections 2 and 3 below we discuss the origin of the present theory as an approximation to the theory of nonlinearly elastic slender bodies, we present various alternative formulations of the theory, and we give a list of conservation laws implied by the system (1.3).

The remainder of the paper is devoted to the theory of traveling waves. The present study of such waves, like that presented in [1992, 1], was much influenced by the very general investigation of ANTMAN & LIU [1979, 1]. Here, as in [1992, 1], the emphasis is laid on explicit solutions. In Sections 5, 6, and 7 we derive expressions, in terms of elliptic functions and integrals, for the axial curve in each type of traveling-wave solution of (1.3): solitary, inflexional, and noninflexional. In the Appendix we consider a limiting process in which the magnitude of tension decreases to zero, and we discuss the manner in which traveling-wave solutions of (1.3) approach solutions of the system of equations,

$$(\cos\theta)_{tt} = (T\cos\theta - G\sin\theta)_{SS}, \tag{1.4a}$$

$$(\sin\theta)_{tt} = (T\sin\theta + G\cos\theta)_{SS}, \tag{1.4b}$$

$$\theta_{tt} - \theta_{SS} = G, \tag{1.4c}$$

governing planar motions in the Kirchhoff-Clebsch theory when inextensibility is assumed.

In the system (1.4), which is the subject of a growing literature (cf. [1984, 1], [1992, 1, 3], [1994, 1–5]), both the tension T and the resultant shear force G are reactive parameters. The system can be obtained from (1.1), (1.2), and Hamilton's principle by imposing, at each S and t, the constraint $\varepsilon = 0$, which can be written:

$$(x_S)^2 + (y_S)^2 = 1. \tag{1.5}$$

Such a variational formulation of (1.4) and the theory of traveling waves governed by that system are discussed in detail in [1992, 1].

If we were using conventional units, such as centimeters, grams, and seconds, instead of the present dimensionless units, the characteristic speed of unity exhibited by (1.3c) and (1.4c) would appear as $\sqrt{E/\rho}$ with E the tensile modulus and ρ the mass density. Both (1.3) and (1.4) have traveling-wave solutions with wave speeds greater than this characteristic speed. When such solutions were found in the theory of (1.4), although they were difficult to interpret physically, there was not available an argument for their out-of-hand rejection. Here, in Section 4, we show that for each such "supersonic" traveling-wave solution of (1.3) there are places at which $\varepsilon < -1$. It follows that if we are granted the principle that a smooth solution of the dynamical equations of a theory of extensible rods is physically meaningful only if it is free from a change in the ordering of the material points on the rod axis, i.e., only if it is free from intervals of values of S on which $\partial s/\partial S < 0$, then we may reject as inadmissible the supersonic traveling-wave solutions of (1.3). (At places in a rod where $\varepsilon = -1$, the mass density per unit length is infinite; hence in the discussion below we shall use a strengthened form of the principle just stated, and we shall assume that a solution in which ε attains the value -1 should be rejected, even if in it ε does not drop below -1.)

We remark here in passing that in the present study we consider only briefly traveling-wave solutions of (1.3) with characteristic speed; for, as is observed in Section 4, for such waves θ is constant and ε obeys the familiar linear wave equation, $\varepsilon_{tt} = \varepsilon_{SS}$.

The system (1.3) is a refinement of (1.4) that takes into account effects of axial elongation and compression. If we are granted, in addition to the principle asserted above, that each physically meaningful solution of the system (1.4) of equations for inextensible rods should be interpretable as a limit for small tension of solutions of (1.3), then results given in Section 4 and in the Appendix tell us that not only for (1.3), but also for (1.4), we can reject supersonic traveling waves as being without physical meaning.

2. Dynamical Equations for Planar Motions of Rods

We are concerned here with the nonlinear dynamical equations governing planar, twist-free, motions of elastic rods in the limit of small strain, but possibly large displacement. We assume the rods to be homogeneous, naturally straight, and subject to external loads only at their end points. When such a rod, $\mathscr{R}$, is in an undistorted, stress-free, configuration, $\mathfrak{C}^{\mathrm{u}}$, it has the form of a solid cylinder with not necessarily circular cross sections, $\mathscr{S}^{\mathrm{u}}$. We write A and h for the area and diameter of the sets $\mathscr{S}^{\mathrm{u}}$. The locus $\mathscr{C}^{\mathrm{u}}$ of the centroids of the sets $\mathscr{S}^{\mathrm{u}}$ is a straight line or a line segment of length L. In discussions of traveling waves, $\mathscr{C}^{\mathrm{u}}$ is a full line and hence L is infinite; in general, whenever a three-dimensional body $\mathscr{R}$ is called a *rod*, it is presupposed that $h/L \ll 1$. The set $\mathscr{C}^{\mathrm{u}}$ is parametrized with S, the distance from a point on $\mathscr{C}^{\mathrm{u}}$ to a preassigned point, $\mathbf{o}$, of $\mathscr{C}^{\mathrm{u}}$. Thus, for each S in an interval $\mathscr{I}$ of length L, $\boldsymbol{x}^{\mathrm{u}}(S)$ is a point on $\mathscr{C}^{\mathrm{u}}$, and $\mathscr{S}^{\mathrm{u}}(S)$ is the (undistorted) cross section that has $\boldsymbol{x}^{\mathrm{u}}(S)$ as its centroid. For simplicity of exposition, we assume that $\boldsymbol{x}^{\mathrm{u}}(S)$ is in $\mathscr{S}^{\mathrm{u}}(S)$.

At each time t in a motion of $\mathscr{R}$, $\mathscr{C}^{\mathrm{u}}$ is taken into a curve $\mathscr{C}(t)$, called the *axial curve at time t*. The point $\boldsymbol{x}(S,t)$ of $\mathscr{C}(t)$ is the present location in space of the material point of $\mathscr{R}$ that is at $\boldsymbol{x}^{\mathrm{u}}(S)$ in $\mathfrak{C}^{\mathrm{u}}$.

For the motions we consider, $\mathscr{C}(t)$ lies in a fixed plane $\mathscr{P}$ which we coordinatize with a Cartesian system x, y, with natural basis, $\boldsymbol{i}, \boldsymbol{j}$. In order to have a broad class of such planar motions possible, we assume that each cross section $\mathscr{S}^{\mathrm{u}}$ has a line of symmetry which we take to lie in $\mathscr{P}$, and we write I for the moment of inertia of a $\mathscr{S}^{\mathrm{u}}$ about its principal axis normal to $\mathscr{P}$. We denote by $\boldsymbol{F} = \boldsymbol{F}(S,t)$ the resultant of the stress vector field acting at time t on $\mathscr{S}(S,t)$, the surface formed at that time t by the material points that are on $\mathscr{S}^{\mathrm{u}}(S)$ in the configuration $\mathfrak{C}^{\mathrm{u}}$.

As the notation $\boldsymbol{x} = \boldsymbol{x}(S,t)$, or equivalently,

$$x = x(S,t), \quad y = y(S,t), \tag{2.1}$$

indicates, we generally parametrize the curve $\mathscr{C}(t)$ not with its own arc length s, but with the parameter S that gives, for each point on $\mathscr{C}(t)$, the distance it would be from the point $\mathbf{o}$ on $\mathscr{C}^{\mathrm{u}}$ if $\mathscr{R}$ were deformed back to its straight, undistorted, configuration $\mathfrak{C}^{\mathrm{u}}$. One may call S the *material coordinate* or *arc-length coordinate in* $\mathfrak{C}^{\mathrm{u}}$ and call s the *arc-length coordinate at time t*. For the *stretch* $\lambda = \lambda(S,t) = \partial s(S,t)/\partial S$ and the *tangent angle*, i.e., the counterclockwise angle $\theta = \theta(S,t)$ from the x-axis to the vector tangent to $\mathscr{C}(t)$ at the material coordinate S, we have

$$x_S = \lambda\cos\theta, \quad y_S = \lambda\sin\theta \tag{2.2}$$

where $x_S = \partial x(S,t)/\partial S$, etc. Let $\boldsymbol{i}$ and $\boldsymbol{j}$ be unit vectors along the coordinate axes, and let $\boldsymbol{t}$ and $\boldsymbol{n}$ be the unit tangent and normal vectors of $\mathscr{C}(t)$:

$$\boldsymbol{t} = \boldsymbol{i}\cos\theta + \boldsymbol{j}\sin\theta, \quad \boldsymbol{n} = -\boldsymbol{i}\sin\theta + \boldsymbol{j}\cos\theta. \tag{2.3}$$

For the components of $\boldsymbol{F}$ relative to the bases $\boldsymbol{i}, \boldsymbol{j}$ and $\boldsymbol{t}, \boldsymbol{n}$, we use the notation

$$\boldsymbol{F} = F^x\boldsymbol{i} + F^y\boldsymbol{j} = T\boldsymbol{t} + G\boldsymbol{n}; \tag{2.4}$$

G is called the *shear resultant* and T the axial force or *tension*. For the magnitude of the resultant moment of the stress field acting on $\mathscr{S}(S,t)$, i.e., for the *bending moment*, we write $M = M(S,t)$.

The rod $\mathscr{R}$ is assumed to be composed of a homogeneous and isotropic and, in general, nonlinear elastic material. We write ρ for the mass density of this material in the undistorted, stress-free, configuration $\mathfrak{C}^{\mathrm{u}}$, and E for the tensile modulus (i.e., Young's modulus) for small deformations from the configuration $\mathfrak{C}^{\mathrm{u}}$.

For each configuration of $\mathscr{R}$, we may define a number δ by

$$\delta = \sup_S \max\{h/L, |\theta_S|h, |\varepsilon|\}, \tag{2.5}$$

where $\varepsilon = \lambda - 1$ is the *elongational strain* at the axis of $\mathscr{R}$. For the planar twist-free motions under consideration, a more general analysis given by KIRCHHOFF [1859, 1], [1876, 1] and CLEBSCH [1862, 1], [1883, 1] and recently discussed from a modern point of view by DILL [1992, 2] (see also [1992, 1]), here yields the conclusion that the dynamical equations expressing the laws of balance linear and angular momentum for the three-dimensional body $\mathscr{R}$ give, to within an error $O(\delta^2)$,

$$\rho A \boldsymbol{x}_{tt} = \boldsymbol{F}_S, \tag{2.6}$$

$$\rho I \theta_{tt} = M_S + \lambda G, \tag{2.7}$$

where, in (2.6), for $T = \boldsymbol{F} \cdot \boldsymbol{t}$ one has

$$T = EA\varepsilon \tag{2.8}$$

and, in (2.7), for M,

$$M = EI\theta_S. \tag{2.9}$$

By differentiating (2.6) with respect to S and making use of (2.8) and (2.9), we obtain the following system of three partial differential equations for θ, λ, and G as functions of S and t:

$$\rho A(\lambda \cos\theta)_{tt} = (EA[\lambda - 1]\cos\theta - G\sin\theta)_{SS}, \tag{2.10a}$$

$$\rho A(\lambda \sin\theta)_{tt} = (EA[\lambda - 1]\sin\theta + G\cos\theta)_{SS}, \tag{2.10b}$$

$$\rho I\theta_{tt} - EI\theta_{SS} = \lambda G. \tag{2.10c}$$

The shear resultant G, whose time-derivative does not occur in the system (2.10), is a reactive parameter not related by a constitutive equation to kinematical variables but rather to be found by solving the dynamical equations for appropriate boundary data.

Since we restrict our attention to motions in which the local stretch ratio λ never attains the value zero, we may use equation (2.10c) to eliminate G from (2.10a) and (2.10b) and thus obtain a system of two equations in two unknown fields, one of which, $\theta = \theta(S, t)$, through its S-derivative, characterizes the flexure of the rod, and the other, $\lambda = \lambda(S, t)$, gives the stretch; that new system is of order two in t-derivatives and of order four in S-derivatives.

As $(I\rho/AE)^{1/2}$ has the dimensions of time, AE the dimension of force, and $(I/A)^{1/2}$ the dimension of length, we now replace t by $(I\rho/AE)^{1/2}\, t$, $\boldsymbol{F}$ by $AE\, \boldsymbol{F}$, and the list $(\boldsymbol{x}, S, s)$ by $(I/A)^{1/2}$ $(\boldsymbol{x}, S, s)$. This replacement leaves λ invariant and induces a change of (T, G) into $AE(T, G)$. In terms of the dimensionless variables so obtained, the system (2.10) reduces to (1.3), i.e.,

$$(\lambda \cos\theta)_{tt} = ([\lambda - 1]\cos\theta - G\sin\theta)_{SS}, \tag{2.11a}$$

$$(\lambda \sin\theta)_{tt} = ([\lambda - 1]\sin\theta + G\cos\theta)_{SS}, \tag{2.11b}$$

$$\theta_{tt} - \theta_{SS} = \lambda G. \tag{2.11c}$$

The first two of these equations, (2.11a) and (2.11b), are equivalent to the vector equation,

$$\boldsymbol{x}_{tt} = \boldsymbol{F}_S, \tag{2.12}$$

whose components in terms of moving basis **t**, **n** are [1992, 1]

$$x_{tt}\cos\theta + y_{tt}\sin\theta = T_S - G\theta_S, \tag{2.13a}$$

$$y_{tt}\cos\theta - x_{tt}\sin\theta = G_S + T\theta_S, \tag{2.13b}$$

where, in the present dimensionless units,

$$T = \lambda - 1 = \varepsilon. \tag{2.14}$$

3. Conservation Laws and Variational Principles

As is common practice, a relation of the form,

$$\frac{\partial}{\partial t} h(S, t) = \frac{\partial}{\partial S} g(S, t), \tag{3.1}$$

is here called a *conservation law* with h a conserved quantity and $-g$ the associated flux.

Equation (2.12) obviously is a conservation law, for it can be written,

$$\frac{\partial}{\partial t} x_t = \frac{\partial}{\partial S} F^x, \quad \frac{\partial}{\partial t} y_t = \frac{\partial}{\partial S} F^y, \tag{3.2}$$

where (in our dimensionless units) x_t and y_t are the x- and y-components of the density (with respect to S) of linear momentum. It is easily verified that, given equation (2.12), equation (2.11c) is equivalent the conservation law,

$$\frac{\partial}{\partial t}[\theta_t + xy_t - yx_t] = \frac{\partial}{\partial S}[\theta_S + xF^y - yF^x]. \tag{3.3}$$

In the forms (3.2) and (3.3) of balance of linear and angular momentum, F^x and F^y are functions of θ, λ, and G; for by (2.3), (2.4):

$$F^x = T\cos\theta - G\sin\theta, \quad F^y = T\sin\theta + G\cos\theta. \tag{3.4}$$

As equation (2.9) for the bending moment here reduces to $M = \theta_S$, and equation (2.8) reduces to (2.14), we may consider the quantity

$$\psi = \tfrac{1}{2}[(\theta_S)^2 + (\lambda - 1)^2] \tag{3.5}$$

to be a density of free energy. Similarly,

$$k = \tfrac{1}{2}[(\theta_t)^2 + \boldsymbol{x}_t \cdot \boldsymbol{x}_t] \tag{3.6}$$

is a density of kinetic energy, and $w = M\theta_t + F^x x_t + F^y y_t$, i.e.,

$$w = \theta_S\theta_t + F^x x_t + F^y y_t, \tag{3.7}$$

is a density for the rate of working of resultant forces and moments. Indeed, by taking the inner product of equation (2.12) with $\boldsymbol{x}_t$ and multiplying equation (2.11c) by θ_t, one easily verifies that those equations imply the conservation law

$$\frac{\partial}{\partial t}[\psi + k] = \frac{\partial w}{\partial S}, \tag{3.8}$$

which is the expected form of the equation of balance of energy.

One further conservation law may be obtained by taking the inner product of equation (2.12) with $\boldsymbol{x}_S = \lambda \boldsymbol{t}$, multiplying equation (2.11c) by θ_S, and following a calculation first carried out for planar motions of inextensible rods by DICHMANN, MADDOCKS & PEGO [1994, 3], for motions of such rods with torsion and twist by COLEMAN, DILL, LEMBO, LU & TOBIAS [1993, 1], and in a very general context by MADDOCKS & DICHMANN [1994, 5]. This new conservation law here takes the form

$$\frac{\partial}{\partial t}[\theta_t \theta_S + \boldsymbol{x}_t \cdot \boldsymbol{x}_S] = \frac{\partial}{\partial S}[\tfrac{1}{2}((\theta_S)^2 + \boldsymbol{x}_S \cdot \boldsymbol{x}_S + (\theta_t)^2 + \boldsymbol{x}_t \cdot \boldsymbol{x}_t)] \tag{3.9}$$

in which, of course, $\boldsymbol{x}_S \cdot \boldsymbol{x}_S = \lambda^2$. For appropriate boundary conditions, equation (3.9) implies that the quantity

$$\int [\theta_t \theta_S + x_t x_S + y_t y_S]\, dS,$$

called the *linear impulse*, is constant in time. We note that the density whose S-derivative appears on the right in (3.9) exceeds the energy density, $\psi + k$, by an amount, $(2\lambda - 1)/2$, whose S-derivative, $\lambda_S = \varepsilon_S$, is zero in the limiting case of an inextensible rod.

Hamilton's principle here becomes the assertion that for each quadruple (S_1, S_2, t_1, t_2) with $S_1 < S_2$ and $t_1 < t_2$, $\delta\mathscr{A}$, the first variation of

$$\mathscr{A} = \int_{t_1}^{t_2} [K - \Psi]\, dt \tag{3.10}$$

must vanish for all smooth variations $\delta\boldsymbol{x}$ of the basic kinetic field $\boldsymbol{x} = \boldsymbol{x}(S, t)$ for which $\delta\boldsymbol{x} = \boldsymbol{0}$ at $t = t_1$ and $t = t_2$. Here

$$K = K(S_1, S_2, t) = \int_{S_1}^{S_2} k\, dS, \tag{3.11a}$$

$$\Psi = \Psi(S_1, S_2, t) = \int_{S_1}^{S_2} \psi\, dS - [\bar{M}\theta + \bar{\boldsymbol{F}} \cdot \boldsymbol{x}]_{S_1}^{S_2}; \tag{3.11b}$$

ψ and k are as in (3.5) and (3.6); $\bar{M}(S_i, t)$ and $\bar{\boldsymbol{F}}(S_i, t)$ are the bending moment and force applied to the cross section $\mathscr{S}^{\mathrm{u}}(S_i)$, $i = 1, 2$. An argument similar to that given in detail for inextensible rods in [1992, 1] here tells us that if we put $\lambda = [\boldsymbol{x}_S \cdot \boldsymbol{x}_S]^{1/2}$, i.e., employ the relations (2.2) relating $\theta(S, t)$ to $\boldsymbol{x}(S, t)$, and *define* $G = G(S, t)$, $T = T(S, t)$ [and hence $\boldsymbol{F}$ in equation (2.4)] by the relations

$$\lambda G = \theta_{tt} - \theta_{SS}, \quad T = \lambda - 1, \tag{3.12}$$

then Hamilton's principle is satisfied if and only if, for all S_1, S_2 with $S_1 < S_2$ and all t, the equation,

$$\boldsymbol{x}_{tt} = \boldsymbol{F}_S \tag{3.13}$$

holds on the interval (S_1, S_2), while at S_1 and S_2

$$\bar{M} = \theta_S, \quad \boldsymbol{F} = \bar{\boldsymbol{F}}. \tag{3.14}$$

Thus, the equations (2.11c) and (2.12), together, are *equivalent* to Hamilton's principle with the action $\mathscr{A}$ given by (3.10) and (3.11) with k and ψ as in (3.6) and (3.5).

We emphasize this equivalence of the differential equations of the present theory to a natural form of Hamilton's principle for the following reason. One may conjecture that solutions of the full equations of elasticity for $\mathscr{R}$, as a three-dimensional body, are such that G is $O(\delta)$ and hence that the term λG in equation (2.11c) differs by terms $O(\delta^2)$ from G itself. If this conjecture be granted, one may consider replacing equation (2.11c) by the equation

$$\theta_{tt} - \theta_{SS} = G, \tag{2.11c)*}$$

familiar in the theory of inextensible rods, albeit we here employ the relation, $T = \lambda - 1$, as a constitutive equation for the tension T, to obtain the form of (2.12) shown in (2.11a), (2.11b). However, the system (2.11a), (2.11b), (2.11c)* does not appear to have a natural variational formulation; nor does it give rise to a conservation law for energy in which ψ has the expected form (3.5) or an acceptable modification of that form.

4. General Theory of Traveling Waves

A solution of the dynamical equations (2.11a, b, c) or, equivalently, of (2.12) and (2.11c), is called a *traveling wave* if for it there is a Cartesian coordinate system on $\mathscr{P}$ for which y and $u = x - S$ are functions of $\xi = S - ct$ with c a constant called the *wave velocity in the material description.* In that coordinate system,

$$y_S(S - ct) = dy(\xi)/d\xi = y'(\xi), \quad y_t(S - ct) = -cy'(\xi), \quad x_t(S, t) = -cu'(\xi), \text{ etc.};$$

by (2.2), $\lambda(S, t) = \lambda(\xi)$, $\theta(S, t) = \theta(\xi)$; by (2.11c) and (2.14), $G(S, t) = G(\xi)$, $T(S, t) = T(\xi)$; and (2.13a) and (2.13b) reduce to

$$c^2[u'' \cos\theta + y'' \sin\theta] = \lambda' - G\theta', \tag{4.1a}$$

$$c^2[y'' \cos\theta - u'' \sin\theta] = G' + (\lambda - 1)\theta'. \tag{4.1b}$$

As we here have $u' + 1 = \lambda \cos\theta$ and $y' = \lambda \sin\theta$, equations (4.1a, b) and (2.11c) yield the following system of three ordinary differential equations for λ, θ, and G:

$$(1 - c^2)\lambda' = G\theta', \tag{4.2a}$$

$$G' = [1 - \lambda(1 - c^2)]\theta', \tag{4.2b}$$

$$(1 - c^2)\theta'' = -\lambda G. \tag{4.2c}$$

When $|c| = 1$, these last equations become $G\theta' = 0$, $G' = \theta'$, $\lambda G = 0$, and, as λ is nowhere zero, G and θ' vanish everywhere, which implies that (2.11a) and (2.11b) both reduce to $\lambda_{tt} = \lambda_{SS}$. Thus, we make the following observation.

Remark 1. When, in our dimensionless units, $c = \pm 1$ [or, in the units of equations (2.8)–(2.10), $c = \pm\sqrt{E/\rho}$], the only traveling waves are such that the shear resultant G vanishes and the rod remains straight (with θ a constant which we may take to be zero); for such waves both the tensile strain ε and the displacement $u = x - S$ obey the linear wave equation, i.e., $u_{tt} = u_{SS}$ [or $\rho u_{tt} = Eu_{SS}$], which governs longitudinal waves in the linear theory of elasticity.

As waves with $c^2 = 1$ show no flexure and obey a familiar and highly developed theory, we now confine our attention to traveling waves with $c^2 \neq 1$.

In general $\boldsymbol{x}_S = \lambda \boldsymbol{t}$, and, for traveling waves, $\boldsymbol{x}_{tt} = c^2 \boldsymbol{x}_{SS}$ and $\boldsymbol{F}' = \boldsymbol{F}_S$. Therefore, equation (2.12) here becomes $c^2(\lambda \boldsymbol{t})' = \boldsymbol{F}'$, and we have $\boldsymbol{F} = c^2 \lambda \boldsymbol{t} + \boldsymbol{v}$ with $\boldsymbol{v}$ a constant vector. We assume $\boldsymbol{v} \neq \boldsymbol{0}$, because it is not difficult to show that in a solution of the system (4.2) with $\boldsymbol{v} = \boldsymbol{0}$ the angle θ either is a constant or is linear in $\xi = s - ct$; in the former case the rod is straight; in the latter its axis is a (degenerate) helix flattened onto the (x, y)-plane in such a way that it lies on a circle and rotates about the center of that circle.

Let $\Omega = \cos^{-1}(\boldsymbol{v}\cdot\boldsymbol{i}/v)$ with $v = |\boldsymbol{v}|$. From this point on we consider traveling waves with $\Omega = 0$, i.e., for which $\boldsymbol{v} = v\boldsymbol{i}$. This is justified because the values, y_Ω, u_Ω, G_Ω, T_Ω, λ_Ω, θ_Ω, of y, u, G, T, λ, θ in a traveling wave with $\Omega \neq 0$ can be obtained from the values, y_0, u_0, G_0, T_0, θ_0, of these quantities in a traveling wave with the same velocity c and the same magnitude v of $\boldsymbol{v}$, but with $\Omega = 0$ (in the same Cartesian system), by use of the formulae:

$$y_\Omega(\xi) = (u_0(\xi) + \xi)\sin\Omega + y_0(\xi)\cos\Omega, \tag{4.3a}$$

$$u_\Omega(\xi) = (u_0(\xi) + \xi)\cos\Omega - y_0(\xi)\sin\Omega - \xi, \tag{4.3b}$$

$$G_\Omega(\theta_\Omega) = G_0(\theta_0), \tag{4.3c}$$

$$T_\Omega(\theta_\Omega) = T_0(\theta_0), \tag{4.3d}$$

$$\lambda_\Omega(\theta_\Omega) = \lambda_0(\theta_0), \tag{4.3e}$$

$$\theta_\Omega(\xi) = \theta_0(\xi) + \Omega. \tag{4.3f}$$

It is not difficult to verify that a derivation of the relations (4.3) can be constructed by extending the analysis given below for the case $\Omega = 0$ to the case of $\Omega \neq 0$ and comparing the generalization so obtained with the results we give for $\Omega = 0$.

For $\Omega = 0$ the first integral of the equation of balance of linear momentum here takes the form

$$\boldsymbol{F} = c^2\lambda\boldsymbol{t} + \beta^2\boldsymbol{i}, \quad \beta = \sqrt{v} > 0. \tag{4.4}$$

The parameter β plays an important role in our theory. By (2.4) and (4.4),

$$T = c^2\lambda + \beta^2\cos\theta, \tag{4.5a}$$

$$G = -\beta^2\sin\theta. \tag{4.5b}$$

As $T = \lambda - 1$, we have, for each traveling wave, the following relation between the stretch λ and the local tangent angle θ:

$$\lambda = \frac{1 + \beta^2 \cos\theta}{1 - c^2}. \tag{4.6}$$

Hence, for λ to vanish at a place in the rod it is necessary and sufficient that at that place $\theta = \theta^\dagger$, where

$$\cos\theta^\dagger = -\beta^{-2}. \tag{4.7}$$

Wherever λ does vanish, the density of mass per unit of arc length is infinite, and hence we shall reject as inadmissible solutions in which there are places where $\lambda = 0$.

For traveling waves, equation (2.11c), expressing balance of angular momentum, becomes (4.2c), which, by (4.5b) and (4.6), is

$$(1 - c^2)\theta'' = \frac{\beta^2}{1 - c^2}\sin\theta + \frac{\beta^4}{1 - c^2}\sin\theta\cos\theta. \tag{4.8}$$

Thus (2.11c) has the first integral

$$\frac{1}{2}\frac{(1 - c^2)^2}{\beta^2}(\theta')^2 = b - \cos\theta + \frac{1}{2}\beta^2\sin^2\theta. \tag{4.9}$$

The phase plane for the second-order differential equation (4.8) has at least two equivalence classes, C^* and C^{**}, of critical points, i.e., points where $\theta' = \theta'' = 0$. In C^*, $\theta = \theta^* = 0, \ \pm 2\pi$, etc. and the critical points are saddle points. In C^{**}, $\theta = \theta^{**} = \ \pm\pi, \ \pm 3\pi$, etc., and the critical points are centers. When $\beta > 1$, equation (4.8) has another class $C^\dagger$ of critical points, namely those for which $\theta = \theta^\dagger$ where $\theta^\dagger$ obeys (4.7). When $\beta = 1$, $C^\dagger$ reduces to C^{**}.

An elementary analysis of the phase-plane diagrams for equation (4.8), i.e., the (θ', θ)-curves determined by (4.9), yields the following remark.

Remark 2. If $\beta \geqq 1$, the class $C^\dagger$ critical points of (4.8) at which (4.7) holds is not empty, each point in $C^\dagger$ is a center, and *every* orbit of equation (4.8) contains a point (θ, θ') at which (4.6) yields $\lambda = 0$. If $\beta < 1$, $C^\dagger$ is empty.

In view of (4.6), the inequalities $c^2 > 1$ and $\beta < 1$ yield $\lambda(\xi) < 0$ for all ξ. Hence, by Remark 2, independently of the value of β we can assert that if $c^2 > 1$, there is at least one place in the rod at which $\lambda \leqq 0$. If $0 \leqq c^2 < 1$ and $\beta < 1$, there is no place at which $\lambda \leqq 0$.

Let us agree to accept the principle that a solution of (2.11) is inadmissible, because it is physically unattainable, if for it there is a pair (S, t) with $\lambda(S, t) \leqq 0$. Then we can assert

Remark 3. In the present theory there are no admissible traveling waves with $|c| > 1$.

Moreover, we have

Remark 4. The traveling waves for which $\lambda(\xi) > 0$ for all ξ correspond to the following set of values of the pair (c^2, β):

$$\mathrm{R} = \{(c^2, \beta) | 0 \leqq c^2 < 1, 0 < \beta < 1\}. \tag{4.10}$$

From this point on we assume that (c^2, β) is in R. In view of Remark 4, equation (4.9) can now be written

$$\tfrac{1}{2}(1 - c^2)(a\theta')^2 = b - \cos\theta + \tfrac{1}{2}\beta^2 \sin^2\theta, \tag{4.11}$$

where a is the positive number defined by

$$a^2 = (1 - c^2)\beta^{-2}. \tag{4.12}$$

As $\beta < 1$, the number b in equation (4.11) can be no less than -1. The solutions of (4.8) with $b = 1$ lie on heteroclinic orbits and correspond to solitary waves; those with $-1 < b < 1$ lie on cyclic orbits and correspond to periodic waves that are inflexional in that they have points at which θ' vanishes; and those with $b > 1$, as they lie on orbits outside the region bounded by heteroclinic separatrices, correspond to waves that are noninflexional, i.e., for which there is no value or limit of ξ at which θ' vanishes.

In Figure 1 are shown phase planes for equation (4.8). For Figure 1A, $c^2 = 0.04$, $\beta^2 = 0.092$, and the values of λ lie in the interval $[0.9, 1.1]$ for each choice of b. For Figure 1B, (c^2, β) is not in the set R of (4.10), because $\beta = \sqrt{2}$; in such a case, i.e., for $\beta \geqq 1$, there are critical points in each of the classes C^*, C^{**}, and $C^\dagger$, and each orbit has at least two points with θ in $C^\dagger$ and hence with $\lambda = 0$.

Let

$$\sigma = \frac{\xi - S_0}{a}, \tag{4.13}$$

where S_0 is chosen as explained after equation (4.18) below. It follows from (4.11) that

$$\sigma = \pm\sqrt{1 - c^2} \int_{\theta_0}^{\theta} [2b - 2\cos\psi + \beta^2 \sin^2\psi]^{-1/2} \, d\psi. \tag{4.14}$$

Here one chooses the symbol $\pm$ to be $+$ ($-$) when θ increases (decreases) with σ. If $b \geqq 1$, θ' does not change sign, and the wave is said to be positive or negative in accord with whether θ' is everywhere positive or negative. If $|b| < 1$, θ is periodic in σ and the choice of $+$ or $-$ in (4.14) varies with position on the cyclic orbit determined by (4.11).

From (4.2c), (4.5b), and (2.2) it follows that $(1 - c^2)\theta'' = \beta^2 y'$, and hence for each traveling wave there is a constant η such that

$$\beta^2 y = (1 - c^2)\theta' + \eta \tag{4.15}$$

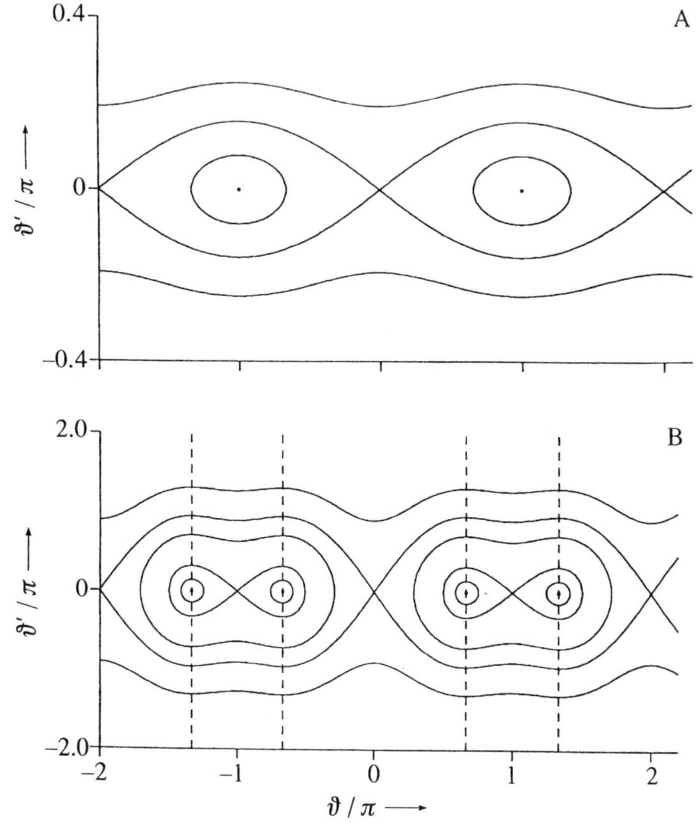

Fig. 1. Phase planes for equation (4.9). In A, $c^2 = 0.04$, $\beta^2 = 0.092$ and hence (c^2, β) is in the set R. In B, $c^2 = 0$, $\beta^2 = 2$; the vertical dashed lines correspond to values of θ in $C^{\dagger}$.

or, by (4.11) and (4.12),

$$\beta y = \pm [2b - 2\cos\theta + \beta^2 \sin^2\theta]^{1/2} + \eta/\beta. \tag{4.16}$$

Moreover, as $x = S + u$, by (2.2) we have $1 + u' = \lambda \cos\theta$, and, as $u' = \theta' du/d\theta$, (4.6) yields

$$\frac{d\xi}{d\theta} + \frac{du}{d\theta} = \frac{(1 + \beta^2 \cos\theta)\cos\theta}{(1 - c^2)\theta'}, \tag{4.17}$$

and hence, by (4.11) and (4.14),

$$u = -a\sigma \pm \frac{1}{\beta}\int_{\theta_0}^{\theta} \frac{(1 + \beta^2 \cos\psi)\cos\psi}{[2b - 2\cos\psi + \beta^2 \sin^2\psi]^{1/2}}\, d\psi. \tag{4.18}$$

Let S_0 be the arc-length coordinate of a point on the axis chosen so that $u(S_0 - ct) = 0$ at $t = 0$; θ_0 in (4.14) and (4.18) is the value of $\theta(S - ct)$ at that point and time. When the symbol $\pm$ is given a value, $+$ or $-$, in one of the equations (4.14), (4.16), (4.18), it must be given that value in all three.

Let $\lambda^{\#}$ be the value of the stretch, λ, at any point at which $\theta = \pm 2n\pi$, i.e., at any point where $\cos\theta = 1$ and hence $T = F^x$ and $G = 0$. By (4.6) and (4.12),

$$\lambda^{\#} = \frac{1+\beta^2}{1-c^2} = \frac{1}{1-c^2} + a^{-2}; \tag{4.19}$$

for the elongational strain $\varepsilon = \lambda - 1$ at such a place we have

$$\varepsilon^{\#} = \frac{\beta^2 + c^2}{1-c^2}. \tag{4.20}$$

It is not difficult to show that a pair (c^2, β) is in the set R of (4.10) if and only if $0 < \beta < 1$ and

$$0 \leqq c^2 < \frac{\varepsilon^{\#}}{1+\varepsilon^{\#}} \tag{4.21}$$

with $\varepsilon^{\#}$ as in (4.20). Moreover, (4.20) and the condition that β be less than 1 imply that if $\varepsilon^{\#}$ exceeds 1, then c^2 must be greater than $(\varepsilon^{\#} - 1)/(1 + \varepsilon^{\#})$.

As $T = \varepsilon$ in our present dimensionless units, $\varepsilon^{\#}$ equals the tension at the places where the tangent $\boldsymbol{t}$ is parallel or antiparallel to the x-axis. We note in passing that equation (4.6) yields

$$\lambda = \frac{1}{1-c^2} + \left(\lambda^{\#} - \frac{1}{1-c^2}\right)\cos\theta. \tag{4.22}$$

When $\lambda^{\#}$ and $\varepsilon^{\#}$ are attained, they equal the maximum stretch and tension in the rod.

Only for noninflexional waves and solitary waves are there places or limits where $\theta = \pm 2n\pi$ and hence where $\lambda = \lambda^{\#}$, $\varepsilon = \varepsilon^{\#}$. However, for all traveling waves, including inflexional waves, there are places where $\theta = \pm(2n+1)\pi$, and at those places λ and ε attain their minimum values,

$$\lambda_{\#} = \frac{1-\beta^2}{1-c^2} = \frac{1}{1-c^2} - a^{-2}, \tag{4.23a}$$

$$\varepsilon_{\#} = \frac{c^2 - \beta^2}{1-c^2}. \tag{4.23b}$$

For noninflexional and solitary waves, the average of the supremum and infimum of the stretches is

$$\frac{\lambda^{\#} + \lambda_{\#}}{2} = \frac{1}{1-c^2}. \tag{4.24}$$

We conclude this section with the following remark which follows forthwith from (4.6) and (4.16).

Remark 5. In each traveling wave, $y(\xi)$ and $\lambda(\xi)$ obey the following formal analogue of a relation between $x(\xi)$ and $y(\xi)$ found by COLEMAN & DILL [1992, 1; eq. (101)] for solitary waves in inextensible rods governed by (1.4):

$$\left(\frac{\lambda}{\beta}\right)^2 + \left(\frac{y - \eta/\beta^2}{a}\right)^2 = \frac{\beta^4 + 2b\beta^2 + 1}{\beta^2(1 - c^2)^2}. \tag{4.25}$$

5. Solitary Waves

A traveling wave is called a *solitary wave* if $\theta'(\xi) \to 0$ in the two limits $\xi \to \infty$, $\xi \to -\infty$. As we observed in Remark 6, the solitary waves lie on heteroclinic orbits and correspond to the choice $b = 1$ in (4.11). As the saddle points of equation (4.8) lie in the set C^*, and hence have $\theta = \pm 2\pi n$, for solitary waves not only does $\theta'(\pm\infty) = 0$, but also

$$\sin\theta(\pm\infty) = 0, \quad \cos\theta(\pm\infty) = 1, \tag{5.1}$$

and equation (4.6) yields

$$\sup\lambda = \lim_{|\xi|\to\infty} \lambda = \lambda^{\#}, \tag{5.2}$$

with $\lambda^{\#}$ as in (4.19). As $\varepsilon = T$ in the units we are using and $\varepsilon^{\#}$ is equal to $\lambda^{\#} - 1$, in each solitary wave the tension in the limits $\pm\infty$, given by (4.20), is the supremum $\varepsilon^{\#}$ of the tension T in the rod and is positive.

The nonconstant solutions of equation (4.11) with $b = 1$ and $\theta(-\infty) = 0$ have the form

$$\theta = -2\tan^{-1}(\sqrt{1+\beta^2}\,\mathrm{csch}(\pm\sigma\sqrt{\lambda^{\#}})) \tag{5.3}$$

with σ as in (4.13) and $\lambda^{\#}$ as in (4.19). Here the choice $+\sigma$ for the expression $\pm\sigma$ gives a *positive solitary wave* in which θ increases monotonically to 2π as $\xi \to \infty$. The choice $-\sigma$ gives a *negative solitary wave* in which θ decreases to -2π in the same limit. For the former type of wave, $\theta = \pi$ at $\sigma = 0$; for the latter $\theta = -\pi$ at $\sigma = 0$. In both cases the minimum value of λ, i.e., $\lambda_{\#}$, is attained at $\sigma = 0$.

Equation (5.3) yields

$$\sin\theta = \frac{-2\sqrt{1+\beta^2}\sinh(\pm\sigma\sqrt{\lambda^{\#}})}{\cosh^2(\sigma\sqrt{\lambda^{\#}}) + \beta^2}, \tag{5.4}$$

$$\cos\theta = \frac{\sinh^2(\sigma\sqrt{\lambda^{\#}}) - (1+\beta^2)}{\sinh^2(\sigma\sqrt{\lambda^{\#}}) + 1 + \beta^2}, \tag{5.5}$$

which, in view of (4.16), yield, in turn,

$$y = \pm 2\frac{1+\beta^2}{\beta}\cdot\frac{\cosh(\sigma\sqrt{\lambda^{\#}})}{\cosh^2(\sigma\sqrt{\lambda^{\#}}) + \beta^2}. \tag{5.6}$$

We here have put $\eta = 0$, which places the x-axis at such a height that $y \to 0$ as $|\xi| \to \infty$. For a positive solitary wave, $y > 0$, and for a negative solitary wave $y < 0$; in both cases the maximum value of $|y|$ is $2a/\sqrt{1 - c^2}$ and occurs at $\sigma = 0$.

For the present case in which $b = 1$, by use of the substitution $z = \operatorname{ctn} \psi/2$, one may evaluate the integral in (4.18) in terms of elementary functions. Upon doing this with $\theta_0 = \pi$ for the positive wave and $-\pi$ for the negative wave, we obtain

$$u = (\xi - S_0)\varepsilon^\# - \Gamma(\sigma; \beta). \tag{5.7a}$$

Here

$$\Gamma(\sigma; \beta) = \frac{\sqrt{1 + \beta^2}}{\beta} \cdot \frac{\sinh(2\sigma\sqrt{\lambda^\#})}{\cosh^2(\sigma\sqrt{\lambda^\#}) + \beta^2}; \tag{5.7b}$$

the coefficient $[1 + \beta^2]^{1/2}/\beta$ equals $a\sqrt{\lambda^\#}$. The term $(\xi - S_0)\varepsilon^\#$ is the value the x-displacement would have at $\xi - S_0 = S - S_0 - ct = \sigma a$ if there were no flexure and the stretch were everywhere equal to the stretch in the limits $\pm\infty$. The remaining term, $-\Gamma(\sigma; \beta)$, is the interesting one, for it gives the effect on the x-displacement of the variation of stretch and flexure with ξ. The total loss of span in the x-direction due to the dependence on ξ of stretch and flexure is

$$\lim_{\sigma \to \infty} [\Gamma(\sigma; \beta) - \Gamma(-\sigma; \beta)] = 4\frac{[1 + \beta^2]^{1/2}}{\beta} = 4a\sqrt{\lambda^\#}. \tag{5.8}$$

The relations (4.22) and (5.5) yield the following expression for $T = T(\sigma) = \lambda(\sigma) - 1$ in a solitary wave:

$$T = \frac{(\cosh^2(\sigma\sqrt{\lambda^\#}) - \beta^2)\varepsilon^\# - 2\beta^2}{\cosh^2(\sigma\sqrt{\lambda^\#}) + \beta^2}. \tag{5.9}$$

Examples of positive and negative solitary waves are shown in Figure 2. The dotted curves illustrate the manner in which graphs of y/a versus x/a approach those corresponding to solutions of (1.4) in the limit in which c^2 and β approach zero, a matter discussed in the Appendix. In that limit, $\varepsilon^\# \to 0$ and T approaches zero uniformly in σ.

The graphs of y versus x seen in Figure 3 illustrate the following property of solitary waves with a specified value of $\varepsilon^\# > 0$: For c^2 in its appropriate range [see, e.g., the paragraph containing (4.21)], the larger c^2, the larger the loop in the axial curve.

A difference between the present theory and the theory of inextensible rods governed by (1.4) is the following: in the present theory the loops corresponding to solitary waves with different speeds $|c|$, or with different limiting tensions $T(\pm\infty)$, are not related by similarity transformations. The reader will notice that the solitary waves shown in Figure 3 for fixed $\varepsilon^\# = T(\pm\infty)$ have loops that are broader the higher the speed.

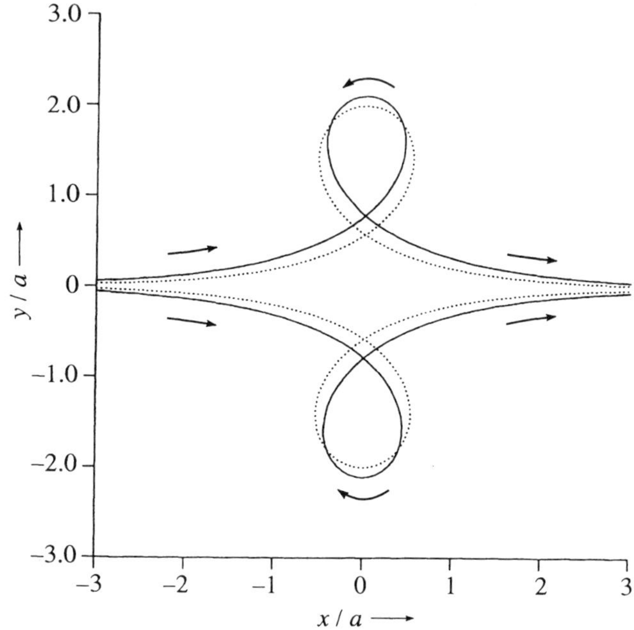

Fig. 2. Graphs of y/a versus x/a for solitary waves described by (5.6) and (5.7). The upper curves are for positive and the lower curves for negative waves. Arrows show the direction of increasing S. Solid curves: $\varepsilon^{\#} = 0.4$, $c^2 = 0.1$, i.e., $\beta^2 = 0.26$. Dotted curves: limit of y/a versus limit of x/a as $c^2 \to 0$ and $\beta \to 0$.

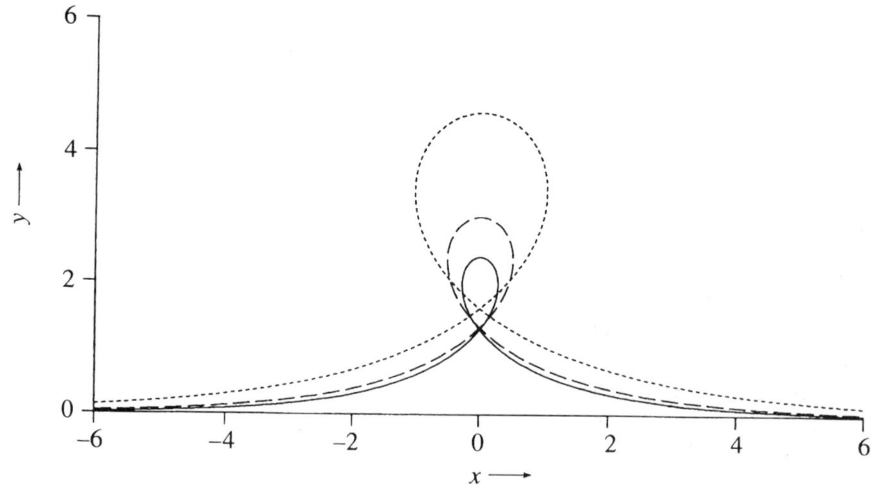

Fig. 3. Graphs of y versus x for positive solitary waves ($b = 1$) with $\varepsilon^{\#} = 0.7$. Short dashes: $c^2 = 0.3$. Long dashes: $c^2 = 0.15$. Solid curve: $c^2 = 0$.

6. Noninflexional Waves

A traveling wave is called *noninflexional* if $(\theta')^2 > 0$ for all ξ and θ' does not have a limit as $\xi \to \pm\infty$. For such waves b in equation (4.11) exceeds 1, and by putting $z = \text{ctn}\,\psi/2$, equation (4.14) (with $\theta_0 = \pi$) can be written

$$\sigma = \mp\sqrt{\frac{2(1-c^2)}{b-1}}\int_0^w [(z^2+d_1)(z^2+d_2)]^{-1/2}\,dz, \tag{6.1}$$

where

$$w = \text{ctn}\,\theta/2, \tag{6.2}$$

$$d_1 = \frac{b+\beta^2-\Delta}{b-1}, \quad d_2 = \frac{b+\beta^2+\Delta}{b-1}, \tag{6.3}$$

$$\Delta = +\sqrt{\beta^4+2b\beta^2+1}. \tag{6.4}$$

As $b > 1$, we have $d_2 > d_1 > 0$, and for

$$m = \frac{d_2-d_1}{d_2} = \frac{2\Delta}{b+\beta^2+\Delta} \tag{6.5}$$

we have $0 < m < 1$. By standard methods we obtain the following implications of (6.1):

$$\sin\theta = \frac{-2\sqrt{d_1}\,\text{sn}(\pm\sigma\gamma|m)\text{cn}(\sigma\gamma|m)}{(d_1-1)\text{sn}^2(\sigma\gamma|m)+1}, \tag{6.6}$$

$$\cos\theta = \frac{(d_1+1)\text{sn}^2(\sigma\gamma|m)-1}{(d_1-1)\text{sn}^2(\sigma\gamma|m)+1}. \tag{6.7}$$

Here γ is the positive number obeying

$$\gamma^2 = \frac{b+\beta^2+\Delta}{2(1-c^2)} = \frac{(b-1)d_2}{2(1-c^2)}; \tag{6.8}$$

sn and cn are Jacobi elliptic functions. (We use the notation of [1964, 1].) On putting (6.6) and (6.7) into (4.16) with η chosen so that $\min|y| = 0$, we find that

$$\frac{y}{a} = \pm\sqrt{\frac{2(b+1)}{1-c^2}}\left[\frac{\text{dn}(\sigma\gamma|m)}{(d_1-1)\text{sn}^2(\sigma\gamma|m)+1} - \sqrt{\frac{b-1}{b+1}}\right]. \tag{6.9}$$

For a positive wave, $\mp$ has the value $-$ in (6.1) while $\pm$ has the value $+$ in (6.6) and (6.9). With z, w, d_1 and d_2 as in (6.1)–(6.3), one can write (4.18) as

$$\xi + S_0 + u = \frac{-a\sqrt{2}}{\sqrt{(1-c^2)(b-1)}}\int_0^w \frac{[(1+\beta^2)z^2+1-\beta^2](z^2-1)}{(1+z^2)^2\sqrt{(z^2+d_1)(z^2+d_2)}}\,dz, \tag{6.10}$$

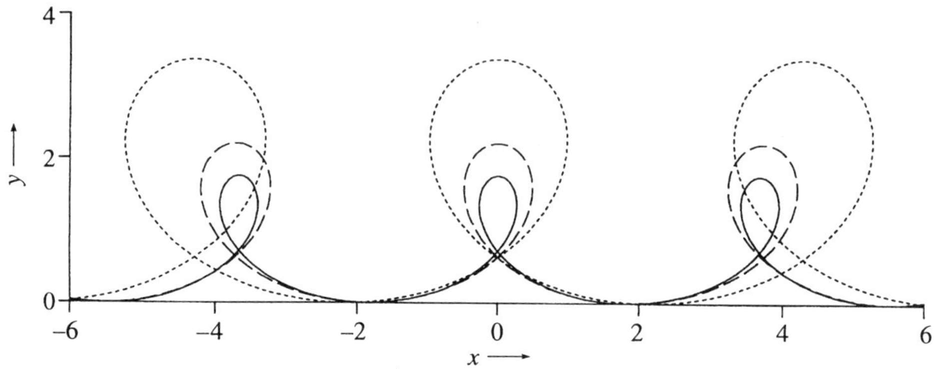

Fig. 4. Graphs of y versus x for positive noninflexional waves with $b = 1.2$ and $\varepsilon^{\#} = 0.7$. Short dashes: $c^2 = 0.3$. Long dashes: $c^2 = 0.15$. Solid curve: $c^2 = 0$.

and, by straightforward but tedious calculation, this can be cast into the form

$$\frac{u}{a} = \left(\frac{b+\beta^2}{1-c^2} - 1\right)\sigma - 2\gamma \mathrm{E}(\sigma\gamma|m)$$

$$- (d_1 - 1)\sqrt{\frac{d_2 2(b-1)}{1-c^2}}\left[\frac{\mathrm{sn}(\sigma\gamma|m)\mathrm{cn}(\sigma\gamma|m)\mathrm{dn}(\sigma\gamma|m)}{(d_1-1)\mathrm{sn}^2(\sigma\gamma|m)+1}\right], \qquad (6.11)$$

where E is an elliptic integral of the second kind (cf. [1964, 1, p. 589]):

$$\mathrm{E}(v|m) = \int_0^u \mathrm{dn}^2(v|m)dv. \qquad (6.12)$$

Placement of (6.7) in (4.6) yields an expression for λ as a function of σ in noninflexional waves. For such waves, as for solitary waves, $|\varepsilon|$ is everywhere less than 1 if and only if $2c^2 < 1 - \beta^2$. The maximum tension T is attained at the place where $\theta = \pm 2n\pi$ and equals the number $\varepsilon^{\#}$ in (4.20) and (4.21).

Figure 4 contains graphs of y versus x for noninflexional waves for one value of b; in that figure, as in Figure 3, as $|c|$ varies, β^2 is chosen so that $\varepsilon^{\#}$ equals a fixed value, 0.7.

7. Inflexional Waves

We here consider traveling waves for which there are values of σ at which $\theta' = 0$. For such a wave, θ is periodic in σ and lies on a cyclic orbit about the critical point $(\theta, \theta') = (0, \pi)$. We write θ_* for the smallest value of θ attained on that orbit. Thus the range of θ on the orbit is the interval $[\theta_*, 2\pi - \theta_*]$, and $\theta' = 0$ where $\theta = \theta_*$ and where $\theta = 2\pi - \theta_*$. For such a wave the value of b in (4.11) is in

$(-1, 1)$. Moreover,

$$\cos\theta_* = \frac{\Delta + \beta^2 + 2b - 1}{\Delta + \beta^2 + 1}, \tag{7.1}$$

where Δ is given by (6.4). Let w, d_1, d_2 be as in (6.2) and (6.3), and note that here $d_1 > 0$ and $d_2 < 0$. Equation (4.19), with θ_0 again set equal to π, can be written

$$\sigma = \mp\sqrt{\frac{2(1-c^2)}{1-b}}\int_0^w [(z^2 + d_1)(-d_2 - z^2)]^{-1/2}\,dz. \tag{7.2}$$

By evaluating the integral in (7.2) on the interval in which w increases from 0 to $\sqrt{-d_2} = \operatorname{ctn}\theta_*/2$ (i.e., in which θ decreases from π to θ_*), one obtains a formula for σ as a function of θ as σ varies over one quarter period of the periodic function $\sigma \mapsto \theta$. Upon doing this and putting, in place of (6.5),

$$\hat{m} = \frac{d_2}{d_2 - d_1} = \frac{\Delta + b + \beta^2}{2\Delta}, \tag{7.3}$$

we find that

$$\sin\theta = \frac{-2\sqrt{\hat{m}d_1}\,\operatorname{sn}(\pm\sigma\hat{\gamma}|\hat{m})\operatorname{dn}(\sigma\hat{\gamma}|\hat{m})}{(d_1 - 1)\hat{m}\operatorname{sn}^2(\sigma\hat{\gamma}|\hat{m}) + 1}, \tag{7.4}$$

$$\cos\theta = \frac{(d_1 + 1)\hat{m}\operatorname{sn}^2(\sigma\hat{\gamma}|\hat{m}) - 1}{(d_1 - 1)\hat{m}\operatorname{sn}^2(\sigma\hat{\gamma}|\hat{m}) + 1}, \tag{7.5}$$

where

$$\hat{\gamma} = +\sqrt{\frac{\Delta}{1-c^2}} \tag{7.6}$$

and $0 < \hat{m} < 1$.

Equations (7.4), (7.5), and (4.16) imply that

$$\frac{y}{a} = \pm\sqrt{\frac{2(b+1)}{1-c^2}}\left[\frac{\operatorname{cn}(\sigma\hat{\gamma}|\hat{m})}{(d_1 - 1)\hat{m}\operatorname{sn}^2(\sigma\hat{\gamma}|\hat{m}) + 1} + \mu\right], \tag{7.7}$$

where $\mu = 1$ when $\min|y| = 0$, as in Figures 5 and 8. The method that gave us equation (6.11) here yields, for inflexional waves,

$$\frac{u}{a} = (\hat{\gamma}^2 - 1)\sigma - 2\hat{\gamma}\,\mathrm{E}(\sigma\hat{\gamma}|\hat{m})$$
$$-(d_1 - 1)\sqrt{\frac{-d_2 2(1-b)\hat{m}}{1-c^2}}\left[\frac{\operatorname{sn}(\sigma\hat{\gamma}|\hat{m})\operatorname{cn}(\sigma\hat{\gamma}|\hat{m})\operatorname{dn}(\sigma\hat{\gamma}|\hat{m})}{(d_1 - 1)\hat{m}\operatorname{sn}^2(\sigma\hat{\gamma}|\hat{m}) + 1}\right]. \tag{7.8}$$

To obtain λ as a function of σ for inflexional waves, one places (7.5) in (4.6). For inflexional waves, the condition that $|\varepsilon| < 1$ (for all σ) is complicated. If $2c^2 < 1 - \beta^2$, then $|\varepsilon| < 1$ for all b. If $1 - \beta^2 \leqq 2c^2 < 1 + \beta^2$, then $|\varepsilon| < 1$ only for b sufficiently near to -1.

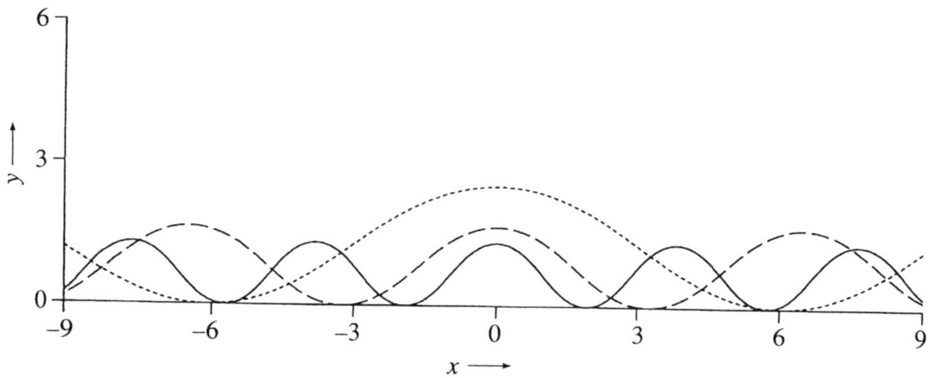

Fig. 5. Graphs of y versus x for positive inflexional waves with $b = -0.85$ and $\varepsilon^{\#} = 0.7$. Short dashes: $c^2 = 0.3$, i.e., $\varepsilon_{\#} = 0.157$. Long dashes: $c^2 = 0.15$, i.e., $\varepsilon_{\#} = -0.347$. Solid curve: $c^2 = 0$, i.e., $\varepsilon_{\#} = -0.7$.

For inflexional waves, the maximum value of the tension T is attained where $\theta = \theta_*$ and $\theta = 2\pi - \theta_*$, and is *less than* $\varepsilon^{\#}$. The minimum tension is attained where $\theta = \pi$ and equals $\varepsilon_{\#}$.

Although ε does not reach the value $\varepsilon^{\#}$ in an inflexional wave, in drawing the graphs shown in Figure 5 for inflexional waves with one value of b and three distinct speeds we have chosen β^2 so that $\varepsilon^{\#}$, here taken to be *defined* by (4.20), is held constant at a value employed for corresponding graphs shown in Figures 3 and 4.

"Figure Eight" Waves

Let $\mathrm{K}(\hat{m})$ and $\mathrm{E}(\hat{m})$ be complete elliptic integrals of the first and second kind defined by

$$\mathrm{K}(\hat{m}) = \int_0^{\pi/2} (1 - \hat{m}\sin^2\varphi)^{-1/2}\, d\varphi, \tag{7.9a}$$

$$\mathrm{E}(\hat{m}) = \int_0^{\pi/2} (1 - \hat{m}\sin^2\varphi)^{1/2}\, d\varphi, \tag{7.9b}$$

and let m_p be the unique solution in (0, 1) of

$$\mathrm{K}(m_{\mathrm{p}}) = 2\mathrm{E}(m_{\mathrm{p}}), \tag{7.10}$$

i.e.,

$$m_{\mathrm{p}} = 0.8261148, \quad k_{\mathrm{p}} = \sqrt{m_{\mathrm{p}}} = 0.9089086. \tag{7.11}$$

For a traveling wave with velocity c, it is occasionally of interest to consider a change of coordinates from x, y to x_*, y_*, with

$$x_* = x - ct, \quad y_* = y. \tag{7.12}$$

The relations $x = S + u(\xi)$, $y = y(\xi)$, $\xi = S - ct$ imply that x_* and y_* are functions of ξ, i.e.,

$$x_* = x_*(\xi) = u(\xi) + \xi, \quad y_* = y_*(\xi) = y(\xi), \tag{7.13}$$

and a graph of y_* versus x_* shows the axial curve as seen by an observer moving in the x-direction with velocity c. For an inflexional wave with $\hat{m} = m_{\rm p}$, it follows from (7.7), (7.8), and (7.13) that not only y_*, but also x_*, is a periodic function of ξ; the minimum period $\xi_{\rm p}$ of y_* is twice the minimum period of x_* and thus is a common period for y_* and x_*. For this common period we have

$$\xi_{\rm p} = \frac{4a{\rm K}(m_{\rm p})}{\hat{\gamma}} \tag{7.14}$$

with a as in (4.12), $\hat{\gamma}$ as in (7.6), and

$${\rm K}(m_{\rm p}) = 2.321049733. \tag{7.15}$$

In view of (7.3), once we know that $\hat{m} = m_{\rm p}$, i.e., that y_* and x_* are both periodic in ξ, then β^2 determines b by

$$b = 4\beta^2 m_{\rm p}(m_{\rm p} - 1) + (2m_{\rm p} - 1)\sqrt{4m_{\rm p}(m_{\rm p} - 1)\beta^4 + 1}\,, \tag{7.16}$$

and β^2 and c^2 determine $\xi_{\rm p}$ by

$$\xi_{\rm p} = \frac{4{\rm K}(m_{\rm p})(1 - c^2)}{[\beta^4(2m_{\rm p} - 1) + \beta^2\sqrt{4m_{\rm p}(m_{\rm p} - 1)\beta^4 + 1}]^{1/2}}. \tag{7.17}$$

Clearly such a wave is possible in a rod of length $L = \xi_{\rm p}$ which is closed, i.e., which, albeit straight in its stress-free reference configuration, has had its ends joined. As the minimum period of y_* is twice the minimum period of x_*, such waves have the form of a "figure eight". The minimum value of ε, $\varepsilon_\# = (c^2 - \beta^2)/(1 - c^2)$, is negative and is attained where $\cos\theta = -1$, that is, at the highest and lowest points of the "figure eight". As we reject, as physically unattainable, configuration with $\varepsilon_\# \leqq -1$, these waves have $\beta^2 < 1$ and, by (7.17), a value of $\xi_{\rm p}/(1 - c^2)$, i.e., of $L/(1 - c^2)$, greater than

$$\left(\frac{L}{1 - c^2}\right)_{\rm inf} = \frac{2\sqrt{2}\,{\rm K}(m_{\rm p})}{\sqrt{2m_{\rm p} - 1}} = 8.1288519. \tag{7.18}$$

Examples of stationary "figure eight" waves, i.e., equilibrium configurations determined by (7.7), (7.8), and (7.13) with $c = 0$, are shown in Figure 6.

"Figure eight" waves in inextensible rods obeying the system (1.4) were discussed by Coleman & Dill [1992, 1]. Their results yield graphs of y_*/L versus x_*/L that for each L are the same as the dotted curve in Figure 6, which is a limit of graphs of y_*/L versus x_*/L as $L \to \infty$ in the present theory.

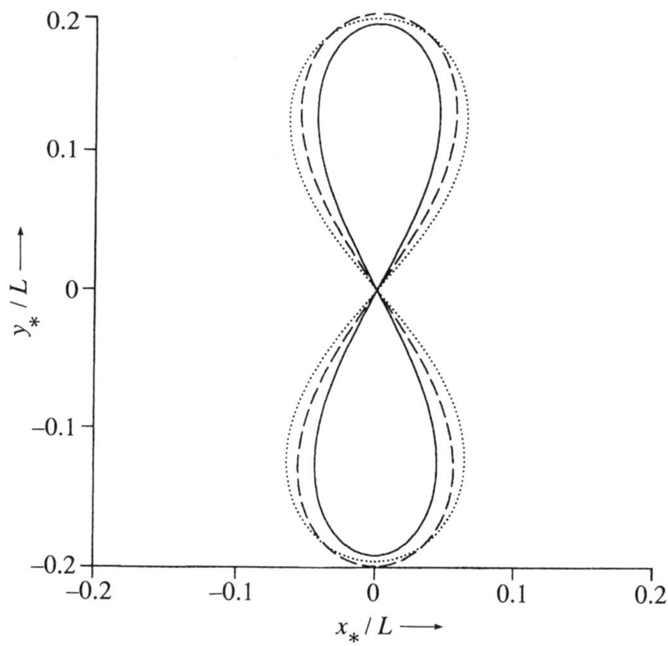

Fig. 6. Graphs y_*/L versus x_*/L for inflexional waves with $\hat{m} = m_p$ and $c = 0$. Here μ is chosen to be zero in (7.7). Solid curve: $\beta^2 = 0.6$, i.e., $b = 0.236109$. Dashed curve: $\beta^2 = 0.3$, i.e., $b = 0.462762$. Dotted curve: limit as $\beta \to 0$, i.e., $L \to \infty$, which corresponds to $b = 0.652230$.

Appendix: Theory of Waves in Inextensible Rods as an Approximation for Small Tension

If, as is customary in the Euler-Kirchhoff-Clebsch theory of rods, one treats the rod as if it were inextensible, then, at each time and place, $s = S$. The constitutive equation (2.8) then does not hold, and both the tension T and the shear resultant G are reactive parameters not given by constitutive equations. As $\lambda = 1$, the relations (2.2) then reduce to

$$x_S = \cos\theta, \quad y_S = \sin\theta, \tag{A.1}$$

and hence, at each time and place

$$(x_S)^2 + (y_S)^2 = 1. \tag{A.2}$$

The constitutive equation (2.9) continues to hold, and the dynamical equations (2.6) and (2.7) become

$$\rho A x_{tt} = F^x_S, \tag{A.3a}$$

$$\rho A y_{tt} = F^y_S, \tag{A.3b}$$

$$\rho I \theta_{tt} = EI\theta_{SS} + G. \tag{A.3c}$$

As $\boldsymbol{F}$ obeys (2.4) and θ is related to $\boldsymbol{x}$ by (A.1), the equations (A.3) may be considered differential equations for the unknowns x, y, F^x, F^y with x and y subject to the constraint (A.2); i.e., we have here four equations for four unknowns, x, y, F^x, F^y, or x, y, T, G. By differentiating (A.3a) and (A.3b) with respect to S, employing (A.1) and the dimensionless variables introduced at the beginning of the paragraph containing (2.11a)–(2.11c), one obtains the system (1.4) shown in the introduction. That system of three equations for three unknowns, θ, T, G, was studied by CAFLISCH & MADDOCKS [1984, 1], COLEMAN & DILL [1992, 1], XU [1992, 3], FALK & XU [1994, 4], COLEMAN & XU [1994, 1], and DICHMANN, MADDOCKS & PEGO [1994, 3]. For (1.4) subject to boundary data appropriate to rods of finite length, CAFLISCH & MADDOCKS [1984, 1] obtained existence and uniqueness theorems and examined the implications of an energy criterion for the stability of equilibrium solutions. COLEMAN & DILL [1992, 1] discussed in detail the relation of the system (1.4) to a form of Hamilton's principle and found the traveling-wave solutions of the system. As we mentioned in Section 3, DICHMANN, MADDOCKS & PEGO [1994, 3] derived the analogue of equation (3.9) for inextensible rods; the set of conservation laws for such rods is discussed in several recent papers [1993, 1], [1994, 3, 4, 5]. An energy preserving second-order finite-difference scheme for numerical solution of (1.4) was developed by XU [1992, 3] and FALK & XU [1994, 4] and employed by COLEMAN & XU [1994, 1] to study the interaction of solitary waves; the numerical evidence indicates that the system (A.4) is not completely integrable in the sense of soliton theory.

We summarize below the results of COLEMAN & DILL [1992, 1] for traveling waves governed by (1.4). For such motions there is a Cartesian system for which θ, y, $u = x - S$, T, and G are functions of $\xi = S - ct$, and (1.4) yields

$$T = c^2 + (\alpha - c^2)\cos\theta, \tag{A.4a}$$

$$G = -(\alpha - c^2)\sin\theta, \tag{A.4b}$$

$$\tilde{a}^2\theta'' = \sin\theta, \tag{A.4c}$$

with $\tilde{a} > 0$ given by

$$\tilde{a}^2 = \frac{1 - c^2}{\alpha - c^2}. \tag{A.5}$$

The integration constant α, which must be such that the right-hand side of (A.5) is positive, equals T at the values (or limits) of ξ for which $\theta = \pm 2n\pi$. When, as is the case for inflexional waves, there are no values of ξ for which $\theta = \pm 2n\pi$, there are values $\xi_\#$ of ξ at which $\theta = \pm(2n+1)\pi$ and

$$\alpha = 2c^2 - T(\xi_\#). \tag{A.6}$$

The first integral of (A.4c) is

$$\tfrac{1}{2}(\tilde{a}\theta')^2 = b - \cos\theta. \tag{A.7}$$

In analogy with equation (4.13), we now put

$$\tilde{\sigma} = (\xi - S_0)/\tilde{a}, \tag{A.8}$$

and note that (A.7) yields

$$\tilde{\sigma} = \pm \int_{\theta_0}^{\theta} [2b - \cos\psi]^{-1/2} d\psi.$$

We are interested in determining the sense in which traveling-wave solutions of the equations of motion (1.4) for inextensible rods are related to asymptotic limits of traveling-wave solutions of the corresponding equations (2.11) for extensible rods for small values of the magnitude of the tension, which for extensible rods means small values of $|\varepsilon| = |\lambda - 1|$. To this end let us agree to write

$$\bar{y} = y/a, \quad \bar{u} = u/a, \tag{A.9}$$

with a as in (4.12), when y and u are determined by (2.11) and (2.2), i.e., obey (4.16) and (4.18). For comparison, we shall write

$$\tilde{y} = y/\tilde{a}, \quad \tilde{u} = u/\tilde{a}, \tag{A.10}$$

with $\tilde{a}$ as in (A.5), when y and u are determined by (A.4) and (A.1).

The parameter b plays the same role in the equations (4.11) and (A.7). For both the extensible rods governed by (2.11) and the inextensible rods governed by (1.4), traveling waves with b in $(-1, 1)$ are inflexional, those with $b = 1$ are solitary, and those with $b > 1$ are noninflexional. In the theory of the inextensible rods the following expressions for θ, $\tilde{y}$, and $\tilde{u}$ as functions of $\tilde{\sigma}$ hold for these three types of waves:

I. For solitary waves [1992, 1; eqs. (97), (99)]:

$$\theta(\tilde{\sigma}) = 4\tan^{-1}(\exp(\pm\tilde{\sigma})) = -2\tan^{-1}(\operatorname{csch}(\pm\tilde{\sigma})), \tag{A.11}$$

$$\tilde{y}(\tilde{\sigma}) = \pm 2\operatorname{sech}\tilde{\sigma}, \tag{A.12}$$

$$\tilde{u}(\tilde{\sigma}) = -2\tanh\tilde{\sigma}. \tag{A.13}$$

II. For noninflexional waves [1992, 1; eqs. (108), (109)]:

$$\sin\theta = -2\operatorname{sn}(\pm\tilde{\sigma}/\tilde{k}|\tilde{m})\operatorname{cn}(\tilde{\sigma}/\tilde{k}|\tilde{m}), \tag{A.14}$$

$$\tilde{y} = \frac{2}{\tilde{k}}[\operatorname{dn}(\tilde{\sigma}/\tilde{k}|\tilde{m}) - (1-\tilde{m})^{1/2}], \tag{A.15}$$

$$\tilde{u} = 2\left[\frac{1-\tilde{m}}{\tilde{m}}\right]\tilde{\sigma} - \frac{2}{\tilde{k}}\mathrm{E}(\tilde{\sigma}/\tilde{k}|\tilde{m}), \tag{A.16}$$

where

$$\tilde{m} = \tilde{k}^2 = 2/(b+1). \tag{A.17}$$

III. For inflexional waves [1992, 1; eqs. (120), (121)]:

$$\sin\theta = -2\tilde{\tilde{k}}\operatorname{sn}(\pm\tilde{\sigma}|\tilde{\tilde{m}})\operatorname{dn}(\pm\sigma|\tilde{\tilde{m}}), \tag{A.18}$$

$$\tilde{y} = \pm 2\tilde{\tilde{k}}[1 + \operatorname{cn}(\tilde{\sigma}|\tilde{\tilde{m}})], \tag{A.19}$$

$$\tilde{u} = -2\mathrm{E}(\tilde{\sigma}|\tilde{\tilde{m}}), \tag{A.20}$$

where

$$\tilde{\tilde{m}} = \tilde{\tilde{k}}^2 = (b + 1)/2. \tag{A.21}$$

For each traveling wave of an extensible rod governed by (2.11) let

$$\tau = \sup\{|T(\sigma)|| - \infty < \sigma < \infty\}, \tag{A.22}$$

and note that, by (4.6),

$$\tau = \sup_\sigma |\varepsilon(\sigma)| = \sup_\sigma \left|\frac{\beta^2 \cos\theta(\sigma) + c^2}{1 - c^2}\right|. \tag{A.23}$$

A natural question that arises is whether, for each fixed value of b, the functions $\tilde{\sigma} \mapsto \theta, \tilde{\sigma} \mapsto \tilde{y}, \tilde{\sigma} \mapsto \tilde{u}$, given above for inextensible rods obeying (1.4), are limits as $\tau \to 0$ of the corresponding expressions for $\sigma \mapsto \theta$, $\sigma \mapsto \bar{y}$, $\sigma \mapsto \bar{u}$, derived in Sections 5, 6, and 7, for extensible rods obeying (2.11).

For $b \geqq 1$, i.e., for a solitary wave and a noninflexional wave, as σ varies over $(-\infty, \infty)$, $\cos\theta$ takes on all values in $(-1, 1)$. Hence when such a wave propagates in a rod obeying (2.11), τ in (A.23) is given by

$$\tau = \varepsilon^{\#} = \frac{\beta^2 + c^2}{1 - c^2}. \tag{A.24}$$

It follows that for each sequence of traveling waves with a fixed $b \geqq 1$,

$$\tau \to 0 \Leftrightarrow \beta \to 0 \quad \text{and} \quad c^2 \to 0. \tag{A.25}$$

For an inflexional wave, i.e., for b in $(-1, 1)$, as σ varies over $(-\infty, \infty)$ the values of $\cos\theta(\sigma)$ cover the interval $[-1, \cos\theta_*]$, where $\cos\theta_*$ is given by (7.1). It follows from (7.1) and (6.4) that for each such b, $\cos\theta_* > -1$ whenever $\beta \neq 0$. As the open interval $[-1, \cos\theta_*]$ is nonempty if $\beta \neq 0$, we may conclude from (A.23) that (A.25) holds also for each sequence of traveling waves with a fixed b in $(-1, 1)$. Thus, by taking the limit of the right-hand side of (4.14) as β and $c^2 \to 0$, we obtain the following conclusion.

Remark 6. For each fixed $b > -1$ and fixed θ_0, as $\tau \to 0$ the functions $\tilde{\sigma} \mapsto \theta$, determined in the theory of (2.11) by (4.14), converge uniformly on compact sets to the function $\tilde{\sigma} \mapsto \theta$, given by (A.9).

In view of this observation and the obvious fact that as $\tau \to 0$ the stretch λ approaches 1 uniformly in σ, we can make the following assertion: as $\tau \to 0$, for solitary waves the expressions (5.3) for θ, (5.6) for $\bar{y}$ and (5.7) for $\bar{u}$, as functions of σ, converge to the expressions (A.11)–(A.13) for θ, $\tilde{y}$, and $\tilde{u}$, as functions of $\tilde{\sigma}$; similarly, for noninflexional waves at fixed b, the expressions (6.6), (6.9), and (6.11) have the expressions (A.14)–(A.16) as limits, while for inflexional waves at fixed b, (7.4), (7.7), (7.8) have (A.18)–(A.20) as limits. Moreover, for each value b, the graphs of $\bar{y}$ versus $\bar{x} = x/a$ approach the graphs of $\tilde{y}$ versus $\tilde{x} = x/\tilde{a}$ as $\tau \to 0$.

Graphs of $\bar{y}$ versus $\bar{x}$ at fixed b and the corresponding limits as $\tau \to 0$ are shown in Figures 2, 7, 8, and 9.

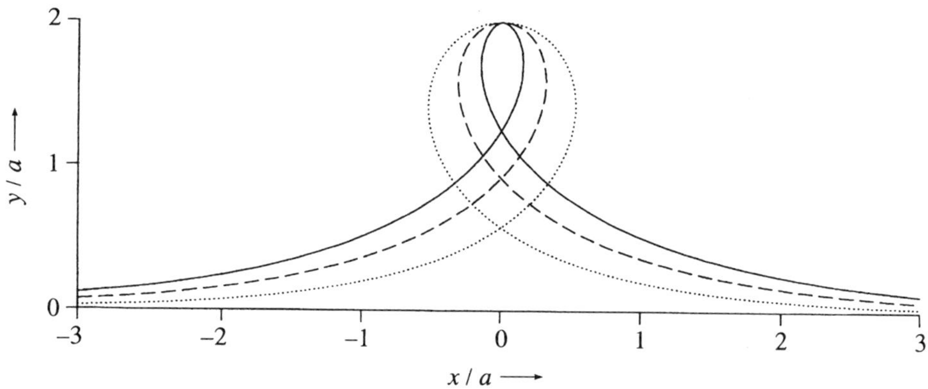

Fig. 7. Graphs of y/a versus x/a for solitary waves that are stationary, i.e., for which $c = 0$. Solid curve: $\varepsilon^{\#} = 0.9$. Dashed curve: $\varepsilon^{\#} = 0.5$. Dotted curve: limit as $\tau = \varepsilon^{\#} \to 0$.

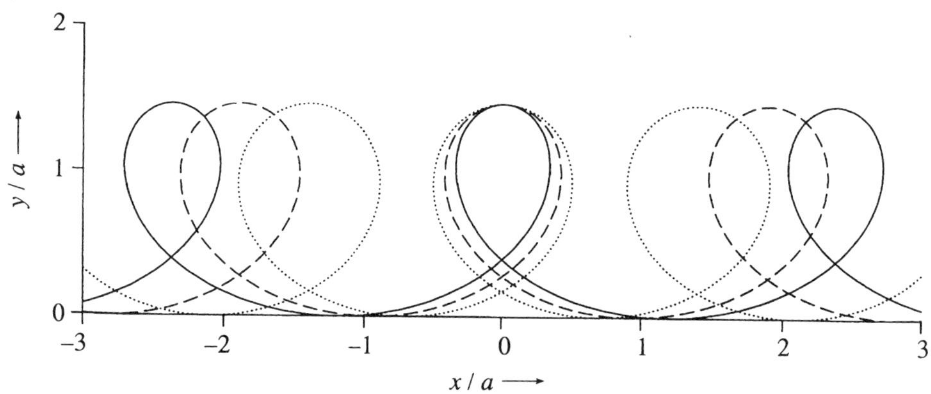

Fig. 8. Graphs of y/a versus x/a for stationary noninflexional waves with $b = 1.2$. Solid curve: $\varepsilon^{\#} = 0.4$. Dashed curve: $\varepsilon^{\#} = 0.2$. Dotted curve: limit as $\tau = \varepsilon^{\#} \to 0$.

In the theory of inextensible rods obeying (1.4) there are traveling waves with $|c| > 1$, and within the confines of that theory there does not appear to be a reason to reject them. However, as we observed in our discussion of Remark 3, traveling waves with $|c| > 1$ in rods obeying (2.11) may be considered to be physically unattainable, because for them there are values of ξ with $\lambda(\xi) < 0$.

The limit of small τ is the only limit in which we expect the theory of inextensible rods to be a useful approximation to the theory of extensible rods, and we have seen that for traveling waves in extensible rods governed by (2.11) the limit of small τ does indeed give expressions for the axial curve which, after appropriate rescaling, have the expected agreement with expressions holding for that curve in the theory of inextensible rods obeying (1.4). As we saw in (A.25), small τ implies small wave speed $|c|$. Thus, if we accept the principle that solutions in the theory of

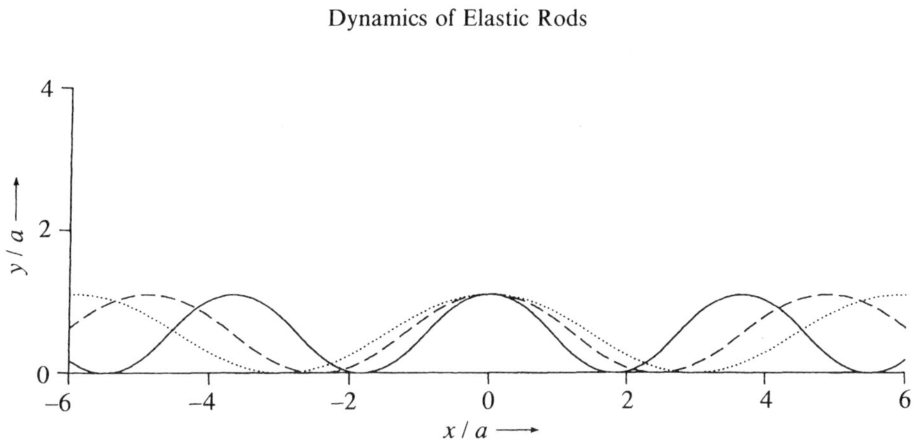

Fig. 9. Graphs of y/a versus x/a for stationary inflexional waves with $b = -0.85$. Solid curve: $\varepsilon^{\#} = 0.6$. Dashed curve: $\varepsilon^{\#} = 0.3$. Dotted curve: limit as $\tau \to 0$.

(1.4) are to be accepted only if they are limits for small τ of solutions in the theory of (2.11), then we have reason to reject as inadmissible supersonic traveling waves not only in the theory of (2.11) but also in the theory of (1.4).

Acknowledgement. This research was supported by the National Science Foundation under Grant DMS-91-07010 and by the Donors of the Petroleum Research Fund, administered by the American Chemical Society.

References

1859 [1] KIRCHHOFF, G., *Über das Gleichgewicht und die Bewegung eines unendlich dünnen elastischen Stabes*, J. f. reine angew. Math. (Crelle), **56**, 285–313.

1862 [1] CLEBSCH, A., *Theorie der Elasticität fester Körper*, B. G. Teubner, Leipzig.

1876 [1] KIRCHHOFF, G., *Vorlesungen über mathematische Physik, Mechanik*, Vorl. 28, B. G. Teubner, Leipzig.

1883 [1] CLEBSCH, A., *Théorie de l'Elasticité des Corps Solides* (translation of [1862, 1] by SAINT-VENANT & FLAMANT), Dunod, Paris.

1964 [1] MILNE-THOMSON, L. M., §16, *Jacobian Elliptic Functions and Theta Functions*; §17, *Elliptic Integrals*, in Handbook of Mathematical Functions, Appl. Math Series, Vol. 55, pp. 567–626, eds: M. ABRAMOWITZ & L. A. STEGUN, National Bureau of Standards, Washington.

1979 [1] ANTMAN, S. S. & LIU, T.-P., *Travelling Waves in Hyperelastic Rods*, Quart. Appl. Math., **36**, 377–399.

1984 [1] CAFLISCH, R. E. & MADDOCKS, J. H., *Nonlinear Dynamical Theory of the Elastica*, Proc. Royal Soc. Edinburgh, **99A**, 1–23.

1992 [1] COLEMAN, B. D. & DILL, E. H., *Flexure Waves in Elastic Rods*, J. Acoustical Soc. Amer., **91**, 2663–2673.

[2] DILL, E. H., *Kirchhoff's Theory of Rods*, Arch. Hist. Exact Sci., **44**, 1–23.

[3] XU, J.-M., *An Analysis of the Dynamical Equations of Elastic Rods and their Numerical Approximation*, Doctoral Dissertation, Rutgers University, New Brunswick, New Jersey.

1993 [1] COLEMAN, B. D. DILL, E. H. LEMBO, M., LU, Z., & TOBIAS, I., *On the Dynamics of Rods in the Theory of Kirchhoff and Clebsch*, Arch. Rational Mech. Anal., **121**, 339–359.

1994 [1] COLEMAN, B. D. & XU, J.-M., *On the Interaction of Solitary Waves of Flexure in Elastic Rods*, Acta Mechanica; in press.

[2] COLEMAN, B. D., LU, Z., & TOBIAS, I., *Traveling Waves in Elastic Rods*. Forthcoming.

[3] DICHMANN, D. J., MADDOCKS, J. H., & PEGO, R. L. *Hamiltonian Dynamics of an Elastica and the Stability of Solitary Waves*, preprint.

[4] FALK, R. S. & XU, J.-M., *Numerical Approximation of the Dynamical Equations of Elastic Rods*. SIAM J. Num. Anal., in press.

[5] MADDOCKS, J. H. & DICHMANN, D. J., *Conservation Laws for the Motion of Rods*, J. Elasticity **34**, 83–96 (1994).

Program in Mechanics
Rutgers University
Piscataway, New Jersey 08855-0909

(Accepted June 24, 1994)

Travelling waves in a nonlinearly suspended beam: some computational results and four open questions

BY Y. CHEN AND P. J. MCKENNA

Department of Mathematics, University of Connecticut, 196 Auditorium Road, U-9, Room 111, Storrs, CT 06269-3009, USA

A new nonlinear differential equation for a nonlinearly supported beam is proposed. Numerical evidence is presented on the existence, interaction, and spontaneous decomposition properties of travelling waves in such a beam. In particular, these waves seem to possess the ability to pass through each other like solitons, and more complicated shapes seem to decompose into simpler ones. At present, there is no mathematical explanation for these phenomena, and several open mathematical questions are posed.

1. Introduction

The purpose of this paper is to bring to the attention of the mathematical community a dramatic gap between a rich and complex collection of computational results on travelling waves in a simple model of a nonlinearly supported beam, and an almost complete absence of mathematical explanation for these results.

McKenna & Walter (1990) found travelling wave solutions to nonlinear beam equations on the real line of the form

$$u_{tt} + u_{xxxx} + u^+ - 1 = 0.$$

The search for travelling wave solutions of the form $u(x,t) = 1 + y(x - ct)$ then led to the ordinary differential equation

$$y'''' + c^2 y'' + (y+1)^+ - 1 = 0. \tag{1.1}$$

In McKenna & Walter (1990), this was shown to have solutions decaying to zero exponentially by 'brute force', namely, by solving the two *linear* ordinary differential equations explicitly for $y > -1$ and $y < -1$ and then matching them at $y = -1$. This purely calculus approach had serious limitations. The slightest perturbation in the nonlinearity rendered all these calculations invalid.

In Chen & McKenna (1997), this problem was solved in the sense that a qualitative proof based on the Mountain Pass lemma (Rabinowitz 1986; Brezis & Nirenberg 1991) was given which showed that for nonlinearities 'like' $(y+1)^+ - 1$, the ordinary differential equation

$$y'''' + c^2 y'' + f(y) = 0 \tag{1.2}$$

had at least one non-trivial solution. This was done by looking for critical points of

Phil. Trans. R. Soc. Lond. A (1997) **355**, 2175–2184

Printed in Great Britain

TEX Paper

the functional

$$I(y) = \int_R (\tfrac{1}{2}|y''|^2 - \tfrac{1}{2}c^2|y'|^2)\,\mathrm{d}x + \int_{y\leqslant -1} (|y| - \tfrac{1}{2})\,\mathrm{d}x + \int_{y>-1} (\tfrac{1}{2}|y|^2)\,\mathrm{d}x. \qquad (1.3)$$

It was shown that $y \equiv 0$ was a local minimum for I and that it was possible to choose other functions in $H^2(R^n)$ which had a lower value for I than $I(0)$. Modulo some technical details, this enabled us to use the Mountain Pass lemma with concentrated compactness to prove the existence of non-trivial solutions for a larger class of nonlinearities f.

Already, we come to the first problem.

Problem 1.1. *Can one prove that there are many solutions of the differential equation (1.3)?*

The calculations of McKenna & Walter (1990) suggest that there are many, possibly infinitely many, solutions, but all that can be proven is that there is at least one non-trivial solution.

Having found travelling wave solutions of (1.3), the next natural question was how these solutions behaved when used as initial data for the full initial value problem (1.1). In particular, it was natural to ask whether some or all of the travelling wave solutions were stable and persisted under slight perturbation in the initial data. If one was daring, one might hope for interesting interaction behaviour when two of these waves collided.

As described in Chen & McKenna (1997), we commenced such an investigation. However, almost immediately, we recognized a significant problem: the nonlinearity u^+ is not smooth and therefore any numerical procedure which treats this function may have large errors. This seemed to be the case in computations.

Accordingly, we decided to substitute for the nonlinearity $u^+ - 1$ a function which shared some of the qualitative behaviour but was analytic. The function we chose was $f(u) = \mathrm{e}^{u-1} - 1$ in the beam equation,

$$u_{tt} + u_{xxxx} + f(u) = 0,$$

giving rise to the nonlinear ordinary differential equation

$$y'''' + c^2 y'' + \mathrm{e}^y - 1 = 0. \qquad (1.4)$$

Again, $y \equiv 0$ is a solution and when one writes the corresponding variational problem, it corresponds to a local minimum of the functional. Again, it is easy to find functions at a lower elevation than the local minimum. However, carrying out the technical details of the previous existence proof has so far proven too difficult.

This brings us to our second open problem.

Problem 1.2. *Can one prove the existence of any non-trivial solutions of (2.1)?*

The next difficulty we faced with this nonlinearity was that of numerically finding solutions of (2.1). There were two solutions to this difficulty. The first one used was a numerical version for the Mountain Pass lemma, which has now been used for a variety of different nonlinear problems.

A second and much more efficient method was suggested to us by Champneys & Spence (1993), whose shooting method proved much faster in finding solutions of (2.1). Experiments described in this paper use solutions obtained by both methods, but primarily the second.

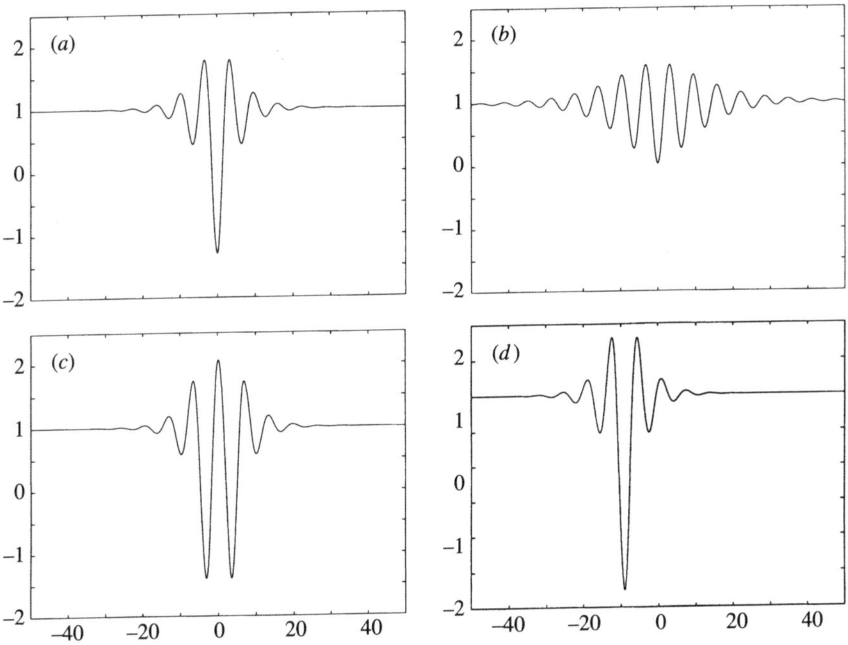

Figure 1. Travelling wave solutions of $u_{tt} + u_{xxxx} + e^{u-1} - 1 = 0$ with speed.

The rest of this paper will consist of three sections. The first will briefly discuss the solutions of (2.1) found by the two methods, with a brief description of the Mountain Pass algorithm.

The next will discuss our observations of the stability properties of the waves when solved as an initial value problem for the beam equation. Finally, we describe some curious interaction properties of these waves, as yet entirely unexplained, including the facts that the waves appear to pass through each other, that their stability properties remain mysterious and that they occasionally can spontaneously decompose to simpler travelling waves of different speeds and magnitudes.

We should emphasize that these properties are of primarily mathematical interest, indicative of a deep and complex underlying structure. We do not expect to see such behaviour in modern-day suspension bridges.

2. Mountain pass algorithm

Based on the Mountain Pass theory, a numerical algorithm was developed in Choi & McKenna (1993).

It has been shown in Chen & McKenna (1997) that if a functional $I \in C^1(H^2, R)$ has a local minimum point e_1 and another point e_2 whose altitude $I(e_2)$ is lower than that of the minimum, then with some additional technicalities, we may conclude that the infimum of the maxima of I along all paths joining e_1 and e_2 is a critical value of I.

On a finite-dimensional approximating subspace, we take a piecewise linear path from e_1 to e_2 and find the maximum of I along that path. Then we deform the path

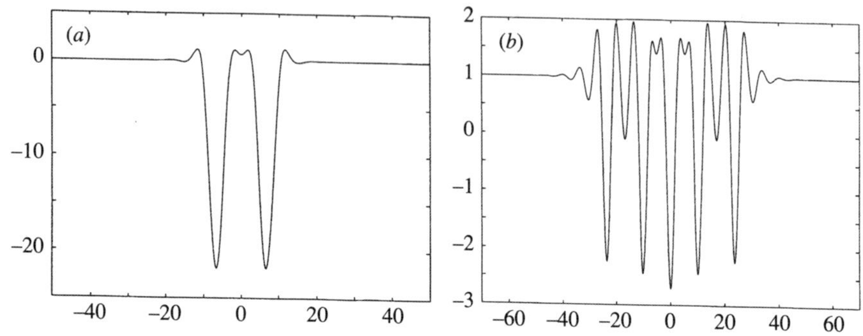

Figure 2. Travelling wave solutions of $u_{tt} + u_{xxxx} + e^{u-1} - 1 = 0$ found by continuation methods.

by pushing the maximum point in the direction of steepest descent. This step will be repeated until the critical point is reached.

Because of the fast growth of the function $e^{u-1} - 1$, we have encountered some difficulty in proving the existence of the travelling wave solutions for the equation

$$u_{tt} + u_{xxxx} + e^{u-1} - 1 = 0, \quad -\infty < x < \infty. \tag{2.1}$$

But by applying the Mountain Pass algorithm on a finite subinterval of R, we did obtain numerical solutions. The functional defined on $H^2(-L, L)$, whose critical points correspond to weak solutions of (2.1), is given by

$$I(y) = \frac{1}{2}\left[\int_{-L}^{L} (|y''|^2 - c^2|y'|^2)\,dx\right] + \int_{-L}^{L} F(y)\,dx, \tag{2.2}$$

where $F(y) = e^y - y - 1$.

The algorithm is similar to that for finding the Mountain Pass type critical points in the direct variational formulation of semilinear elliptic equations (Choi & McKenna 1993), the major difference being that the steepest descent direction is being sought in the H^2 norm in our case.

The conclusions of the search for solutions of (2.1) seem to indicate that the situation is rather similar to that of McKenna & Walter (1990). Solutions seem to exist in the range $0 < c < \sqrt{2}$. As the wave speed approaches $\sqrt{2}$, the solutions become highly oscillatory in nature, whereas when c approaches 0, they appear to go to infinity in amplitude.

There appear to be many different solutions for each wave speed and these can be obtained either by the Mountain Pass algorithm and varying the choice of the initial path by choosing different points for e_2, or by varying the shooting parameters.

We show some of the solutions obtained in figures 1 and 2. Notice that figures 1c and 1d are two different waves obtained from the Mountain Pass algorithm with the same wave speed.

3. Stability properties

Taking one of the numerical solutions we have obtained by applying the Mountain Pass algorithm as the initial function $u(x, 0)$, we solve the initial-boundary value problem

$$u_{tt} + u_{xxxx} + f(u) = 0$$

by using a standard explicit central finite difference scheme (Strikwerda 1989) with the following non-reflecting boundary conditions:

$$\left.\begin{aligned}\left(\frac{\partial}{\partial t}-c\frac{\partial}{\partial x}\right)u(-L,t)=0,\quad \left(\frac{\partial}{\partial t}+c\frac{\partial}{\partial x}\right)u(L,t)=0\\ \left(\frac{\partial}{\partial t}-c\frac{\partial}{\partial x}\right)^2u(-L,t)=0,\quad \left(\frac{\partial}{\partial t}+c\frac{\partial}{\partial x}\right)^2u(L,t)=0.\end{aligned}\right\}\tag{3.1}$$

We then follow the solution of the initial value problem for some time in the hope of observing the stability properties of these waves. Figure 3 shows one of such travelling wave solutions.

We have also tested the stability of the solutions by disturbing the initial function $u(x,0)$ or the speed c of the wave, and the waves maintained the same basic shape over time. The results show that some of the solutions are quite stable in a sense that the nearby states remain nearby for future times (Strauss 1989).

The surprising conclusion seems to be that the stability of the simple wave is related to the shape. If a simple solitary wave is concave up about the centre, then it will be stable over a large time interval. On the other hand, if it is concave down, then solving the same initial value problem results in the wave rapidly losing its shape. Some of the tests are shown in figures 4 and 5.

In the case that a simple wave is unstable, an even more interesting phenomenon is observed in figure 5 (with periodic boundary conditions). The unstable wave seems to break into two simple stable waves, both concave up about their centres, one moves slower with larger amplitude, the other moves faster with smaller amplitude.

There are also multi-bump solutions—solutions which appear to be the linear combination of two waves that are separated apart. Two multi-bump solutions obtained by the continuation method are shown in figure 6.

So far it is not clear whether or not the stability properties of this kind of solutions are related to their shapes. Shown in figure 7 is a solution which appears to be a combination of two stable waves. This solution is stable for a long period of time, but is concave down at the central point.

Problem 3.1. *Prove that some solutions of (2.2) are stable. Account for why some of the solutions appear unstable. What is the connection between the shape of the solutions and their stability properties?*

To the best of the authors' knowledge, the only results available on stability which might be related are those of Levandosky (1998) and Sandstede (this volume), which are for a different type of nonlinearity.

4. Interaction properties

To observe the interaction of two travelling waves, we first obtain two solutions $y_1(t)$ and $y_2(t)$ by the Mountain Pass algorithm, then take the initial function $u(x,0)$ to be the linear combination of these two solutions:

$$u(x,0)=y_1(x+L)+y_2(x-L)+1,\tag{4.1}$$

where L is quater the length of the experimental interval.

We can make the two waves travel toward each other by taking opposite signs for their wave speeds. It is shown in figure 8 that the two waves emerge apparently intact after the collision.

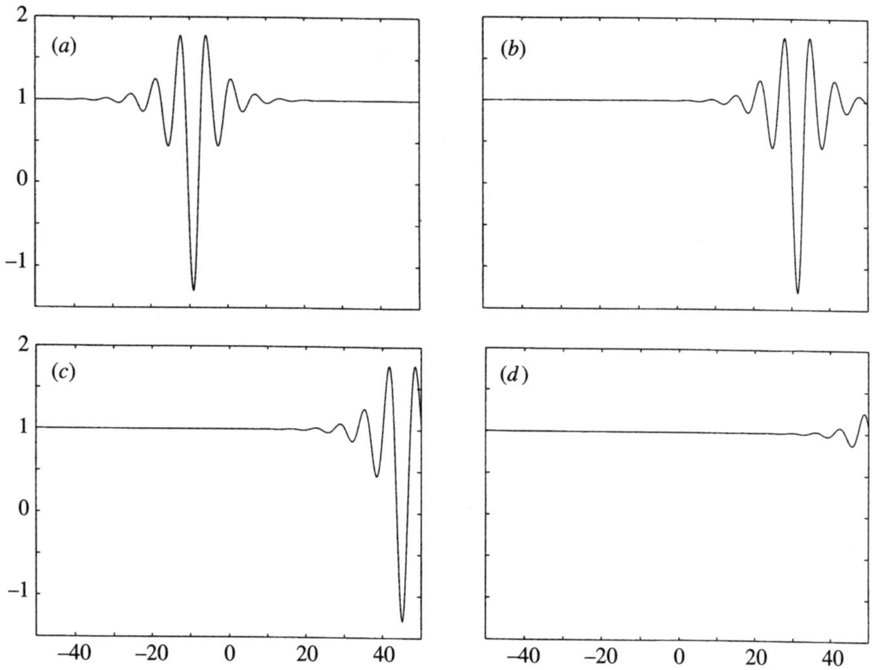

Figure 3. A travelling wave solution of $u_{tt} + u_{xxxx} + e^{u-1} - 1 = 0$ with speed $c = 1.354$.

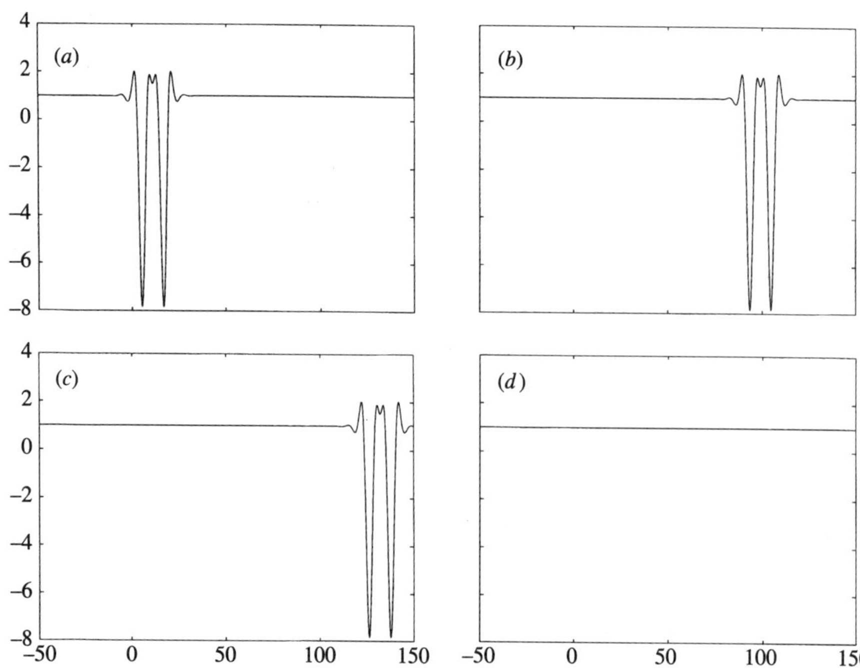

Figure 4. A stable solution with speed $c = 1.10$.

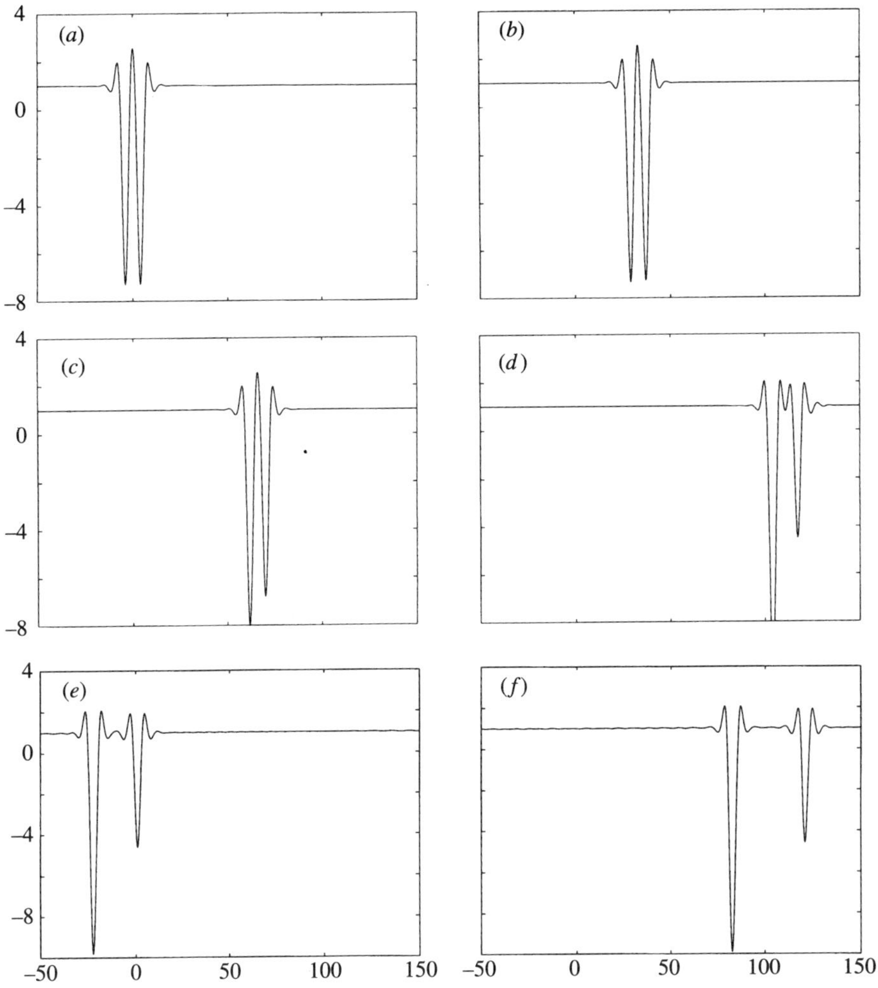

Figure 5. An unstable solution with speed $c = 1.10$.

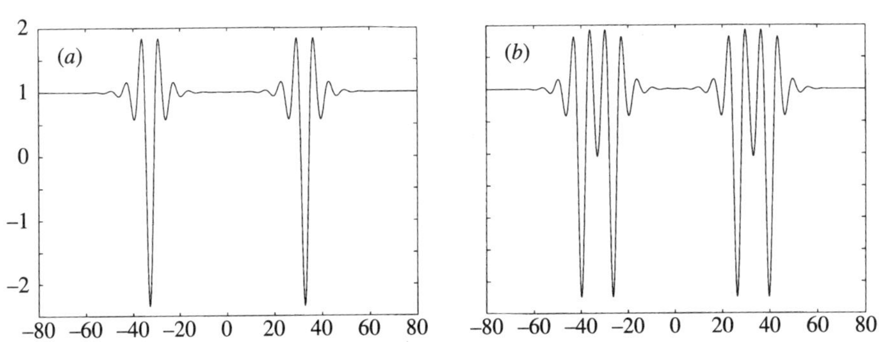

Figure 6. Solutions with speed $c = 1.30$.

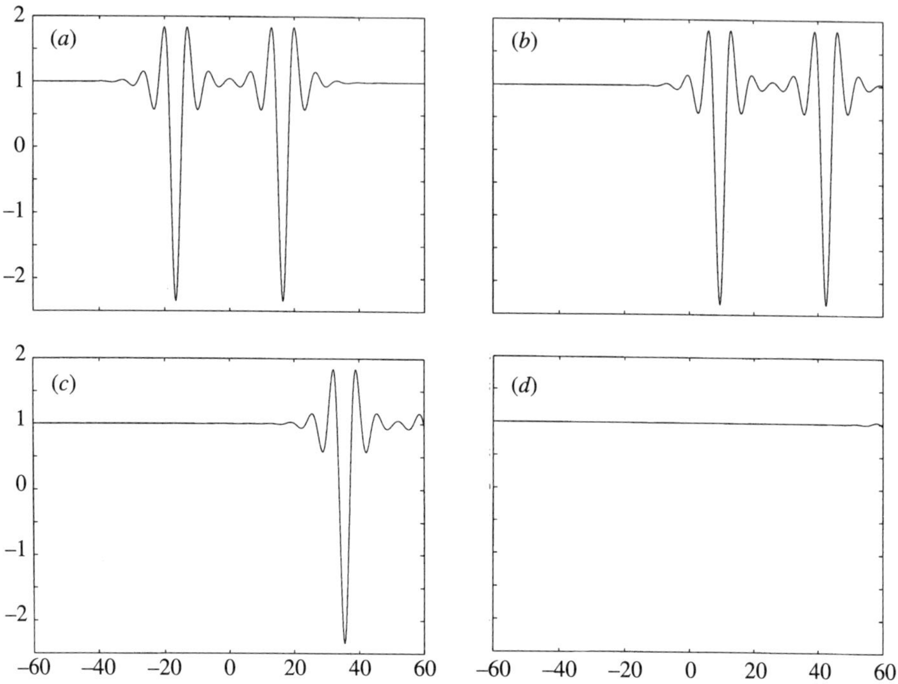

Figure 7. Solutions with speed $c = 1.30$.

We have not been able to observe this behaviour at wave speeds that are not close to $\sqrt{2}$. This may be because that as the wave speed gets smaller, the amplitude gets larger and that makes the initial value problem harder to solve accurately on a relatively coarse grid.

Usually, when one sees two nonlinear waves interacting in this manner, one expects to find additional conservation laws that are obeyed by solutions of the equation.

Problem 4.1. *Are there additional conserved quantities (apart from energy) for solutions of equation (2.2)? Can one give an explanation for the apparent property that solutions pass through each other?*

The authors are grateful to the referee who pointed out the connection with two other papers in the literature, namely Iooss & Pérouème (1993) and Grimshaw *et al.* (1994). An analysis of what happens in the neighbourhood for $c = \sqrt{2}$ was carried out using normal form analysis for some related problems including the Korteweg–de Vries equation.

5. Summary

We summarize the results of this paper. Probably the most striking result is the observation that these travelling waves pass through each other without losing their stability. This type of interaction of nonlinear waves has only previously been seen in completely integrable systems.

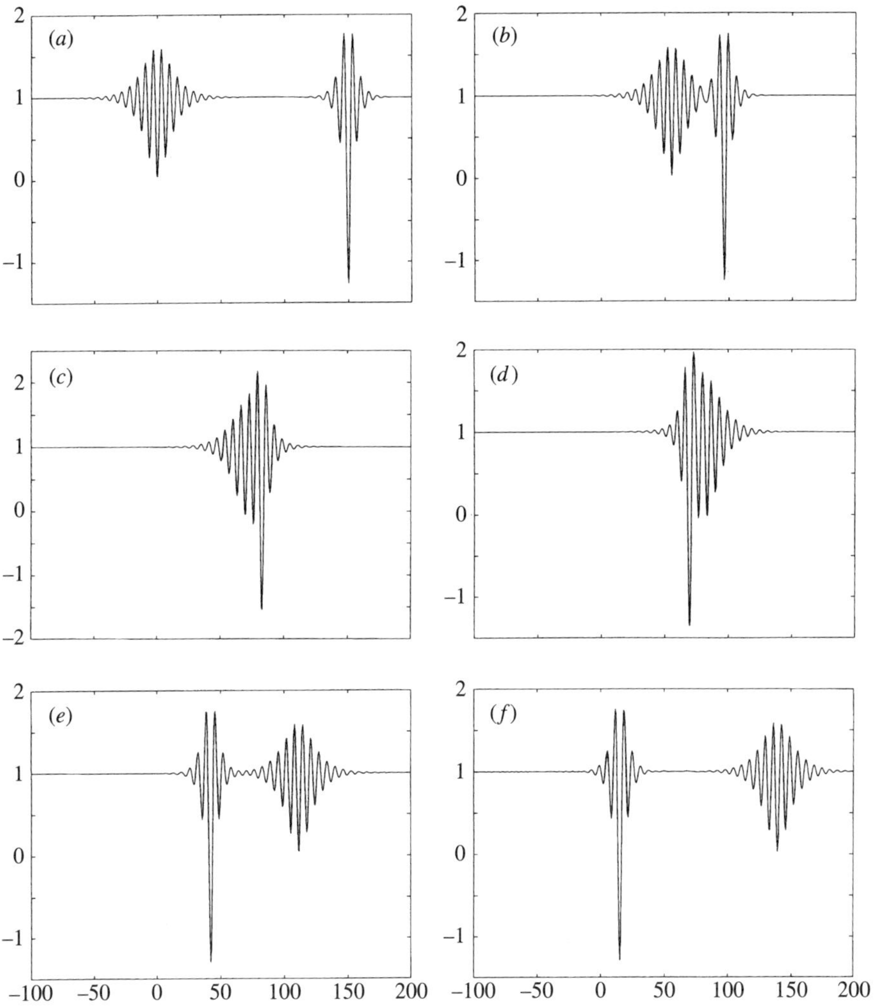

Figure 8. Interaction of two traveling waves of $u_{tt} + u_{xxxx} + \mathrm{e}^{u-1} - 1 = 0$.

This leads us to wonder whether we are in the presence of such a system†. Among experts, there seems to be little agreement about how to go about determining whether this is the case, other than be searching for additional conservation laws, in which we have so far been unsuccessful. Is there a methodical way to test for additional conservation laws?

Alternatively, this remarkable property of passing through each other may be related to some entirely different deep underlying structure, whose nature is still not apparent, but not of the completely integrable variety.

The experiment shown in figure 5 is also puzzling. Apparently, the more complex many-noded wave can exist for a while, and then spontaneously decompose into two

† One internationally known figure conjectured that equation (2.1) is connected to the Toda lattice, but the nature of the connection is as yet not clear.

simpler waves with less structure. While we are not proposing that a nonlinear beam equation should model elementary particle interactions, this result nonetheless also hints at some deeper fundamental property of these waves.

Just as earlier results on completely integrable systems were preceded by a period where the numerical results were unexplained, we believe this may be the case here, and that in the near future we shall have significant mathematical progress on the existence, stability, interactions and decompositions of these remarkable waves.

References

Brezis, H., Nirenberg, L. 1991 Remarks on finding critical points. *Commun. Pure Appl. Math.* **44**, 939–963.

Champneys, A. R. & Spence, A. 1993 Hunting for homoclinic orbits in reversible systems; a shooting technique. *Adv. Comp. Maths* **1**, 81–108.

Chen, Y. & McKenna, P. J. 1997 Traveling waves in a nonlinearly suspended beam: theoretical results and numerical observations. *J. Differ. Equat.* **136**, 325–355.

Choi, Y. S. & McKenna, P. J. 1993 A mountain pass method for the numerical solution of semilinear elliptic problems. *Nonlinear Analys. TMA* **20**, 417–437.

Grimshaw, R., Malomed, B. & Benilov, E. 1994 Solitary waves with damped oscillatory tails: an analysis of the fifth-order Korteweg–de Vries equation. *Physica* D **77**, 473–485.

Iooss, G. & Pérouème, M.-C. 1993 Periodic homoclinic solutions in reversible 1:1 resonance vector fields. *J. Differ. Equat.* **102**, 62–88.

Levandosky, S. 1998 Stability and instability for fourth-order solitary waves. Preprint.

McKenna, P. J. & Walter, W. 1990 Traveling waves in a suspension bridge. *SIAM J. Appl. Math.* **50**, 703–715.

Rabinowitz, P. H. 1986 Minimax methods in critical point theory with applications to differential equations. *CBMS Reg. Conf. Ser. Math.* **65**.

Strauss, W. A. 1989 Nonlinear wave equations. *CBMS Reg. Conf. Ser. Math.* **73**.

Strikwerda, H. C. 1989 *Finite difference schemes and partial differential equations.* Wadsworth & Brooks/Cole Advanced Books & Software.